Computer Algebra in Quantum Field Theory

Texts and Monographs in Symbolic Computation

A Series of the Research Institute
for Symbolic Computation,
Johannes Kepler University, Linz, Austria

For further volumes:
http://www.springer.com/series/3073

Carsten Schneider • Johannes Blümlein
Editors

Computer Algebra in Quantum Field Theory

Integration, Summation and Special Functions

Springer

Editors
Carsten Schneider
RISC, Johannes Kepler University
Linz
Austria

Johannes Blümlein
DESY
Zeuthen
Germany

ISSN 0943-853X
ISBN 978-3-7091-1615-9 ISBN 978-3-7091-1616-6 (eBook)
DOI 10.1007/978-3-7091-1616-6
Springer Wien Heidelberg New York Dordrecht London

Library of Congress Control Number: 2013950385

Printed on acid-free paper

Springer is part of Springer Science+Business Media (www.springer.com)

Preface

The research topics of computer algebra, special functions and quantum field theory have been deriving outstanding achievements from computational, algorithmic and theoretical point of view. As it turns out, there is a strong overlap of common interests concerning mathematical, physical and computer science aspects, and in the last years, the topics started a vital and promising interaction in the field of the automated computation of multi-loop and multi-leg Feynman diagrams in precision calculations. This observation has led, e.g. to an intensive cooperation between RISC (Research Institute for Symbolic Computation) of the Johannes Kepler University Linz and DESY (Deutsches Elektronen-Synchrotron). In order to push forward the interaction of the three research fields, the summer school and conference "Integration, Summation and Special Functions in Quantum Field Theory" organized by the European Network LHCPhenonet in cooperation with RISC and DESY was held at Hagenberg/Austria. Here central topics have been introduced with the special emphasis to present the current developments and to point out further possible connections.

This book collects the presented work in form of survey articles for a general readership. It aims at pushing forward the interdisciplinary ties between the very active research areas of computer algebra, special functions and quantum field theory. The driving questions of this book can be summarized as follows:

- How do special functions, such as generalized hypergeometric series, Appell functions, nested harmonic sums, nested multiple polylogarithms and multiple zeta values, emerge in quantum field theories?
- What properties do these functions and constants have and how are they related to each other?
- How can one extract information from such functions or how can one simplify voluminous expressions in terms of such functions with computer algebra, in particular with the help of symbolic summation and symbolic integration?
- What is the irreducible analytic and algebraic structure of multi-loop and multi-leg Feynman integrals?

This book tries to throw light to the underlying problems and to work out possible future cooperations between the different fields:

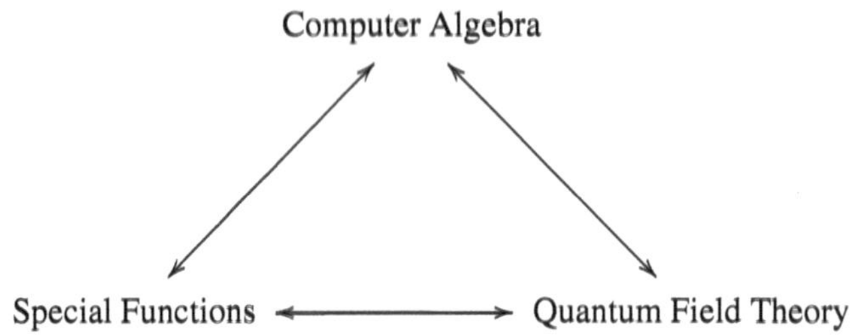

We emphasize that the interdisciplinary aspects are also reflected in the spirit of the articles. The authors have different backgrounds concerning mathematics, computer science and theoretical physics, and their different approaches bring in new aspects that shall push forward the presented topics of this book.

In this regard, we highlight the following rising aspects:

In Harmonic Sums, Polylogarithms, Special Numbers, and their Generalizations (J. Ablinger, J. Blümlein), *special functions* such as nested sums, associated iterated integrals and special constants which hierarchically appear in the evaluation of massless and massive Feynman diagrams at higher loops are discussed. In particular, the properties of harmonic sums and their generalizations of cyclotomic sums, generalized harmonic sums and sums containing binomial and inverse-binomial weights are worked out that give rise to the simplification of such sums by means of *computer algebra*.

In Multiple Zeta Values and Modular Forms in Quantum Field Theory (D. Broadhurst), properties of *special functions* like multiple zeta values and alternating Euler sums are worked out, and it is indicated where they arise in *quantum field theory*. In particular, the article deals with massive Feynman diagrams whose evaluations yield polylogarithms of the sixth root of unity, products of elliptic integrals and L-functions of modular forms inside their critical strips.

In Computer-Assisted Proofs of Some Identities for Bessel Functions of Fractional Order (S. Gerhold, M. Kauers, C. Koutschan, P. Paule, C. Schneider, B. Zimmermann), big parts of the *computer algebra software* of the combinatorics group of RISC are used to prove a collection of identities involving Bessel functions and other *special functions*. These identities appear in the famous Handbook of Mathematical Functions by Abramowitz and Stegun, as well as in its successor, the DLMF, but their proofs were lost. Here generating functions and symbolic summation techniques are utilized to produce new proofs for them.

In Conformal Methods for Massless Feynman Integrals and Large N_f Methods (J. A. Gracey), the large N method based on *conformal integration methods* is presented that calculates high-order information on the renormalization group functions in a *quantum field theory*. The possible future directions for the large N methods are discussed in light of the development of more recent techniques such as the Laporta algorithm.

In The Holonomic Toolkit (M. Kauers), an overview over standard techniques for *holonomic functions* is given covering, e.g. big parts of Feynman integrals coming from *quantum field theory*. It gives a collection of standard examples and states several fundamental properties of holonomic objects. Two techniques which are most useful in applications are explained in some more detail: *closure properties*, which can be used to prove identities among holonomic functions, and *guessing*, which can be used to generate plausible conjectures for equations satisfied by a given function.

In Orthogonal Polynomials (T. H. Koornwinder), an introduction to *orthogonal polynomials* is presented. It works out the general theory and properties of such special functions, and it is concerned with constructive aspects on how certain formulas can be derived. Special classes, such as *Jacobi polynomials*, *Laguerre polynomials* and *Hermite polynomials*, are discussed in details. It ends with some remarks about the usage of *computer algebra* for this theory.

In Creative Telescoping for Holonomic Functions (C. Koutschan), a broad overview of the available *summation and integration algorithms* for *holonomic functions* is presented. In particular, it is worked out how the underlying algorithms can be executed within the Mathematica package HolonomicFunctions. Special emphasis is put on concrete examples that are of particular relevance for problems coming, e.g. from special functions and *physics*.

In Renormalization and Mellin transforms (D. Kreimer and E. Panzer), the Hopf algebraic framework is utilized to study renormalization in a kinetic scheme. Here a direct *combinatorial description* of renormalized amplitudes in terms of *Mellin transform* coefficients is given using the universal property of rooted trees. The application to scalar *quantum field theory* reveals the scaling behaviour of individual Feynman graphs.

In Relativistic Coulomb Integrals and Zeilberger's Holonomic Systems Approach I (P. Paule, S. K. Suslov), *symbolic summation algorithms* such as Zeilberger's extension of Gosper's algorithm and a parameterized variant are utilized to calculate *recurrence relations* and transformation formulas for generalized hypergeometric series. More precisely, the basic facts within the theory of relativistic Coulomb integrals are presented, and the presented summation technology is used to tackle open problems there.

In Hypergeometric Functions in Mathematica® (O. Pavlyk), a short introduction to the constructive theory of *generalized hypergeometric functions* is given dealing, e.g. with differential equations, Mellin transforms and Meijer's G-functions. Special emphasis is put on concrete examples and notes on the implementation in the *computer algebra system Mathematica*.

In Solving Linear Recurrence Equations with Polynomial Coefficients (M. Petkovšek, H. Zakrajšek), *computer algebra algorithms* for finding polynomial, rational, hypergeometric, d'Alembertian and Liouvillian *solutions of linear recurrences* with polynomial coefficients are described. In particular, an alternative proof of a recent result of Reutenauer's is given that Liouvillian sequences are precisely the interlacing of d'Alembertian ones. In addition, algorithms for factoring linear recurrence operators and finding the minimal annihilator of a given holonomic sequence are presented.

In Generalization of Risch's Algorithm to Special Functions (C. G. Raab), *indefinite integration algorithms* in differential fields are presented. In particular, the basic ideas of Risch's algorithm for *elementary functions* and generalizations thereof are introduced. These algorithms give rise to more general algorithms dealing also with *definite integration*, i.e. calculating linear recurrences and differential equations integrals involving extra parameters.

In Multiple Hypergeometric Series – Appell Series and Beyond (M. J. Schlosser), a collection of basic material on *multiple hypergeometric series* of Appell type is presented covering *contiguous relations*, *recurrences*, *partial differential equations*, *integral representations* and *transformations*. More general series and related types such as Horn functions, Kampé de Fériet series and Lauricella series are introduced.

In Simplifying Multiple Sums in Difference Fields (C. Schneider), difference field algorithms for *symbolic summation* are presented. This includes the *simplification of indefinite nested sums*, *computing recurrence relations* of definite sums and *solving recurrence relations*. Special emphasis is put on new aspects in how the summation problems are rephrased in terms of difference fields, how the problems are solved there and how the derived results can be reinterpreted as solutions of the input problem. In this way, large-scale summation problems for the evaluation of Feynman diagrams in *quantum field theories* can be solved completely automatically.

In Potential of FORM 4.0 (J. A. M. Vermaseren), the *computer algebra system* FORM is presented that is heavily used in *quantum field theory* for large-scale calculations. Special emphasis is put on the main new features concerning factorization algorithms, polynomial arithmetic, special functions and code simplification.

Finally, in Feynman Graphs (S. Weinzierl), *Feynman graphs* and the associated *Feynman integrals* are discussed. It presents four different definitions from the mathematical and physical point of view. In particular, the most prominent class of *special functions*, the multiple polylogarithms, with their algebraic properties are worked out, which appear in the evaluation of Feynman integrals. The final part is devoted to Feynman integrals, which cannot be expressed in terms of multiple polylogarithms. Methods from algebraic geometry provide tools to tackle these integrals.

In addition, we want to emphasize the following fascinating presentations that are not part of this book, but which contributed substantially to our summer school and conference "Integration, Summation and Special Functions in Quantum Field Theory": the key note lecture Mate is Meta by Bruno Buchberger, Hypergeometric Functions and Loop Integrals by Nigel Glover, Polynomial GCDs and Factorization by Jürgen Gerhard and Holonomic Summation and Integration by Frédéric Chyzak.

The present project has been supported in part by the EU Network LHCPhenonet PITN-GA-2010-264564; the Austrian Science Fund (FWF) grants (P20347-N18, DK W1214); the Research Institute for Symbolic Computation, RISC; and Deutsches Elektronen-Synchrotron, DESY, which are kindly acknowledged.

We would like to thank I. Brandner-Foissner, T. Guttenberger, R. Oehme-Pöchinger (RISC) and M. Mende (DESY) for their help in organizing this meeting.

Zeuthen, Germany
Hagenberg, Austria
May, 2013

Johannes Blümlein
Carsten Schneider

Contents

List of Contributors

Jakob Ablinger Research Institute for Symbolic Computation (RISC), Johannes Kepler University, Linz, Austria

Johannes Blümlein Deutsches Elektronen-Synchrotron, DESY, Zeuthen, Germany

David Broadhurst Department of Physical Sciences, Open University, Milton Keynes, UK

Stefan Gerhold Financial and Actuarial Mathematics, Vienna University of Technology, Vienna, Austria

John A. Gracey Theoretical Physics Division, Department of Mathematical Sciences, University of Liverpool, Liverpool, UK

Manuel Kauers Research Institute for Symbolic Computation, Johannes Kepler University, Linz, Austria

Tom H. Koornwinder Korteweg-de Vries Institute, University of Amsterdam, Amsterdam, The Netherlands

Christoph Koutschan Johann Radon Institute for Computational and Applied Mathematics (RICAM), Austrian Academy of Sciences (ÖAW), Vienna, Austria

Dirk Kreimer Departments of Mathematics and Physics, Humboldt University, Berlin, Germany

Erik Panzer Departments of Mathematics and Physics, Humboldt University, Berlin, Germany

Peter Paule Research Institute for Symbolic Computation (RISC), Johannes Kepler University Linz, Linz, Austria

Oleksandr Pavlyk Wolfram Research, Inc., Champaign, IL, USA

Marko Petkovšek Faculty of Mathematics and Physics, University of Ljubljana, Ljubljana, Slovenia

Clemens G. Raab Deutsches Elektronen-Synchrotron, DESY, Zeuthen, Germany

Michael J. Schlosser Fakultät für Mathematik, Universität Wien, Vienna, Austria

Carsten Schneider Research Institute for Symbolic Computation (RISC), Johannes Kepler University Linz, Linz, Austria

Sergei K. Suslov School of Mathematical and Statistics Sciences, Mathematical, Computational and Modeling Sciences Center, Arizona State University, Phoenix, AZ, USA

Jos A.M. Vermaseren Nikhef, Amsterdam, The Netherlands

Stefan Weinzierl PRISMA Cluster of Excellence, Institut für Physik, Johannes Gutenberg-Universität Mainz, Mainz, Germany

Helena Zakrajšek Faculty of Mechanical Engineering, University of Ljubljana, Ljubljana, Slovenia

Burkhard Zimmermann Research Institute for Symbolic Computation, Johannes Kepler University, Linz, Austria

Harmonic Sums, Polylogarithms, Special Numbers, and Their Generalizations

Jakob Ablinger and Johannes Blümlein

Abstract In these introductory lectures we discuss classes of presently known nested sums, associated iterated integrals, and special constants which hierarchically appear in the evaluation of massless and massive Feynman diagrams at higher loops. These quantities are elements of stuffle and shuffle algebras implying algebraic relations being widely independent of the special quantities considered. They are supplemented by structural relations. The generalizations are given in terms of generalized harmonic sums, (generalized) cyclotomic sums, and sums containing in addition binomial and inverse-binomial weights. To all these quantities iterated integrals and special numbers are associated. We also discuss the analytic continuation of nested sums of different kind to complex values of the external summation bound N.

1 Introduction

In the solution of physical problems very often new classes of special functions have been created during the last three centuries, cf. [1–5]. This applies especially also to the analytic calculation of Feynman-parameter integrals [6] for massless and massive two- and more-point functions, also containing local operator insertions and corresponding quantities, cf. [7, 8]. In case of zero mass-scale quantities the associated integrals map to special numbers, lately having been called periods [9], see also [10]. In case of single-scale quantities, expressed as a ratio $x \in [0, 1]$ to

J. Ablinger
Research Institute for Symbolic Computation (RISC), Johannes Kepler University, Altenbergerstraße 69, A-4040 Linz, Austria
e-mail: jablinge@risc.uni-linz.ac.at

J. Blümlein (✉)
Deutsches Elektronen-Synchrotron, DESY, Platanenallee 6, D-14738 Zeuthen, Germany
e-mail: Johannes.Bluemlein@desy.de

C. Schneider and J. Blümlein (eds.), *Computer Algebra in Quantum Field Theory*, Texts & Monographs in Symbolic Computation, DOI 10.1007/978-3-7091-1616-6_1,

the defining mass scale, the integrals are Poincaré iterated integrals [11, 12] or they emerge as a Mellin-transform at $N \in \mathbb{N}$ [13] in terms of multiply nested sums. A systematic way to these structures has been described in [14, 15]. Here an essential tool consists in representations by Mellin–Barnes [16] integrals. They are applicable also for integrals of multi-scale more-loop and multi-leg Feynman integrals [17], which are, however, less explored at present.

In the practical calculations dimensional regularization in $D = 4 + \varepsilon$ space-time dimensions [18] is used, which is essential to maintain conservation laws due to the Noether theorem and probability. It provides the singularities of the problem in terms of poles in ε. However, the Feynman parameter integrals are not performed over rational integrands but hyperexponential ones. Thus one passes through higher transcendental functions [4, 5] from the beginning. The renormalization is carried out in the $\overline{\text{MS}}$-scheme, chosen as the standard. In new calculations various ingredients as anomalous dimensions and expansion coefficients of the β-functions needed in the renormalization can thus be used referring to results given in the literature. At higher orders the calculation of these quantities requests a major investment and is not easily repeated at present within other schemes in a short time.

With growing complexity of the perturbative calculations in Quantum Field Theories the functions emerging in integration and summation had to be systematized. While a series of massless two-loop calculations, cf. [19], during the 1980s and 1990s initially still could be performed referring to the classical polylogarithms [12, 20–23] and Nielsen-integrals [24], the structure of the results became readily involved. In 1998 a first general standard was introduced [25, 26] by the nested harmonic sums, and shortly after the harmonic polylogarithms [27]. Further extensions are given by the generalized harmonic sums, the so-called S-sums [28, 29] and the (generalized) cyclotomic sums [30], see Fig. 1. Considering problems at even higher loops and a growing number of legs, also associated with more mass scales, one expects various new levels of generalization to emerge. In particular, also elliptic integrals will contribute [31]. These structures can be found systematically by applying symbolic summation, cf. [32], and integration formalisms, cf. [33, 34], which also allow to proof the relative transcendence of the basis elements found and are therefore applied in the calculation of Feynman diagrams.

In this survey we present an introduction to a series of well-studied structures which have been unraveled during the last years. The paper is organized as follows. In Sect. 2 a survey is given on polylogarithms, Nielsen integrals and harmonic polylogarithms. In Sect. 3 harmonic sums are discussed. Both harmonic polylogarithms and harmonic sums obey algebraic and structural relations on which a survey is given in Sect. 4. In Sect. 5 we discuss properties of the multiple zeta values which emerge as special constants in the context of harmonic sums and polylogarithms. The S-sums, associated iterated integrals, and special numbers are considered in Sect. 6. The generalization of harmonic sums and S-sums to (generalized) cyclotomic sums, polylogarithms and numbers is given in Sect. 7. A further generalization, which appears in massive multi-loop calculations, to nested binomial and inverse-binomial harmonic sums and polylogarithms is outlined in

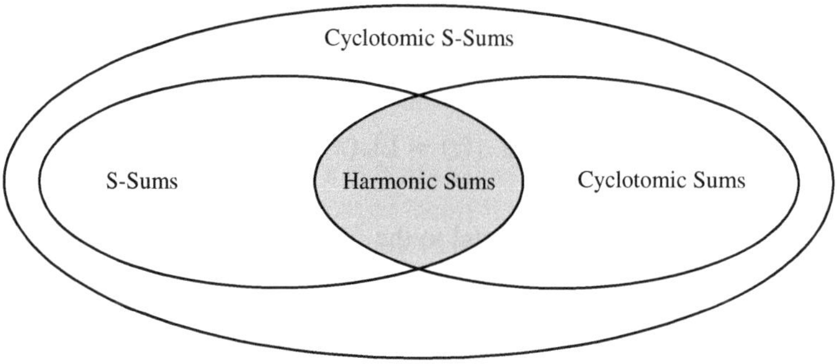

Fig. 1 Relations between the different extensions of harmonic sums

Sect. 8. Finally, we discuss in Sect. 9 the analytic continuation of the different kind of nested sums in the argument N to complex numbers, which is needed in various physical applications. Section 10 contains the conclusions. The various mathematical relations between the different quantities being discussed in the present article are implemented in the Mathematica package `HarmonicSums.m` [29, 35].

2 Polylogarithms, Nielsen Integrals, Harmonic Polylogarithms

Different particle propagators $1/A_k(p_i, m_i)$ can be linked using Feynman's integral representation [36]

$$\frac{1}{A_1^{\nu_1}\dots A_n^{\nu_n}} = \frac{\Gamma(\sum_{k=1}^n \nu_k)}{\prod_{k=1}^n \Gamma(\nu_k)} \int_0^1 \prod_{k=1}^n dx_k \frac{\prod_{k=1}^n x^{\nu_k - 1}}{(\sum_{k=1}^n x_k A_k)^{\sum_{k=1}^n \nu_k}} \delta\left(1 - \sum_{k=1}^n x_k\right), \nu_i \in \mathbb{R}. \quad (1)$$

While the momentum integrals over p_i can be easily performed, the problem consists in integrating the Feynman parameters x_k. In the simplest cases the associated integrand is a multi-rational function. In the first integrals one obtains multi-rational functions, but also logarithms [33]. The logarithms [37] have to be introduced as new functions being transcendental to the rational functions

$$\int_0^x \frac{dz}{1-z} = -\ln(1-x), \text{ etc.} \quad (2)$$

Iterating this integral by

$$\int_0^x \frac{dz_1}{z_1} \int_0^{z_1} \frac{dz_2}{1-z_2} = \mathrm{Li}_2(x) \quad (3)$$

In physics applications they appeared early in [44, 45]. Their systematic use dates back to Refs. [25, 26]. They can be represented as a Mellin transform

$$S_{\mathbf{a}}(N) = \mathbf{M}[f(x)](N) = \int_0^1 dx\, x^{N-1} f(x), \quad N \in \mathbb{N}\backslash\{0\}, \tag{20}$$

where $f(x)$ denotes a linear combination of HPLs. For example,

$$S_{-2,1,1}(N) = (-1)^{N+1} \int_0^1 dx \frac{\mathrm{H}_{0,1,1}(x) - \zeta_3}{x+1} - \mathrm{Li}_4\left(\frac{1}{2}\right) - \frac{\ln^4(2)}{24} + \frac{\ln^2(2)\zeta_2}{4} - \frac{7\ln(2)\zeta_3}{8} + \frac{\zeta_2^2}{8} \tag{21}$$

holds. Harmonic sums possess algebraic and structural relations, cf. Sect. 4. In the limit $N \to \infty$ they define the multiple zeta values, cf. Sect. 5. They are originally defined at integer argument N. In physical applications they emerge in the context of the light–cone expansion [46]. The corresponding operator matrix elements are analytically continued to complex values of N either from the even *or* the odd integers, cf. Sect. 9.

4 Algebraic and Structural Relations

4.1 Algebraic Relations

Algebraic relations of harmonic polylogarithms and harmonic sums, respectively, are implied by their products and depend on their index structure only, i.e. they are a consequence of the associated shuffle or quasi–shuffle (stuffle) algebras [47]. These properties are widely independent of the specific realization of these algebras. To one of us (JB) it appeared as a striking surprise, when finding the determinant-formula for harmonic sums of equal argument [26] Eqs. (157,158)

$$S_{\underbrace{\mathrm{a},\ldots,\mathrm{a}}_{k}}(N) = \frac{1}{k}\sum_{l=0}^{k} S_{\underbrace{\mathrm{a},\ldots,\mathrm{a}}_{l}}(N) S_{\wedge_{m=1}^{k-l} a}(N), \quad a \wedge b = \mathrm{sign}(ab)(|a|+|b|) \tag{22}$$

also in Ramanujan's notebook [48], but for *integer sums*, which clearly differ in value from the former ones. Related relations to again different quantities were given by Faá die Bruno [49].

Iterated integrals with the same argument x obey shuffle relations w.r.t. their product,

$$\mathrm{H}_{a_1,\ldots,a_k}(x) \cdot \mathrm{H}_{b_1,\ldots,b_l}(x) = \sum_{\mathbf{c} \in \mathbf{a} \,\sqcup\!\sqcup\, \mathbf{b}} \mathrm{H}_{c_1,\ldots c_{k+l}}(x). \tag{23}$$

The shuffle-operation runs over all combinations of the sets **a** and **b** leaving the order of these sets unchanged. Likewise, the (generalized) harmonic sums obey quasi-shuffle or stuffle-relations, which are found recursively using [28, 29]

$$\begin{aligned}
& \mathrm{S}_{a_1,\ldots,a_k}(x_1,\ldots,x_k;n)\,\mathrm{S}_{b_1,\ldots,b_l}(y_1,\ldots,y_l;n) = \\
& \qquad \sum_{i=1}^{n} \frac{x_1^i}{i^{a_1}} \mathrm{S}_{a_2,\ldots,a_k}(x_2,\ldots,x_k;i)\,\mathrm{S}_{b_1,\ldots,b_l}(y_1,\ldots,y_l,i) \\
& \qquad + \sum_{i=1}^{n} \frac{y_1^i}{i^{b_1}} \mathrm{S}_{a_1,\ldots,a_k}(x_1,\ldots,x_k,i)\,\mathrm{S}_{b_2,\ldots,b_l}(y_2,\ldots,y_l;i) \\
& \qquad - \sum_{i=1}^{n} \frac{(x_1\cdot y_1)^i}{i^{a_1+b_1}} \mathrm{S}_{a_2,\ldots,a_k}(x_2,\ldots,x_k,i)\,\mathrm{S}_{b_2,\ldots,b_l}(y_2,\ldots,y_l;i), \\
& \qquad x_i, y_i \in \mathbb{C}, a_i, b_i \in \mathbb{N}\backslash\{0\}\,.
\end{aligned} \tag{24}$$

The presence of trace terms in form of lower weight products in addition to the shuffled terms, cf. [50], leads to the name stuffle relations. In case the corresponding values exist, both (23,24) can be applied to the multiple zeta values or other special numbers applying the integral and sum-representations at $x = 1$ and $N \to \infty$, cf. [51]. The basis elements applying the (quasi) shuffle relations in case of the harmonic sums and polylogarithms at a given weight w can be identified by the Lyndon words [52,53]. Let $\mathfrak{A} = \{a, b, c, d, \ldots\}$ be an *ordered* alphabet and $\mathfrak{A}^*(\mathfrak{A})$ the set of words w given as concatenation products. Under the ordering of $\mathfrak{A}$ a Lyndon word is smaller than any of its suffixes. For example, the set $\{a, a, a, b, b, b\}$, $a < b$ is associated to the Lyndon words $\{aaabbb, aababb, aabbab\}$. Radford showed [54] that a shuffle algebra is freely generated by the Lyndon words. The number of Lyndon words can be counted using Witt formulae [55]. Let $\mathfrak{M}$ be a set of letters q in which the letter a_k emerges n_k times, and $n = \sum_{k=1}^{q} n_k$. The number of Lyndon words associated to this set is given by

$$l_n(n_1,\ldots,n_q) = \frac{1}{n} \sum_{d|n_k} \mu(d) \frac{(n/d)!}{(n_1/d)!\ldots(n_q/d)!}\,. \tag{25}$$

Similarly one may count the basis elements occurring for all combinations at a given weight, if the alphabet has m letters:

$$N_A(w) = \frac{1}{w} \sum_{d|w} \mu\left(\frac{w}{d}\right) m^d\,, \tag{26}$$

where μ denotes Möbius' function [56]. In case of the harmonic sums and polylogarithms one has $m = 3$. The original number of harmonic polylogarithms is 3^w and in case of the harmonic sums $2 \cdot 3^{w-1}$. Algebraic relations for the harmonic polylogarithms and harmonic sums are implemented in the FORM-codes `summer` [25] and `harmpol` [27], `HPL` [57], and also `HarmonicSums.m` [29, 35].

4.2 *Structural Relations*

Structural relations of harmonic polylogarithms and harmonic sums are implied by operations on their arguments x and N, respectively.

4.2.1 Harmonic Polylogarithms

Harmonic polylogarithms satisfy argument-relations, as has been illustrated in (8–14) for some examples in case of the Nielsen integrals. Not all argument relations map inside the harmonic polylogarithms, however, cf. [27]. Some of them are valid only for the sub-alphabet $\{\omega_0, \omega_1\}$. While the transformation $x \rightarrow -x$ is general

$$\mathrm{H}_{\mathbf{a}}(-x) = (-1)^p \mathrm{H}_{-\mathbf{a}}(x), \tag{27}$$

with the last letter in **a** different from 0 and p the number of non-zero letters in **a**. The transformations

$$x \rightarrow 1-x, \qquad x \rightarrow x^2 \tag{28}$$

apply to subsets only. Examples are:

$$\mathrm{H}_{1,0,1}(1-x) = -\mathrm{H}_0(x)\mathrm{H}_{0,1}(x) + 2\mathrm{H}_{0,0,1}(x) - \zeta_2\mathrm{H}_0(x) - 2\zeta_3 \tag{29}$$

$$\mathrm{H}_{1,0,0,1}\left(x^2\right) = 4\left[\mathrm{H}_{1,0,0,1}(x) - \mathrm{H}_{1,0,0,-1}(x) - \mathrm{H}_{-1,0,0,1}(x) + \mathrm{H}_{-1,0,0,-1}(x)\right]. \tag{30}$$

One may transform arguments by $x \rightarrow 1/y + i\varepsilon$,

$$\begin{aligned}\mathrm{H}_{1,0,1}\left(\frac{1}{x}\right) = {} & \mathrm{H}_0(x)\left[\mathrm{H}_{0,1}(x) + i\pi\mathrm{H}_1(x) - 4\zeta_2 + \pi^2\right] - 2[\mathrm{H}_{0,0,1}(x) - \mathrm{H}_{0,1,1}(x) \\ & +\zeta_3] + [-\mathrm{H}_1(x) - i\pi]\,\mathrm{H}_{0,1}(x) + 2\zeta_2\mathrm{H}_1(x) - \frac{1}{6}\mathrm{H}_0^3(x) \\ & +\frac{1}{2}i\,[\pi + i\mathrm{H}_1(x)]\,\mathrm{H}_0^2(x)\,.\end{aligned} \tag{31}$$

An important general transformation is

$$x \rightarrow \frac{1-t}{1+t} \tag{32}$$

which acts on the HPLs but not on the subset of Nielsen-integrals. An example is:

$$\mathrm{H}_{1,-1,0}\left(\frac{1-x}{1+x}\right) = \frac{1}{6}\mathrm{H}_{-1}^{3}(x) + \mathrm{H}_{-1,-1,1}(x) - \mathrm{H}_{0,-1,-1}(x) - \mathrm{H}_{0,-1,1}(x) + \frac{15\zeta_3}{8}$$
$$-\frac{1}{2}\zeta_2\left[\mathrm{H}_{-1}(x) - \mathrm{H}_0(x)\right] - 2\left[\frac{\zeta_3}{8} - \frac{\ln(2)\zeta_2}{2}\right] - 2\ln(2)\zeta_2\,. \tag{33}$$

In most of these relations also HPLs at argument $x = 1$ contribute, cf. Sect. 5. Structural relations of HPLs are implemented in the packages `harmpol` [27], `HPL` [57], and `HarmonicSums.m` [29, 35].

4.2.2 Harmonic Sums

Harmonic sums obey the duplication relation

$$S_{i_1,\ldots,i_n}(N) = 2^{i_1+\ldots+i_n-n} \sum_{\pm} S_{\pm i_1,\ldots,\pm i_n}(2N), \quad i_k \in \mathbb{N}\backslash\{0\}\,. \tag{34}$$

This allows to define harmonic sums at half-integer, i.e. rational, values. Ultimately, one would like to derive expressions for $N \in \mathbb{C}$, cf. Sect. 9. Another extension is to $N \in \mathbb{R}$ [14, 26]. The representation of harmonic sums through Mellin-transforms (20) implies analyticity for a finite range around a given value of N. The Mellin-transform of a harmonic polylogarithm can thus be differentiated for N

$$\frac{d}{dN}\int_0^1 dx x^{N-1}\mathrm{H}_{\mathbf{a}}(x) = \int_0^1 dx x^{N-1}\mathrm{H}_0(x)\mathrm{H}_{\mathbf{a}}(x)\,. \tag{35}$$

In turn, the shuffling relation (23) allows to represent the r.h.s. in (35) as the Mellin-transform of other HPLs. It turns out that differentiation of harmonic sums for N is closed under additional association of the multiple zeta values [14]. The number of basis elements by applying the duplication relation (H), resp. its combination with the algebraic relations is [58]

$$N_H(w) = 2\cdot 3^{w-1} - 2^{w-1}, \qquad N_{AH}(w) = \frac{1}{w}\sum_{d|w}\mu\left(\frac{w}{d}\right)\left[2^2 - 3^d\right]\,. \tag{36}$$

Differentiation in combination with the other relation yields

$$N_D(w) = 4\cdot 3^{w-1}, N_{DH}(w) = 4\cdot 3^{w-2} - 2^{w-2}, N_{ADH}(w) = N_{AH}(w) - N_{AH}(w-1). \tag{37}$$

Let us close with a remark on observables or related quantities in physics which are calculated to a certain loop level and can be thoroughly expressed in terms of harmonic sums. As a detailed investigation of massless and single mass two-loop quantities showed [40] seven basic functions of up to weight $w = 4$, cf. [14], are sufficient to express all quantities. The three-loop anomalous dimensions [41] contributing to the $1/\varepsilon$ poles of the corresponding matrix elements require 15 functions of up to weight $w = 5$ and further 20 basic functions are needed to also express the massless Wilson coefficients [7] in deep-inelastic scattering [59], cf. Ref. [60]. Despite of the complexity of these calculations finally a rather compact structure is obtained for the representation of the results. Structural relations of harmonic sums are implemented in the package `HarmonicSums.m` [29,35].

5 Multiple Zeta Values

The multiple zeta values (MZVs) [61,62][1] are obtained by the limit $N \to \infty$ of the harmonic sums

$$\lim_{N\to\infty} S_{\mathbf{a}}(N) = \sigma_{\mathbf{a}} \tag{38}$$

and may also be represented in terms of linear combinations of harmonic polylogarithms $\mathrm{H}_{\mathbf{b}}(1)$ over the alphabet $\{\omega_0, \omega_1, \omega_{-1}\}$.[2] In the former case one usually includes the divergent harmonic sums since all divergent contributions are uniquely represented in terms of polynomials in $\sigma_1(\infty) \equiv \sigma_0$ due to the algebraic relations. Likewise, not all harmonic polylogarithms can be calculated at $x = 1$, requiring their re-definition in terms of distributions. Some examples for MZVs, which already appear in case of Nielsen integrals, are:

$$\mathrm{Li}_n(1) = \zeta_n \tag{39}$$

$$\mathrm{Li}_n(-1) = -\left(1 - \frac{1}{2^{n-1}}\right)\zeta_n \tag{40}$$

$$\mathrm{S}_{1,p}(1) = \zeta_{p+1} \tag{41}$$

$$\mathrm{S}_{1,2}(-1) = \frac{1}{8}\zeta_3 \tag{42}$$

$$\mathrm{Li}_2\left(\frac{1}{2}\right) = \frac{1}{2}\left[\zeta_2 - \ln^2(2)\right] \tag{43}$$

[1] For a detailed account on the literature on MZVs see [63,64] and the surveys Ref. [65].

[2] The numbers associated with this alphabet are sometimes also called Euler-Zagier values and those of the sub-alphabet $\{\omega_0, \omega_1\}$ multiple zeta values.

$$\mathrm{Li}_3\left(\frac{1}{2}\right) = \frac{7}{8}\zeta_3 - \frac{1}{2}\zeta_2 \ln(2) + \frac{1}{6}\ln^3(2) \tag{44}$$

$$\mathrm{S}_{2,2}(1) = \frac{1}{10}\zeta_2^2 \tag{45}$$

$$\mathrm{S}_{2,2}(-1) = -\frac{3}{4}\zeta_2^2 + 2\mathrm{Li}_4\left(\frac{1}{2}\right) + \frac{7}{4}\zeta_3 \ln(2) - \frac{1}{2}\zeta_2 \ln^2(2) + \frac{1}{12}\ln^4(2) \tag{46}$$

$$\mathrm{S}_{1,3}(-1) = -\frac{2}{5}\zeta_2^2 + \mathrm{Li}_4\left(\frac{1}{2}\right) + \frac{7}{8}\zeta_3 \ln(2) - \frac{1}{4}\zeta_2 \ln^2(2) + \frac{1}{24}\ln^4(2)\,. \tag{47}$$

In case of physics applications, MZVs played a role in loop calculations rather early, cf. [66]. Since for $\mathrm{Li}_m(1/2)$ for $m = 2, 3$ these numbers are not elementary, (46, 47) seem to fail to provide a corresponding relation for $m = 4$. Similarly, for larger values of m also no reduction has been observed.

A central question concerns the representation of harmonic sums in terms of polynomial bases. This has been analyzed systematically in [63, 67].[3] For MZVs over $\{0, 1\}$ a proof on the maximum of basis elements at fixed weight w has been given in Refs. [68]. At the lowest weights the shuffle and stuffle relations imply all relations for the MZVs. Starting with weight $w = 8$ one also needs the duplication relation (34), and from weight $w = 12$ also the generalized duplication relations Sect. 4.1 in [63]. The latter are closely related to the conformal transformation relations of the HPLs at $x = 1$, see (32). Let us give one example for the combined use of the shuffle and stuffle relation for illustration, [51][4]:

$$\begin{aligned} \text{shuffle}: \quad & \zeta_{2,1}\zeta_2 = 6\zeta_{3,1,1} + 2\zeta_{2,2,1} + \zeta_{2,1,2} \\ \text{stuffle}: \quad & \zeta_{2,1}\zeta_2 = 2\zeta_{2,2,1} + \zeta_{41} + \zeta_{2,3} + \zeta_{2,1,2} \\ \Longrightarrow \quad & \zeta_{3,1,1} = \frac{1}{6}\left[\zeta_{4,1} + \zeta_{2,3} - \zeta_{2,2,1}\right]\,. \end{aligned} \tag{48}$$

Finally one derives a basis for the MZVs using the above relations. Up to weight $w = 7$ reads, cf. [25],

$$\Big\{ (\sigma_1(\infty), \ln(2))\,; \zeta_2; \zeta_3; \mathrm{Li}_4\left(\tfrac{1}{2}\right); \left(\zeta_5, \mathrm{Li}_5\left(\tfrac{1}{2}\right)\right); \left(\mathrm{Li}_6\left(\tfrac{1}{2}\right), \sigma_{-5,-1}\right); \\ \left(\zeta_7, \mathrm{Li}_7\left(\tfrac{1}{2}\right), \sigma_{-5,1,1}, \sigma_{-5,1,1}\right); \Big\} \tag{49}$$

---

[3]For some aspects of the earlier development including results by the Leuven-group, Zagier, Broadhurst, Vermaseren and the Lille-group, see [63].

[4]Here the $\zeta_{\mathbf{a}}$-values are defined $\zeta_{a_1,\ldots,a_m} = \sum_{n_1>n_2>\ldots>n_m}^{\infty} \prod_{k=1}^{m} n_k^{-a_1}$.

In [63] bases were calculated up to $w = 12$ for the alphabet $\{0, 1, -1\}$ and to $w = 22$ for the alphabet $\{0, 1\}$ in explicit form resp. for $w = 24$ restricting to basis elements only. In the latter case the conjecture by Zagier [62] that the shuffle, stuffle and duplication relations were the only ones was confirmed up to the weights quoted. For these cases counting relations were conjectured in Refs. [69,70]. One may represent the basis for the MZVs over the alphabets $\{0, 1\}$ resp. $\{0, 1, -1\}$ by polynomial bases or count just the factors appearing in these polynomials of special numbers occurring newly in the corresponding weight, which is called Lyndon-basis [63]. In the first case the basis for the MZVs[$\{0, 1\}$] is conjectured to be counted by the Padovan numbers $\hat{P}_k$ [71] generated by

$$\frac{1+x}{1-x^2-x^3} = \sum_{k=0}^{\infty} x^k \hat{P}_k, \quad \hat{P}_1 = \hat{P}_2 = \hat{P}_3 = 1. \tag{50}$$

In case of the Lyndon basis the Perrin numbers P_k appear [72]

$$\frac{3-x^2}{1-x^2-x^3} = \sum_{k=0}^{\infty} x^k \hat{P}_k, \quad P_1 = 0, P_2 = 2, P_3, = 3\,. \tag{51}$$

Both the above sequences obey the Fibonacci-recurrence [73]

$$P_d = P_{d-2} + P_{d-3}, \quad d \geq 3\,. \tag{52}$$

The length of the Lyndon basis at weight w is given by

$$l(w) = \frac{1}{w} \sum_{d|w} \mu\left(\frac{w}{d}\right) P_d\,. \tag{53}$$

Hoffman [74] conjectured that all MZVs over the alphabet $\{0, 1\}$ can be represented over a basis of MZVs carrying 2 and 3 as indices only. This has been confirmed up to $w = 24$. An explicit proof has been given in [75].

The polynomial basis of the MZVs[$\{0, 1, -1\}$] is conjectured to be counted by the Fibonacci numbers [76]

$$f_d = \frac{1}{\sqrt{5}}\left[\left(\frac{1+\sqrt{5}}{2}\right)^d - \left(\frac{1-\sqrt{5}}{2}\right)^d\right] \tag{54}$$

which obey

$$\frac{x}{1-x-x^2} = \sum_{k=0}^{\infty} x^k f_k\,. \tag{55}$$

For the corresponding Lyndon basis the counting relation

$$l(w) = \frac{1}{w} \sum_{d|w} \mu\left(\frac{w}{d}\right) L_d \tag{56}$$

is conjectured [69, 70], where L_d are the Lucas numbers [77],

$$L_d = \left(\frac{1+\sqrt{5}}{2}\right)^d + \left(\frac{1-\sqrt{5}}{2}\right)^d \tag{57}$$

$$\frac{2-x}{1-x-x^2} = \sum_{k=0}^{\infty} x^k L_k, \tag{58}$$

$$L_d = L_{d-1} + L_{d-2}, d \geq 4, \quad L_1 = 1, L_2 = 3, L_3 = 4. \tag{59}$$

There is a series of Theorems proven on the MZVs, see also [63], which can be verified using the data base [63]. The **duality theorem** [62] in case of the alphabet $\{\omega_0, \omega_1\}$ states

$$\mathrm{H}_{\mathbf{a}}(1) = \mathrm{H}_{\mathbf{a}^\dagger}(1), \quad \mathbf{a}^\dagger = \mathbf{a}^T_{0\longleftrightarrow 1}\,. \tag{60}$$

In case of the alphabet $\{\omega_0, \omega_1, \omega_{-1}\}$ it is implied by the transformation (32), see [63]. Another relation is the **sum theorem**, Refs. [78, 79],

$$\sum_{i_1+\ldots+i_k=n, i_1>1} \zeta_{i_1,\ldots,i_k} = \zeta_n\,. \tag{61}$$

The sum-theorem was conjectured in [80], cf. [81]. For its derivation using the Euler connection formula for polylogarithms, cf. [82].

Further identities are given by the **derivation theorem**, [80, 83]. Let $I = (i_1, \ldots, i_k)$ any sequence of positive integers with $i_1 > 1$. Its derivation $D(I)$ is given by

$$D(I) = (i_1+1, i_2, \ldots, i_k) + (i_1, i_2+1, \ldots, i_k) + \ldots (i_1, i_2, \ldots, i_k+1)$$

$$\zeta_{D(I)} = \zeta_{(i_1+1, i_2, \ldots, i_k)} + \ldots + \zeta_{(i_1, i_2, \ldots, i_k+1)}\,. \tag{62}$$

The derivation theorem states

$$\zeta_{D(I)} = \zeta_{\tau(D(\tau(I)))}\,, \tag{63}$$

where τ denotes the duality-operation, cf. (60). An index-word w is called admissible, if its first letter is not 1. The words form the set $\mathfrak{H}^0$. $|w| = \mathrm{w}$ is the weight and $d(w)$ the depth of w. For the MZVs the words w are build in terms of concatenation products $x_0^{i_1-1}x_1x_0^{i_2-1}x_0 \ldots x_0^{i_k-1}x_1$. The height of a word, $\mathrm{ht}(w)$, counts the number of (non-commutative) factors $x_0^a x_1^b$ of w. The operator D and its dual $\overline{D}$ act as follows [84],

$$D x_0 = 0, \quad D x_1 = x_0 x_1, \quad \overline{D} x_0 = x_0 x_1, \quad \overline{D} x_1 = 0 .$$

Define an anti-symmetric derivation

$$\partial_n x_0 = x_0 (x_0 + x_1)^{n-1} x_1 .$$

A generalization of the derivation theorem was given in [83, 85]. The identity

$$\zeta(\partial_n w) = 0 \tag{64}$$

holds for any $n \geq 1$ and any word $w \in \mathfrak{H}^0$. Further theorems are the Le–Murakami theorem, [86], the Ohno theorem, [87], which generalizes the sum- and duality theorem, the Ohno–Zagier theorem, [88], which covers the Le–Murakami theorem and the sum theorem, which generalizes a theorem by Hoffman [80, 81], and the cyclic sum theorem, [89].

There are also relations for MZVs at repeated arguments, cf. [51, 84, 90], on which examples are:

$$\zeta(\{2\}_n) = \frac{2(2\pi)^{2n}}{(2n+1)!} \frac{1}{2} \tag{65}$$

$$\zeta(2, \{1\}_n) = \zeta(n+2) \tag{66}$$

$$\zeta(\{3,1\}_n) = \frac{1}{4^n} \zeta(\{4\}_n) = \frac{2\pi^{4n}}{(4n+2)!} \tag{67}$$

$$\zeta(\{10\}_n) = \frac{10(2\pi)^{2n}}{(10n+5)!} \left[1 + \left(\frac{1+\sqrt{5}}{2} \right)^{10n+5} + \left(\frac{1-\sqrt{5}}{2} \right)^{10n+5} \right] . \tag{68}$$

Finally, we mention a main conjecture for the MZVs over $\{0, 1\}$. Consider tuples $\mathbf{k} = (k_1, \ldots, k_r) \in \mathbb{N}^r, k_1 \geq 1$. One defines

$$\begin{aligned} \mathscr{Z}_0 &:= \mathbb{Q} \\ \mathscr{Z}_1 &:= \{0\} \\ \mathscr{Z}_w &:= \sum_{|\mathbf{k}|=w} \mathbb{Q} \cdot \zeta(\mathbf{k}) \subset \mathbb{R} . \end{aligned} \tag{69}$$

If further

$$\mathscr{Z}^{\text{Go}} := \sum_{w=0}^{\infty} \mathscr{Z}_w \subset \mathbb{R} \qquad \text{(Goncharov)} \tag{70}$$

$$\mathscr{Z}^{\mathrm{Ca}} := \bigoplus_{w=0}^{\infty} \mathscr{Z}_w \qquad \text{(Cartier)} \tag{71}$$

the conjecture states

(a) $\mathscr{Z}^{\mathrm{Go}} \cong \mathscr{Z}^{\mathrm{Ca}}$. There are no relations over $\mathbb{Q}$ between the MZVs of different weight w.
(b) $\dim \mathscr{Z}_w = d_w$, with $d_0 = 1, d_1 = 0, d_2 = 1, d_w = d_{w-2} + d_{w-3}$.
(c) All algebraic relations between MZVs are given by the extended double-shuffle relations [91], cf. also [92]. If this conjecture turns out to be true all MZVs are irrational numbers.

Let us also mention a few interesting relations for $\mathrm{Li}_2(z)$ for special arguments found by Ramanujan [93], which are of use e.g. in massive calculations at three-loops [94]. These numbers relate to constants beyond the MZVs, which occur for generalized sums and their extension allowing for binomial and inverse binomial weights:

$$\mathrm{Li}_2\left(\frac{1}{3}\right) - \mathrm{Li}_2\left(\frac{1}{9}\right) = \frac{\pi^2}{18} - \frac{1}{6}\ln^2(3) \tag{72}$$

$$\mathrm{Li}_2\left(-\frac{1}{2}\right) + \frac{1}{6}\mathrm{Li}_2\left(\frac{1}{9}\right) = -\frac{\pi^2}{18} + \ln(2)\ln(3) - \frac{1}{2}\ln^2(2) - \frac{1}{3}\ln^2(3) \tag{73}$$

$$\mathrm{Li}_2\left(\frac{1}{4}\right) + \frac{1}{3}\mathrm{Li}_2\left(\frac{1}{9}\right) = \frac{\pi^2}{18} + 2\ln(2)\ln(3) - 2\ln^2(2) - \frac{2}{3}\ln^2(3) \tag{74}$$

$$\mathrm{Li}_2\left(-\frac{1}{3}\right) - \frac{1}{3}\mathrm{Li}_2\left(\frac{1}{9}\right) = -\frac{\pi^2}{18} + \frac{1}{6}\ln^2(3) \tag{75}$$

$$\mathrm{Li}_2\left(-\frac{1}{8}\right) + \frac{1}{3}\mathrm{Li}_2\left(\frac{1}{9}\right) = -\frac{1}{2}\ln^2\left(\frac{9}{8}\right)\,. \tag{76}$$

For further specific numbers, which occur in the context of Quantum Field Theory calculations see also Sects. 6.3, 7, and 9.

6 Generalized Harmonic Sums and Polylogarithms

6.1 Generalized Harmonic Sums

Generalized harmonic sums, also called S-sums, are defined by [28, 29, 95]

$$S_{a_1,\ldots,a_k}(x_1,\ldots,x_k;N) = \sum_{i_1=1}^{N} \frac{x_1^{i_1}}{i_1^{a_1}} S_{a_2,\ldots,a_k}(x_2,\ldots,x_k;i_1),$$
$$S_\emptyset = 1\,, \quad x_i \in \mathbb{R}\backslash\{0\}, \quad a_i \in \mathbb{N}\backslash\{0\} \tag{77}$$

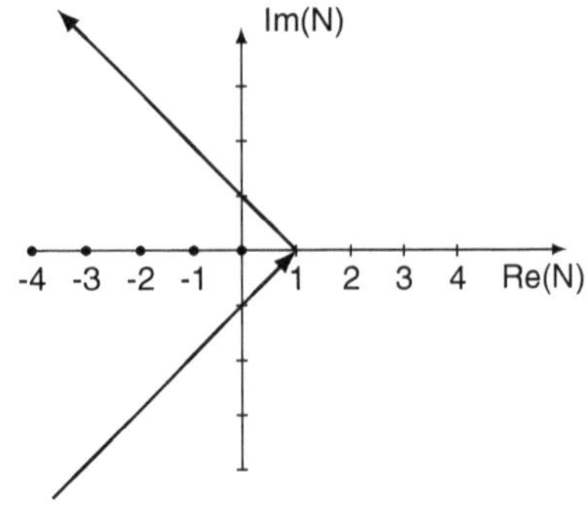

Fig. 2 Path of the contour integral (112)

9 Analytic Continuation of Harmonic Sums

The loop-corrections to various physical quantities take a particular simple form in Mellin-space being expressed in terms of harmonic sums and their generalizations. Moreover, in this representation the renormalization group equations can be solved analytically, cf. [59,112]. For a wide variety of non-perturbative parton distributions Mellin-space representations can be given as well, see e.g. [113].

Thus one obtains complete representations for observables in N-space. In case of the perturbative part, the singularities are situated at the integers $N \leq N_0$, with usually $N_0 = 1$, see e.g. [114]. The harmonic sums possess a unique polynomial representation in terms of the sum $S_1(N)$ and harmonic sums which can be represented as Mellin transforms having a representation by factorial series [115, 116]. They are transformed to x-space by a single precise numerical contour integral around the singularities of the problem to compare with the data measured in experiment. The analytic continuation of the perturbative evolution kernels and Wilson coefficients from even or odd integers to complex values of N is unique [117].[5] To perform this integral a representation of the harmonic sums for $N \in \mathbb{C}$ is required. Accurate numeric representations up to $w = 5$ have been given in [118], see also [119]. Arbitrary precise representations can be obtained using the analytic expressions for the asymptotic representation [14, 60] together with the recursion relations given in Sect. 3. These are given up to $w = 8$ in [58]. A path to perform the inverse Mellin transform

$$f(x) = \frac{1}{\pi} \int_0^\infty dz \, \mathrm{Im}\left[e^{i\phi}\, x^{-C} \mathbf{M}[f](N = C)\right], \quad C = c + ze^{i\phi} \qquad (112)$$

is shown in Fig. 2. The asymptotic representations can also be obtained in analytic form for the S-sums [29], cyclotomic (S)-sums [30], as well as for the nested binomial cyclotomic S-sums [94, 110].

[5] For a detailed proof also in case of generalized harmonic sums see [29].

In case the expressions in N-space result from Mellin transforms of functions $f(x) \sim x^\alpha, \alpha \in]0,1[$ the singularities are shifted by α. This is usually the case for the non-perturbative parton distribution functions, but also in case of some root-valued harmonic polylogarithms considered in Sect. 8.

The inverse Mellin-transform cannot be performed in the above way for integrands which do not vanish sufficiently fast enough as $|N| \to \infty$. Contributions of this kind are those leading to distribution-valued terms in x-space as to $\delta(1-x)$, $[\ln^k(1-x)/(1-x)]_+$ the Dirac δ-distribution and the $+$-distribution defined by

$$\int_0^1 dx\ [f(x)]_+\, g(x) = \int_0^1 dx\ [g(x) - g(1)]\, f(x)\ . \tag{113}$$

Also for terms which grow like $a^N, a \in \mathbb{R}, a > 1$ in N-space, the Mellin transform cannot be performed numerically in general. They are not supposed to emerge in physical observables. The physical quantities in hadronic scattering contain the parton distribution functions, which, however, damp according contributions occurring in the evolution kernels sufficiently. On the other hand, the inverse Mellin transform can always be performed analytically changing form nested sum-representations in N-space to iterated integral representations in x-space as has been outlined before.

10 Conclusions

Feynman integrals in Quantum Field Theories generate a hierarchic series of special functions, which allow their unique representation. They emerge in terms of special nested sums, iterated integrals and numbers. Their variety gradually extends enlarging the number of loops and legs, as well as the associated mass scales. The systematic exploration of these structures has been started about 15 years ago and several levels of complexity have been unraveled since. The relations of the various associated sums and integrals are schematically illustrated in Fig. 3 and are widely explored. Many relations are implied by the shuffle resp. stuffle algebras, others are structural relations. The number of relations grows with the number of admissible operations.

A large amount of transformations and relations between the different quantities being discussed in this article are encoded in the package `HarmonicSums` [29,35] for public use. For newly emerging structures the algebraic relations are easily generalized but they will usually apply structural relations of a new kind. With the present programme revealing their strict (atomic) structure, they are fully explored analytically and Feynman's original approach to completely organize the calculation of observables in Quantum Field Theory is currently extended to massive calculations at the three-loop level in Quantum Electrodynamics and Quantum Chromodynamics at the perturbative side. For these quantities efficient numerical

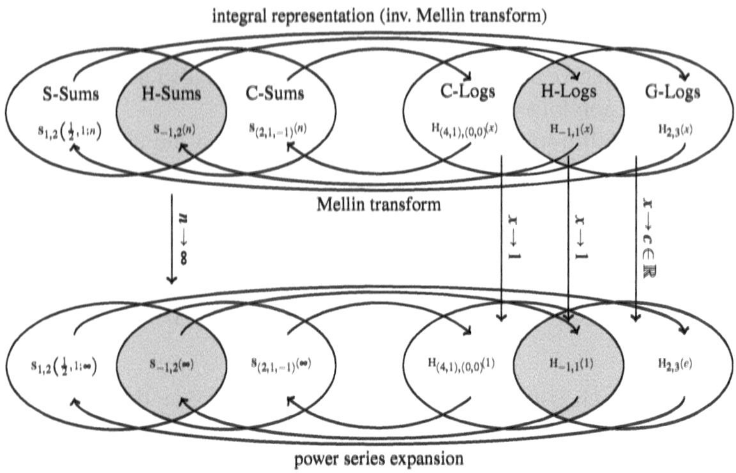

Fig. 3 Connection between harmonic sums (H-Sums), S-sums (S-Sums) and cyclotomic harmonic sums (C-Sums), their values at infinity and harmonic polylogarithms (H-Logs), generalized harmonic polylogarithms (G-Logs) and cyclotomic harmonic polylogarithms (C-Logs) and their values at special constants

representations have to be derived. Working in Mellin space the treatment may even remain *completely analytic*, in a very elegant way, up to a single final numerical contour integral around the singularities of the problem, cf. Sect. 9.

Despite of the achievements being obtained many more physical classes still await their systematic exploration in the future. It is clear, however, that the various concrete structures are realized as combinations of words over certain alphabets, which may be called the *genetic code of the microcosm* [120].

Acknowledgements We would like to thank D. Broadhurst, F. Brown, A. De Freitas, E.W.N. Glover, A. Hasselhuhn, D. Kreimer, C. Raab, C. Schneider, S. Weinzierl, F. Wißbrock, and J. Vermaseren for discussions. This work has been supported in part by DFG Sonderforschungsbereich Transregio 9, Computergestützte Theoretische Teilchenphysik, by the Austrian Science Fund (FWF) grants P20347-N18, P22748-N18, SFB F50 (F5009-N15) and by the EU Network LHCPHENOnet PITN-GA-2010-264564.

References

1. Lense, J.: Reihenentwicklungen in der Mathematischen Physik. Walter de Gryter, Berlin (1953)
2. Sommerfeld, A.: Partielle Differentialgleichungen der Physik. In: Vorlesungen über Theoretische Physik, vol. VI. Akademische Verlagsgesellschaft Geest and Prtig, Leipzig (1958)
3. Forsyth, A.R.: Theory of Differential Equations, pp. 2–4. Cambridge University Press, Cambridge (1900–1902);
   Kamke, E.: Differentialgleichungen, Lösungsmethoden und Lösungen. Akademische Verlagsgesellschaft Geest and Portig, Leipzig (1967)

4. Kratzer, A., Franz, W.: Transzendente Funktionen. Akademische Verlagsgesellschaft Geest and Portig, Leipzig (1963)
5. Andrews, G.E., Askey, R., Roy, R.: Special Functions. Cambridge University Press, Cambridge (2006)
6. Nakanishi, N.: Parametric integral formulas and analytic properties in perturbation theory. Suppl. Progr. Theor. Phys. **18**, 1–125 (1961); Graph Theory and Feynman Integrals. Gordon and Breach, New York (1970);
Bogner, C., Weinzierl, S.: Feynman graph polynomials. Int. J. Mod. Phys. A **25**, 2585–2618 (2010). [arXiv:1002.3458 [hep-ph]];
Weinzierl, S.: (this volume)
7. Vermaseren, J.A.M., Vogt, A., Moch, S.: The third-order QCD corrections to deep-inelastic scattering by photon exchange. Nucl. Phys. B **724**, 3–182 (2005). [arXiv:hep-ph/0504242] and refences therein
8. Bierenbaum, I., Blümlein, J., Klein, S.: Mellin moments of the $O(\alpha_s^3)$ heavy flavor contributions to unpolarized deep-inelastic scattering at $Q^2 \gg m^2$ and anomalous dimensions. Nucl. Phys. B **820**, 417–482 (2009). [arXiv: 0904.3563 [hep-ph]]
9. Kontsevich, M., Zagier, D.: Periods. IMHS/M/01/22. In: Engquist, B., Schmid, W. (eds.) Mathematics Unlimited – 2001 and Beyond, pp. 771–808. Springer, Berlin (2011)
10. Bogner, C., Weinzierl, S.: Periods and Feynman integrals. J. Math. Phys. **50**, 042302 (2009). [arXiv:0711.4863 [hep-th]]
11. Poincaré, H.: Sur les groupes des équations linéaires. Acta Math. **4**, 201–312 (1884);
Lappo-Danilevsky, J.A.: Mémoirs sur la Théorie des Systèmes Différentielles Linéaires. Chelsea Publishing Company, New York (1953);
Chen, K.T.: Algebras of iterated path integrals and fundamental groups. Trans. A.M.S. **156**(3), 359–379 (1971)
12. Jonquière, A.: Über einige Transcendente, welche bei der wiederholten Integration rationaler Funktionen auftreten. Bihang till Kongl. Svenska Vetenskaps-Akademiens Handlingar **15**, 1–50 (1889)
13. Mellin, H.: Über die fundamentale Wichtigkeit des Satzes von Cauchy für die Theorien der Gamma- und hypergeometrischen Funktionen. Acta Soc. Fennicae **21**, 1–115 (1886); Über den Zusammenhang zwischen den linearen Differential- und Differenzengleichungen. Acta Math. **25**, 139–164 (1902)
14. Blümlein, J.: Structural relations of harmonic sums and Mellin transforms up to weight w = 5. Comput. Phys. Commun. **180**, 2218–2249 (2009). [arXiv:0901.3106 [hep-ph]]
15. Blümlein, J., Klein, S., Schneider, C., Stan, F.: A symbolic summation approach to Feynman integral calculus. J. Symb. Comput. **47**, 1267–1289 (2012). [arXiv:1011.2656 [cs.SC]]
16. Barnes, E.W.: A new development of the theory of the hypergeometric functions. Proc. Lond. Math. Soc. **6**(2), 141 (1908); A transformation of generalized hypergeometric series. Quart. Journ. Math. **41**, 136–140 (1910);
Mellin, H.: Abriß einer einheitlichen Theorie der Gamma- und der hypergeometrischen Funktionen. Math. Ann. **68**, 305–337 (1910)
17. Gluza, J., Kajda, K., Riemann, T.: AMBRE: a mathematica package for the construction of Mellin-Barnes representations for Feynman integrals. Comput. Phys. Commun. **177**, 879–893 (2007). [arXiv:0704.2423 [hep-ph]]
18. 't Hooft, G., Veltman, M.J.G.: Regularization and renormalization of gauge fields. Nucl. Phys. B **44**, 189–213 (1972);
Bollini, C.G., Giambiagi, J.J.: Dimensional renormalization: the number of dimensions as a regularizing parameter. Nuovo Cim. B **12**, 20–26 (1972);
Ashmore, J.F.: A method of gauge invariant regularization. Lett. Nuovo Cim. **4**, 289–290 (1972);
Cicuta, G.M., Montaldi, E.: Analytic renormalization via continuous space dimension. Lett. Nuovo Cim. **4**, 329–332 (1972)
19. Berends, F.A., van Neerven, W.L., Burgers, G.J.H.: Higher order radiative corrections at LEP energies. Nucl. Phys. B **297**, 429–478 (1988). [Erratum-ibid. B **304**, 921 (1988)];

Blümlein, J., De Freitas, A., van Neerven, W.: Two-loop QED operator matrix elements with massive external fermion lines. Nucl. Phys. B **855**, 508–569 (2012). [arXiv:1107.4638 [hep-ph]];
Hamberg, R., van Neerven, W.L., Matsuura, T.: A complete calculation of the order α_s^2 correction to the Drell-Yan K-factor. Nucl. Phys. B **359**, 343–405 (1991). [Erratum-ibid. B **644**, 403–404 (2002)];
Zijlstra, E.B., van Neerven, W.L.: Contribution of the second order gluonic Wilson coefficient to the deep inelastic structure function. Phys. Lett. B **273**, 476–482 (1991); O(α_s^2) contributions to the deep inelastic Wilson coefficient. Phys. Lett. B **272**, 127–133 (1991); O(α_s^2) QCD corrections to the deep inelastic proton structure functions F_2 and F_L. Nucl. Phys. B **383**, 525–574 (1992); O(α_s^2) corrections to the polarized structure function $g_1(x, Q^2)$. Nucl. Phys. B **417**, 61–100 (1994). [Erratum-ibid. B **426**, 245 (1994)], [Erratum-ibid. B **773**, 105–106 (2007)];
Kazakov, D.I., Kotikov, A.V.: Totalas correction to deep-inelastic scattering cross section ratio $R = \sigma_L/\sigma_T$ in QCD. Calculation of the longitudinal structure function. Nucl. Phys. B **307**, 721–762 (1988). [Erratum-ibid. B **345**, 299 (1990)];
Kazakov, D.I., Kotikov, A.V., Parente, G., Sampayo, O.A., Sanchez Guillen, J.: Complete quartic (α_s^2) correction to the deep inelastic longitudinal structure function F_L in QCD. Phys. Rev. Lett. **65**, 1535–1538 (1990). [Erratum-ibid. **65**, 2921 (1990)];
Sanchez Guillen, J., Miramontes, J., Miramontes, M., Parente, G., Sampayo, O.A.: Next-toleading order analysis of the deep inelastic $R = \sigma_L/\sigma_{total}$. Nucl. Phys. B **353**, 337–345 (1991)

20. Leibniz, G.W.: Mathematische Schriften. In: Gerhardt C.J. (ed.) vol. III, p. 351. Verlag H.W.Schmidt, Halle (1858);
Euler, L.: Institutiones calculi integralis, vol. I, pp. 110–113. Impensis Academiae Imperialis Scientiarum, Petropoli (1768); Mémoires de l'Académie de Sint-P'etersbourg (1809–1810), vol. 3, pp. 26–42 (1811);
Landen, J.: A new method of computing sums of certain series. Phil. Trans. R. Soc. Lond. **51**, 553–565 (1760); Mathematical Memoirs, p. 112 (Printed for the Author, Nourse, J., London, 1780);
Lewin, L.: Dilogarithms and Associated Functions. Macdonald, London (1958);
Kirillov, A.N.: Dilogarithm identities. Prog. Theor. Phys. Suppl. **118**, 61–142 (1995). [hep-th/9408113];
Maximon, L.C.: The dilogarithm function for complex argument. Proc. R. Soc. A **459**, 2807–2819 (2003);
Zagier, D.: The remarkable dilogarithm. J. Math. Phys. Sci. **22**, 131–145 (1988); In: Cartier, P., Julia, B., Moussa et al., P. (eds.) Frontiers in Number Theory, Physics, and Geometry II – On Conformal Field Theories, Discrete Groups and Renormalization, pp. 3–65. Springer, Berlin (2007)
21. Spence, W.: An Essay of the Theory of the Various Orders of Logarithmic Transcendents. John Murray, London (1809)
22. Kummer, E.E.: Ueber die Transcendenten, welche aus wiederholten Integrationen rationaler Formeln entstehen. J. für Math. (Crelle) **21**, 74–90 (1840)
23. Lewin, L.: Polylogarithms and Associated Functions. North Holland, New York (1981);
Devoto, A., Duke, D.W.: Table of integrals and formulae for Feynman diagram calculations. Riv. Nuovo Cim. **7**(6), 1–39 (1984)
24. Nielsen, N.: Der Eulersche Dilogarithmus und seine Verallgemeinerungen. Nova Acta Leopold. **XC** Nr. 3, 125–211 (1909);
Kölbig, K.S., Mignoco, J.A., Remiddi, E.: On Nielsen's generalized polylogarithms and their numerical calculation. BIT **10**, 38–74 (1970);
Kölbig, K.S.: Nielsen's generalized polylogarithms. SIAM J. Math. Anal. **17**, 1232–1258 (1986)
25. Vermaseren, J.A.M.: Harmonic sums, Mellin transforms and integrals. Int. J. Mod. Phys. A **14**, 2037–2076 (1999). [hep-ph/9806280]

26. Blümlein, J., Kurth, S.: Harmonic sums and Mellin transforms up to two loop order. Phys. Rev. D **60**, 014018 (1999). [hep-ph/9810241]
27. Remiddi, E., Vermaseren, J.A.M.: Harmonic polylogarithms. Int. J. Mod. Phys. A **15**, 725–754 (2000). [hep-ph/9905237]
28. Moch, S.-O., Uwer, P., Weinzierl, S.: Nested sums, expansion of transcendental functions and multiscale multiloop integrals. J. Math. Phys. **43**, 3363–3386 (2002). [hep-ph/0110083]
29. Ablinger, J., Blümlein, J., Schneider, C.: Analytic and algorithmic aspects of generalized harmonic sums and polylogarithms. [arXiv:1302.0378 [math-ph]]
30. Ablinger, J., Blümlein, J., Schneider, C.: Harmonic sums and polylogarithms generated by cyclotomic polynomials. J. Math. Phys. **52**, 102301 (2011). [arXiv:1105.6063 [math-ph]]
31. Laporta, S.: High precision ε-expansions of massive four loop vacuum bubbles Phys. Lett. B **549**, 115–122 (2002). [hep-ph/0210336]; Analytical expressions of 3 and 4-loop sunrise Feynman integrals and 4-dimensional lattice integrals. Int. J. Mod. Phys. A **23**, 5007–5020 (2008). [arXiv:0803.1007 [hep-ph]];
Bailey, D.H., Borwein, J.M., Broadhurst, D., Glasser, M.L.: Elliptic integral evaluations of Bessel moments. [arXiv:0801.0891 [hep-th]];
Müller-Stach, S., Weinzierl, S., Zayadeh, R.: Picard-Fuchs equations for Feynman integrals. [arXiv:1212.4389 [hep-ph]];
Adams, L., Bogner, C., Weinzierl, S.: The two-loop sunrise graph with arbitrary masses. [arXiv:1302.7004 [hep-ph]]
32. Schneider, C.: The summation package sigma: underlying principles and a rhombus tiling application. Discret. Math. Theor. Comput. Sci. **6**, 365–386 (2004); Solving parameterized linear difference equations in terms of indefinite nested sums and products. Differ, J.: Equ. Appl. **11**(9), 799–821 (2005); A new sigma approach to multi-summation. Adv. Appl. Math. **34**(4), 740–767 (2005); Product representations in $\Pi\Sigma$-fields. Ann. Comb. **9**(1), 75–99 (2005); Symbolic summation assists combinatorics. Sem. Lothar. Combin. **56**, 1–36 (2007); A refined difference field theory for symbolic summation. J. Symb. Comp. **43**(9), 611–644 (2008). arXiv:0808.2543 [cs.SC]; Parameterized telescoping proves algebraic independence of sums. Ann. Comb. **14**(4), 533–552 (2010). [arXiv:0808.2596 [cs.SC]]; Structural theorems for symbolic summation. Appl. Algebra Eng. Comm. Comput. **21**(1), 1–32 (2010); A symbolic summation approach to find optimal nested sum representations. In: Carey, A., Ellwood, D., Paycha, S., Rosenberg, S. (eds.) Motives, Quantum Field Theory, and Pseudodifferential Operators, vol. 12, pp. 285–308. Clay Mathematics Proceedings. American Mathematical Society (2010). [arXiv:0904.2323 [cs.SC]]; and this volume
33. Bronstein, M.: Symbolic Integration I: Transcendental Functions. Springer, Berlin (1997)
34. Raab, C.: Definite integration in differential fields. PhD thesis, Johannes Kepler University Linz (2012); and this volume
35. Ablinger, J.: A computer algebra toolbox for harmonic sums related to particle physics. Master's thesis, Johannes Kepler University (2009). [arXiv:1011.1176 [math-ph]]; Computer algebra algorithms for special functions in particle physics. PhD thesis, Johannes Kepler University Linz (2012)
36. Feynman, R.P.: Space - time approach to quantum electrodynamics. Phys. Rev. **76**, 769–789 (1949)
37. Napier, J.: Mirifici logarithmorum canonis descriptio, ejusque usus, in utraque trigonometria; ut etiam in omni logistica mathematica, amplissimi, facillimi, & expeditissimi explacatio. Andrew Hart, Edinburgh (1614)
38. Racah, G.: Sopra l'rradiazione nell'urto di particelle veloci. Nuovo Com. **11**, 461–476 (1934)
39. Fleischer, J., Kotikov, A.V., Veretin, O.L.: Analytic two loop results for selfenergy type and vertex type diagrams with one nonzero mass. Nucl. Phys. B **547**, 343–374 (1999). [hep-ph/9808242]
40. Blümlein, J., Ravindran, V.: Mellin moments of the next-to-next-to leading order coefficient functions for the Drell-Yan process and hadronic Higgs-boson production. Nucl. Phys. B **716**, 128–172 (2005). [hep-ph/0501178]; O(α_s^2) timelike Wilson coefficients for parton-fragmentation functions in Mellin space. Nucl. Phys. B **749**, 1–24 (2006). [hep-ph/0604019];

Blümlein, J., Klein, S.: Structural relations between harmonic sums up to w=6. PoS ACAT **084** (2007). [arXiv:0706.2426 [hep-ph]];
Bierenbaum, I., Blümlein, J., Klein, S., Schneider, C.: Two-loop massive operator matrix elements for unpolarized heavy flavor production to O(ε). Nucl. Phys. B **803**, 1–41 (2008). [arXiv:0803.0273 [hep-ph]];
Czakon, M., Gluza, J., Riemann, T.: Master integrals for massive two-loop Bhabha scattering in QED. Phys. Rev. D **71**, 073009 (2005). [hep-ph/0412164]
41. Moch, S., Vermaseren, J.A.M., Vogt, A.: The three loop splitting functions in QCD: the non-singlet case. Nucl. Phys. B **688**, 101–134 (2004). [hep-ph/0403192]; The three-loop splitting functions in QCD: the Singlet case. Nucl. Phys. B **691**, 129–181 (2004). [hep-ph/0404111]
42. Gehrmann, T., Remiddi, E.: Numerical evaluation of harmonic polylogarithms. Comput. Phys. Commun. **141**, 296–312 (2001). [arXiv:hep-ph/0107173]
43. Vollinga, J., Weinzierl, S.: Numerical evaluation of multiple polylogarithms. Comput. Phys. Commun. **167**, 177–194 (2005). [arXiv:hep-ph/0410259]
44. Gonzalez-Arroyo, A., Lopez, C., Yndurain, F.J.: Second order contributions to the structure functions in deep inelastic scattering. 1. Theoretical calculations. Nucl. Phys. B **153**, 161–186 (1979);
Floratos, E.G., Kounnas, C., Lacaze, R.: Higher order QCD effects in inclusive annihilation and deep inelastic scattering. Nucl. Phys. B **192**, 417–462 (1981)
45. Mertig, R., van Neerven, W.L.: The calculation of the two loop spin splitting functions $P_{ij}^{(1)}(x)$. Z. Phys. C **70**, 637–654 (1996). [hep-ph/9506451]
46. Wilson, K.G.: Non-lagrangian models of current algebra. Phys. Rev. **179**, 1499–1512 (1969); Zimmermann, W.: Lecture on Elementary Particle Physics and Quantum Field Theory, Brandeis Summer Institute, vol. 1, p. 395. MIT Press, Cambridge (1970); Brandt, R.A., Preparata, G.: The light cone and photon-hadron interactions. Fortsch. Phys. **20**, 571–594 (1972); Frishman, Y.: Operator products at almost light like distances. Ann. Phys. **66**, 373–389 (1971); Blümlein, J., Kochelev, N.: On the twist-2 and twist-three contributions to the spin dependent electroweak structure functions. Nucl. Phys. B **498**, 285–309 (1997). [hep-ph/9612318]
47. Hoffman, M.E.: Quasi-shuffle products. J. Algebraic Combin. **11**, 49–68 (2000). [arXiv:math/9907173 [math.QA]]; The Hopf algebra structure of multiple harmonic sums. Nucl. Phys. (Proc. Suppl.) **135**, 215 (2004). [arXiv:math/0406589]
48. Berndt, B.C.: Ramanujan's Notebooks, Part I. Springer, Berlin (1985)
49. Faà di Bruno, F.: Einleitung in die Theorie dier Binären Formen, deutsche Bearbeitung von Th. Walter. Teubner, Leipzig (1881)
50. Blümlein, J.: Algebraic relations between harmonic sums and associated quantities. Comput. Phys. Commun. **159**, 19–54 (2004). [hep-ph/0311046]
51. Borwein, J.M., Bradley, D.M., Broadhurst, D.J., Lisonek, P.: Special values of multiple polylogarithms. Trans. Am. Math. Soc. **353**, 907–941 (2001). [math/9910045 [math-ca]]
52. Lyndon, R.C.: On Burnsides problem. Trans. Am. Math. Soc. **77**, 202–215 (1954); On Burnsides problem II. Trans. Amer. Math. Soc. **78**, 329–332 (1955)
53. Reutenauer, C.: Free Lie algebras. University Press, Oxford (1993)
54. Radford, D.E.: J. Algebra **58**, 432–454 (1979)
55. Witt, E.: Treue Darstellung Liescher Ringe. J. Reine und Angew. Math. **177**, 152–160 (1937); Die Unterringe der freien Lieschen Ringe. Math. Zeitschr. **64**, 195–216 (1956)
56. Möbius, A.F.: Über eine besondere Art von Umkehrung der Reihen. J. Reine Angew. Math. **9**, 105–123 (1832);
Hardy, G.H., Wright, E.M.: An Introduction to the Theory of Numbers, 5th edn. Calendron Press, Oxford (1978)
57. Maitre, D.: HPL, a mathematica implementation of the harmonic polylogarithms. Comput. Phys. Commun. **174**, 222–240 (2006). [hep-ph/0507152]
58. Ablinger, J., Blümlein, J., Schneider, C.: DESY 13-064

59. Blümlein, J.: The theory of deeply inelastic scattering. Prog. Part. Nucl. Phys. **69**, 28 (2013). [arXiv:1208.6087 [hep-ph]]
60. Blümlein, J.: Structural relations of harmonic sums and Mellin transforms at weight w = 6. In: Carey, A., Ellwood, D., Paycha, S., Rosenberg, S. (eds.) Motives, Quantum Field Theory, and Pseudodifferential Operators, vol. 12, pp. 167–186. Clay Mathematics Proceedings, American Mathematical Society (2010). [arXiv:0901.0837 [math-ph]].
61. Euler, L.: Meditationes circa singulare serium genus. Novi Commentarii academiae scientiarum Petropolitanae **20**, 140–186 (1776)
62. Zagier, D.: Values of zeta functions and their applications. In: First European Congress of Mathematics (Paris, 1992), vol. II. Prog. Math. **120**, 497–512 (Birkhäuser, Basel–Boston) (1994)
63. Blümlein, J., Broadhurst, D.J., Vermaseren, J.A.M.: The multiple zeta value data mine. Comput. Phys. Commun. **181**, 582–625 (2010). [arXiv:0907.2557 [math-ph]] and references therein
64. Hoffman, M.E.: Hoffman's page. http://www.usna.edu/Users/math/~meh/biblio.html
65. Fischler, S.: Irrationalité de valeurs de zéta. Sém. Bourbaki, Novembre 2002, exp. no. **910**. Asterisque **294**, 27–62 (2004) http://www.math.u-psud.fr /~fischler/publi.html;
Colmez, P.: Arithmetique de la fonction zêta. In: Journées X-UPS 2002. La fontion zêta. Editions de l'Ecole polytechnique,
pp. 37–164. Paris (2002). http://www.math.polytechnique.fr/xups/vol02.html;
Waldschmidt, M.: Multiple polylogarithms: an introduction. In: Agarwal, A.K., Berndt, B.C., Krattenthaler, C.F., Mullen, G.L., Ramachandra, K., Waldschmidt, M. (eds.) Number Theory and Discrete Mathematics, pp. 1–12. Hindustan Book Agency, New Delhi (2002);
Waldschmidt, M.: Valeurs zêta multiples. Une introduction. Journal de théorie des nombres de Bordeaux **12**(2), 581–595 (2000);
Huttner, M., Petitot, M.: Arithmeétique des fonctions d'zetas et Associateur de Drinfel'd. UFR de Mathématiques, Lille (2005)
Hertling, C.: AG Mannheim-Heidelberg, SS2007;
Cartier, P.: Fonctions polylogarithmes, nombres polyzêtas et groupes pro-unipotents. Sém. Bourbaki, Mars 2001, 53*e* année, exp. no. **885**. Asterisque **282**, 137–173 (2002)
66. Barbieri, R., Mignaco, J.A., Remiddi, E.: Electron form-factors up to fourth order. 1. Nuovo Cim. A **11**, 824–864 (1972);
Levine, M.J., Remiddi, E., Roskies, R.: Analytic contributions to the G factor of the electron in sixth order. Phys. Rev. D **20**, 2068–2076 (1979)
67. Kuipers, J., Vermaseren, J.A.M.: About a conjectured basis for multiple zeta values. [arXiv:1105.1884 [math-ph]]
68. Goncharov, A.B.: Multiple polylogarithms and mixed Tate motives. [arxiv:math.AG/0103059];
Terasoma, T.: Mixed Tate motives and multiple zeta values. Invent. Math. **149**(2), 339–369 (2002). arxiv:math.AG/010423;
Deligne, P., Goncharov, A.B.: Groupes fondamentaux motiviques de Tate mixtes. Ann. Sci. Ecole Norm. Sup. Série IV **38**(1), 1–56 (2005)
69. Broadhurst, D.J.: On the enumeration of irreducible k-fold Euler sums and their roles in knot theory and field theory. [arXiv:hep-th/9604128]
70. Broadhurst, D.J., Kreimer, D.: Association of multiple zeta values with positive knots via Feynman diagrams up to 9 loops. Phys. Lett. B **393**, 403–412 (1997). [arXiv:hep-th/9609128]
71. http://en.wikipedia.org/wiki/Padovan_sequence;
http://www.emis.de/journals/NNJ/conferences/N2002-Padovan.html
72. Perrin, R.: Item 1484, L'Intermédiare des Math. **6**, 76–77 (1899);
Williams, A., Shanks, D.: Strong primality tests that are not sufficient. Math. Comput. **39**(159), 255–300 (1982)
73. de filiis Bonaccij, L.P.: Liber abaci, Cap. 12.7. Pisa (1202);
Sigler, L.E.: Fibonacci's Liber Abaci. Springer, Berlin (2002)
74. Hoffman, M.E.: The algebra of multiple harmonic series. J. Algebra **194**, 477–495 (1997)

75. Brown, F.: Mixed tate motives over Z. Ann. Math. **175**(1), 949–976 (2012)
76. The formula goes back to de Moivre, Bernoulli, Euler, and later Binet, see: Beutelspacher, A., Petri, B.: Der Goldene Schnitt. Spektrum, Heidelberg (1988)
77. Lucas, E.: Théorie des fonctions numériques simplement périodiques. Am. J. Math. **1**, 197–240 (1878)
78. Euler, L.: Meditationes circa singulare serium genus. Novi Comm. Acad. Sci. Petropol. **20**, 140–186 (1775). (reprinted in Opera Omnia Ser I, vol. 15, pp. 217–267. B.G. Teubner, Berlin (1927))
79. Hoffman, M. E., Moen, C.: Sums of triple harmonic series. J. Number Theory **60**, 329–331 (1996);
Granville, A.: A decomposition of Riemann's zeta-function. In: Motohashi, Y. (ed.) Analytic Number Theory, London Mathematical Society. Lecture Note Series, vol. 247, pp. 95–101. Cambridge University Press, Cambridge (1997);
Zagier, D.: Multiple Zeta Values. (preprint)
80. Hoffman, M.E.: Multiple harmonic series. Pac. J. Math. **152**, 275–290 (1992)
81. Hoffman, M.E.: Algebraic aspects of multiple zeta values. In: Aoki et al., T. (eds.) Zeta Functions, Topology and Quantum Physics. Developments in Mathematics, vol. 14, pp. 51–74. Springer, New York (2005). [arXiv:math/0309452 [math.QA]]
82. Okuda, J., Ueno, K.: The sum formula of multiple zeta values and connection problem of the formal Knizhnik–Zamolodchikov equation. In: Aoki et al., T. (eds.) Zeta Functions, Topology and Quantum Physics. Developments in Mathematics, vol. 14, pp. 145–170. Springer, New York (2005). [arXiv:math/0310259 [math.NT]]
83. Hoffmann, M.E., Ohno, Y.: Relations of multiple zeta values and their algebraic expressions. J. Algebra **262**, 332–347 (2003)
84. Zudilin, V.V.: Algebraic relations for multiple zeta values. Uspekhi Mat. Nauk **58**(1), 3–22
85. Ihara, K., Kaneko, M.: A note on relations of multiple zeta values (preprint)
86. Le, T.Q.T., Murakami, J.: Kontsevich's integral for the Homfly polynomial and relations between values of multiple zeta functions. Topol. Appl. **62**, 193–206 (1995)
87. Ohno, Y.: A generalization of the duality and sum formulas on the multiple zeta values. J. Number Theory **74**, 189–209 (1999)
88. Ohno, Y., Zagier, D.: Indag. Math. (N.S.) **12**, 483–487 (2001)
89. Ohno, Y., Wakabayashi, N.: Cyclic sum of multiple zeta values. Acta Arithmetica **123**, 289–295 (2006)
90. Borwein, J.M., Bradley, D.M., Broadhurst, D.J.: Evaluations of K fold Euler/Zagier sums: a compendium of results for arbitrary k. [hep-th/9611004]
91. Ihara, K., Kaneko, M., Zagier, D.: Derivation and double shuffle relations for multiple zeta values. Compositio Math. **142**, 307–338 (2006); Preprint MPIM2004-100
92. Écalle, J.: Théorie des moules, vol. 3, prépublications mathématiques d'Orsay (1981, 1982, 1985); La libre génération des multicêtas et leur d'ecomposition canonico-explicite en irréductibles, automne (1999); Ari/gari et la décomposition des multizêtas en irréductibles. Prépublication, avril (2000)
93. Berndt, B.C.: Ramanujan's Notebooks, Part IV, pp. 323–326. Springer, New York (1994)
94. Ablinger, J., Blümlein, J., Raab, C., Schneider, C., Wißbrock, F.: DESY 13–063
95. Weinzierl, S.: Feynman Graphs. [arXiv:1301.6918 [hep-ph]]
96. Hopf, H.: Über die Topologie der Gruppen-Mannigfaltigkeiten und ihrer Verallgemeinerungen. Ann. Math. **42**, 22–52 (1941);
Milner, J., Moore, J.: On the structure of Hopf algebras. Ann. Math. **81**, 211–264 (1965);
Sweedler, M.E.: Hopf Algebras. Benjamin, New York (1969)
97. Goncharov, A.B.: Multiple polylogarithms, cyclotomy and modular complexes. Math. Res. Lett. **5**, 497–516 (1998). [arXiv:1105.2076 [math.AG]]
98. Gehrmann, T., Remiddi, E.: Two loop master integrals for $\gamma^* \to 3$ jets: the planar topologies. Nucl. Phys. B **601**, 248–286 (2001). [hep-ph/0008287]; Numerical evaluation of twodimensional harmonic polylogarithms. Comput. Phys. Commun. **144**, 200–223 (2002). [hep-ph/0111255]

99. Weinzierl, S.: Symbolic expansion of transcendental functions. Comput. Phys. Commun. **145**, 357–370 (2002). [math-ph/0201011]
100. Moch, S.-O., Uwer, P.: XSummer: transcendental functions and symbolic summation in FORM. Comput. Phys. Commun. **174**, 759–770 (2006). [math-ph/0508008]
101. Appell, P.: Sur Les Fonctions Hypérgéometriques de Plusieurs Variables. Gauthier-Villars, Paris (1925);
Appell, P., Kampé de Fériet, J.: Fonctions Hypérgéometriques; Polynômes d'Hermite. Gauthier-Villars, Paris (1926);
Bailey, W.N.: Generalized Hypergeometric Series. Cambridge University Press, Cambridge (1935);
Erdélyi, A. (ed.): Higher Transcendental Functions, Bateman Manuscript Project, vol. I. McGraw-Hill, New York (1953);
Exton, H.: Multiple Hypergeometric Functions and Applications. Ellis Horwood Limited, Chichester (1976); Handbook of Hypergeometric Integrals. Ellis Horwood Limited, Chichester (1978);
Slater, L.J.: Generalized Hypergeometric Functions. Cambridge University Press, Cambridge (1966);
Schlosser, H.: (this volume);
Anastasiou, C., Glover, E.W.N., Oleari, C.: Application of the negative dimension approach to massless scalar box integrals. Nucl. Phys. B **565**, 445–467 (2000). [hep-ph/9907523]; Scalar one loop integrals using the negative dimension approach. Nucl. Phys. B **572**, 307–360 (2000). [hep-ph/9907494];
Glover, E.W.N.: (this volume)
102. Ablinger, J., Blümlein, J., Hasselhuhn, A., Klein, S., Schneider, C., Wißbrock, F.: Massive 3-loop ladder diagrams for quarkonic local operator matrix elements. Nucl. Phys. B **864**, 52–84 (2012). [arXiv:1206.2252 [hep-ph]]
103. Ablinger, J., et al.: New results on the 3-loop heavy Flavor Wilson coefficients in deep-inelastic scattering. [arXiv:1212.5950 [hep-ph]]; Three-loop contributions to the gluonic massive operator matrix elements at general values of N. PoS LL **2012**, 033 (2012). [arXiv:1212.6823 [hep-ph]]
104. Lang, S.: Algebra, 3rd edn. Springer, New York (2002)
105. Euler, L.: Theoremata arithmetica nova methodo demonstrata. Novi Commentarii academiae scientiarum imperialis Petropolitanae, vol. 8, pp. 74–104 (1760/1, 1763); Opera Omnia, Ser. I, vol. 2, pp. 531–555. Takase, M.: Euler's Theory of Numbers In: Baker, R. (ed.) Euler Reconsidered, pp. 377–421. Kedrick Press, Heber City (2007). leonhardeuler.web.fc2.com/eulernumber_en.pdf
106. Catalan, E.: Recherches sur la constant G, et sur les integrales euleriennes. Mémoires de l'Academie imperiale des sciences de Saint-Pétersbourg, Ser. **7**(31), 1–51 (1883);
Adamchik, V. http://www-2.cs.cmu.edu/~adamchik/articles/catalan/catalan.htm
107. Broadhurst, D.J.: Massive three - loop Feynman diagrams reducible to SC^* primitives of algebras of the sixth root of unity. Eur. Phys. J. C **8** 311–333 (1999). [hep-th/9803091]
108. Racinet, G.: Torseurs associes a certaines relations algebriques entre polyzetas aux racines de l'unite. Comptes rendus de l'Académie des sciences. Série 1, Mathématique **333**(1), 5–10 (2001). [arXiv:math.QA/0012024]
109. Davydychev, A.I., Kalmykov, M.Y.: Massive Feynman diagrams and inverse binomial sums. Nucl. Phys. B **699**, 3–64 (2004). [hep-th/0303162];
Weinzierl, S.: Expansion around half integer values, binomial sums and inverse binomial sums. J. Math. Phys. **45**, 2656–2673 (2004). [hep-ph/0402131];
Kalmykov, M.Y.: Gauss' hypergeometric function: Reduction, ε-expansion for integer/halfinteger parameters and Feynman diagrams. JHEP **0604**, 056 (2006). [hep-th/0602028];
Huber, T., Maitre, D.: Expanding hypergeometric functions about half-integer parameters. Comput. Phys. Commun. **178**, 755–776 (2008). [arXiv:0708.2443 [hep-ph]]

The first irreducible MZV of depth 2, namely $\zeta(5,3)$, occurs in a counterterm coming from the most symmetric six-loop diagram for the ϕ^4 coupling, in which each of the 4 vertices connected to an external line is connected to each of the 3 other vertices, giving 12 internal propagators (or edges, as mathematicians prefer to call them). It hence diverges, at large loop momenta, in the manner of $\int \mathrm{d}^{24}k/k^{24}$. Its contribution to the β-function of ϕ^4-theory is scheme-independent and may be computed to high accuracy by using Gegenbauer polynomial expansions in x-space, which give the counterterm as a four-fold sum that is far from obviously a MZV. Accelerated convergence of truncations of this sum gave an empirical **Q**-linear of combination of $\zeta(5)\zeta(3)$ with

$$\zeta(5,3) - \frac{29}{12}\zeta(8)$$

and the latter combination was found to occur in another six-loop counterterm. I shall attempt to demystify the multiple of $\zeta(8)$ after discussing alternating Euler sums.

At seven loops, Dirk Kreimer and I found the combination

$$\zeta(3,5,3) - \zeta(3)\zeta(5,3)$$

in three different counterterms, where it occurs in combination with rational multiples of $\zeta(11)$ and $\zeta^2(3)\zeta(5)$.

4 Alternating Euler Sums

This second topic is closely related to the first, namely alternating sums of the form

$$\sum_{n_1>n_2>\ldots>n_k>0}^{\infty} \frac{\varepsilon_1^{n_1}\ldots\varepsilon_k^{n_k}}{n_1^{s_1}\ldots n_k^{s_k}}$$

with positive integers s_j and signs $\varepsilon_j = \pm 1$. We may compactly indicate the presence of an alternating sign, when $\varepsilon_j = -1$, by placing a bar over the corresponding integer exponent s_j. Thus we write

$$\zeta(\overline{3},\overline{1}) = \sum_{m>n>0} \frac{(-1)^{m+n}}{m^3 n}$$

$$\zeta(3,\overline{6},3,\overline{6},3) = \sum_{j>k>l>m>n>0} \frac{(-1)^{k+m}}{j^3 k^6 l^3 m^6 n^3}$$

using the same symbol ζ as we did for the MZVs. Such sums may be studied using `EZFace` and the `DataMine`.

4.1 Three-Letter Alphabet

Alternating sums have a stuffle algebra, from their representation as nested sums, and a shuffle algebra, from their representation as iterated integrals. In the integral representation we need a third letter, C, in our alphabet, corresponding to the differential form $\mathrm{d}x/(1+x)$. Consider

$$Z(ABC) = \int_0^1 \frac{\mathrm{d}x}{x} \int_0^x \frac{\mathrm{d}y}{1-y} \int_0^y \frac{\mathrm{d}z}{1+z}.$$

The z-integral gives $\log(1+y) = -\sum_{j>0}(-y)^j/j$ and hence

$$Z(ABC) = -\sum_{j>0} \int_0^1 \frac{\mathrm{d}x}{x} \int_0^x \frac{\mathrm{d}y}{1-y} \frac{(-y)^j}{j}.$$

Expanding $1/(1-y) = \sum_{k>0} y^{k-1}$ and integrating over y we obtain

$$Z(ABC) = -\sum_{k>0} \sum_{j>0} \int_0^1 \frac{\mathrm{d}x}{x} \frac{x^{j+k}}{j+k} \frac{(-1)^j}{j}$$

and the final integration gives

$$Z(ABC) = -\sum_{k>0} \sum_{j>0} \frac{1}{(j+k)^2} \frac{(-1)^j}{j}.$$

Finally, the substitution $k = m - j$ gives

$$Z(ABC) = -\sum_{m>j>0} \frac{(-1)^j}{m^2 j} = -\zeta(2, \overline{1}).$$

It takes a bit of practice to translate between words and sums. Here's another example:

$$Z(ACCAC) = (-1)^3 \sum_{l>0} \sum_{k>0} \sum_{j>0} \frac{(-1)^l}{(j+k+l)^2} \frac{(-1)^k}{j+k} \frac{(-1)^j}{j^2}$$

gives

$$Z(ACCAC) = -\sum_{m>n>j>0} \frac{(-1)^m}{m^2 n j^2} = -\zeta(\overline{2}, 1, 2)$$

after the substitutions $l = m - n$ and $k = n - j$.

Going from sums to words is quite tricky. For example, try to find the word W and the sign $\varepsilon(W)$ such that

$$\zeta(3,\overline{6},3,\overline{6},3) = \varepsilon(W)Z(W).$$

Note that $\varepsilon(W)$ is $+1$ or -1 according as whether there is an odd or even number of letters C in the word W. The word W begins $AABAAAAACAA\ldots$. The next letter is either B or C, but which is it?

4.2 *Shuffles and Stuffles for Alternating Sums*

The six shuffles in

$$\mathcal{S}(ab, xy) = \{abxy, axby, xaby, axyb, xayb, xyab\}$$

give six different words, with $a = A, b = B, x = y = C$:

$$\begin{aligned} Z(AB)Z(CC) = \; & Z(ABCC) + Z(ACBC) + Z(CABC) \\ & + Z(ACCB) + Z(CACB) + Z(CCAB) \end{aligned}$$

which translates to

$$\zeta(2)\zeta(\overline{1},1) = \zeta(2,\overline{1},1) + \zeta(\overline{2},\overline{1},\overline{1}) + \zeta(\overline{1},\overline{2},\overline{1}) + \zeta(\overline{2},1,\overline{1}) + \zeta(\overline{1},2,\overline{1}) + \zeta(\overline{1},1,\overline{2}).$$

The stuffles for this product are

$$\zeta(2)\zeta(\overline{1},1) = \zeta(2,\overline{1},1) + \zeta(\overline{3},1) + \zeta(\overline{1},2,1) + \zeta(\overline{1},3) + \zeta(\overline{1},1,2).$$

4.3 *Transforming Words*

The transformation $x = (1-y)/(1+y)$ gives

$$\begin{aligned} \mathrm{d}\log(x) &= \mathrm{d}\log(1-y) - \mathrm{d}\log(1+y) \\ \mathrm{d}\log(1-x) &= \mathrm{d}\log(y) - \mathrm{d}\log(1+y) \\ \mathrm{d}\log(1+x) &= -\mathrm{d}\log(1+y) \end{aligned}$$

and maps $x = 0$ and $x = 1$ to $y = 1$ and $y = 0$. Thus, if we take a word W, write it backwards, and make the transformations

$$A \to (B + C)$$
$$B \to (A - C)$$

we may obtain an expression for $Z(W)$ by expanding the brackets.

For example the transformation

$$AB \to (A - C)(B + C) = AB + AC - CB - CC$$

gives

$$Z(AB) = Z(AB) + Z(AC) - Z(CB) - Z(CC).$$

Combining this with the shuffle

$$Z(C)Z(C) = Z(CC) + Z(CC)$$

we obtain

$$0 = Z(AC) - Z(CB) - \frac{1}{2}Z(C)Z(C) = -\zeta(\overline{2}) + \zeta(\overline{1}, \overline{1}) - \frac{1}{2}\zeta(\overline{1})\zeta(\overline{1}).$$

Combining this with the stuffle

$$\zeta(\overline{1})\zeta(\overline{1}) = \zeta(\overline{1}, \overline{1}) + \zeta(2) + \zeta(\overline{1}, \overline{1})$$

we obtain

$$\zeta(\overline{2}) = -\frac{1}{2}\zeta(2)$$

which is also obtainable as follows.

4.4 Doubling Relations

For $a > 1$ we have

$$\zeta(a) + \zeta(\overline{a}) = \sum_{n>0} \frac{1 + (-1)^n}{n^a} = \sum_{k>0} \frac{2}{(2k)^a} = 2^{1-a}\zeta(a)$$

by the substitution $n = 2k$. Hence

$$\zeta(\overline{a}) = (2^{1-a} - 1)\zeta(a).$$

At $a = 2$, we obtain $\zeta(\overline{2}) = -\zeta(2)/2$, as above. Note also that $\zeta(\overline{1}) = -\log(2)$.

We may take any MZV and convert it into a combination of MZVs and alternating sums, by doubling the summation variables. For example, we obtain

$$\begin{aligned} 2^{2-a-b}\zeta(a,b) &= \sum_{m>n>0} \frac{2}{(2m)^a}\frac{2}{(2n)^b} \\ &= \sum_{j>k>0} \frac{1+(-1)^j}{j^a}\frac{1+(-1)^k}{k^a} \\ &= \zeta(a,b) + \zeta(\overline{a},b) + \zeta(a,\overline{b}) + \zeta(\overline{a},\overline{b}) \end{aligned}$$

by the transformations $j = 2m$ and $k = 2n$.

More complicated doubling relations were used in constructing the `DataMine`. With these, it was possible to avoid using the time-consuming transformations $A \to (B + C)$ and $B \to (A - C)$ as algebraic input. It was verified that the output, obtained by shuffling, stuffling and doubling, satisfied the relations that follow from word transformation.

4.5 Conjectured Enumeration of Irreducible Alternating Sums

Before considering the enumeration of irreducible MZVs, in the (A, B) alphabet, I already had a rather simple conjecture for the generator of the number, $E_{w,d}$, of irreducible sums of weight w and depth d in the (A, B, C) alphabet, namely

$$\prod_{w>2}\prod_{d>0}(1 - x^w y^d)^{E_{w,d}} \stackrel{?}{=} 1 - \frac{x^3 y}{(1-xy)(1-x^2)}. \tag{9}$$

If this be true, it is easy to obtain $E_{w,d}$ by Möbius transformation of the binomial coefficients in Pascal's triangle. Let [8]

$$T(a,b) = \frac{1}{a+b}\sum_{c|a,b}\mu(c)\frac{(a/c + b/c)!}{(a/c)!(b/c)!} \tag{10}$$

where the sum is over all positive integers c that divide both a and b and the Möbius function is defined by

$$\mu(c) = \begin{cases} 1 & \text{when } c = 1 \\ 0 & \text{when } c \text{ is divisible by the square of a prime} \\ (-1)^k & \text{when } c \text{ is the product of } k \text{ distinct primes.} \end{cases} \tag{11}$$

When w and d have the same parity, and $w > d$, one obtains from (9)

$$E_{w,d} = T\left(\frac{w-d}{2}, d\right). \tag{12}$$

The DataMine now provides extensive evidence to support this conjecture. It was verified at depth 6 up to weight 12, solving the algebraic input in rational arithmetic, and then up to weight 18, using arithmetic modulo a 31-bit prime. At depth 5, the corresponding weights are 17 and 21. At depth 4, they are 22 and 30.

5 Pushdown from MZVs to Alternating Sums

Now consider the integers $M_{w,d}$ generated by an even simpler process:

$$\prod_{w>2}\prod_{d>0}(1-x^w y^d)^{M_{w,d}} = 1 - \frac{x^3 y}{1-x^2}. \tag{13}$$

But what is the question, to which this is the answer?

I conjectured that $M_{w,d}$ is the number of irreducible sums of weight w and depth d in the (A, B, C) alphabet that suffice for the evaluation of MZVs in the (A, B) alphabet.

5.1 *Pushdown at Weight 12*

As already hinted, the first place that this conjecture becomes non-trivial is at weight 12, where the enumerations $M_{12,4} = 0$ and $M_{12,2} = 2$ are to be contrasted with the enumerations $D_{12,4} = 1$ and $D_{12,2} = 1$ of irreducible MZVs. The conjecture requires that

$$\zeta(6,4,1,1) = \sum_{k>l>m>n>0} \frac{1}{k^6 l^4 mn}$$

be reducible to sums of lesser depth, if we include an alternating double sum in the basis.

In 1996, I found such a "pushdown" empirically, using the integer-relation search routine PSLQ [10]. It took another decade to prove such an integer relation, by the laborious process of solving all the known algebraic relations in the (A, B, C) alphabet at weight 12 and depths up to 4. Jos Vermaseren derived this proven identity from the DataMine:

$$\begin{aligned}\zeta(6,4,1,1) = &-\frac{64}{27}A(7,5) - \frac{7{,}967}{1{,}944}\zeta(9,3) + \frac{1}{12}\zeta^4(3) + \frac{11{,}431}{1{,}296}\zeta(7)\zeta(5)\\ &-\frac{799}{72}\zeta(9)\zeta(3) + 3\zeta(2)\zeta(7,3) + \frac{7}{2}\zeta(2)\zeta^2(5) + 10\zeta(2)\zeta(7)\zeta(3)\\ &+\frac{3}{5}\zeta^2(2)\zeta(5,3) - \frac{1}{5}\zeta^2(2)\zeta(5)\zeta(3) - \frac{18}{35}\zeta^3(2)\zeta^2(3)\\ &-\frac{5{,}607{,}853}{6{,}081{,}075}\zeta^6(2)\end{aligned}$$

where

$$A(7,5) = Z(AAAAAA(B-C)AAAAB) = \zeta(7,5) + \zeta(\overline{7},\overline{5}).$$

It is now proven that all MZVs of weight up to 12 are reducible to **Q**-linear combinations of $\zeta(5,3)$, $\zeta(7,3)$, $\zeta(3,5,3)$, $\zeta(9,3)$, $\zeta(\overline{7},\overline{5})$, single zeta values, and products of these terms.

5.2 *Enumeration of MZVs Revisited*

I can now explain the rather simple-minded procedure that Dirk Kreimer and I used in 1996 to arrive at the conjecture [8]

$$\prod_{w>2}\prod_{d>0}(1-x^w y^d)^{D_{w,d}} \stackrel{?}{=} 1 - \frac{x^3 y}{1-x^2} + \frac{x^{12}y^2(1-y^2)}{(1-x^4)(1-x^6)}$$

for the number $D_{w,d}$ of irreducible sums in the (A, B) alphabet of pure MZVs. We added the third term to the much simpler conjectured generator for the much complicated question answered by $M_{w,d}$, namely the number of irreducibles in the (A, B, C) alphabet that suffice for reductions of MZVs. The numerator, $x^{12}y^2$ $(1-y^2)$, of this term was determined by the single pushdown observed at weight 12, from an MZV of depth 4 to an alternating sum of depth 2. The denominator, $(1-x^4)(1-x^6)$, was chosen to agree with the empirical number $D_{2n,2} = \lceil n/3 \rceil - 1$ of double non-alternating irreducible sums of weight $2n$. Then we assumed that the enumeration of all other pushdowns would be generated by filtration. It was possible to check this, in a few cases, using PSLQ in 1996.

The list of explicit pushdowns that have now been obtained, in accord with the conjecture, has grown since then.

At weights 15, 16, 17, we have found pushdowns from MZVs to these alternating sums: $\zeta(\overline{6},3,\overline{6})$, $\zeta(\overline{13},\overline{3})$, $\zeta(\overline{6},5,\overline{6})$.
At weight 18, there were pushdowns to $\zeta(\overline{15},\overline{3})$ and $\zeta(6,\overline{5},\overline{4},3)$.
At weight 19, to $\zeta(\overline{8},3,\overline{8})$ and $\zeta(\overline{6},7,\overline{6})$.
At weight 20, to $\zeta(\overline{17},\overline{3})$, $\zeta(8,\overline{5},\overline{4},3)$ and $\zeta(6,\overline{5},\overline{6},3)$.

Our most ambitious efforts were at weight 21, where 3 MZVs of depth 5 are pushed down to the alternating sums $\zeta(\overline{8},5,\overline{8})$, $\zeta(\overline{6},9,\overline{6})$and $\zeta(\overline{8},3,\overline{10})$. Moreover the first pushdown from an MZV of depth 7 to an alternating sum of depth 5 is predicted at weight 21. A demanding PSLQ computation gave a relation of the form

$$\zeta(6,2,3,3,5,1,1) = -\frac{326}{81}\zeta(3,\overline{6},3,\overline{6},3) + \ldots \tag{14}$$

where the remaining 150 terms are formed by MZVs with depth no greater than 5, and their products. At such weight and depth, it becomes rather non-trivial to decide on a single alternating sum that might replace a MZV of greater depth. It took several attempts to discover that the alternating sum

$$\zeta(3,\overline{6},3,\overline{6},3) = \sum_{j>k>l>m>n>0} \frac{(-1)^{k+m}}{(jk^2lm^2n)^3}$$

is an "honorary MZV" that performs this task.

5.3 Suppression of π in Massless Diagrams

Now I can demystify, somewhat, the combination

$$\zeta(5,3) - \frac{29}{12}\zeta(8)$$

that occurs in scheme-independent counterterms of ϕ^4 theory at six loops. Dirk Kreimer and I discovered that the combinations [8]

$$N(a,b) = \zeta(\overline{a},b) - \zeta(\overline{b},a),$$

with distinct odd integers a and b, simplify the results for counterterms. In particular, the use of

$$N(3,5) = \frac{27}{80}\left(\zeta(5,3) - \frac{29}{12}\zeta(8)\right) + \frac{45}{64}\zeta(3)\zeta(5)$$

removes all powers of π from both subdivergence-free diagrams that contribute to the six-loop β-function. In each case, the contribution is a $\mathbf{Z}$-linear combination of $N(3,5)$ and $\zeta(3)\zeta(5)$.

At higher loop numbers, Oliver Schnetz has found that $N(3,7)$ suppresses the appearance π^{10}. However, at 8 loops he found that $N(3,9)$ and $N(5,7)$ are not sufficient to remove π^{12}. Like the maths, the physics becomes different at weight 12.

6 Magnetic Moment of the Electron

The magnetic moment of an electron, with charge $-e$ and mass m, is slightly greater than the Bohr magneton

$$\frac{e\hbar}{2m} = 9.274 \times 10^{-24}\ \mathrm{J\,T^{-1}}$$

which was the value predicted by Dirac. Here I included $\hbar = h/(2\pi)$, which we usually set to unity in QFT.

Using perturbation theory, we may expand in powers of the fine structure constant:

$$\alpha = \frac{e^2}{4\pi\varepsilon_0\hbar c} = \frac{1}{137.035999\ldots}.$$

In QFT, we usually set $\varepsilon_0 = 1$ and $c = 1$ and expand in powers of $\alpha/\pi = e^2/(4\pi^2)$, obtaining a perturbation expansion

$$\frac{\text{magnetic moment}}{\text{Bohr magneton}} = 1 + A_1\frac{\alpha}{\pi} + A_2\left(\frac{\alpha}{\pi}\right)^2 + A_3\left(\frac{\alpha}{\pi}\right)^3 + \ldots$$

which is known up to three loops.

In 1947, Schwinger [11] found the first correction term $A_1 = \frac{1}{2}$. In 1950, Karplus and Kroll [12] claimed the value

$$28\zeta(3) - 54\zeta(2)\log(2) + \frac{125}{6}\zeta(2) - \frac{2{,}687}{288} = -2.972604271\ldots$$

for the coefficient of the next correction. It turned out that they had made a mistake in this rather difficult calculation. The correct result

$$A_2 = \frac{3}{4}\zeta(3) - 3\zeta(2)\log(2) + \frac{1}{2}\zeta(2) + \frac{197}{144} = -0.3284789655\ldots$$

was not obtained until 1957 [13, 14]. Not until 1996 was the next coefficient

$$\begin{aligned} A_3 = {} & -\frac{215}{24}\zeta(5) + \frac{83}{12}\zeta(3)\zeta(2) - \frac{13}{8}\zeta(4) - \frac{50}{3}\zeta(\overline{3},\overline{1}) \\ & + \frac{139}{18}\zeta(3) - \frac{596}{3}\zeta(2)\log(2) + \frac{17{,}101}{135}\zeta(2) + \frac{28{,}259}{5{,}184} \\ = {} & 1.181241456\ldots \end{aligned} \tag{15}$$

found, by Stefano Laporta and Ettore Remiddi [15]. The irrational numbers appearing on the second line are those already seen in A_2. On the first line we see zeta values and a new number, namely the alternating double sum

$$\zeta(\overline{3},\overline{1}) = \sum_{m>n>0} \frac{(-1)^{m+n}}{m^3 n} \approx -0.11787599965050932684101395083413761\ldots$$

I visited Stefano and Ettore in Bologna when they were working on this formidable calculation and recommended to them a method of integration by parts, in D dimensions, that I had found useful for related calculations in the quantum field theory of electrons and photons [16]. Here $D = 4 - 2\varepsilon$ is eventually set to 4, the number of dimensions of space-time. But it turns out to be easier if we keep it as a variable until the final stage of the calculation. Then if we find parts of the result that are singular at $\varepsilon = 0$ we need not worry: all that matters is that the complete result is finite. Based on D-dimensional experience, I expected their final result to look simplest when written in terms of $\zeta(\overline{3},\overline{1})$.

The D-dimensional calculation that informed this intuition involved three-loop massive diagrams contributing to charge renormalization in QED [16]. These yielded Saalschützian F_{32} hypergeometric series, with parameters differing from $\frac{1}{2}$ by multiples of ε, namely

$$W(a_1, a_2; a_3, a_4) = \sum_{n=0}^{\infty} \frac{(\frac{1}{2} - a_1\varepsilon)_n(\frac{1}{2} - a_2\varepsilon)_n}{(\frac{1}{2} + a_3\varepsilon)_{n+1}(\frac{1}{2} + a_4\varepsilon)_{n+1}}$$

with $(\alpha)_n \equiv \Gamma(\alpha+n)/\Gamma(\alpha)$. In particular, I needed the expansions of $W(1,1;1,0)$ and $W(1,0;1,1)$ in ε. The result for the most difficult three-loop diagram had the value $\pi^2\log(2) - \frac{3}{2}\zeta(3)$ at $\varepsilon = 0$. Noting that this also occurs in the two-loop contribution to the magnetic moment, I expanded the charge-renormalization result to $O(\varepsilon)$, where I found only $\zeta(\overline{3},\overline{1})$ and $\zeta(4)$. I thus hazarded the guess that these two sums would exhaust the weight-4 contributions to the magnetic moment at three loops, which happily is the case.

One may also write (15) in terms of a polylog that is not evaluated on the unit circle, such as

$$\mathrm{Li}_4(1/2) = \sum_{n=1}^{\infty} \frac{1}{n^4}\left(\frac{1}{2}\right)^n = -\frac{1}{24}\log^4(2) + \frac{1}{4}\zeta(2)\log^2(2) + \frac{1}{4}\zeta(4) - \frac{1}{2}\zeta(\overline{3},\overline{1}),$$

but then the result for A_3 will acquire extra terms, involving powers of $\log^2(2)$.

7 Three-Loop Massive Bubble Diagrams

Here we consider three-loop diagrams with a massive particle in at least one of the internal lines. If this mass is much larger than the scales set by external momenta, we may set the latter to zero, and obtain vacuum bubbles.

7.1 *Tetrahedral Bubbles from Two-Loop Propagators*

There are ten distinct colourings of a tetrahedron by mass, shown in Fig. 1.

The massive lines in V_{2A} and V_{2N} are adjacent and non-adjacent, respectively; in the dual cases, V_{4A} and V_{4N}, it is the massless lines that are adjacent and non-adjacent; in cases V_{3T}, V_{3S} and V_{3L}, the massive lines form a triangle, star and line, and hence the massless lines form a star, triangle and line.

Defining the finite two-point function (with space-like p^2)

$$I(r_1 \dots r_5; p^2/m^2) := \frac{p^2}{\pi^4} \int d^4k \int d^4l \quad P_1(k)P_2(p+k)P_3(k-l)P_4(l)P_5(p+l) \tag{16}$$

with $P_j(k) := 1/(k^2 + m^2 r_j)$, in 4 dimensions, we obtain

$$V(r_1 \dots r_5, 0) - V(r_1 \dots r_5, 1) = \int_0^\infty dx\, I(r_1 \dots r_5; x) \left\{ \frac{1}{x} - \frac{1}{x+1} \right\} + O(\varepsilon) \tag{17}$$

for the difference of vacuum diagrams with a massless and massive sixth propagator. This difference is finite in 4 dimensions.

Suppressing the parameters $r_1 \dots r_5$, temporarily, we exploit the dispersion relation

$$I(x) = \int_{s_0}^\infty ds\, \sigma(s) \left\{ \frac{1}{s+x} - \frac{1}{s} \right\} \tag{18}$$

where $-2\pi i \sigma(s) = I(-s + i0) - I(-s - i0)$ is the discontinuity across the cut $[-\infty, -s_0]$ on the negative axis. Integration by parts then gives

$$I(x) = \int_{s_0}^\infty ds\, \sigma'(s) \left\{ -\log\left(1 + \frac{x}{s}\right) + \log\left(1 + \frac{x}{s_0}\right) \right\} \tag{19}$$

where the constant term in the logarithmic weight function may be dropped if $\sigma(s_0) = 0$, as occurs when $s_0 = 0$. As $x \to \infty$, we obtain the universal asymptotic value

$$6\zeta(3) = I(\infty) = \int_{s_0}^\infty ds\, \sigma'(s) \{\log(s) - \log(s_0)\} \tag{20}$$

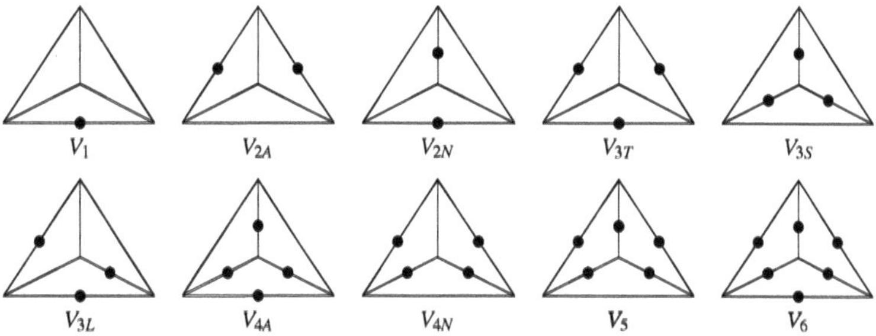

Fig. 1 Colourings of a tetrahedron by mass, denoted by a blob

with the $\log(s_0)$ term dropped when $s_0 = 0$. The finite difference in (17) is obtained from (19) as

$$\int_0^\infty dx\, I(x) \left\{\frac{1}{x} - \frac{1}{x+1}\right\} = \int_{s_0}^\infty ds\, \sigma'(s) \left\{L_2(s) - L_2(s_0)\right\} \tag{21}$$

with a dilogarithmic weight function

$$L_2(s) := \int_0^\infty \frac{dx}{x(x+1)} \log\left(\frac{1+x}{1+x/s}\right) = \mathrm{Li}_2(1-1/s) = -\frac{1}{2}\log^2(s) - \mathrm{Li}_2(1-s) \tag{22}$$

that is chosen to satisfy $L_2(1) = 1$, thus enabling one to drop $L_2(s_0)$ for $s_0 = 0$ and $s_0 = 1$, which covers all the cases with $N \le 3$ massive particles in the two-point function, and hence $N + 1 \le 4$ massive particles in vacuum diagrams.

We now prove that the two terms in the weight function (22) can be separated to yield the finite parts of the vacuum diagrams combined in (17), as follows:

$$F(r_1 \ldots r_5, 0) \equiv \frac{1}{2}\int_{s_0}^\infty ds\, \sigma'(r_1 \ldots r_5; s) \left\{\log^2(s) - \log^2(s_0)\right\} \tag{23}$$

$$F(r_1 \ldots r_5, 1) \equiv -\int_{s_0}^\infty ds\, \sigma'(r_1 \ldots r_5; s) \left\{\mathrm{Li}_2(1-s) - \mathrm{Li}_2(1-s_0)\right\} \tag{24}$$

with constant terms in the weight functions that are inert when $s_0 = 0$ and when $s_0 = 1$. The proof uses the representation

$$I(x) = 6\zeta(3) + \int_{s_0}^\infty ds\, \sigma'(s) \left\{-\log(x+s) + \log(x+s_0)\right\} \tag{25}$$

in which the asymptotic value (20) is subtracted. Then one obtains

$$\int_0^\infty dx \frac{I(\infty) - I(x)}{x+1} = -\int_{s_0}^\infty ds\, \sigma'(s) \left\{\mathrm{Li}_2(1-s) - \mathrm{Li}_2(1-s_0)\right\}. \tag{26}$$

Specializing the analysis to cases with $r_j = 0$ or 1, we obtain from [17]

$$\begin{aligned}
\sigma'(r_1 \ldots r_5; s) &= \left\{\sigma'_a(r_1 \ldots r_5; s)\, \Theta\left(s - (r_1 + r_2)^2\right) + (1 \leftrightarrow 4, 2 \leftrightarrow 5)\right\} \\
&\quad + \left\{\sigma'_b(r_1 \ldots r_5; s)\, \Theta\left(s - (r_2 + r_3 + r_4)^2\right) + (1 \leftrightarrow 2, 4 \leftrightarrow 5)\right\}
\end{aligned}$$

$$\sigma'_a(r_1 \ldots r_5; s) := 2\,\Re \int_{(r_4+r_5)^2}^{\infty} dx\, \frac{T(x, r_1, r_2, r_3, r_4, r_5)}{\Delta(s, r_1, r_2)} \frac{\partial}{\partial x}\left(\frac{\Delta(x, r_1, r_2)}{x - s + i0}\right) \tag{27}$$

$$\sigma'_b(r_1 \ldots r_5; s) := 2\,\Re \int_{(r_3+r_4)^2}^{(\sqrt{s}-r_2)^2} dx\, \frac{\partial}{\partial s}\left(\frac{T(x, s, r_2, r_5, r_4, r_3)}{x - r_1 + i0}\right) \tag{28}$$

$$T(s, a, b, c, d, e) := \operatorname{arctanh}\left(\frac{\Delta(s, a, b)\Delta(s, d, e)}{x^2 - x(a + b - 2c + d + e) + (a - b)(d - e)}\right) \tag{29}$$

$$\Delta(a, b, c) := \sqrt{a^2 + b^2 + c^2 - 2ab - 2bc - 2ca} \tag{30}$$

with integration by parts in (27) giving a logarithmic result, in all cases, and differentiation in (28) giving a logarithmic result when $r_1 r_3 r_5 = r_2 r_3 r_4 = 0$, i.e. when there is no intermediate state with three massive particles.

7.2 The Totally Massive Case

We were able to hand nine cases by methods that avoided intermediate states with three massive particles. Now there is no option, since

$$F_6 = -\int_4^{\infty} ds\, \overline{\sigma}'(s)\, \mathrm{Li}_2(1 - s) \tag{31}$$

involves intermediate states with two and three massive particles in

$$\overline{\sigma}'(s) = \overline{\sigma}'_a(s)\, \Theta(s - 4) + \overline{\sigma}'_b(s)\, \Theta(s - 9)\,. \tag{32}$$

We may, however, simplify matters by separating these contributions in

$$F_6 - F_5 = \int_4^{\infty} ds\, \overline{\sigma}'(s)\, \mathrm{Li}_2(1 - 1/s) = F_a + F_b \tag{33}$$

$$F_a := \int_4^{\infty} ds\, \overline{\sigma}'_a(s)\, \{\mathrm{Li}_2(1 - 1/s) - \zeta(2)\} \tag{34}$$

$$F_b := \int_9^{\infty} ds\, \overline{\sigma}'_b(s)\, \{\mathrm{Li}_2(1 - 1/s) - \zeta(2)\} \tag{35}$$

where F_5 may be evaluated without encountering elliptic integrals.

The two-particle cut gives a logarithm in

$$\overline{\sigma}'_a(s) = \frac{2}{s-3}\left\{\operatorname{arccosh}(s/2-1) - \frac{2\pi}{\sqrt{3s(s-4)}}\right\} \tag{36}$$

while the three-particle cut gives the elliptic[3] integral

$$\overline{\sigma}'_b(s) = -2\int_4^{(\sqrt{s}-1)^2} \frac{dx}{x-1}\,\frac{\Delta(x,1,1)}{\Delta(x,s,1)}\,\frac{x+s-1}{\Delta^2(x,s,1)+xs}\,. \tag{37}$$

At large s, contributions (36) and (37) are each $O(\log(s)/s)$, while their sum is $O(\log(s)/s^2)$. The integrals (34) and (35) converge separately, thanks to the $\zeta(2)$ in their weight functions, to which the combination (33) is blind.

It appears that we need to integrate the product of a dilog and an elliptic integral. To avoid this, we may we reverse the order of integration. Setting $x = 1/u^2 \in [4,\infty]$ in (37), which now becomes the outer integration, and $s = (1/u+v)(1/u+1/v) \in [(1/u+1)^2,\infty]$ in the inner, we then integrate by parts on $v \in [0,1]$ to convert the dilog to a product of logs, with the result [18]

$$F_b = 2\int_0^{\frac{1}{2}} du\left(\frac{dA(u)}{du}\right)\int_0^1 dv\left(\frac{\partial B(u,v)}{\partial v}\right) C(u,v)\, D(u,v) \tag{38}$$

$$A(u) := \log\left(\frac{u^2}{1-u^2}\right) \tag{39}$$

$$B(u,v) := \log\left(\frac{(1+uv)(u+v)}{u+v+uv^2}\right) \tag{40}$$

$$C(u,v) := \log\left(\frac{(1+uv)(u+v)}{u^2 v}\right) \tag{41}$$

$$D(u,v) := \log\left(\frac{1+2uv+v^2+(1-v^2)\sqrt{1-4u^2}}{1+2uv+v^2-(1-v^2)\sqrt{1-4u^2}}\right) \tag{42}$$

which establishes that F_b is the integral of a trilogarithm.

The NAG routine D01FCF is notably efficient at evaluating rectangular double integrals in double-precision FORTRAN, which was ample to to discover the remarkable relation

$$F_6 = F_{3S} + F_{4N} - F_{2N} = 4\left(\mathrm{Cl}_2^2(\pi/3) + 4\zeta(4) + 2\zeta(\overline{3},\overline{1})\right) \tag{43}$$

[3] I am told that Källén was disappointed to find that the two-loop electron propagator involves an elliptic integral, unlike the simpler photon propagator.

where $\mathrm{Cl}_2(\pi/3) = \sum_{n>0} \sin(n\pi/3)/n^2$. This corresponds to a direct relation between diagrams

$$V_6 + V_{2N} = V_{3S} + V_{4N} + O(\varepsilon) \tag{44}$$

verified to 15 digits. It stands as testament to the oft remarked fact that results in quantum field theory have a simplicity that tends to increase with the labour expended.

8 Massive Banana Diagrams

To progress beyond diagrams that yield polylogs and elliptic integrals we now turn attention to vacuum diagrams with merely two vertices. I shall call these "banana" diagrams. The L-loop banana diagram has $L+1$ edges, each representing a massive propagator with unit mass. To avoid ultra-violate divergences, let us consider these in two space-time dimensions.

8.1 Schwinger's Bananas

Let A be the diagonal $N \times N$ matrix with entries $A_{i,j} = \delta_{i,j}\alpha_i$. Let U be the column vector of length N with unit entries, $U_i = 1$. Then $B = U\tilde{U}$ is the $N \times N$ matrix with unit entries, $B_{i,j} = 1$. The banana diagram with $N+1$ edges of unit mass, in two space-time dimensions, may be evaluated by Schwinger's trick as a multiple of the N-dimensional integral

$$\overline{V}_{N+1} = \int_{\alpha_i>0} \frac{\mathrm{d}\alpha_1 \ldots \mathrm{d}\alpha_N}{\mathrm{Det}(A+B)(\mathrm{Tr}(A)+1)} \tag{45}$$

where

$$\mathrm{Det}(A+B) = \sum_{i=0}^{N} \frac{1}{\alpha_i} \prod_{j=0}^{N} \alpha_j$$

is the first Symanzik polynomial, with $\alpha_0 = 1$ fixed by momentum conservation, and the second Symanzik polynomial

$$\mathrm{Tr}(A) + 1 = \sum_{i=0}^{N} \alpha_i$$

results from the fact that the $N+1$ edges are propagators with unit mass.

8.2 Bessels's Bananas

We may also evaluate banana diagrams in x-space, since the two-dimensional Fourier transform of the p-space Euclidean propagator $1/(p^2+m^2)$, with $p^2 = p_0^2 + p_1^2$, yields the Bessel function $K_0(mx)$, with $x^2 = x_0^2 + x_1^2$. The normalization in (45) corresponds to

$$\overline{V}_{N+1} = 2^N \int_0^\infty [K_0(t)]^{N+1} t \, \mathrm{d}t \tag{46}$$

which differs by a power of 2 from the Bessel moments that I studied with Bailey, Borwein and Glasser [19].

Hence I but a bar over V and use the subscript $N+1$ to indicate the number of Bessel functions.

8.3 Known Bananas

It is proven that [19]

$$\overline{V}_1 = 1 \tag{47}$$

$$\overline{V}_2 = 1 \tag{48}$$

$$\overline{V}_3 = 3L_{-3}(2) \tag{49}$$

$$\overline{V}_4 = 7\zeta(3) \tag{50}$$

where

$$L_{-3}(s) = \sum_{n\geq 0}\left(\frac{1}{(3n+1)^s} - \frac{1}{(3n+2)^s}\right)$$

is the Dirichlet L function with conductor -3.

The zero-loop evaluation (47) merely checks our normalization.

The one-loop evaluation

$$\overline{V}_2 = \int_0^\infty \frac{\mathrm{d}\alpha_1}{(\alpha_1+1)^2} = 1$$

follows neatly from (45), since with $N = 1$ we have $\mathrm{Det}(A+B) = \mathrm{Tr}(A) + 1 = \alpha_1 + 1$.

I shall now use $\{a, b, c, \ldots\}$ for the Schwinger parameters $\{\alpha_1, \alpha_2, \alpha_3, \ldots\}$.

8.4 Three-Edge Banana and Sixth Root of Unity

At two loops, the Schwinger method gives the banana diagram with 3 edges as

$$\overline{V}_3 = \int_0^\infty \int_0^\infty \frac{\mathrm{d}a\,\mathrm{d}b}{(ab+a+b)(a+b+1)}.$$

To proceed we may take partial fractions with respect to b. Then

$$\frac{a^2+a+1}{(ab+a+b)(a+b+1)} = \frac{a+1}{ab+a+b} - \frac{1}{a+b+1} = \frac{\partial}{\partial b}\log\left(\frac{ab+a+b}{a+b+1}\right)$$

enables integration over b. Hence we obtain

$$\overline{V}_3 = \int_0^\infty \frac{G(a)\,\mathrm{d}a}{a^2+a+1} \qquad (51)$$

with contributions to

$$G(a) = \log(1+a) + \log(1+1/a) \qquad (52)$$

at $b=\infty$ and $b=0$. It is apparent from (51) that the sixth root of unity $\lambda = (1+\mathrm{i}\sqrt{3})/2$ is implicated, since $a^2+a+1 = (a+\lambda)(a+\overline{\lambda})$, where $\overline{\lambda} = (1-\mathrm{i}\sqrt{3})/2 = 1-\lambda$ is the conjugate root. Working out the corresponding dilogarithms we obtain

$$\overline{V}_3 = \frac{4}{\sqrt{3}}\Im\mathrm{Li}_2(\lambda) = 3L_{-3}(2)$$

in agreement with the well known result (49).

8.5 Four-Edge Banana and ζ(3)

To evaluate

$$\overline{V}_4 = \int_0^\infty \int_0^\infty \int_0^\infty \frac{\mathrm{d}a\,\mathrm{d}b\,\mathrm{d}c}{(abc+ab+bc+ca)(a+b+c+1)}$$

we take partial fractions with respect to c and then integrate over c, to obtain

$$\overline{V}_4 = \int_0^\infty \int_0^\infty \frac{L(a,b)\,\mathrm{d}a\,\mathrm{d}b}{(a+1)(b+1)(a+b)}$$

with

$$L(a,b) = \log\left(\frac{(ab+a+b)(a+b+1)}{ab}\right).$$

Hence with

$$F(a) = \int_0^\infty \frac{(a-1)L(a,b)\,\mathrm{d}b}{(b+1)(a+b)}$$

we have

$$\overline{V}_4 = \int_0^\infty \frac{F(a)\,\mathrm{d}a}{a^2-1} = \int_0^1 \frac{(F(a)-F(1/a))\,\mathrm{d}a}{a^2-1}. \tag{53}$$

I shall need only the derivative of $F(a)$. Let

$$K(a,b) = \frac{b\,L(a,b)}{a+b} + \log(ab+a+b) - 2\log(a+b+1).$$

Then, by construction,

$$\frac{\partial}{\partial b}K(a,b) = a\frac{\partial}{\partial a}\left(\frac{(a-1)L(a,b)}{(b+1)(a+b)}\right)$$

and hence

$$a\frac{\mathrm{d}}{\mathrm{d}a}F(a) = K(a,\infty) - K(a,0) = 2G(a)$$

where $G(a)$ was given in (52). We now integrate (53) by parts, to obtain

$$\overline{V}_4 = \int_0^1 \frac{\mathrm{d}a}{a}\log\left(\frac{1+a}{1-a}\right)(G(a)+G(1/a))$$

and use Nielsen's evaluations

$$-\int_0^1 \frac{\mathrm{d}a}{a}\log(1-a)\log(1+a) = \frac{5}{8}\zeta(3)$$

$$-\int_0^1 \frac{\mathrm{d}a}{a}\log(a)\log(1+a) = \frac{3}{4}\zeta(3)$$

$$\int_0^1 \frac{\mathrm{d}a}{a}\log^2(1+a) = \frac{1}{4}\zeta(3)$$

$$\int_0^1 \frac{\mathrm{d}a}{a}\log(a)\log(1-a) = \zeta(3)$$

to obtain

$$\overline{V}_4 = \left(4 \times \frac{5}{8} + 2 \times \frac{3}{4} + 4 \times \frac{1}{4} + 2\right) \zeta(3) = 7\zeta(3)$$

in agreement with the previously known result (50).

8.6 *Unknown Banana*

The next diagram has 5 edges and hence 4 loops. After an easy first integration, we obtain

$$\overline{V}_5 = \int_0^\infty \int_0^\infty \int_0^\infty \frac{M(a,b,c)\,\mathrm{d}a\,\mathrm{d}b\,\mathrm{d}c}{(ab+a+b)c^2 + (ab+a+b)(a+b)c + (a+b)ab}$$

with

$$M(a,b,c) = \log(a+b+c+1) + \log\left(1 + \frac{1}{a} + \frac{1}{b} + \frac{1}{c}\right).$$

But then integration over c will produce complicated dilogarithms with arguments involving the square root of the discriminant

$$D(a,b) = (ab+a+b)(a+b)(ab(a+b) + (a-b)^2)$$

of the quadratic in c. The result will have the form

$$\overline{V}_5 = \int_0^\infty \int_0^\infty \frac{L_2(a,b)\mathrm{d}a\,\mathrm{d}b}{\sqrt{D(a,b)}}$$

with undisclosed dilogs in $L_2(a,b)$. Integration by parts, to reduce the dilogs to logs, would require us to introduce an elliptic function, since $D(a,b)$ is a quartic in b.

We know nothing about the number theory of $\overline{V}_5$. Its value is known to 1,000 decimal places.

9 Cut Bananas: On-Shell Sunrise Diagrams

For $N > 2$ we may cut an edge in $\overline{V}_N$ and set the two external half edges on the unit mass shell, which is at $p^2 = -1$. I call the result $\overline{S}_N$. It has $N-1$ internal edges and hence $N-2$ loops. Thus $\overline{V}_3$ and $\overline{S}_4$ correspond to the two-loop diagrams of Fig. 2, with the "sunrise" diagram $\overline{S}_4$ obtained by cutting an edge of $\overline{V}_4$.

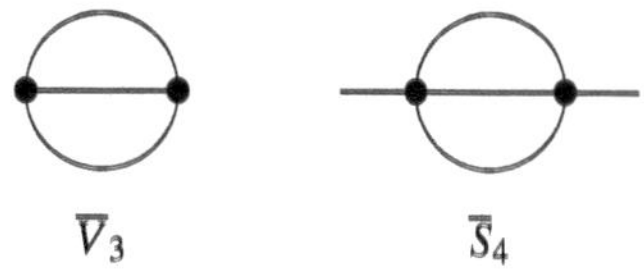

Fig. 2 Two-loop banana and sunrise diagrams

9.1 Schwinger's Cut Bananas

At N loops, the integral over Schwinger parameters is

$$\overline{S}_{N+2} = \int_{\alpha_i > 0} \frac{\mathrm{d}\alpha_1 \ldots \mathrm{d}\alpha_N}{\mathrm{Det}(A+B)\mathrm{Tr}(A) + \tilde{U}CU}. \tag{54}$$

where C is the adjoint of $A + B$, with

$$(A+B)C = \mathrm{Det}(A+B)I$$

where I is the unit matrix with $I_{i,j} = \delta_{i,j}$. The denominator in (54) is the second Symanzik polynomial.

9.2 Bessels's Cut Bananas

In x-space, cutting an edge and putting it on the mass shell corresponds to replacing one instance of the Bessel function $K_0(t)$ by $I_0(t)$, to obtain

$$\overline{S}_{N+2} = 2^N \int_0^\infty I_0(t)[K_0(t)]^{N+1} t\, \mathrm{d}t \tag{55}$$

at N loops. Note that $\overline{S}_2$ is divergent, since

$$I_0(t) = \sum_{k \geq 0} \left(\frac{t^k}{2^k k!} \right)^2$$

grows exponentially, with

$$I_0(t) = \frac{\exp(t)}{\sqrt{2\pi t}} \left(1 + \frac{1}{8t} + O(1/t^2) \right)$$

as $t \to \infty$, while

$$K_0(t) = \sqrt{\frac{\pi}{2t}} \exp(-t) \left(1 - \frac{1}{8t} + O(1/t^2)\right)$$

is exponentially damped.

9.3 Known Cut Bananas

It is proven that [19]

$$\overline{S}_3 = 2L_{-3}(1) = \frac{2\pi}{3\sqrt{3}} \tag{56}$$

$$\overline{S}_4 = \mathrm{Li}_2(1) - \mathrm{Li}_2(-1) = \frac{\pi^2}{4} \tag{57}$$

and it is conjectured that [19]

$$\overline{S}_5 \stackrel{?}{=} \frac{1}{30\sqrt{5}} \Gamma\left(\frac{1}{15}\right) \Gamma\left(\frac{2}{15}\right) \Gamma\left(\frac{4}{15}\right) \Gamma\left(\frac{8}{15}\right) \tag{58}$$

which holds to at least 1,000 decimal places.

9.4 Cut Banana with Sixth Root of Unity

The Schwinger formula (54) at one loop gives

$$\overline{S}_3 = \int_0^\infty \frac{\mathrm{d}a}{a^2 + a + 1} = \frac{\log(\lambda) - \log(\overline{\lambda})}{\lambda - \overline{\lambda}} = \frac{2\arctan(\sqrt{3})}{\sqrt{3}} = \frac{2\pi}{3\sqrt{3}}$$

as claimed in (56).

9.5 Cut Banana with π^2

At two loops, we have

$$\overline{S}_4 = \int_0^\infty \int_0^\infty \frac{\mathrm{d}a\, \mathrm{d}b}{(a+b)(a+1)(b+1)}$$

with a convenient factorization of the second Symanzik polynomial. Hence

$$\overline{S}_4 = \int_0^\infty \frac{\log(a)\,\mathrm{d}a}{a^2-1} = 2\int_0^1 \frac{\log(a)\,\mathrm{d}a}{a^2-1}$$

yields dilogs at square roots of unity, namely

$$\overline{S}_4 = \mathrm{Li}_2(1) - \mathrm{Li}_2(-1) = \frac{\pi^2}{6} + \frac{\pi^2}{12} = \frac{\pi^2}{4}$$

as claimed in (57).

9.6 Cut Banana at the 15th Singular Value

At three loops, we have

$$\overline{S}_5 = \int_0^\infty \int_0^\infty \int_0^\infty \frac{\mathrm{d}a\,\mathrm{d}b\,\mathrm{d}c}{P(a,b,c)}$$

where

$$P(a,b,c) = (abc + ab + bc + ca)(a + b + c) + (ab + bc + ca)$$

with the final term, $(ab + bc + ca)$, resulting from the adjoint matrix. Grouping powers of c, we see that

$$P(a,b,c) = (ab + a + b)c^2 + (ab + a + b)(a + b + 1)c + (a + b + 1)ab$$

yields a discriminant

$$\Delta(a,b) = (ab + a + b)(a + b + 1)((ab + a + b)(a + b + 1) - 4ab)$$

and the integral over c gives

$$\overline{S}_5 = \int_0^\infty \int_0^\infty \frac{\mathrm{d}a\,\mathrm{d}b}{\sqrt{\Delta(a,b)}} \log\left(\frac{1 + X(a,b)}{1 - X(a,b)}\right)$$

with

$$X(a,b) = \sqrt{1 - \frac{4ab}{(ab + a + b)(a + b + 1)}}\,.$$

Conjecture (58) was stimulated by a proven result for

$$\overline{T}_5 \equiv 4\int_0^\infty [I_0(t)]^2[K_0(t)]^3 t\, dt = \int_0^\infty \int_0^\infty \frac{da\, db}{\sqrt{\Delta(a,b)}}$$

namely

$$\overline{T}_5 = \frac{\sqrt{3}}{120\pi}\Gamma\left(\frac{1}{15}\right)\Gamma\left(\frac{2}{15}\right)\Gamma\left(\frac{4}{15}\right)\Gamma\left(\frac{8}{15}\right) \tag{59}$$

obtained at the 15th singular value, by diamond mining [19].

Numerical evaluation suggested that

$$\frac{\overline{S}_5}{\overline{T}_5} \stackrel{?}{=} \frac{4\pi}{\sqrt{15}}$$

and this has been confirmed at 1,000-digit precision. Yet it remains to be proved that

$$\int_0^\infty \int_0^\infty \frac{da\, db}{\sqrt{\Delta(a,b)}}\left(\log\left(\frac{1+X(a,b)}{1-X(a,b)}\right) - \frac{4\pi}{\sqrt{15}}\right) \tag{60}$$

vanishes. It has been shown that its magnitude is smaller than $10^{-1,000}$.

10 Diagrams Evaluating as L-Series of Modular Forms

Finally, I indicate how sunrise diagrams lead to evaluations in terms of the Dirichlet L-functions of modular forms, evaluated at integers inside their critical strips.

10.1 L-Series of a K3 Surface

For $s > 2$ let

$$L(s) = \prod_p \frac{1}{1 - \frac{A_p}{p^s} + \left(\frac{p}{15}\right)\frac{p^2}{p^{2s}}}$$

where $\left(\frac{\cdot}{15}\right)$ is a Kronecker symbol and the product is over all primes p, with integers

$$A_3 = -3,$$
$$A_5 = 5,$$
$$A_p = 0, \text{ for } \left(\frac{p}{15}\right) = -1,$$

$$A_p = 2x^2 + 2xy - 7y^2, \text{ for } x^2 + xy + 4y^2 = p \equiv 1, 4 \pmod{15}, \quad (61)$$

$$A_p = x^2 + 8xy + y^2, \text{ for } 2x^2 + xy + 2y^2 = p \equiv 2, 8 \pmod{15}, \quad (62)$$

with pairs of integers (x, y) defined, for $x > 0$, by the quadratic forms in (61) and (62).

As shown by Peters, Top and van der Vlugt [20], the L-series

$$L(s) = \sum_{n>0} \frac{A_n}{n^s}$$

is generated by the weight-3 modular form

$$f_3(q) = \eta(q)\eta(q^3)\eta(q^5)\eta(q^{15})R(q) = \sum_{n>0} A_n q^n \quad (63)$$

where

$$\frac{\eta(q)}{q^{1/24}} = \prod_{j>0}(1 - q^j) = \sum_{n\in\mathbf{Z}}(-1)^n q^{n(3n+1)/2}, \quad (64)$$

$$R(q) = \sum_{m,n\in\mathbf{Z}} q^{m^2+mn+4n^2}. \quad (65)$$

Note that $A_1 = 1$, since $1 + 3 + 5 + 15 = 24$. If $q = p^r$ is a prime power, then

$$A_{pq} = A_p A_q - \left(\frac{p}{15}\right) p^2 A_{q/p}.$$

If $n = \prod_j q_j$, with prime powers $q_j = p_j^{r_j}$, then $A_n = \prod_j A_{q_j}$. Thus (61) and (62) suffice to compute A_n and are easily programmed using the `qfbsolve` command of `Pari-GP`.

I now describe how I was able to evaluate 20,000 good digits of the conditionally convergent series $L(2) = \sum_{n>0} A_n/n^2$. Let

$$\Lambda(s) = \frac{\Gamma(s)}{c^s} L(s), \text{ with } c = \frac{2\pi}{\sqrt{15}}.$$

Then the functional equation $\Lambda(s) = \Lambda(3 - s)$ may be used to extend the Mellin transform

$$\Lambda(s) = \sum_{n>0} A_n \int_0^\infty \frac{dx}{x} x^s \exp(-cnx) \quad (66)$$

throughout the complex s-plane, as follows

$$\Lambda(s) = \sum_{n>0} A_n \left(\frac{\Gamma(s, cn\lambda)}{(cn)^s} + \frac{\Gamma(3-s, cn/\lambda)}{(cn)^{3-s}} \right) \tag{67}$$

where

$$\Gamma(s, y) = \int_y^\infty \frac{\mathrm{d}x}{x} x^s \exp(-x)$$

is the incomplete Γ function and $\lambda \geq 0$ is an arbitrary real parameter. To establish (67), I remark that it agrees with (66), at $\lambda = 0$, and that its derivative with respect to λ vanishes by virtue of the inversion symmetry

$$M(\lambda) \equiv \lambda^{3/2} \sum_{n>0} A_n \exp(-cn\lambda) = M(1/\lambda).$$

Optimal convergence is achieved at $\lambda = 1$, where

$$\Lambda(s) = \sum_{n>0} A_n \int_1^\infty \frac{\mathrm{d}x}{x} \left(x^s + x^{3-s}\right) \exp\left(-\frac{2\pi n x}{\sqrt{15}}\right) \tag{68}$$

makes the relation $\Lambda(s) = \Lambda(3-s)$ explicit. Zeros on the critical line $\Re s = 3/2$ occur when

$$\Lambda(3/2 + \mathrm{i}s_0) = 2 \sum_{n>0} A_n \int_1^\infty \mathrm{d}x\, x^{1/2} \cos(s_0 \log(x)) \exp\left(-\frac{2\pi n x}{\sqrt{15}}\right)$$

vanishes. I have computed 100 good digits of the first zero, obtaining

$$\begin{aligned} s_0 = {} & 4.841925814229962588045533711247175448399945840634 7 \\ & 6693950953608563348168047411353721585251883775250 05\ldots \end{aligned}$$

At $s = 2$, the integral in (68) is elementary and we have dramatically improved convergence for

$$L(2) \equiv \sum_{n>0} \frac{A_n}{n^2} = \sum_{n>0} \frac{A_n}{n^2} \left(1 + \frac{4\pi n}{\sqrt{15}}\right) \exp\left(-\frac{2\pi n}{\sqrt{15}}\right) \tag{69}$$

from which I obtained more than 20,000 good digits in less than a minute, by computing the first 30,000 terms, with the aid of (61) and (62). The result is consistent with the conjecture

$$\begin{aligned}
3L(2) &\stackrel{?}{=} \overline{T}_5 && (70)\\
&\equiv 4\int_0^\infty [I_0(t)]^2[K_0(t)]^3 t\,\mathrm{d}t && (71)\\
&= \int_0^\infty\int_0^\infty \frac{\mathrm{d}a\,\mathrm{d}b}{\sqrt{(ab+a+b)(a+b+1)((ab+a+b)(a+b+1)-4ab)}} && (72)\\
&= \frac{\pi^2}{8}\left(\sqrt{15}-\sqrt{3}\right)\left(1+2\sum_{n>0}\exp\left(-\sqrt{15}\pi n^2\right)\right)^4 && (73)\\
&= \frac{\sqrt{3}}{120\pi}\Gamma\left(\frac{1}{15}\right)\Gamma\left(\frac{2}{15}\right)\Gamma\left(\frac{4}{15}\right)\Gamma\left(\frac{8}{15}\right) && (74)\\
&\stackrel{?}{=} \frac{\sqrt{15}}{4\pi}\overline{S}_5 && (75)
\end{aligned}$$

where $\overline{T}_5$ is defined in (71) as a Bessel moment, with a proven integral representation over Schwinger parameters in (72), a proven evaluation at the 15th singular value in (73), a proven reduction to Γ values in (74) and a conjectural relation to $\overline{S}_5$ in (75).

Unfortunately, I did not succeed in relating $\overline{V}_5$ to $L(3)$ and/or $L(4)$.

10.2 L-Series for 6 Bessel Functions

We are interested in relating Bessel moments of the form

$$\overline{V}_N = 2^{N-1}\int_0^\infty [K_0(t)]^N t\,\mathrm{d}t,\ \text{for } N>0, \tag{76}$$

$$\overline{S}_N = 2^{N-2}\int_0^\infty I_0(t)[K_0(t)]^{N-1} t\,\mathrm{d}t,\ \text{for } N>2, \tag{77}$$

$$\overline{T}_N = 2^{N-3}\int_0^\infty I_0^2(t)[K_0(t)]^{N-2} t\,\mathrm{d}t,\ \text{for } N>4, \tag{78}$$

$$\overline{U}_N = 2^{N-4}\int_0^\infty I_0^3(t)[K_0(t)]^{N-3} t\,\mathrm{d}t,\ \text{for } N\ge 6, \tag{79}$$

$$\overline{W}_N = 2^{N-5}\int_0^\infty I_0^4(t)[K_0(t)]^{N-4} t\,\mathrm{d}t,\ \text{for } N\ge 8, \tag{80}$$

to L-series derived from modular forms. In [19] it was conjectured that

$$\overline{S}_5 \overset{?}{=} \frac{4\pi}{\sqrt{15}} \overline{T}_5 \tag{81}$$

$$\overline{S}_6 \overset{?}{=} \frac{4\pi^2}{3} \overline{U}_6 \tag{82}$$

$$\overline{T}_8 \overset{?}{=} \frac{18\pi^2}{7} \overline{W}_8 \tag{83}$$

with a notable appearance of 7 in the denominator on the right hand side of (83).

Francis Brown suggested that the weight-4 modular form

$$f_4(q) = \left[\eta(q)\eta(q^2)\eta(q^3)\eta(q^6)\right]^2 = \sum_{n>0} A_{4,n} q^n \tag{84}$$

of Hulek, Spandaw, van Geemen, and van Straten [21] might yield an L-series

$$L_4(s) = \sum_{n>0} \frac{A_{4,n}}{n^s} = \frac{1}{1+2^{1-s}} \frac{1}{1+3^{1-s}} \prod_{p>3} \frac{1}{1 - \frac{A_{4,p}}{p^s} + \frac{p^3}{p^{2s}}}$$

with values related to the problem with 6 Bessel functions. Note that $A_{4,1} = 1$, since $2(1+2+3+6) = 24$.

The Mellin transform

$$\Lambda_4(s) = \frac{\Gamma(s)}{(2\pi/\sqrt{6})^s} L_4(s) = \sum_{n>0} A_{4,n} \int_0^\infty \frac{\mathrm{d}x}{x} x^s \exp\left(-\frac{2\pi n x}{\sqrt{6}}\right)$$

may be analytically continued to give

$$\Lambda_4(s) = \sum_{n>0} A_{4,n} \int_1^\infty \frac{\mathrm{d}x}{x} \left(x^s + x^{4-s}\right) \exp\left(-\frac{2\pi n x}{\sqrt{6}}\right)$$

by virtue of the inversion symmetry

$$M_4(\lambda) \equiv \lambda^2 \sum_{n>0} A_{4,n} \exp\left(-\frac{2\pi n \lambda}{\sqrt{6}}\right) = M_4(1/\lambda)$$

that gives the reflection symmetry $\Lambda_4(s) = \Lambda_4(4-s)$.

Then, at $s = 2$ and $s = 3$, we obtain the very convenient formulas

$$L_4(2) = \sum_{n>0} \frac{A_{4,n}}{n^2} \left(2 + \frac{4\pi n}{\sqrt{6}}\right) \exp\left(-\frac{2\pi n}{\sqrt{6}}\right) \tag{85}$$

$$L_4(3) = \sum_{n>0} \frac{A_{4,n}}{n^3} \left(1 + \frac{2\pi n}{\sqrt{6}} + \frac{2\pi^2 n^2}{3}\right) \exp\left(-\frac{2\pi n}{\sqrt{6}}\right) \tag{86}$$

without resort to incomplete Γ functions that entail exponential integrals. By this means, I was able to compute 20,000 good digits of (85) and (86) in less than 100 s. Then the conjectural evaluations

$$\overline{S}_6 \stackrel{?}{=} 48\zeta(2)L_4(2) \tag{87}$$

$$\overline{T}_6 \stackrel{?}{=} 12L_4(3) \tag{88}$$

$$\overline{U}_6 \stackrel{?}{=} 6L_4(2) \tag{89}$$

were discovered and checked at 1,000-digit precision.

I remark that Francis Brown had expected a result of form (88), for $\overline{T}_6$, with an unknown rational coefficient, which I here evaluate as 12. The existence of a relation of the form (89), for $\overline{U}_6$, had not been predicted, since I had been unable to provide an expression for this Bessel moment as an integral over Schwinger parameters of an algebraic or polylogarithmic function. However, it was quite natural to guess that a reduction of $\overline{T}_6$ to $L_4(3)$ would be accompanied by a reduction of $\overline{U}_6$ to $L_4(2)$. Then the reduction of $\overline{S}_6$ to $\zeta(2)L_4(2)$ follows from conjecture (82), which I had already checked at 1,000-digit precision in [19].

10.3 L-Series for 8 Bessel Functions

Next, Francis Brown provided the first 100 Fourier coefficients of a weight-6 modular form $f_6(q) \equiv \sum_{n>0} A_{6,n}q^n$, whose L-series

$$L_6(s) \equiv \sum_{n\geqslant 0} \frac{A_{6,n}}{n^s} = \frac{1}{1-2^{2-s}}\,\frac{1}{1+3^{2-s}}\prod_{p>3}\frac{1}{1-\frac{A_{6,p}}{p^s}+\frac{p^5}{p^{2s}}}$$

was expected to yield values related to the problem with 8 Bessel functions. His data may be condensed down to the values

```
-66, 176, -60, -658, -414, 956, 600, 5574, -3592, -8458,
19194, 13316, -19680, -31266, 26340, -31090, -16804, 6120,
-25558, 74408, -6468, -32742, 166082
```

of $A_{6,p}$ for the primes $p = 5, 7, \ldots, 97$.

From this I inferred that the explicit modular form is given by

$$f_6(q) = g(q)g(q^2) \tag{90}$$

$$g(q) = \left[\eta(q)\eta(q^3)\right]^2 \sum_{m,n\in\mathbf{Z}} q^{m^2+mn+n^2} \tag{91}$$

with $f_6(q)/f_4(q)$ given by the θ function of the strongly 6-modular lattice [22] indexed by `QQF.4.g`[4] with expansion coefficients in entry `A125510` of Neil Sloane's On-Line Encyclopedia of Integer Sequences.[5]

Proceeding along the lines of the previous section, I accelerated the convergence of

$$\Lambda_6(s) = \frac{\Gamma(s)}{(2\pi/\sqrt{6})^s} L_6(s) = \sum_{n>0} A_{6,n} \int_0^\infty \frac{\mathrm{d}x}{x} x^s \exp\left(-\frac{2\pi n x}{\sqrt{6}}\right)$$

by using the functional relation $\Lambda_6(s) = \Lambda_6(6-s)$ to obtain

$$\Lambda_6(s) = \sum_{n>0} A_{6,n} \int_1^\infty \frac{\mathrm{d}x}{x} \left(x^s + x^{6-s}\right) \exp\left(-\frac{2\pi n x}{\sqrt{6}}\right)$$

and hence the convenient formulas

$$L_6(3) = \sum_{n>0} \frac{A_{6,n}}{n^3} \left(2 + \frac{4\pi n}{\sqrt{6}} + \frac{2\pi^2 n^2}{3}\right) \exp\left(-\frac{2\pi n}{\sqrt{6}}\right), \tag{92}$$

$$L_6(4) = \sum_{n>0} \frac{A_{6,n}}{n^4} \left(1 + \frac{2\pi n}{\sqrt{6}} + \frac{4\pi^2 n^2}{9} + \frac{4\pi^3 n^3}{9\sqrt{6}}\right) \exp\left(-\frac{2\pi n}{\sqrt{6}}\right), \tag{93}$$

$$L_6(5) = \sum_{n>0} \frac{A_{6,n}}{n^5} \left(1 + \frac{2\pi n}{\sqrt{6}} + \frac{\pi^2 n^2}{3} + \frac{2\pi^3 n^3}{9\sqrt{6}} + \frac{\pi^4 n^4}{27}\right) \exp\left(-\frac{2\pi n}{\sqrt{6}}\right). \tag{94}$$

The resulting fits

$$\overline{T}_8 \stackrel{?}{=} 216 L_6(5) \tag{95}$$

$$\overline{U}_8 \stackrel{?}{=} 36 L_6(4) \tag{96}$$

$$\overline{W}_8 \stackrel{?}{=} 8 L_6(3) \tag{97}$$

are rather satisfying. They leave the conjectural relation

$$L_6(5) \stackrel{?}{=} \frac{4}{7} \zeta(2) L_6(3) \tag{98}$$

as a restatement of the notable conjecture (83) given in [19].

---

[4] http://www2.research.att.com/~njas/lattices/QQF.4.g.html

[5] http://oeis.org/A125510

Thanks to the explicit formula (90) for the weight-6 modular form, conjecture (98) has now been checked to 20,000-digit precision.

11 Open Questions

This article has provided examples of single-scale Feynman diagrams that evaluate to five types of number: multiple zeta values, alternating sums, polylogarithms of the sixth root of unity, products of elliptic integrals, and L-functions of modular forms. In each case, I indicate an open question concerning the physics and an open puzzle concerning the mathematics.

Q1: At which loop-number do the counterterms of QFT cease to evaluate to MZVs?
There remains a single subdivergence-free counterterm in ϕ^4 theory at seven loops that has not been reduced to MZVs, but might be expected to evaluate to polylogs. At eight loops there is a diagram for which there is good reason [23] to suppose that no reduction to polylogs will be possible, yet there is no concrete guess of the type of new number that might emerge.
On the mathematical side, the conjectural enumeration [8] of irreducible MZVs by weight and depth, in (8), is still unproven. Might it be that the conjecture fails at weights higher than those in the table of Sect. 2.4, notwithstanding the notable agreement so far achieved?
Q2: At which loop-number do the diagrams for the magnetic moment of the electron case to evaluate in terms of alternating sums?
One might guess that this will happen at four loops, since there one has diagrams with five electrons in an intermediate state and the corresponding on-shell sunrise diagram in two dimensions, with 6 Bessel functions, evaluates to the L function of a modular form, as seen in Eq. (87).
On the mathematical side, one would like to understand why a depth-5 alternating sum like $\zeta(3,\overline{6},3,\overline{6},3)$ in (14) is an honorary MZV of depth 7.
Q3: Does a polylogarithm of the sixth root of unity appear in the seven-loop beta-function of ϕ^4 theory?
It has been argued [24] that this may happen, for one special diagram. However, comparable arguments suggested the appearance of alternating sums from a pair of simpler seven-loop diagrams and these were found to evaluate to MZVs.
On the mathematical side, one would like to have an economical basis for polylogs of the sixth root of unity up to weight 11, so as to tackle the seven-loop problem in QFT. However, that seems to be a daunting task.
Q4: What type of number results from the four-loop banana diagram $\overline{V}_5$, with 5 Bessel functions?
We have seen that the three-loop on-shell sunrise diagram $\overline{S}_5$ evaluates, empirically, to the square of an elliptic integral at the 15th singular value. Yet the

simplest result so far achieved for $\overline{V}_5$ is the integral of the product of a dilogarithm and a complete elliptic integral [19].

On the mathematical side, one would like to be able to prove the vanishing of the remarkable integral (60).

Q5: Is there a modular form whose L-function gives an evaluation of the on-shell five-loop sunrise diagram $\overline{S}_7$?

It is frustrating to have identified modular forms for problems with 5, 6 and 8 Bessel functions, yet to have failed to do so for any 7-Bessel problem.

On the mathematical side, one would like to understand the relation between integrals of powers of Bessel functions and Kloosterman sums [25, 26] that evaluate to rational numbers.

In conclusion, these open questions arose from fertile meetings of number theory, algebraic geometry and quantum field theory, reported in part by this article. While much remains to be understood, we may still rejoice that mathematicians and physicists continue to learn how to share their understanding and their puzzles at the work-face of perturbative quantum field theory.

References

1. Zagier, D.: Values of zeta functions and their applications. In: First European Congress of Mathematics, Paris, 1992, vol. II, pp. 497–512. Birkhäuser, Basel (1994)
2. Hoffman, M.E: Multiple harmonic series. Pac. J. Math. **152**, 275–290 (1992)
3. Borwein, J.M., Bradley, D.M., Broadhurst, D.J.: Evaluations of k-fold Euler/Zagier sums: a compendium of results for arbitrary k. Electron. J. Comb. **4**(2), R5 (1997)
4. Borwein, J.M., Bradley, D.M., Broadhurst, D.J., Lisonek, P.: Special values of multiple polylogarithms. Trans. Am. Math. Soc. **353**, 907–941 (2001)
5. Espinosa, E., Moll, V.H.: The evaluation of Tornheim double sums. J. Number Theory **116**, 200–229 (2006)
6. Borwein, J.M., Girgensohn, R.: Evaluation of triple Euler sums, with appendix Euler sums in quantum field theory by D.J. Broadhurst. Electron. J. Comb. **3**, R23 (1996)
7. Blümlein, J., Broadhurst, D.J., Vermaseren, J.A.M.: The multiple zeta value data mine. Comput. Phys. Commun. **181**, 582–625 (2010)
8. Broadhurst, D.J., Kreimer, D.: Association of multiple zeta values with positive knots via Feynman diagrams up to 9 loops. Phys. Lett. **B393**, 403–412 (1997)
9. Broadhurst, D.J., Kreimer, D.: Knots and numbers in ϕ^4 theory to 7 loops and beyond. Int. J. Mod. Phys. **C6**, 519–524 (1995)
10. Bailey, D.H., Broadhurst, D.J.: Parallel integer relation detection: techniques and applications. Math. Comput. **70**, 1719–1736 (2000)
11. Schwinger, J.: On quantum-electrodynamics and the magnetic moment of the electron. Phys. Rev. **73**, 416 (1948)
12. Karplus, R., Kroll, N.M.: Fourth-order corrections in quantum electrodynamics and the magnetic moment of the electron. Phys. Rev. **77**, 536–549 (1950)
13. Sommerfield, C.M.: Magnetic dipole moment of the electron. Phys. Rev. **107**, 328–329 (1957)
14. Petermann, A.: Fourth order magnetic moment of the electron. Helv. Phys. Acta. **30**, 407 (1957)
15. Laporta, S., Remiddi, E.: The analytical value of the electron $(g-2)$ at order α^3 in QED. Phys. Lett. **B379**, 283–291 (1996)

16. Broadhurst, D.J.: Three-loop on-shell charge renormalization without integration: $\Lambda_{\mathrm{QED}}^{\overline{\mathrm{MS}}}$ to four loops. Z. Phys. **C54**, 599–606 (1992)
17. Broadhurst, D.J.: The master two-loop diagram with masses. Z. Phys. **C47**, 115–124 (1990)
18. Broadhurst, D.J.: Massive 3-loop Feynman diagrams reducible to SC* primitives of algebras of the sixth root of unity. Eur. Phys. J. **C8**, 311–333 (1999)
19. Bailey, D.H., Borwein, J.M., Broadhurst, D.J., Glasser, M.L.: Elliptic integral evaluations of Bessel moments and applications. J. Phys. A Math. Theor. **41**, 205203 (2008)
20. Peters, C., Top, J., van der Vlugt, M.: The Hasse zeta function of a K3 surface related to the number of words of weight 5 in the Melas codes. J. Reine Angew. Math. **432**, 151–176 (1992)
21. Hulek, K., Spandaw, J., van Geemen, B., van Straten, D.: The modularity of the Barth-Nieto quintic and its relatives. Adv. Geom. **1**, 263–289 (2001)
22. Rains, E.M., Sloane, N.J.A.: The shadow theory of modular and unimodular lattices. J. Number Theory **73**, 359–389 (1998)
23. Brown, F., Schnetz, O.: A K3 in ϕ^4. Duke Math. J. **161**, 1817–1862 (2012)
24. Schnetz, O.: Quantum field theory over F_q. Electron. J. Comb. **18**, P102 (2011)
25. Evans, R.: Seventh power moments of Kloosterman sums. Isr. J. Math. **175**, 349–362 (2010)
26. Fu, L., Wan, D.: Functional equations of L-functions for symmetric products of the Kloosterman sheaf. Trans. Am. Math. Soc. **362**, 5947–5695 (2010). (S0002-9947, 05172-4)

Computer-Assisted Proofs of Some Identities for Bessel Functions of Fractional Order

Stefan Gerhold, Manuel Kauers, Christoph Koutschan, Peter Paule, Carsten Schneider, and Burkhard Zimmermann

In memory of Frank W.J. Olver (1924–2013)

Abstract We employ computer algebra algorithms to prove a collection of identities involving Bessel functions with half-integer orders and other special functions. These identities appear in the famous Handbook of Mathematical Functions, as well as in its successor, the DLMF, but their proofs were lost. We use generating functions and symbolic summation techniques to produce new proofs for them.

1 Introduction

The Digital Library of Mathematical Functions [15] is the successor of the classical Handbook of Mathematical Functions [2] by Abramowitz and Stegun. Beginning of June 2005 Peter Paule was supposed to meet Frank Olver, the mathematics editor of [15], at the NIST headquarters in Gaithersburg (Maryland, USA). On May 18, 2005, Olver sent the following email to Paule:

M. Kauers · P. Paule · C. Schneider · B. Zimmermann
Research Institute for Symbolic Computation (RISC), Johannes Kepler University Linz, Linz, Austria
e-mail: Manuel.Kauers@risc.jku.at; Peter.Paule@risc.jku.at; Carsten.Schneider@risc.jku.at; Burkhard.Zimmermann@risc.jku.at

S. Gerhold
Financial and Actuarial Mathematics, Vienna University of Technology, Vienna, Austria
e-mail: sgerhold@fam.tuwien.ac.at

C. Koutschan (✉)
Johann Radon Institute for Computational and Applied Mathematics (RICAM), Austrian Academy of Sciences (ÖAW), Linz, Austria
e-mail: christoph.koutschan@ricam.oeaw.ac.at

C. Schneider and J. Blümlein (eds.), *Computer Algebra in Quantum Field Theory*, Texts & Monographs in Symbolic Computation, DOI 10.1007/978-3-7091-1616-6_3,

> The writing of DLMF Chapter BS[1] by Leonard Maximon and myself is now largely complete [...] However, a problem has arisen in connection with about a dozen formulas from Chapter 10 of Abramowitz and Stegun for which we have not yet tracked down proofs, and the author of this chapter, Henry Antosiewiecz, died about a year ago. Since it is the editorial policy for the DLMF not to state formulas without indications of proofs, I am hoping that you will be willing to step into the breach and supply verifications by computer algebra methods [...] I will fax you the formulas later today.

In view of the upcoming trip to NIST, Paule was hoping to be able to provide at least some help in this matter. But the arrival of Olver's fax chilled the enthusiasm quite a bit. Despite containing some identities with familiar pattern, the majority of the entries involved Bessel functions of fractional order or with derivatives applied with respect to the order.

Let us now display the bunch of formulas we are talking about. Here, $J_\nu(z)$ and $Y_\nu(z)$ denote the Bessel functions of the first and second kind, respectively, $I_\nu(z)$ and $K_\nu(z)$ the modified Bessel functions, $j_n(z)$ and $y_n(z)$ the spherical Bessel functions, $P_n(z)$ the Legendre polynomials, and $\mathrm{Si}(z)$ and $\mathrm{Ci}(z)$ the sine and cosine integral, respectively. Unless otherwise specified, all parameters are arbitrary complex numbers.

$$\frac{1}{z}\sin\sqrt{z^2+2zt} = \sum_{n=0}^{\infty}\frac{(-t)^n}{n!}y_{n-1}(z) \qquad (2|t|<|z|, |\Im(z)|\le\Re(z)) \tag{10.1.39}$$

$$\frac{1}{z}\cos\sqrt{z^2-2zt} = \sum_{n=0}^{\infty}\frac{t^n}{n!}j_{n-1}(z) \qquad z\neq 0 \tag{10.1.40}$$

$$\left[\frac{\partial}{\partial\nu}j_\nu(z)\right]_{\nu=0} = \frac{1}{z}(\mathrm{Ci}(2z)\sin z-\mathrm{Si}(2z)\cos z) \qquad (z\in\mathbb{C}\setminus]-\infty,0]) \tag{10.1.41}$$

$$\left[\frac{\partial}{\partial\nu}j_\nu(z)\right]_{\nu=-1} = \frac{1}{z}(\mathrm{Ci}(2z)\cos z+\mathrm{Si}(2z)\sin z) \qquad (z\in\mathbb{C}\setminus]-\infty,0]) \tag{10.1.42}$$

$$\left[\frac{\partial}{\partial\nu}y_\nu(z)\right]_{\nu=0} = \frac{1}{z}(\mathrm{Ci}(2z)\cos z+[\mathrm{Si}(2z)-\pi]\sin z) \qquad (z\in\mathbb{C}\setminus]-\infty,0]) \tag{10.1.43}$$

$$\left[\frac{\partial}{\partial\nu}y_\nu(z)\right]_{\nu=-1} = -\frac{1}{z}(\mathrm{Ci}(2z)\sin z-[\mathrm{Si}(2z)-\pi]\cos z) \qquad (z\in\mathbb{C}\setminus]-\infty,0]) \tag{10.1.44}$$

$$J_0(z\sin\theta) = \sum_{n=0}^{\infty}(4n+1)\frac{(2n)!}{2^{2n}n!^2}j_{2n}(z)P_{2n}(\cos\theta) \tag{10.1.48}$$

[1] Finally Chap. 10 Bessel Functions.

$$j_n(2z) = -n!z^{n+1}\sum_{k=0}^{n}\frac{2n-2k+1}{k!(2n-k+1)!}j_{n-k}(z)y_{n-k}(z) \quad (n=0,1,2,\ldots) \tag{10.1.49}$$

$$\sum_{n=0}^{\infty} j_n^2(z) = \frac{\mathrm{Si}(2z)}{2z} \tag{10.1.52}$$

$$\frac{1}{z}\sinh\sqrt{z^2-2izt} = \sum_{n=0}^{\infty}\frac{(-it)^n}{n!}\sqrt{\tfrac{1}{2}\pi/z}\,I_{-n+\frac{1}{2}}(z) \qquad (2|t|<|z|, |\Im(z)|\le\Re(z)) \tag{10.2.30}$$

$$\frac{1}{z}\cosh\sqrt{z^2+2izt} = \sum_{n=0}^{\infty}\frac{(it)^n}{n!}\sqrt{\tfrac{1}{2}\pi/z}\,I_{n-\frac{1}{2}}(z) \qquad z\neq 0 \tag{10.2.31}$$

$$\left[\frac{\partial}{\partial\nu}I_\nu(z)\right]_{\nu=1/2} = -\frac{1}{\sqrt{2\pi z}}(\mathrm{Ei}(2z)\mathrm{e}^{-z}+\mathrm{E}_1(2z)\mathrm{e}^{z}) \qquad (z\in\mathbb{C}\setminus]-\infty,0]) \tag{10.2.32}$$

$$\left[\frac{\partial}{\partial\nu}I_\nu(z)\right]_{\nu=-1/2} = \frac{1}{\sqrt{2\pi z}}(\mathrm{Ei}(2z)\mathrm{e}^{-z}-\mathrm{E}_1(2z)\mathrm{e}^{z}) \qquad (z\in\mathbb{C}\setminus]-\infty,0]) \tag{10.2.33}$$

$$\left[\frac{\partial}{\partial\nu}K_\nu(z)\right]_{\nu=\pm 1/2} = \pm\sqrt{\frac{\pi}{2z}}\mathrm{E}_1(2z)\mathrm{e}^{z} \qquad (z\in\mathbb{C}\setminus]-\infty,0]) \tag{10.2.34}$$

The numbering follows that in Abramowitz and Stegun [2], and Olver remarked on the fax: "Irene Stegun left a record (without proofs) that (10.1.41)–(10.1.44) have errors: the factor $\frac{1}{2}\pi$ should not be there, and (10.1.44) also has the wrong sign. Equations (10.2.32)–(10.2.34) have similar errors. Their correct versions are given by [...]".

In view of these unfamiliar objects and of the approaching trip to NIST, Paule asked his young collaborators for help. Within 2 weeks, all identities succumbed to the members of the algorithmic combinatorics group of RISC. Moreover, in addition to the typos in [2] mentioned by Olver, further typos in (10.1.39) and (10.2.30) were found. Above we have listed the corrected versions of the formulas, and when we use the numbering from [2], we refer to the corrected versions of the formulas here and throughout the paper.

At this place we want to relate the numbering from [2] to the one used in [15]: (10.1.39) and (10.1.40) are DLMF entries 10.56.2 and 10.56.1, respectively. With the help of the rewriting rule DLMF 10.47.3, (10.1.41) and (10.1.42) are DLMF entries 10.15.6 and 10.15.7, respectively; using the rule DLMF 10.47.4, (10.1.43) and (10.1.44) are DLMF entries 10.15.8 and 10.15.9, respectively. Entry (10.1.48) is DLMF 10.60.10, (10.1.49) is DLMF 10.60.4, and (10.1.52) is DLMF 10.60.11. With the help of DLMF 10.47.8, entry (10.2.30) turns into DLMF 10.56.4; and with the help of DLMF 10.46.7, entry (10.2.31) turns into DLMF 10.56.3. Formulas

(10.2.32) and (10.2.33) are bundled in DLMF entry 10.38.6; formula (10.2.34) is DLMF 10.38.7.

The goal of our exposition is to convince the reader that only a very limited amount of techniques has to be mastered to be able to prove such special function identities with computer algebra.

Our computer proofs are based on the algorithmic theory of holonomic functions and sequences, and symbolic summation algorithms. In the following two sections, we do purely algebraic manipulations; where necessary, analytical justifications (convergence of series, etc.) are given in Sect. 4. In general, we rely on the following computer algebra toolbox; underlying ideas are described in [9, 10].

Holonomic closure properties. The packages gfun [18] (for Maple) and GeneratingFunctions [13] (for Mathematica) are useful for the manipulation of functions $f(x)$ that satisfy linear ordinary differential equations (LODEs) with polynomial coefficients, as well as for sequences f_n satisfying linear recurrence equations (LOREs) with polynomial coefficients. Such objects are called *holonomic.* It can be shown that whenever $f(x)$ and $g(x)$ (resp. f_n and g_n) are holonomic, then so are $f(x) \cdot g(x)$ and $f(x) + g(x)$ (resp. $f_n \cdot g_n$ and $f_n + g_n$). Furthermore, if $f(x) = \sum_{n=0}^{\infty} f_n x^n$, then $f(x)$ is holonomic if and only if f_n is holonomic as a sequence. The packages gfun and GeneratingFunctions provide procedures for "executing closure properties," i.e., from given differential equations for $f(x)$ and $g(x)$ they can compute differential equations for $f(x) \cdot g(x)$ and $f(x) + g(x)$, and likewise for sequences. Also several further closure properties can be executed in this sense, and there are procedures for obtaining a recurrence equation for f_n from a differential equation for its generating function $f(x) = \sum_{n=0}^{\infty} f_n x^n$, and vice versa.

Symbolic summation tools. The package Zb [16] (for Mathematica) and the more general and powerful packages Mgfun [4] (for Maple), HolonomicFunctions [11] and Sigma [19, 20] (both for Mathematica) provide algorithms to compute for a given definite sum $S(n,z) = \sum_{k=0}^{n} f(n,z,k)$ recurrences (in n) and/or differential equations (in z). Here the essential assumption is that the summand $f(n,z,k)$ satisfies certain types of recurrences or differential equations; see Sect. 3.

Subsequently, we restrict our exposition to the Mathematica packages GeneratingFunctions, Zb, HolonomicFunctions, and Sigma. In the Appendix, for the reader's convenience we list all formulas from Abramowitz and Stegun [2] that we apply in our proofs.

As for applications of differentiating Bessel functions w.r.t. order, we mention maximum likelihood estimation for the generalized hyperbolic distribution, and calculating moments of the Hartman-Watson distribution. Both distributions have applications in mathematical finance [6, 17]. Prause's PhD thesis [17] in fact cites formulas (9.6.42)–(9.6.46).

2 Basic Manipulations of Power Series

Let us now show how to apply these computer algebra tools for proving identities. The basic strategy is to determine algorithmically a differential equation (LODE) or a recurrence (LORE) for both sides of an identity and check initial conditions.

First we load the package GeneratingFunctions in the computer algebra system Mathematica.

In[1]:= **<< GeneratingFunctions.m**

GeneratingFunctions Package by Christian Mallinger – © RISC Linz

2.1 *LODE and Initial Conditions for (10.1.39)*

We show that both sides of the equation satisfy the same differential equation in t, and then check a suitable number of initial values.

First we compute a differential equation for the left hand side $\frac{1}{z}\sin\sqrt{z^2+2zt}$. We view this function as the composition of $\frac{1}{z}\sin(t)$ with $\sqrt{z^2+2zt}$ and compute a differential equation for it from defining equations of the components, by using the command AlgebraicCompose. (The last argument specifies the function under consideration. This symbol is used both in input and output.)

In[2]:= $\mathbf{AlgebraicCompose}[f''[t] == -f[t], f[t]^2 == z^2 + 2zt, f[t]]$

Out[2]= $zf[t] + f'[t] + (2t+z)f''[t] == 0$

In order to obtain a differential equation for the right hand side, we first compute a recurrence equation for the coefficient sequence $c_n := (-1)^n/n!\, y_{n-1}(z)$ from the recurrences of its factors (using (10.1.19)). (The coefficient-wise product of power series is called *Hadamard product*, which explains the name of the command REHadamard.)

In[3]:= $\mathbf{REHadamard}[c[n+1] == -c[n]/(n+1), c[n-1] + c[n+1] ==$
$(2(n-1)+1)/z\, c[n], c[n]]$

CanRE::denom : Warning. The input equation will be multiplied by its denominator.

Out[3]= $zc[n] + (1+n)(1+2n)c[n+1] + (1+n)(2+n)zc[n+2] == 0$

Then we convert the recurrence equation for c_n into a differential equation for its generating function $\sum_{n=0}^{\infty} c_n t^n$, which is the right hand side.

In[4]:= $\mathbf{RE2DE}[\%, c[n], f[t]]$

Out[4]= $zf[t] + f'[t] + (2t+z)f''[t] == 0$

This agrees with output 2. To complete the proof, we need to check two initial values.

In[5]:= **Series[1/z Sin[$\sqrt{z^2 + 2zt}$], {t, 0, 1}]**

Out[5]= $\frac{\mathrm{Sin}[\sqrt{z^2}]}{z} + \frac{\sqrt{z^2}\mathrm{Cos}[\sqrt{z^2}]}{z^2} t + O[t]^2$

By (10.1.12) and (10.1.19), this agrees with the initial values of the right hand side for $z \in \mathbb{R}_{\geq 0}$. The extension to complex z will be discussed in Sect. 4.

Alternatively, we could have derived a differential equation only for the right hand side and then check with Mathematica that the left hand side satisfies this equation:

In[6]:= **Out[4] /. f → (1/z Sin[$\sqrt{z^2 + 2z\#}$]&)**

Out[6]= True

The proofs for (10.1.40), (10.2.30), and (10.2.31) follow the same scheme as the proof above. Both variants of the proof work in each case.

In summary, the most systematic way is to compute a differential equation for the difference of left hand side and right hand side, and then check that an appropriate number of initial values are zero.

2.2 *Proof of (10.1.41)*

This time we will not derive an LODE, but instead a recurrence relation for the Taylor coefficients of the difference of the left and the right hand side. The term $\log(z/2)$ that occurs in the pertinent expansion (9.1.64) is not analytic at $z = 0$, hence we first treat that one "by hand." (Working with Taylor series at $z = 1$, say, promises not much but additional complications.) This will leave us with a rather complicated expression for a holonomic formal power series, for which we have to prove that it is zero. At this point, we will employ the GeneratingFunctions package for computing a recurrence equation for the coefficient sequence of that series. Upon checking a suitable number of initial values, zero equivalence is then established.

One might think that we would not even have to compute the recurrences, since it is known a priori that the sum of two sequences satisfying recurrences of order r_1 and r_2, respectively, satisfies a recurrence of order at most $r_1 + r_2$. The same holds for products, with $r_1 r_2$ instead of $r_1 + r_2$. The catch is that the leading coefficient of the combined recurrence might have roots in the positive integers. It is clear that in order to give an inductive proof there must not be an integer root beyond the places where we check initial values.

Proposition 1. *Identity (10.1.41) holds for $z \in \mathbb{C} \setminus \mathbb{R}_{\leq 0}$.*

Proof. First we consider the left hand side. Using (10.1.1) and (9.1.64) from the Appendix, we get

$$\frac{\partial}{\partial \nu} j_\nu(z) = j_\nu(z) \log\frac{z}{2} - \frac{1}{2}\sqrt{\pi}\sum_{n=0}^{\infty}(-1)^n \frac{\psi(\nu+n+\frac{3}{2})}{\Gamma(\nu+n+\frac{3}{2})}\frac{(\frac{1}{4}z^2)^n}{n!},$$

where $\Gamma(x)$ and $\psi(x) = \frac{\frac{d}{dx}\Gamma(x)}{\Gamma(x)}$ denote the Gamma and digamma function, respectively. Hence, with (10.1.11),

$$\left[\frac{\partial}{\partial \nu} j_\nu(z)\right]_{\nu=0} = \frac{\sin z}{z}\log\frac{z}{2} - \frac{1}{2}\sqrt{\pi}\sum_{n=0}^{\infty}(-1)^n \frac{\psi(n+\frac{3}{2})}{\Gamma(n+\frac{3}{2})}\frac{(\frac{1}{4}z^2)^n}{n!}.$$

For the right hand side, we need (5.2.14), (5.2.16), and the Taylor expansions of $\sin z$ and $\cos z$. We have to show that

$$\begin{aligned}
&\left[\frac{\partial}{\partial \nu} j_\nu(z)\right]_{\nu=0} - \left(\mathrm{Ci}(2z)\sin z - \mathrm{Si}(2z)\cos z\right)/z \\
&= \frac{\sin z}{z}\log\frac{z}{2} - \frac{1}{2}\sqrt{\pi}\sum_{n=0}^{\infty}\frac{(-1)^n\psi(n+\frac{3}{2})}{\Gamma(n+\frac{3}{2})}\frac{(\frac{1}{4}z^2)^n}{n!} + \frac{\mathrm{Si}(2z)\cos z}{z} \\
&\quad - \frac{\sin z}{z}\left(\gamma + \log(2z) + \sum_{n=1}^{\infty}\frac{(-1)^n(2z)^{2n}}{2n(2n)!}\right) \\
&= -\frac{1}{2}\sqrt{\pi}\sum_{n=0}^{\infty}\frac{(-1/4)^n\psi(n+\frac{3}{2})z^{2n}}{\Gamma(n+\frac{3}{2})n!} + 2\sum_{n=0}^{\infty}\frac{(-4)^n z^{2n}}{(2n+1)(2n+1)!}\sum_{n=0}^{\infty}\frac{(-1)^n z^{2n}}{(2n)!} \\
&\quad - (\gamma + 2\log 2)\sum_{n=0}^{\infty}\frac{(-1)^n z^{2n}}{(2n+1)!} + 4z^2\sum_{n=0}^{\infty}\frac{(-4)^n z^{2n}}{2(n+1)(2(n+1))!}\sum_{n=0}^{\infty}\frac{(-1)^n z^{2n}}{(2n+1)!}
\end{aligned} \qquad (*)$$

is identically zero, i.e., $c_n = 0$ for all $n \geq 0$, where c_n is defined as $(*) = \sum_{n=0}^{\infty} c_n z^{2n}$.

To this end, we compute step by step a recurrence equation for c_n from the various coefficient sequences appearing in $(*)$. We suppress some of the output, in order to save space. Recurrences for most of the inner coefficient sequences are easy to obtain. For instance, for

In[7]:= $f[n\_] := \frac{(-4)^n}{(2n+1)(2n+1)!}$

we have

In[8]:= **FullSimplify[f [n + 1]/f [n]]**

Out[8]= $\frac{-2(2n+1)}{(n+1)(2n+3)^2}$

and hence the recurrence $f_{n+1} = \frac{-2(2n+1)}{(n+1)(2n+3)^2} f_n$. Only the series involving $\psi(n+3/2)$ requires a bit more work. Here, we use the package GeneratingFunctions to obtain a recurrence from the recurrence (6.3.5) for $\psi(n+3/2)$ and the first order recurrence of $(-1/4)^n/\Gamma(n+\frac{3}{2})n!$.

In[9]:= $\textbf{recSum} = \textbf{REHadamard}[f[n+1] == f[n] + \frac{1}{n+3/2},$
$f[n+1] == \frac{-1}{2(2n+3)(n+1)} f[n], f[n]];$

Next, we compute recurrence equations for the coefficient sequence of the two series products in (∗).

In[10]:= $\textbf{recSiCos} = \textbf{RECauchy}[f[n+1] == \frac{-2(2n+1)}{(n+1)(2n+3)^2} f[n],$
$f[n+1] == \frac{-1}{2(2n+1)(n+1)} f[n], f[n]];$

In[11]:= $\textbf{recCiSin} = \textbf{RECauchy}[f[n+1] == \frac{-2(n+1)}{(n+2)^2(2n+3)} f[n],$
$f[n+1] == \frac{-1}{2(n+1)(2n+3)} f[n], f[n]];$

The latter recurrence has to be shifted by 1, owing to the factor z^2.

In[12]:= $\textbf{recCiSin} = \textbf{recCiSin} \;/.\; f[n\_] \to f[n+1] \;/.\; n \to n-1;$

The recurrences collected so far can now be combined to a recurrence for c_n.

In[13]:= $\textbf{rec1} = \textbf{REPlus}[\textbf{recSiCos}, \textbf{recSum}, f[n]];$

In[14]:= $\textbf{rec2} = \textbf{REPlus}[\textbf{recCiSin}, f[n+1] == \frac{-1}{2(n+1)(2n+3)} f[n], f[n]];$

In[15]:= $\textbf{rec} = \textbf{REPlus}[\textbf{rec1}, \textbf{rec2}, f[n]]$

Out[15]= $5184(227+60n)f[n]+\cdots$

$\cdots+7600(4+n)(5+n)(6+n)^2(9+2n)(11+2n)(13+2n)^2(167+60n)f[n+6]=0$

The precise shape of the recurrence is irrelevant, it only matters that it has order 6 and that the coefficient of $f[n+6]$ (i.e., of c_{n+6}) does not have roots at nonnegative integers. As this is the case, we can complete the proof by checking that the coefficients of $z^0,\ldots,z^{10}$ in (∗) vanish, which can of course be done with Mathematica.

Alternatively, a similar proof can be obtained more conveniently using the package

In[16]:= **<< HolonomicFunctions.m**

HolonomicFunctions package by Christoph Koutschan, RISC-Linz, Version 1.6 (12.04.2012)

One of the main features of this package is the Annihilator command; it analyzes the structure of a given expression and executes the necessary closure properties automatically, in order to compute a system of differential equations and/or recurrences for the expression. We apply it to (∗):

In[17]:= $\textbf{Annihilator}\Big[-\frac{\textbf{Sin}[z]}{z}\Big(\textbf{EulerGamma} + 2\,\textbf{Log}[2] + \textbf{Sum}\Big[\frac{(-1)^n\,(2z)^{2n}}{2n\,(2n)!}, \{n, 1, \infty\}\Big]\Big)$
$-\frac{\sqrt{\pi}}{2}\textbf{Sum}\Big[\frac{(-1/4)^n\,z^{2n}\,\textbf{PolyGamma}[0, n+3/2]}{n!\,\textbf{Gamma}[n+3/2]}, \{n, 0, \infty\}\Big]$
$+\frac{\textbf{Cos}[z]}{z}\textbf{SinIntegral}[2z], \textbf{Der}[z]\Big]$

Out[17]= $\{(48z^5 + 95z^3)D_z^8 + (864z^4 + 1900z^2)D_z^7 + (576z^5 + 5436z^3 + 10830z)D_z^6 + (7968z^4 + 23684z^2 + 17100)D_z^5 + (1440z^5 + 32442z^3 + 77002z)D_z^4 + (13344z^4 + 59332z^2 + 83448)D_z^3 + (1344z^5 + 33596z^3 + 82858z)D_z^2 + (6240z^4 + 31404z^2 + 46892)D_z + (432z^5 + 6495z^3 + 15150z)\}$

Since the HolonomicFunctions package uses operator notation, the second argument indicates that a differential equation w.r.t. z is desired; instead of an equation the corresponding operator is returned with $D_z = \mathrm{d}/\mathrm{d}z$. As before, the proof is completed by checking a few initial values (see also Sect. 4). □

3 Symbolic Summation Tools

It is not always the case that recurrences for the power series coefficients can be obtained by the package GeneratingFunctions. Sometimes combinatorial identities such as the following one are needed. Its proof gives occasion to introduce the Mathematica package Zb, an implementation of Zeilberger's algorithm for hypergeometric summation [22].

Lemma 1. *For $k \in \mathbb{Z}_{\geq 0}$ we have*

$$\sum_{j=1}^{k} \frac{(-2)^j}{j}\binom{k}{j} = \begin{cases} \mathrm{H}_{n+1} - 2\mathrm{H}_{2n+2} & k = 2n+1 \quad \textit{is odd} \\ \mathrm{H}_n - 2\mathrm{H}_{2n} & k = 2n \quad \textit{is even,} \end{cases}$$

where $\mathrm{H}_n := \sum_{k=1}^{n} \frac{1}{k}$ denotes the harmonic numbers.

It can be a chore to locate such identities in the literature. The closest match that the authors found is the similar identity $\sum_{j=1}^{k}(-1)^{j+1}j^{-1}\binom{k}{j} = \mathrm{H}_k$ [7, p. 281]. Thus, an automatic identity checker like the one we describe now is helpful. We note in passing that we can not only verify such identities, but even compute the right hand side from the left hand side [19].

Proof (of Lemma 1). We denote the sum on the left hand side by a_k. Using the Mathematica package

In[18]:= **<< Zb.m**

Fast Zeilberger Package by Peter Paule and Markus Schorn (enhanced by Axel Riese) – © RISC Linz

we find

In[19]:= **Zb[(−2)^j /j Binomial[2n + 1, j], {j, 1, 2n + 1}, n]**

If '1 + 2 n' is a natural number, then:

Out[19]= $\{(n+1)(2n+3)\mathrm{SUM}[n] - (4n^2 + 14n + 13)\mathrm{SUM}[n+1] + (n+2)(2n+5)\mathrm{SUM}[n+2] == -2\}$

In[20]:= **Zb[(−2)^j /j Binomial[2n, j], {j, 1, 2n}, n]**

If '2n' is a natural number, then:

Out[20]= {(n + 1)(2n + 1)SUM[n] − (4n^2 + 10n + 7)SUM[n + 1]
+ (n + 2)(2n + 3)SUM[n + 2] == −2}

hence the sequence a_k satisfies the recurrences

$$(n+1)(2n+3)a_{2n+1} - (4n^2+14n+13)a_{2n+3} + (n+2)(2n+5)a_{2n+5} = -2$$

and

$$(n+1)(2n+1)a_{2n} - (4n^2+10n+7)a_{2n+2} + (2n+3)(n+2)a_{2n+4} = -2.$$

The right hand side satisfies these recurrences, too:

```
In[21]:= Out[19] /. SUM[n_] → HarmonicNumber[n + 1] − 2HarmonicNumber[2n + 2]
         // ReleaseHold // FullSimplify

Out[21]= {True}

In[22]:= Out[20]/. SUM[n_] → HarmonicNumber[n] − 2HarmonicNumber[2n]
         // ReleaseHold // FullSimplify

Out[22]= {True}
```

Hence the desired result follows by checking the initial conditions $k = 0, 1, 2, 3$.

□

Proposition 2. *Identities (10.2.32) and (10.2.33) follow from Lemma 1. They hold for* $z \in \mathbb{C} \setminus \mathbb{R}_{\leq 0}$.

Proof. We do Taylor series expansion on both sides of (10.2.32), and then compare coefficients. Using the expansions (5.1.10) and (5.1.11) and computing Cauchy products, we find that the right hand side of (10.2.32) equals

$$\sqrt{\frac{2}{\pi z}}\left((\log z + \log 2 + \gamma)\sinh z + \sum_{n=0}^{\infty}\frac{a_{2n+1}}{(2n+1)!}z^{2n+1}\right), \tag{1}$$

where a_k is the sum from Lemma 1. The expansion of the left hand side of (10.2.32) can be done with (9.6.10) and (9.6.42). Since

$$\frac{(z^2/4)^n}{\Gamma(n+\frac{3}{2})n!} = \frac{2z^{2n}}{\sqrt{\pi}(2n+1)!}$$

and

$$\psi(n+\tfrac{3}{2}) = -\gamma - 2\log 2 + 2\mathrm{H}_{2n+2} - \mathrm{H}_{n+1},$$

the left hand side of (10.2.32) turns out to be

$$\sqrt{\frac{2}{\pi z}}\left((\log z + \log 2 + \gamma)\sinh z + \sum_{n=0}^{\infty}(\mathrm{H}_{n+1} - 2\mathrm{H}_{2n+2})\frac{z^{2n+1}}{(2n+1)!}\right). \tag{2}$$

Lemma 1 completes the coefficient comparison.

Identity (10.2.33) can be proved analogously; replace sinh by cosh and $2n+1$ by $2n$ in (1), and sinh by cosh and the summand by $(\mathrm{H}_n - 2\mathrm{H}_{2n})z^{2n}/(2n)!$ in (2). □

We proceed to prove the identities (10.1.48), (10.1.49), and (10.1.52) by the same strategy as above: compute LODEs or LOREs for both sides, and check initial values. Since in these identities definite sums occur for which one cannot derive LOREs or LODEs by using holonomic closure properties, symbolic summation algorithms enter the game. For hypergeometric sums, like in Lemma 1, the package Zb is the perfect choice. Since in the following identities the occurring sums do not have hypergeometric summands, we use more general summation methods [19] and [11] that are available in the packages Sigma and HolonomicFunctions, respectively.

In general, the sums under consideration are of the form

$$S(n,z) = \sum_{k=0}^{\infty} h(n,k)f(n,z,k) \tag{3}$$

with integer parameter n and complex parameter z where h and f have the following properties: $h(n,k)$ is a hypergeometric term in n and k, i.e., $h(n+1,k)/h(n,k)$ and $h(n,k+1)/h(n,k)$ are rational functions in n and k. Furthermore, $f(n,z,k)$ satisfies a recurrence relation of the form

$$\begin{aligned} f(n,z,k+d) &= \alpha_0(n,z,k)f(n,z,k) \\ &\quad + \alpha_1(n,z,k)f(n,z,k+1) + \cdots + \alpha_{d-1}(n,z,k)f(n,z,k+d-1), \end{aligned} \tag{4}$$

and either a recurrence relation

$$\begin{aligned} f(n+1,z,k) &= \beta_0(n,z,k)f(n,z,k) \\ &\quad + \beta_1(n,z,k)f(n,z,k+1) + \cdots + \beta_{d-1}(n,z,k)f(n,z,k+d-1) \end{aligned} \tag{5}$$

or a differential equation

$$\begin{aligned} \frac{\mathrm{d}}{\mathrm{d}z}f(n,z,k) &= \beta_0(n,z,k)f(n,z,k) \\ &\quad + \beta_1(n,z,k)f(n,z,k+1) + \cdots + \beta_{d-1}(n,z,k)f(n,z,k+d-1), \end{aligned} \tag{6}$$

where the α_i, β_i are rational functions in k, n, and z. From recurrences of the forms (4) and (5) we will derive a recurrence relation in n for $S(n,z)$. If, on the other hand, we have (6) instead of (5), we will compute a differential equation for $S(n,z)$ in z.

We note that the HolonomicFunctions package allows more flexible recurrence/differential systems as input specifying the shift/differential behavior of the summand accordingly. However, the input description given above gives rise to rather efficient algorithms implemented in the Sigma package to calculate LOREs and LODEs for $S(n,z)$.

3.1 LORE and Initial Conditions for (10.1.49)

We compute a LORE for the right hand side

$$S(n) := \sum_{k=0}^{n} -n!z^{n+1}\frac{2n-2k+1}{k!(2n-k+1)!}j_{n-k}(z)y_{n-k}(z)$$
$$= \sum_{k=0}^{n} \frac{-n!z^{n+1}(2k+1)}{(n-k)!(n+k+1)!}j_k(z)y_k(z)$$

using

In[23]:= **<< Sigma.m**

Sigma - A summation package by Carsten Schneider © RISC-Linz

First we insert the sum in the form (3) with recurrences of the type (4) and (5). Note that $h(n,k) = \frac{-n!z^{n+1}(2k+1)}{(n-k)!(n+k+1)!}$ is hypergeometric in n and k. Moreover, by (10.1.19) the spherical Bessel functions of the first kind $j(k) := j_k(z)$ (we suppress the parameter z in our Mathematica session) fulfill the recurrence

In[24]:= $\mathbf{recJ = z j[k] - (2k+3) j[k+1] + z j[k+2] == 0;}$

Since the same recurrence holds for $y_k(z)$, see (10.1.19), we obtain with

In[25]:= $\mathbf{recJY = REHadamard[recJ, recJ, j[k]]/.\{j \to f\};}$

Out[25]= $(-2k-5)z^2 f[k] + (2k+3)(4k^2+16k-z^2+15) f[k+1]$
$\qquad -(2k+5)(4k^2+16k-z^2+15) f[k+2] + (2k+3)z^2 f[k+3] = 0$

a recurrence in the form (4) for $f(k) := j_k(z)y_k(z)$. Since $f(k)$ is free of n, we choose $f[n+1,k] == f[k]$ for the required recurrence of the form (5). Given these recurrences we are ready to compute a recurrence for our sum

In[26]:= $\mathbf{mySum = \sum_{k=0}^{n} \frac{-n!z^{n+1}(2k+1)}{(n-k)!(n+k+1)!} f[k];}$

by using the Sigma-function

In[27]:= $\mathbf{GenerateRE[mySum, n, \{recJY, f[k]\}, f[n+1,k] == f[k]]}$

Out[27]= $2z\mathrm{SUM}[n] - (2n+3)\mathrm{SUM}[n+1] + 2z\mathrm{SUM}[n+2] == 0$

Note that $S(n) = \sum_{k=0}^{n} h(n,k) f(k) (= \text{mySum} = \text{SUM}[n])$. Since besides $S(n)$ also $j_n(2z)$ fulfills the computed recurrence and since $S(n) = j_n(2z)$ for $n = 0, 1$, we have $S(n) = j_n(2z)$ for all $n \geq 0$.

A correctness proof. Denote $\Delta_k g(z,k) := g(z,k+1) - g(z,k)$. The correctness of the produced recurrence follows from the computed proof certificate

$$\Delta_k g(n,k) = c_0 h(n,k) f(k) + c_1 h(n+1,k) f(k) + c_2 h(n+2,k) f(k) \tag{7}$$

given by $c_0 = 2z$, $c_1 = -(2n+3)$, $c_2 = 2z$ and

$$g(n,k) = \frac{z^{n+1} n!}{(2k+3)(n+k+2)!(n-k+2)!} \big[g_0 \mathrm{f}(k) + g_1 \mathrm{f}(k+1) + g_2 \mathrm{f}(k+2) \big]$$

with

$$\begin{aligned} g_0 =& 8k^5 - 8(n-1)k^4 - (z^2 + 28n + 30)k^3 + 2(2n^2 + (2z^2 - 9)n + 2z^2 - 19) \\ & k^2 + ((z^2+8)n^2 + (8z^2+15)n + 8z^2 + 1)k + (n^2 + 3n + 2)(2z^2 + 3) \\ g_1 =& (2k+3)(k-n-2)(2k^3 + (3-2n)k^2 - (5n+2)k + (n+1)(z^2-3)), \\ g_2 =& -(k+1)(k-n-2)(k-n-1)z^2. \end{aligned}$$

Namely, one can show that (7) holds for all $n \geq 0$ and $0 \leq k \leq n$ as follows. Express $\Delta_k g(n,k)$ in terms of $f(k)$ and $f(k+1)$ by using the recurrence given in Out[25] and rewrite any factorial in (7) in terms of $(n+k+2)!$ and $(n-k+2)!$. Afterwards verify (7) by polynomial arithmetic. The summation of (7) over k from 0 to n gives the recurrence in Out[27]; here we needed the first evaluations of $f(i) = j_i(z) y_i(z)$, $i = 0, 1, 2$, from (10.1.11) and (10.1.12).

We remark that the underlying algorithms [19] unify the creative telescoping paradigm [22] in the difference field setting [20] and holonomic setting [4]. This general point of view opens up interesting applications, e.g., in the field of combinatorics [3] and particle physics [1].

3.2 *LODE and Initial Conditions for (10.1.48)*

For the proof of (10.1.48) we choose the package HolonomicFunctions. As we have seen before, holonomic closure properties include algebraic substitution; but since $\sin(\theta)$ is not algebraic, we have to transform identity (10.1.48) slightly in order to make it accessible to our software: just replace $\cos(\theta)$ by c and $\sin(\theta)$ by $\sqrt{1-c^2}$. Now it is an easy task to compute a LODE in z for the left hand side:

In[28]:= **Annihilator** $\left[\textbf{BesselJ}\left[0, z\sqrt{1-c^2}\right], \textbf{Der}[z]\right]$

Out[28]= $\{z D_z^2 + D_z + (z - c^2 z)\}$

The sum on the right hand side requires some more work. Similar to identity (10.1.49) above, the technique of creative telescoping [22] is applied and it fits perfectly to the HolonomicFunctions package. The latter can deal with multivariate holonomic functions and sequences, i.e., roughly speaking, mathematical objects that satisfy (for each variable in question) either a LODE or a LORE of arbitrary (but fixed) order. For example, the expression

$$f(n,z,c) = (4n+1)\frac{(2n)!}{2^{2n}n!^2} j_{2n}(z) P_{2n}(c)$$

satisfies a LORE in n of order 4 and LODEs w.r.t. z and c, both of order 2. To derive a LODE in z for the sum we employ the following command (the shift operator S_n, defined by $S_n f(n) = f(n+1)$, is input as S[n], and the derivation D_z, defined by $D_z f(z) = f'(z)$, is input as Der[z]):

```
In[29]:= CreativeTelescoping[(4n+1)(2n)!/(2^(2n) n!^2) SphericalBesselJ[2n,z] LegendreP[2n,c],
         S[n] − 1, Der[z]]
```

Out[29]=
$$\left\{\{zD_z^2 + D_z + (z - c^2 z)\}, \left\{\frac{4(n+1)^2}{4n+5} S_n D_z + \frac{4(n+1)^2(8n^2+18n-z^2+9)}{(4n+3)(4n+5)z} S_n + \frac{4n^2}{4n+1} D_z + \frac{-16c^2n^2z^2 - 16c^2nz^2 - 3c^2z^2 + 32n^4 + 40n^3 + 4n^2z^2 + 12n^2 + 4nz^2 + z^2}{(4n+1)(4n+3)z}\right\}\right\}$$

The output consists of two operators, say P and Q, which are called *telescoper* and *certificate* (note already that P equals Out[28]). They satisfy the relation

$$\big(P + (S_n - 1)Q\big)\, f(n,z,c) = 0, \tag{8}$$

a fact that can be verified using the well-known LODEs and LOREs for spherical Bessel functions and Legendre polynomials. Summing (8) w.r.t. n and telescoping yields

$$P\sum_{n=0}^{\infty} f(n,z,c) - (Qf)(0,z,c) + \lim_{n\to\infty}(Qf)(n,z,c) = 0$$

(P is free of n and S_n and therefore can be interchanged with the summation quantifier). Using (9.3.1) and (10.1.1) it can be shown that the limit is 0, and also the part $(Qf)(0,z,c)$ vanishes.

Consequently, we have established that both sides of (10.1.48) satisfy the same second-order LODE. It suffices to compare the initial conditions at $z = 0$ (see Sect. 4). For the left hand side we have $J_0(0) = 1$. From (10.1.25) it follows that the Taylor expansion of $j_{2n}(z)$ starts with z^{2n} and hence for $z = 0$ all summands are zero except the first one. With (10.1.11) we see that the initial conditions on both sides agree.

Before turning to the next identity, we want to point to [20] where a different computer algebra proof of (10.1.48) has been given. More examples of proving special function identities with the HolonomicFunctions package are collected in [12].

3.3 LODE and Initial Conditions for (10.1.52)

Again we compute a LODE with Sigma. In order to get a LODE of the left hand side of (10.1.52) we compute a LODE of its truncated version

In[30]:= $\mathbf{mySum} = \sum_{k=0}^{a} j[k]^2;$

Note that the summand of our input-sum depends non-linearly on $j_k(z)$. In order to handle this type of summation input, Sigma needs in addition the package [5]

In[31]:= **<< OreSys.m**

OreSys package by Stefan Gerhold © RISC-Linz

for uncoupling systems of LODE-systems. Then using a new feature of Sigma we can continue as "as usual". Given the difference-differential equation of the form (6) for $j(k) := j_k(z)$ and $j^{(0,1)}(k, z) := \frac{d}{dz} j_k(z)$:

In[32]:= $\mathbf{recZ} = j^{(0,1)}[k, z] == \frac{k}{z} j[k] + j[k+1];$

see (10.2.20), and the recurrence In[24] of the form (5), we compute a LODE for mySum(= SUM[n]):

In[33]:= $\mathbf{mySum} = \sum_{k=0}^{a} j[k]^2;$

In[34]:= **GenerateDE[mySum, n, {recJ, $j[k]$}, recZ]**

Out[34]= $z\mathrm{SUM}'[z] + \mathrm{SUM}[z] == (zj[a]j[a+1] - (2a+1)j[a]^2) - (zj[0]j[1] - j[0]^2)$

A correctness proof. The correctness of the LODE can be checked by the computed proof certificate

$$\Delta_k g(z, k) = c_0 j(k)^2 + c_1 j^{(0,1)}(k, z)^2 \tag{9}$$

with $c_0 = 1$, $c_1 = z$ and $g(z, k) = zj(k)j(k+1) - (2k+1)j(k)^2$. Namely, one can easily show that (9) holds for all $0 \le k$ as follows. Express (9) in terms of $j(k)$ and $j(k+1)$ by using the recurrence given in In[24] and the difference-differential equation given in In[32]. Afterwards verify (9) by polynomial arithmetic. Then summing (9) over k from 0 to a gives the recurrence in Out[34]; here we used the initial values (10.1.11).

Next, we let $a \to \infty$. Then $j_a(z)$ tends to zero by (9.3.1). Therefore, the left hand side of (10.1.52) satisfies the LODE

$$S(z) + z\frac{dS(z)}{dz} = \frac{\sin(2z)}{2z}. \tag{10}$$

It is readily checked that the right hand side satisfies it, too, and both sides equal 1 at $z = 0$. This establishes equality of both sides of (10.1.52).

Alternatively, we can derive the inhomogeneous differential equation for the left hand side of (10.1.52) with HolonomicFunctions:

In[35]:= **Annihilator[Sum[SphericalBesselJ[n, z]^2, {n, 0, ∞}], Der[z], Inhomogeneous → True]**

Out[35]= $\{\{zD_z + 1\}, \{\text{Hold}[\text{Limit}[\ldots, n \to \infty]] + \ldots\}\}$

The output consists of a differential operator and an expression that gives the inhomogeneous part (abbreviated above). Without help, Mathematica is not able to simplify the latter (i.e., compute the limit), but using (9.3.1) it succeeds and we get

$$\left(zD_z + 1\right)S(z) - \frac{\sin(z)\cos(z)}{z} = 0$$

which of course agrees with (10).

4 Series Solutions of LODEs and Analyticity

In some proofs we have determined a differential equation that is satisfied by both sides of the identity in question, and then compared initial values. In contrast to the case of recurrences, the validity of this approach needs some non-trivial justification. This procedure can be justified by well-known uniqueness results for solutions of LODEs, to be outlined in this section. In the proofs of (10.1.39), (10.1.40), (10.2.30), and (10.2.31), the point $t = 0$, where we checked initial conditions, is an ordinary point of the LODE (i.e., the leading coefficient of the LODE does not vanish at $t = 0$). Then there is a unique analytic solution, if the number of prescribed initial values equals the order of the equation. The identity then holds (at least) in the domain (containing zero) where we can establish analyticity of both sides.

Proposition 3. *Identity (10.1.40) holds for all complex t and all complex $z \neq 0$. The same is true for (10.2.31).*

Proof. We consider (10.1.40) and omit the analogous considerations for (10.2.31). For $n \in \mathbb{Z}$, the function $j_{n-1}(z)$ is defined for $z \in \mathbb{C}^*$. We fix such a z and consider both sides of (10.1.40) as functions of t. By (9.3.1), the right hand side converges uniformly for all complex t, therefore it is an entire function of t. The left hand side is also entire, since $\cos\sqrt{w} = \sum_{n\geq 0}(-1)^n w^n/(2n)!$ is an entire function of w. Initial values at $t = 0$ and an LODE satisfied by both sides were already presented in Sect. 2, hence, by the above uniqueness property, identity (10.1.40) is proved. □

Proposition 4. *Identity (10.1.39) holds for all complex z and t with $|\Im(z)| \leq \Re(z)$ and $2|t| < |z|$. If $|\Im(z)| \leq -\Re(z)$, then the identity holds with switched sign for all t with $2|t| < |z|$. The same is true for (10.2.30).*

Proof. We give the proof in the case of (10.1.39); (10.2.30) is treated analogously. First we complete the check of initial values from Sect. 2. For $t = 0$, the right hand side is $y_{-1}(z) = (\sin z)/z$, and on the left hand side we have $(\sin \sqrt{z^2})/z$. Thus, at $t = 0$ both sides agree for $|\arg(z)| < \pi/2$, which follows from $|\Im(z)| \leq \Re(z)$; for $\pi/2 < |\arg(z)| < \pi$, which follows from $|\Im(z)| \leq -\Re(z)$, the identity holds at $t = 0$ with switched sign, because the function $w \mapsto \sqrt{w^2}$ changes sign when crossing the branch cut $\mathrm{i}\mathbb{R}$. The first derivatives at $t = 0$ are $(\cos \sqrt{z^2})/\sqrt{z^2} = (\cos z)/\sqrt{z^2}$ and $-y_0(z) = (\cos z)/z$, respectively. The same consideration as for the first initial value completes the check of the initial conditions.

Now we show that both sides of (10.1.39) are analytic functions of t for fixed $z \neq 0$ with $|\Im(z)| \leq |\Re(z)|$. Let us start by determining the radius of convergence of the right hand side. It is an easy consequence of (9.3.1) that

$$y_n(z) \sim -\frac{\sqrt{2}}{z}\left(\frac{2n}{\mathrm{e}z}\right)^n, \quad n \to \infty,\ z \neq 0.$$

Hence, by Stirling's formula, the radius of convergence is $|z|/2$, and so the right hand side is analytic for $2|t| < |z|$.

The left hand side of (10.1.39) has a branch cut along a half-line starting at $t = -z/2$, a point on the circle of convergence of the right hand side. If this half line has no other intersection with this circle, then the left hand side is analytic in the disk $\{t : 2|t| < |z|\}$. Otherwise, the branch cut separates the disk into two segments, and the identity does not necessarily hold in a segment that does not contain $t = 0$. As we will now show, our assumptions exclude the possibility of a second intersection. Once again it is convenient to proceed by computer algebra. Note that the presence of two intersections means that

$$\left(\exists s \neq t \in \mathbb{C}\right)\left(2|s| = |z| \wedge 2|t| = |z| \wedge z^2 + 2zs \in]-\infty, 0] \wedge z^2 + 2zt \in]-\infty, 0]\right)$$

holds. Upon rewriting this formula with real variables, it can be simplified by Mathematica's Reduce command; the result – translated back into complex language – is the equivalent formula $|\Re(z)| < |\Im(z)|$. Summing up, under our assumptions on z the left hand side of (10.1.39) is analytic in the disk $\{t : 2|t| < |z|\}$. □

Now consider the LODE (10), which we want to employ to prove (10.1.52). The point $z = 0$ is not an ordinary point, so the question of uniqueness of the solution is more subtle. The origin is a regular singular point of (10), since the degree of the indicial polynomial

$$[z^0] p_s(z)^{-1} z^{s-\sigma} \mathscr{L} z^\sigma = \sigma + 1$$

agrees with the order $s = 1$ of the LODE. Here, $p_s(z) = z$ denotes the leading coefficient, and $\mathscr{L}$ the differential operator

$$\mathscr{L} := 1 + zD_z.$$

The following classical result [8] describes the structure of a fundamental system at a regular singular point. See also the concise exposition in Meunier and Salvy [14].

Theorem 1. *Let $z = 0$ be a regular singular point of a homogeneous LODE of order s. Denote the roots of the indicial polynomial by $\sigma_1, \dots, \sigma_s$, and let $m_1, \dots, m_s$ be their multiplicities. Then the equation has a basis of s solutions*

$$z^{\sigma_i} \sum_{j=0}^{d_i} \log^j(z) \Phi_{ij}(z), \qquad 1 \le i \le s, \tag{11}$$

where $d_i < s$, and the $\Phi_{ij}(z)$ are convergent power series. Each of these solutions is uniquely defined by the coefficients of the s "monomials"

$$\bigcup_{i=1}^{s} \{z^{\sigma_i}, z^{\sigma_i} \log z, \dots, z^{\sigma_i} \log^{m_i - 1} z\}$$

in the series (11).

Proposition 5. *Identity (10.1.52) holds for all $z \in \mathbb{C}$.*

Proof. We have shown in the preceding section that both sides satisfy the LODE (10). As seen above, the indicial polynomial of the homogeneous equation $\mathscr{L} f = f + zf' = 0$ is $\sigma + 1$. Hence, by Theorem 1, a solution of $\mathscr{L} f = 0$ that has the form (11) is uniquely defined by the coefficient of z^{-1}. Hence the zero function is the only analytic solution of the homogeneous initial value problem $\mathscr{L} f = 0$, $f(0) = 0$. It is a trivial consequence that the inhomogeneous equation (10) cannot have more than one analytic solution with $f(0) = 1$. Therefore, (10.1.52) holds in a neighbourhood of $z = 0$. The left hand side of (10.1.52) is entire since it is a uniform limit of entire functions, and the right hand side is entire by (5.2.14). Thus, the identity holds in the whole complex plane by analytic continuation. □

Proposition 6. *Identity (10.1.48) holds for all complex z and θ.*

Proof. By the Laplace-Heine formula [21, Theorem 8.21.1], $P_{2n}(\cos\theta)$ grows at most exponentially as $n \to \infty$. Together with (9.3.1) and $n(2n)!/(2^{2n} n!^2) = O(\sqrt{n})$, this shows that the right hand side of (10.1.48) is an entire function of z and θ. In Sect. 3 we showed that both sides of (10.1.48) satisfy the differential equation $zf''(z) + f'(z) + z(1 - c^2) f(z) = 0$ (whose indicial equation is $\sigma^2 = 0$) and that the initial condition at z^0 agrees. The result follows from Theorem 1 and the fact that both sides are entire functions. □

5 Non-computer Proofs

Some of our identities can be easily proved from some of the others, without using any software machinery. The computer proofs that we have in hand suffice for establishing the remaining identities (10.1.42)–(10.1.44), and (10.2.34) in this spirit. The reader should by now be convinced that, if desired, all of them can also be proved by the algorithmic methods we have presented.

Proposition 7. *Identities (10.1.42)–(10.1.44) follow from (10.1.41). They hold for* $z \in \mathbb{C} \setminus \mathbb{R}_{\leq 0}$.

Proof. Identities (10.1.42)–(10.1.44) can be done analogously to (10.1.41), but we instead present (non-computer) deductions from (10.1.41). The derivative of Y_ν w.r.t. ν can be expressed in terms of J_ν, $J_{-\nu}$, and Y_ν, see (9.1.65) in the appendix. Note that $\cot(\nu + 1/2)\pi$ vanishes for $\nu = 0, 1$. Equation (9.1.65) thus yields

$$\left[\frac{\partial}{\partial \nu} y_\nu(z)\right]_{\nu=0} = \left[\frac{\partial}{\partial \nu} j_\nu(z)\right]_{\nu=-1} - \frac{\pi \sin z}{z}$$

and

$$\left[\frac{\partial}{\partial \nu} y_\nu(z)\right]_{\nu=-1} = -\left[\frac{\partial}{\partial \nu} j_\nu(z)\right]_{\nu=0} - \frac{\pi \cos z}{z}.$$

Therefore, we have a relation between the left hand sides of (10.1.42) and (10.1.43), and one between the left hand sides of (10.1.41) and (10.1.44). It is easy to verify that the respective right hand sides satisfy the same relations. Hence the assertion will be established once we show that (10.1.42) follows from (10.1.41). To this end, it suffices to show that the left hand sides of these identities satisfy

$$\frac{\partial}{\partial z}\left(z\left[\frac{\partial}{\partial \nu} j_\nu(z)\right]_{\nu=0}\right) - z\left[\frac{\partial}{\partial \nu} j_\nu(z)\right]_{\nu=-1} = -\frac{\sin z}{z}, \tag{12}$$

since once again it is easy to see that the right hand sides of (10.1.41) and (10.1.42) obey the same relation. By (9.1.64), the recurrence relations of Γ and ψ, and the duplication formula of Γ, the left hand side of (12) equals

$$\begin{aligned}
&\frac{\sin z}{z} - \sqrt{\pi}\sum_{k=0}^{\infty}(-\tfrac{1}{4})^k\left(\frac{\psi(k+\frac{3}{2})}{\Gamma(k+\frac{3}{2})/(k+\frac{1}{2})} - \frac{\psi(k+\frac{1}{2})}{\Gamma(k+\frac{1}{2})}\right)\frac{z^{2k}}{k!}\\
=&\frac{\sin z}{z} - \sqrt{\pi}\sum_{k=0}^{\infty}(-\tfrac{1}{4})^k\frac{1}{\Gamma(k+\frac{1}{2})(k+\frac{1}{2})}\frac{z^{2k}}{k!}\\
=&\frac{\sin z}{z} - \sum_{k=0}^{\infty}(-\tfrac{1}{4})^k\frac{2^{2k+1}z^{2k}}{\Gamma(2k+2)} = -\frac{\sin z}{z}.
\end{aligned}$$

□

Proposition 8. *Identity (10.2.34) follows from (10.1.32) and (10.2.33). It holds for* $z \in \mathbb{C} \setminus \mathbb{R}_{\leq 0}$.

Proof. Indeed, by (9.6.43) we have

$$\begin{aligned}\left[\frac{\partial}{\partial \nu} K_\nu(z)\right]_{\nu=\pm 1/2} &= \frac{\pi}{2}\csc(\nu\pi)\left[\frac{\partial}{\partial \nu} I_{-\nu}(z) - \frac{\partial}{\partial \nu} I_\nu(z)\right]_{\nu=\pm 1/2} \\ &= -\frac{\pi}{2}\csc(\nu\pi)\left(\left[\frac{\partial}{\partial \nu} I_\nu(z)\right]_{\nu=\mp 1/2} + \left[\frac{\partial}{\partial \nu} I_\nu(z)\right]_{\nu=\pm 1/2}\right) \\ &= \pm\sqrt{\frac{\pi}{2z}}\mathrm{e}^z \mathrm{E}_1(2z).\end{aligned}$$

□

Finally, we note that (10.2.32), which was proved in Proposition 2, can be proved by hand from (10.1.41). Indeed, replacing z with iz in (9.1.64) makes the k-sum in (9.1.64) equal the k-sum in (9.6.42). Solving both relations for the k-sum allows to express $\frac{\partial}{\partial \nu} I_\nu(z)$ by $I_\nu(z)$, $J_\nu(\mathrm{i}z)$, and $\frac{\partial}{\partial \nu} J_\nu(\mathrm{i}z)$. Plugging in $\nu = \frac{1}{2}$, rewriting $\frac{\partial}{\partial \nu} J_\nu(\mathrm{i}z)$ with (10.1.41), and using the relations (5.2.21) and (5.2.23) between the exponential integral and the sine and cosine integrals gives (10.2.32). Analogously, (10.2.33) follows from (10.1.42).

Acknowledgements This work has been supported by the Austrian Science Fund (FWF) grants P20347-N18, P24880-N25, Y464-N18, DK W1214 (DK6, DK13) and SFB F50 (F5004-N15, F5006-N15, F5009-N15), and by the EU Network LHCPhenoNet PITN-GA-2010-264564.

Appendix: List of Relevant Table Entries

For the reader's convenience, we collect here all identities from Abramowitz, Stegun [2] that we have used.

$$\mathrm{Ei}(x) = \gamma + \ln x + \sum_{n=1}^{\infty} \frac{x^n}{n\, n!} \qquad (x > 0) \qquad (5.1.10)$$

$$E_1(z) = -\gamma - \ln z - \sum_{n=1}^{\infty} \frac{(-1)^n z^n}{n\, n!} \qquad (|\arg z| < \pi) \qquad (5.1.11)$$

$$\mathrm{Si}(x) = \sum_{n=0}^{\infty} \frac{(-1)^n x^{2n+1}}{(2n+1)(2n+1)!} \qquad (5.2.14)$$

$$\mathrm{Ci}(x) = \gamma + \log x + \sum_{n=1}^{\infty} \frac{(-1)^n x^{2n}}{2n(2n)!} \qquad (5.2.16)$$

$$\mathrm{Si}(x) = \frac{1}{2\mathrm{i}}(E_1(\mathrm{i}z) - E_1(-\mathrm{i}z)) + \frac{\pi}{2} \qquad (|\arg z| < \frac{\pi}{2}) \tag{5.2.21}$$

$$\mathrm{Ci}(x) = -\frac{1}{2}(E_1(\mathrm{i}z) + E_1(-\mathrm{i}z)) \qquad (|\arg z| < \frac{\pi}{2}) \tag{5.2.23}$$

$$\psi(z+1) = \psi(z) + \frac{1}{z} \tag{6.3.5}$$

$$\frac{\partial}{\partial\nu} J_\nu(z) = J_\nu(z)\log(\tfrac{1}{2}z) - (\tfrac{1}{2}z)^\nu \sum_{k=0}^{\infty} (-1)^k \frac{\psi(\nu+k+1)}{\Gamma(\nu+k+1)} \frac{(\frac{1}{4}z^2)^k}{k!} \tag{9.1.64}$$

$$\frac{\partial}{\partial\nu} Y_\nu(z) = \cot(\nu\pi)\Big(\frac{\partial}{\partial\nu} J_\nu(z) - \pi Y_\nu(z)\Big) \tag{9.1.65}$$

$$- \csc(\nu\pi)\frac{\partial}{\partial\nu} J_{-\nu}(z) - \pi J_\nu(z) \qquad (\nu \neq 0, \pm 1, \pm 2, \ldots)$$

$$J_\nu(z) \sim \frac{1}{\sqrt{2\pi\nu}}\Big(\frac{\mathrm{e}z}{2\nu}\Big)^\nu, \quad Y_\nu(z) \sim -\sqrt{\frac{2}{\pi\nu}}\Big(\frac{\mathrm{e}z}{2\nu}\Big)^{-\nu} \quad (\nu \to \infty) \tag{9.3.1}$$

$$I_\nu(z) = (\tfrac{1}{2}z)^\nu \sum_{k=0}^{\infty} \frac{(\frac{1}{4}z^2)^k}{k!\,\Gamma(\nu+k+1)} \tag{9.6.10}$$

$$\frac{\partial}{\partial\nu} I_\nu(z) = I_\nu(z)\ln(\tfrac{1}{2}z) - (\tfrac{1}{2}z)^\nu \sum_{k=0}^{\infty} \frac{\psi(\nu+k+1)}{\Gamma(\nu+k+1)} \frac{(\frac{1}{4}z^2)^k}{k!} \tag{9.6.42}$$

$$j_n(z) = \sqrt{\tfrac{1}{2}\pi/z}\,J_{n+\frac{1}{2}}(z), \quad y_n(z) = \sqrt{\tfrac{1}{2}\pi/z}\,Y_{n+\frac{1}{2}}(z) \tag{10.1.1}$$

$$j_0(z) = \frac{\sin z}{z}, \quad j_1(z) = \frac{\sin z}{z^2} - \frac{\cos z}{z}, \tag{10.1.11}$$

$$j_2(z) = \Big(\frac{3}{z^3} - \frac{1}{z}\Big)\sin z - \frac{3}{z^2}\cos z$$

$$y_0(z) = -j_{-1}(z) = -\frac{\cos z}{z}, \quad y_1(z) = j_{-2}(z) = -\frac{\cos z}{z^2}, \tag{10.1.12}$$

$$y_2(z) = -j_{-3}(z) = \Big(\frac{1}{z} - \frac{3}{z^2}\Big)\cos z - \frac{3}{z^2}\sin z$$

$$j_{n-1}(z) + j_{n+1}(z) = (2n+1)z^{-1} j_n(z) \qquad (n \in \mathbb{Z}) \tag{10.1.19}$$

$$y_{n-1}(z) + y_{n+1}(z) = (2n+1)z^{-1} y_n(z) \qquad (n \in \mathbb{Z})$$

$$j_n(z) = z^n\Big(-\frac{1}{z}\frac{\partial}{\partial z}\Big)^n \frac{\sin z}{z} \tag{10.1.25}$$

$$\frac{n+1}{z} j_n(z) + \frac{\mathrm{d}}{\mathrm{d}z} j_n(z) = j_{n-1}(z) \tag{10.2.20}$$

References

1. Ablinger, J., Blümlein, J., Round, M., Schneider, C.: Advanced Computer Algebra Algorithms for the Expansion of Feynman Integrals. In: Loops and Legs in Quantum Field Theory 2012, PoS(2012), Wernigerode, pp. 1–14 (2012)
2. Abramowitz, M., Stegun, I.A.: Handbook of Mathematical Functions. Dover, New York (1973). A reprint of the tenth National Bureau of Standards edition, 1964
3. Andrews, G.E., Paule, P., Schneider, C.: Plane partitions VI: Stembridge's TSPP theorem. Adv. Appl. Math. **34**(4), 709–739 (2005). (Special Issue Dedicated to Dr. David P. Robbins. Edited by D. Bressoud)
4. Chyzak, F.: An extension of Zeilberger's fast algorithm to general holonomic functions. Discrete Math. **217**, 115–134 (2000)
5. Gerhold, S.: Uncoupling systems of linear Ore operator equations. Master's thesis, RISC, J. Kepler University, Linz (2002)
6. Gerhold, S.: The Hartman-Watson distribution revisited: asymptotics for pricing Asian options. J. Appl. Probab. **48**(3), 892–899 (2011)
7. Graham, R.L., Knuth, D.E., Patashnik, O.: Concrete Mathematics, 2nd edn. Addison Wesley, Reading (1994)
8. Ince, E.L.: Ordinary Differential Equations. Dover, New York (1926)
9. Kauers, M.: The holonomic toolkit. In: Blümlein, J., Schneider, C. (eds.) Computer Algebra in Quantum Field Theory: Integration, Summation and Special Functions. Texts and Monographs in Symbolic Computation. Springer, Vienna (2013)
10. Kauers, M., Paule, P.: The Concrete Tetrahedron. Springer, Wien/New York (2011)
11. Koutschan, C.: Advanced applications of the holonomic systems approach. PhD thesis, RISC, J. Kepler University, Linz (2009)
12. Koutschan, C., Moll, V.H.: The integrals in Gradshteyn and Ryzhik. Part 18: some automatic proofs. SCIENTIA Ser. A Math. Sci. **20**, 93–111 (2011)
13. Mallinger, C.: Algorithmic manipulations and transformations of univariate holonomic functions and sequences. Master's thesis, RISC, J. Kepler University, Linz (1996)
14. Meunier, L., Salvy, B.: ESF: an automatically generated encyclopedia of special functions. In: Rafael Sendra, J. (ed.) Proceedings of the ISSAC'03, Philadelphia, pp. 199–206. ACM (2003)
15. Olver, F.W.J., Lozier, D.W., Boisvert, R.F., Clark, C.W. (eds.): NIST digital library of mathematical functions. http://dlmf.nist.gov/. Release 1.0.5 of 2012-10-01
16. Paule, P., Schorn, M.: A mathematica version of Zeilberger's algorithm for proving binomial coefficient identities. J. Symb. Comput. **20**(5–6), 673–698 (1995)
17. Prause, K.: The generalized hyperbolic model: estimation, financial derivatives, and risk measures. PhD thesis, Albert-Ludwigs-Universität, Freiburg i. Br. (1999)
18. Salvy, B., Zimmermann, P.: Gfun: a package for the manipulation of generating and holonomic functions in one variable. ACM Trans. Math. Softw. **20**, 163–177 (1994)
19. Schneider, C.: A new Sigma approach to multi-summation. Adv. Appl. Math. **34**(4), 740–767 (2005). (Special Issue Dedicated to Dr. David P. Robbins. Edited by D. Bressoud)
20. Schneider, C.: Symbolic summation assists combinatorics. Sém. Lothar. Comb. **56**, 1–36, (2007). (Article B56b)
21. Szegö, G.: Orthogonal Polynomials. Volume XXIII of Colloquium Publications, 4th edn. AMS, Providence (1975)
22. Zeilberger, D.: The method of creative telescoping. J. Symb. Comput. **11**, 195–204 (1991)

Conformal Methods for Massless Feynman Integrals and Large N_f Methods

John A. Gracey

Abstract We review the large N method of calculating high order information on the renormalization group functions in a quantum field theory which is based on conformal integration methods. As an example these techniques are applied to a typical graph contributing to the β-function of $O(N)$ ϕ^4 theory at $O(1/N^2)$. The possible future directions for the large N methods are discussed in light of the development of more recent techniques such as the Laporta algorithm.

1 Introduction

One of the main problems in renormalization theory is the construction of the renormalization group functions. These govern how the parameters of a quantum field theory, such as the coupling constant, depend on scale. In situations where one has to compare with precision data, this ordinarily requires knowing the renormalization group functions to very high orders in a perturbative expansion. The quantum field theories we have in mind are not only the gauge theories of particle physics but also the scalar and fermionic ones which arise in condensed matter problems. These are central in understanding phase transitions. To attain such precision in perturbative expansions means that large numbers of Feynman diagrams have to be determined with the number of graphs increasing with the loop order. Moreover, as the order increases the underlying integrals require more sophisticated methods in order to deduce their value analytically. The widely established methods of computing Feynman graphs will be reported elsewhere in this volume. Here we review an alternative approach which complements explicit perturbative techniques. It does so in such a way that for low loop orders there is overlap but at orders

J.A. Gracey (✉)
Theoretical Physics Division, Department of Mathematical Sciences, University of Liverpool, P.O. Box 147, Liverpool, L69 3BX, UK
e-mail: gracey@liv.ac.uk

C. Schneider and J. Blümlein (eds.), *Computer Algebra in Quantum Field Theory*, Texts & Monographs in Symbolic Computation, DOI 10.1007/978-3-7091-1616-6_4,

beyond that already known part of the perturbative series can be deduced at *all* orders within a certain approximation. This is known as the large N or large N_f method where N is a parameter deriving from a symmetry of the theory such as a Lie group or the number of massless quark flavours, N_f, in Quantum Chromodynamics (QCD). In this method the Feynman graphs are related to those of perturbation theory but because of the nature of the expansion parameter, the powers of the propagators appearing in such graphs are not the canonical value of unity but instead differ from unity by $O(\epsilon)$ where ϵ corresponds to the regularizing parameter of dimensional regularization. In addition beyond leading order in the $1/N$ expansion, the propagator powers will include the anomalous dimensions in addition to the leading or canonical dimension. Therefore, standard perturbative techniques such as integration by parts requires care in its use since one may not be able to actually reduce a graph to a simpler topology. Instead a different technique has had to be refined and developed. It is based on a conformal property of Feynman integrals and we review it here in the context of the large N methods. Though it has had some applications in perturbative computations.

The article is organized as follows. We devote the next section to the notation and techniques of computing Feynman graphs using conformal methods in d-dimensions. We focus on the general two loop self energy graph in the subsequent section and review the work of [1, 2], upon which this review is mostly based, and others in the methods of evaluating it. These techniques are then applied to a problem in scalar quantum field theory in Sect. 4 where a graph with 10 internal integrations is evaluated *exactly* in d-dimensions. We conclude in Sect. 5 with thoughts on the direction in which the technique could be developed next given recent advances in the computation of Feynman graphs using conventional perturbative techniques.

2 Notation and Elementary Techniques

We begin by introducing the notation we will use which will be based on [1,2]. There Feynman graphs were represented in coordinate or configuration space notation. By this we mean that in writing a Feynman integral graphically the integration variables are represented as the vertices. By contrast in momentum space representation the integration variables correspond to the momenta circulating around a loop. So in coordinate space representation propagators are denoted by lines between two fixed points, as illustrated in Fig. 1. There the power of the propagator is denoted by a number or symbol beside the line. One can map between coordinate and momentum space representation by using a Fourier transform. In the notation of [1, 2] we have

$$\frac{1}{(x^2)^\alpha} = \frac{a(\alpha)}{2^{2\alpha}\pi^\mu} \int_k d^d k \, \frac{e^{ikx}}{(k^2)^{\mu-\alpha}} \tag{1}$$

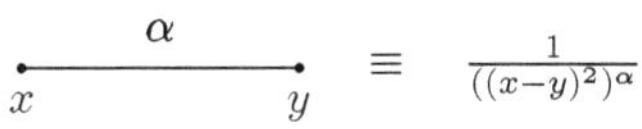

Fig. 1 Coordinate space propagator

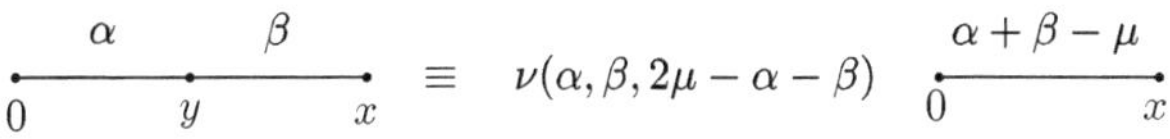

Fig. 2 Chain integration

where x is in coordinate space and k is the conjugate momentum. Also for shorthand we set

$$d \;=\; 2\mu \tag{2}$$

which is used throughout to avoid the appearance of $d/2$ in the Euler Γ-function. This symbol should not be confused with the mass scale appearing in renormalization group equations. Clearly

$$a(\alpha) \;=\; \frac{\Gamma(\mu-\alpha)}{\Gamma(\alpha)} \tag{3}$$

which is singular when $\alpha = \mu + n$ where n is zero or a positive integer. Also $a(\alpha)$ vanishes at the negative integers. The elementary identity

$$a(\alpha)a(\mu-\alpha) \;=\; 1 \tag{4}$$

follows trivially as does

$$a(\alpha) \;=\; \frac{a(\alpha-1)}{(\alpha-1)(\mu-\alpha)} \tag{5}$$

from the Γ-function identity $\Gamma(z+1) = z\Gamma(z)$. With this notation the elementary one loop self energy graph in momentum space is replaced by chain integration in coordinate space representation. This is represented graphically in Fig. 2 where, [1, 2],

$$\nu(\alpha,\beta,\gamma) \;=\; \pi^{\mu}a(\alpha)a(\beta)a(\gamma)\,. \tag{6}$$

However, in practice Feynman graphs have more complicated integration points. In other words in coordinate space representation one has more than two lines intersecting at a point. Therefore, more involved integration techniques are required to evaluate the Feynman integrals. One very useful technique is that of uniqueness or conformal integration which was introduced in three dimensions in [3]. It has

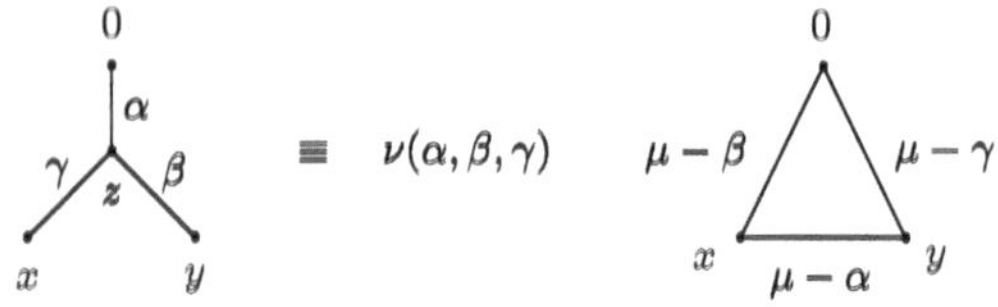

Fig. 3 Conformal integration when $\alpha + \beta + \gamma = 2\mu$

been developed in several ways subsequently and specifically to d-dimensions. For example, see [4]. We follow [1, 2] and use the rule represented in Fig. 3, where z is the integration variable, which follows when the sum of the exponents of the lines intersecting at the 3-point vertex add to the spacetime dimension

$$\alpha + \beta + \gamma = 2\mu . \tag{7}$$

This is known as the uniqueness condition. By the same token if a graph contains a triangle where the lines comprising the triangle sum to μ such as

$$(\mu - \alpha) + (\mu - \beta) + (\mu - \gamma) = \mu \tag{8}$$

as is the case in Fig. 3, then the unique triangle can be replaced by the vertex on the left side. There are several methods to establish the uniqueness integration rule. If one uses standard text book methods such as Feynman parameters then the integral over z can be written as

$$\begin{aligned} &\frac{\pi^{\mu}\Gamma(\mu-\alpha)\Gamma(\alpha+\beta+\gamma-\mu)}{\Gamma(\beta)\Gamma(\gamma)\Gamma(\mu)} \\ &\times \int_0^1 db \frac{b^{\beta-1}(1-b)^{\gamma-1}}{[b(1-b)(x-y)^2]^{\alpha+\beta+\gamma-\mu}} \\ &\times {}_2F_1\left(\alpha+\beta+\gamma-\mu, \alpha; \mu; -\frac{[bx+(1-b)y]}{b(1-b)(x-y)^2}\right) \end{aligned} \tag{9}$$

prior to using, (7). When that condition is set then the hypergeometric function collapses to the geometric series and allows the integration over the Feynman parameter b to proceed which results in

$$\begin{aligned} &\frac{\pi^{\mu}\Gamma(\mu-\alpha)\Gamma(\mu-\beta)\Gamma(\mu-\gamma)}{\Gamma(2\mu-\beta-\gamma)\Gamma(\beta)\Gamma(\gamma)(y^2)^{\alpha}[(x-y)^2]^{\beta+\gamma-\mu}} \\ &\times {}_2F_1\left(\alpha, \mu-\gamma; 2\mu-\beta-\gamma; 1-\frac{x^2}{y^2}\right) \end{aligned} \tag{10}$$

Applying the uniqueness condition a second time produces the right hand side of Fig. 3 since the hypergeometric function again reduces to the geometric series. This is in such a way that the canonical propagators emerge.

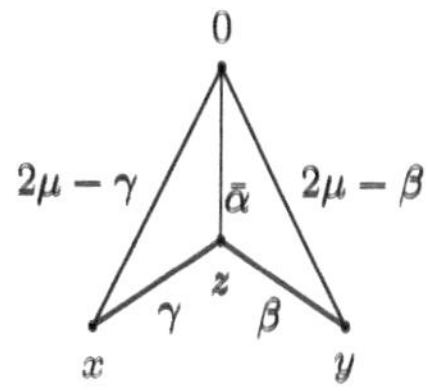

Fig. 4 Vertex of Fig. 3 after a conformal transformation with base at 0 where $\bar{\alpha} = 2\mu - \alpha - \beta - \gamma$

An alternative method is to apply a conformal transformation on the coordinates of the integral, [2]. In this approach, which is applicable to any graph in general, one external point is labelled as an origin and given 0 as a coordinate. The other points are denoted by coordinates x, y and z. The conformal transformation changes the integration coordinate as well as the external points through

$$x_\mu \rightarrow \frac{x_\mu}{x^2} . \tag{11}$$

Thus for two coordinates y and z undergoing such a transformation we have the lemma

$$(y - z)^2 \rightarrow \frac{(y - z)^2}{y^2 z^2} . \tag{12}$$

An integration measure also produces contributions to the lines joining to the origin since

$$d^d z \rightarrow \frac{d^d z}{(z^2)^{2\mu}} . \tag{13}$$

Therefore, for the vertex on the left side of Fig. 3 this transformation produces the intermediate integral of Fig. 4.

To complete the integration requires setting the uniqueness condition (7) which produces a chain integral since the line from 0 to z is absent from the graph. To complete the derivation one undoes the original conformal transformations to produce the right hand side of Fig. 3. If one compares the two derivations, the latter is in fact of more practical use. This is because it avoids the use of writing the original integral in terms of Feynman parameters which would become tedious for higher order cases. Also it is simple to implement graphically.

Having recalled the derivation of the uniqueness rule it is straightforward to see that there is a natural extension. In the first derivation there was not a unique way to collapse the hypergeometric function to an elementary type of propagator. Instead this will happen if the sum of the exponents is $(2\mu + n)$ where n is a positive integer. Although the collapse in this case will not be to the geometric series, it will reduce to simple algebraic functions which are of the propagator type. So, for instance, when $n = 1$ we have the result of Fig. 5, [5],

A similar rule has been constructed and used in [5]. We will use Fig. 5 later in order to simplify various integrals.

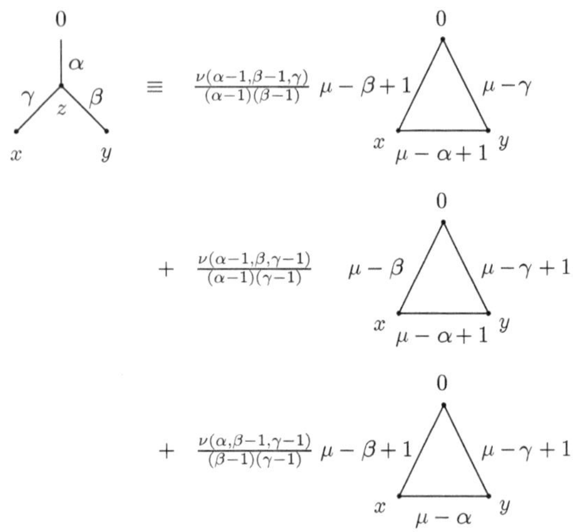

Fig. 5 Conformal integration when $\alpha + \beta + \gamma = 2\mu + 1$

3 Two Loop Self Energy Graph

We can illustrate some of the techniques of conformal integration by considering the massless two loop self energy graph with arbitrary powers, α_i on the propagators. It is illustrated in Fig. 6 where we have used the coordinate space representation. Thus the vertices are integrated over rather than the loop momenta. To clarify, the integral of Fig. 6 is

$$I(\alpha_1, \alpha_2, \alpha_3, \alpha_4, \alpha_5) = \int_{yz} \frac{1}{(y^2)^{\alpha_1}((x-y)^2)^{\alpha_2}((x-z)^2)^{\alpha_3}(z^2)^{\alpha_4}((y-z)^2)^{\alpha_5}} \tag{14}$$

where $\int_y = \int \frac{d^d y}{(2\pi)^d}$. The structure of this integral has been widely studied and we briefly highlight several properties of relevance. The analysis of [6, 7] determined that the symmetry group of the graph was $Z_2 \times S_6$ which has 1,440 elements. Exploiting this the ϵ expansion of the integral in $d = 4 - 2\epsilon$ with propagator powers of order ϵ from unity was determined up to $O(\epsilon^6)$, [6,7]. At $O(\epsilon^5)$ it was discovered that the first multi-zeta value occurred, [7]. Specifically

$$\begin{aligned} I(1, 1, \mu - 1, 1, \mu - 1) &= 6\zeta_3 + 9\zeta_4\epsilon + 7\zeta_5\epsilon^5 \\ &+ \frac{5}{2}\left[\zeta_6 - 2\zeta_3^2\right]\epsilon^3 - \frac{1}{8}\left[91\zeta_7 + 120\zeta_3\zeta_4\right]\epsilon^4 \\ &+ \frac{1}{81{,}920}\left[653{,}440\zeta_5\zeta_3 - 7{,}059{,}417\zeta_8 + 576F_{53}\right]\epsilon^5 \\ &+ O(\epsilon^6) \end{aligned} \tag{15}$$

where ζ_z is the Riemann zeta function and $F_{53} = \sum_{n>m>0} \frac{1}{n^5m^3}$ in the original notation of [8]. Subsequent to this it has been shown that the only numbers which appear in the full series expansion in ϵ are multiple zeta values, [9]. While the work of [6, 7] illustrated the power of group theory to evaluate master integrals explicitly, using conformal integration allows one to relate two loop self energy integrals by exploiting the masslessness of the original diagrams. This was originally developed in [1, 2] and we summarize that here as there appears to be scope nowadays to take this method to three and higher loop order graphs.

The transformations developed in [2] fall into several classes. The first is that derived from the elementary use of the Fourier transform. Writing

$$I(\alpha_1,\alpha_2,\alpha_3,\alpha_4,\alpha_5) = \frac{\Gamma}{(x^2)^{D-2\mu}} \tag{16}$$

where $D = \sum_{i=1}^5 \alpha_i$ and Γ is independent of x and corresponds to the value of the integral, then taking the Fourier transform produces an integral which is also the two loop self energy. Though the propagator powers are different. In this sense one can say that the graph is self-dual which is not a property all Feynman graphs have. Thus, [2],

$$I(\alpha_1,\alpha_2,\alpha_3,\alpha_4,\alpha_5) = \frac{\prod_{i=1}^5 a(\alpha_i)}{a(D-2\mu)} I(\mu-\alpha_2,\mu-\alpha_3,\mu-\alpha_4,\mu-\alpha_1,\mu-\alpha_5)\,. \tag{17}$$

This transformation is known as the momentum representation or MR. It can be easily generalized to other topologies and there is a simple graphical rule for this. Although not immediately apparent from the self energy because of the self-duality, each 3-valent vertex of the original graph has an associated triangle in the dual graph. For other topologies 4-valent vertices are mapped to squares and 5-valent vertices to pentagons with a clear generalization pattern.

A set of less obvious transformations can be deduced from the uniqueness condition. First, we define the shorthand notation, [2],

$$\begin{aligned} s_1 &= \alpha_1+\alpha_2+\alpha_5 \quad , \quad s_2 = \alpha_3+\alpha_4+\alpha_5 \\ t_1 &= \alpha_1+\alpha_4+\alpha_5 \quad , \quad t_2 = \alpha_2+\alpha_3+\alpha_5 \end{aligned} \tag{18}$$

and illustrate the technique for one case. If one considers the central propagator it can be replaced by a chain integral. Although there are an infinite number of ways of doing this one can choose the exponents of the chain so that the top vertex is unique. In other words

$$\begin{aligned} &\frac{1}{((y-z)^2)^{\alpha_5}} \\ &= \frac{1}{\nu(2\mu-\alpha_1-\alpha_2, s_1-\mu, \mu-\alpha_5)} \int_u \frac{1}{((y-u)^2)^{2\mu-\alpha_1-\alpha_2}((u-z)^2)^{s_1-\mu}} \end{aligned} \tag{19}$$

where u is the intermediate integration point. As the y vertex of Fig. 6 is now unique the conformal integration rule can be used to rewrite the integral. This results in, [2],

$$I(\alpha_1,\alpha_2,\alpha_3,\alpha_4,\alpha_5) = \frac{a(\alpha_1)a(\alpha_2)a(\alpha_5)}{a(s_1-\mu)} I(\mu-\alpha_2,\mu-\alpha_1,\alpha_3,\alpha_4,s_1-\mu) \,. \quad (20)$$

In the notation of [2] this transformation is known as $\uparrow$. It is elementary to see that there are five other such transformations which are denoted by $\nearrow$, $\nwarrow$, $\downarrow$, $\searrow$ and $\swarrow$. The syntax is that when an arrow points in a general upwards direction it is a transformation on the y vertex and by contrast in a downwards direction it relates to the z vertex. The propagator which one replaces by a chain to make the vertex unique is in correspondence with the direction of the arrow. While these six transformations operate on the internal vertices there are two which act on each of the external vertices. One can complete the uniqueness of one of these by realizing that the integral itself is a propagator with power $(D-2\mu)$ as indicated in (16), [2]. For example, if the right external point is chosen as the base integration vertex then the appending propagator has power $(2\mu-\alpha_2-\alpha_3)$. This produces

$$I(\alpha_1,\alpha_2,\alpha_3,\alpha_4,\alpha_5) = \frac{a(\alpha_2)a(\alpha_3)}{a(D-2\mu)a(2\mu-t_1)} I(\alpha_1,\mu-\alpha_3,\mu-\alpha_2,\alpha_4,t_2-\mu) \quad (21)$$

and this is denoted by $\leftarrow$. The corresponding transformation on the left external point is called $\rightarrow$.

The final set of transformations are based on the conformal transformations (11) and (12) together with the effect they have on the two vertex measures, [2]. One can choose either of the external vertices as the origin of the transformation. Once decided the result of the conformal transformation is that all propagators joining to the origin have their powers changed to the difference of 2μ and the sum of the exponents at the point at the other end of that propagator. This means all points including those not directly connected to the base point in the first place. For the two loop self energy there are no such points but for higher loop graphs this will be the case. We will give an example of this in Sect. 4. Thus the conformal left transformation is, [2],

$$I(\alpha_1,\alpha_2,\alpha_3,\alpha_4,\alpha_5) = I(2\mu-s_1,\alpha_2,\alpha_3,2\mu-s_2,\alpha_5) \quad (22)$$

where there is no Γ-function factor and this is denoted by CL in contrast to CR which is the transformation based on the right external vertex as the origin of the conformal transformation. The full set of transformations and the result of applying each to the graph of Fig. 6 are summarized in a Table in [2]. However, as brief examples of the transformations the integral of (15) is related as follows

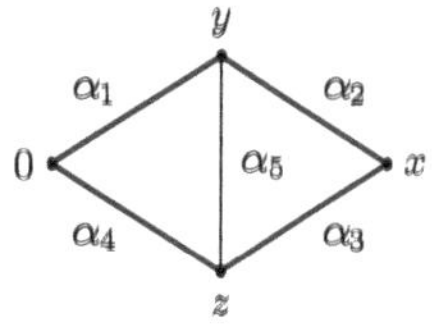

Fig. 6 Two loop self energy graph in coordinate space representation

$$I(1,1,\mu-1,1,\mu-1) \stackrel{\uparrow}{=} I(\mu-1,\mu-1,1,\mu-1,1) \stackrel{\mathrm{CR}}{=} I(\mu-1,1,1,1,\mu-1)\,. \tag{23}$$

Though the latter follows from a simple rotation of the integral as well.

Aside from the transformations there are other techniques which allow one to evaluate the two loop self energy and higher order graphs. Perhaps the most exploited is that of integration by parts which was introduced for (16) in [10]. It determined that the first term in the ϵ expansion of $I(1,1,1,1,1)$ was $6\zeta_3$ and has also been used in other applications, [2]. Indeed more recently the technique has been developed by Laporta in [11] to produce an algorithm which relates all integrals in a Feynman graph to a base set of master integrals. These can then be evaluated by direct methods to complete the overall computation. In the coordinate space representation we use here the basic rule is given in Fig. 7 where the + or − on a line indicates that the power of that propagator is increased or decreased by unity. For example, with this, [10],

$$I(1,1,1,1,1) = \frac{\nu(1,1,2\mu-2)}{(\mu-2)}\left[\nu(1,2,2\mu-3) - \nu(3-\mu,2,3\mu-5)\right] \tag{24}$$

which can be expanded in powers of ϵ. Clearly the series can only involve rationals and ζ_n. Indeed the rule can also be applied to more general cases. In [2] it was shown that

$$I(\alpha,\mu-1,\mu-1,\beta,\mu-1) = \frac{a(2\mu-2)}{\Gamma(\mu-1)}\left[\frac{a(\alpha)a(2-\alpha)}{(1-\beta)(\alpha+\beta-2)} + \frac{a(\alpha+\beta-1)a(3-\alpha-\beta)}{(\alpha-1)(\beta-1)} + \frac{a(\beta)a(2-\beta)}{(1-\alpha)(\alpha+\beta-2)}\right] \tag{25}$$

for arbitrary α and β. However, not all graphs can be integrated by parts. An example of such a case is $I(1,\alpha,\beta,\gamma,1)$ for non-unit α, β and γ. Another example is (15), [7], whose expansion has a non-Riemann zeta value at some point in the expansion. Indeed this is perhaps an indication of an obstruction to integrability.

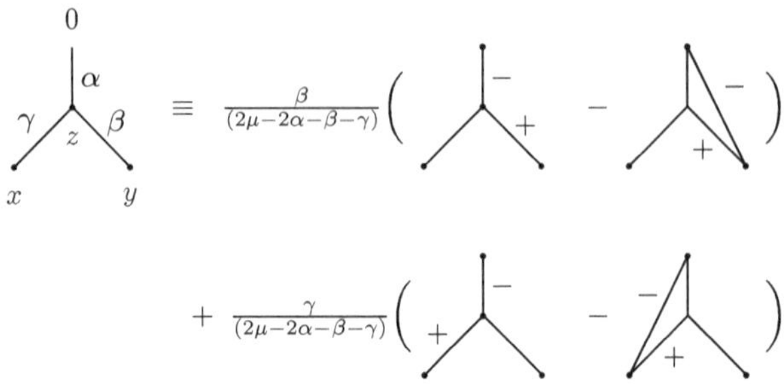

Fig. 7 Integration by parts in coordinate space representation

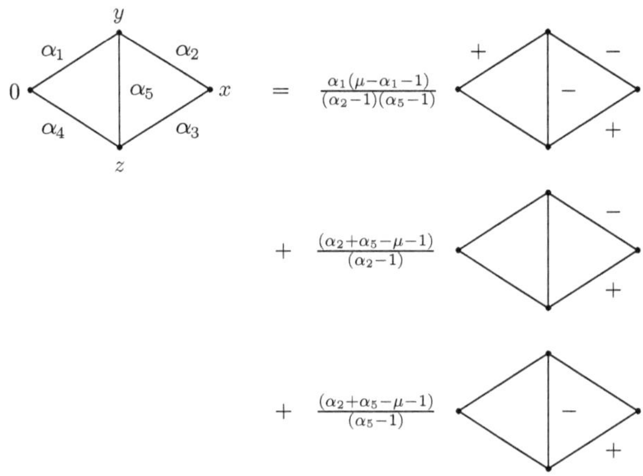

Fig. 8 Reduction formula for two loop self energy based on the generalized $\nearrow$ transformation

While integration by parts allows one to reduce the powers of various propagators by unity within a Feynman diagram it is not the only method to achieve this. A modification of the uniqueness method can be used to derive rules similar to Fig. 7. Specifically if one chooses the exponents of the propagators to be $(2\mu + 1)$ then one finds the extension given in Fig. 5. Using this rule and repeating the analysis of the transformations on the two loop self energy graph provides relations specific to this topology, [5]. For instance, extending $\nearrow$ to have the upper vertex exponents summing to $(2\mu + 1)$ gives the relation in Fig. 8 where the $+$ or $-$ on the right side indicates that the exponent of that line is increased or decreased by unity. In Fig. 8 provided $\alpha_2 \neq 1$ and $\alpha_5 \neq 1$ then the powers of the respective propagators can be reduced by unity. However, this restriction is a drawback if one wishes to reduce graphs which have unit exponents. Instead it is possible to extend the method which produced the relation of Fig. 8. For instance, rather than begin with the general two loop self energy and applying the generalized uniqueness rule, one can use one of the transformations of [2] and then apply a rule like that of Fig. 8 before applying the transformation inverse to the original one. In this way one can build up a suite of relations.

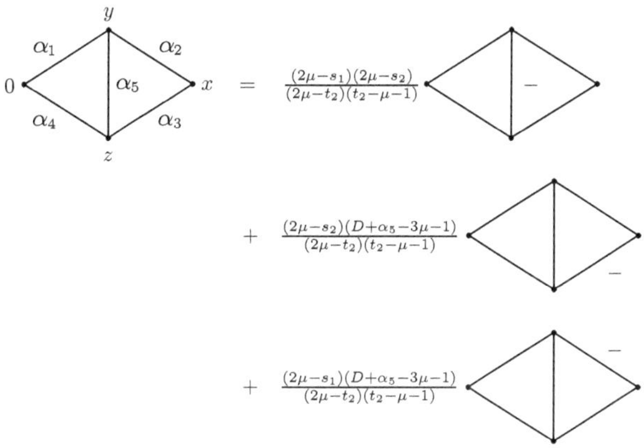

Fig. 9 Another reduction formula for two loop self energy

One such useful relation is illustrated in Fig. 9 which is derived in several stages. The first is to construct a relation similar to that of Fig. 9 by first applying $\leftarrow$ to the graph of Fig. 6 and then undoing it by applying the rule of Fig. 5 to the same external vertex. This produces a relation where t_2 increases by unity in each of the three resulting graphs. The second stage is to apply this rule to the graph of Fig. 6 after a CR transformation has been enacted. To complete the derivation the final step is to undo with another CR transformation. Thus the t_2 value of each graph on the right hand side of Fig. 9 is one less than that of the graph on the left side. This reduction has coefficients on the right hand side which are non-singular for unit propagators. Other rules can be derived by this method and a fuller set are recorded in Appendix B of [12]. It is worth noting that similar rules based on the generalized uniqueness where developed in [4].

4 QFT Application

Having discussed the general techniques for determining massless Feynman integrals using conformal methods, we illustrate their usefulness in a practical problem in a quantum field theory. Specifically we focus on the determination of the critical exponents at a phase transition in various models in the large N expansion. The background which we describe here is based on a series of articles, [1, 2, 13], where exponents were determined in d-dimensions at $O(1/N^2)$ and $O(1/N^3)$. The fact that d-dimensional results are computable means that information on the renormalization group functions can be deduced in various spacetime dimensions. This is due to a special feature of critical point field theories and that is that at a non-trivial fixed point of the renormalization group flow the critical exponents correspond to the associated renormalization group function at that fixed point.

Thus information on the renormalization group functions is encoded in these exponents. Moreover, at a fixed point several quantum field theories can lie in the same universality class despite having different structures. This is invariably as a consequence of a common interaction in the Lagrangian. Thus the same exponents can be used to access the structure of the renormalization group functions of two different theories. Further, as the spacetime dimension d is not used as a regulator, information on the exponents can be deduced simultaneously in several different dimensions such as three and four. For more background to the use of the renormalization group equation at near criticality in quantum field theories see, for example, [14].

For the application of the conformal methods we consider here we concentrate on the $O(N)$ nonlinear σ model which is critically equivalent in d-dimensions to $O(N)$ ϕ^4 theory. For the latter theory the Lagrangian is

$$L = \frac{1}{2}(\partial_\mu \phi^i)^2 + \frac{g}{8}(\phi^i\phi^i)^2 \tag{26}$$

where g is the coupling constant and $1 \leq i \leq N$. Introducing an auxiliary field σ equates this Lagrangian to

$$L = \frac{1}{2}(\partial_\mu \phi^i)^2 + \frac{1}{2}\sigma(\phi^i\phi^i) - \frac{\sigma^2}{2g} . \tag{27}$$

At criticality it is the interaction which drives the dynamics and thus it is straightforward to see that in this formulation the Lagrangian interaction is the same as that of the $O(N)$ nonlinear σ model when the fields are constrained to lie on an $(N-1)$-dimensional sphere. The constraint would have a final term linear in σ rather than a quadratic one together with a different coupling constant. This essentially is the origin of both field theories being in the same universality class. The linear or quadratic terms in σ at criticality serve effectively to define the structure of the propagators. In coordinate space representation these are, [1,2],

$$\langle \phi^i(0)\phi^j(x)\rangle = \frac{\delta^{ij}A}{(x^2)^\alpha} \quad , \quad \langle \sigma(0)\sigma(x)\rangle = \frac{B}{(x^2)^\beta} \tag{28}$$

where A and B are x-independent amplitudes and α and β are the scaling dimensions of the fields. The latter comprise two parts. The first is the canonical dimension and the other is the anomalous dimension. Here

$$\alpha = \mu - 1 + \tfrac{1}{2}\eta \quad , \quad \beta = 2 - \eta - \chi \tag{29}$$

where η is the anomalous dimension of ϕ^i and χ is the vertex anomalous dimension. The former is related to the renormalization group function which is also termed the anomalous dimension, $\gamma(g)$, by

$$\eta = \gamma(g_c) \tag{30}$$

Fig. 10 Two loop self energy for σ

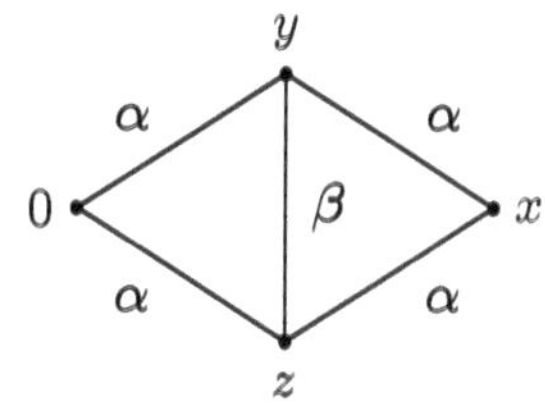

where g_c is the value of the coupling constant at the critical point,

$$\gamma(g) \;=\; \mu \frac{d}{d\mu} \ln Z_\phi \tag{31}$$

and Z_ϕ is the wave function renormalization constant. (In (31) we have temporarily used μ to denote the standard renormalization group scale that underlies any renormalization group equation.) To determine the values of the exponents to a particular order in $1/N$ requires solving the skeleton Schwinger-Dyson equation for the 2-point functions at the same order. We do not discuss that formalism here, which can be found in [1, 2], as our focus is rather on the evaluation of the Feynman graphs contributing to these equations. Though we should say that the presence of the non-zero anomalous dimensions in the propagators means that in 2-point functions there are no self energy corrections on any internal propagator as otherwise there would be double counting. So the number of graphs to consider is smaller than the corresponding perturbative case.

The coupling constant at the critical point is denoted by g_c and is defined as a nontrivial zero of the β-function, $\beta(g_c) = 0$. As we are working in d-dimensions such a non-trivial zero exists in our theories since away from the spacetime dimension where the theory is renormalizable the coupling constant becomes dimensionful. Hence the first term of the d-dimensional β-function depends on d. Moreover, g_c will depend on the parameters of the theory which in our case here is N. Thus $g_c = g_c(d, N)$. Similarly $\eta = \eta(d, N)$ and $\chi = \chi(d, N)$. These can all be expanded in powers of $1/N$ where N is large in such a way that the coefficients of $1/N$ are d-dependent. Thence if one expresses these coefficients in powers of ϵ where $d = 4 - 2\epsilon$ for ϕ^4 theory or $d = 2 + \bar{\epsilon}$ for the nonlinear σ model, then one can deduce the coefficients in the corresponding renormalization group equation to *all* orders in perturbation theory at that order in $1/N$. In this respect it is important to note that in the large N expansion ϵ or $\bar{\epsilon}$ do not play the role of a regulator as they would do in conventional perturbation theory.

Instead to see the origin of where a regulator is required one should consider the simple two loop contribution to the σ self energy graph given in Fig. 10 in coordinate space representation. To use conformal methods one has to check the sum of the exponents at a vertex in coordinate space representation. From (29) one can see that

$$2\alpha + \beta \;=\; 2\mu \;-\; \chi \,. \tag{32}$$

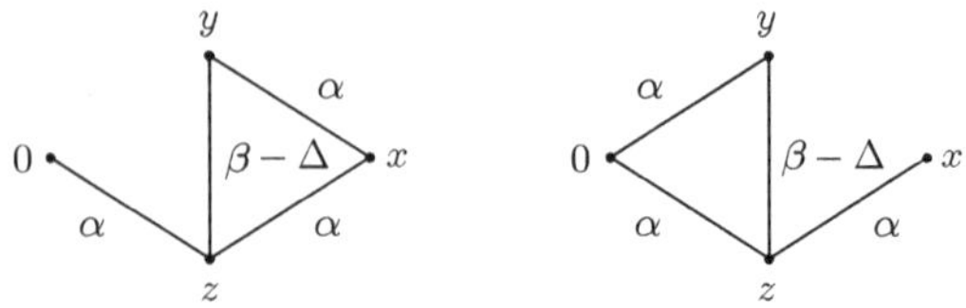

Fig. 11 Subtracted graphs for computation of σ two loop self energy

However, from the structure of the renormalization group equation at criticality the anomalous dimensions η and χ begin as $O(1/N)$. More, specifically

$$\eta = \sum_{i=1}^{\infty} \frac{\eta_i}{N^i} . \tag{33}$$

Thus at leading order in $1/N$ the basic vertex is unique, [2]. Hence at this order one can integrate at either of the vertices and produce the first contribution to the integral which is $\nu(\mu-1, \mu-1, 2)$. The second integration is a simple chain and naively gives $\nu(\mu, \mu, 0)$. This is clearly ill-defined due to the zeroes and singularities deriving from the Γ-function. However, this graph was chosen to illustrate the fact that the graph and indeed the theory requires a regularization in this critical point formulation. The method developed in [1,2] was to use analytic regularization which is introduced by shifting the vertex anomalous dimension by an infinitesimal amount, Δ, via

$$\beta \rightarrow \beta - \Delta . \tag{34}$$

In some respect one is in effect performing a perturbative expansion in the vertex anomalous dimension, [1,2]. Consequently even at leading order the graph of Fig. 10 no longer has a unique vertex due to a non-zero Δ. Therefore, to determine the graph to the finite part in Δ requires the addition and subtraction of the graphs of Fig. 11, [2].

These two graphs have been chosen in such a way that their singularity structure in Δ exactly matches that of Fig. 10, [2]. Clearly they represent simple chain integrals which can be determined as $2\nu(\alpha, \beta-\Delta, 2\mu-\alpha-\beta+\Delta)\nu(\alpha, \mu-\Delta, \mu-\alpha+\Delta)$ where the singularity is clearly regularized. To complete the evaluation introduces another technique, which we will use later, to extract a finite term of a graph. This is a temporary regularization, [2]. If one subtracts the graphs of Fig. 11 from that of Fig. 10, the combination is finite with respect to Δ which is therefore not required and can be set to zero. Thus one can complete the first integration at the upper vertex of each graph. (Without a regularization the point where one integrates in each graph has to be the same and thence the order of integration is important.) This produces $\nu(\alpha, \alpha, \beta)$ for each graph. However, each of the three subsequent chain integrals has a singular exponent, μ. To circumvent this the lower two propagators of all three graphs are temporarily regularized by $\alpha \rightarrow \alpha - \delta$ where δ is arbitrary. Thus the three graphs give

$$
\begin{aligned}
&[\nu(\mu-\delta,\mu-\delta,2\delta)-\nu(\alpha-\delta,\mu-\delta,\mu-\alpha+2\delta) \\
&-\nu(\mu-\delta,\alpha-\delta,\mu-\alpha+2\delta)]\,\nu(\alpha,\alpha,\beta)
\end{aligned} \tag{35}
$$

which is clearly finite as $\delta \to 0$, [2]. Thus to $O(\Delta)$ the graph of Fig. 10 evaluates to, [2],

$$
\frac{2\pi^{2\mu}a^2(\alpha)a(\beta)}{\Gamma(\mu)}\left[\frac{1}{\Delta}+B(\beta)-B(\alpha)+O(\Delta)\right] \tag{36}
$$

where $B(z) = \psi(z) + \psi(\mu - z)$ for z and $(\mu - z)$ not equal to zero or a negative integer and $\psi(z)$ is derivative of the logarithm of the Γ-function.

A more involved example which uses many of the techniques of the previous section occurs in the computation of the $O(1/N^2)$ correction to the β-function in $O(N)$ ϕ^4 theory. The relevant critical exponent is ω which is related through the critical renormalization group equation to the β-function slope at criticality. In this case it has the form

$$
\omega = 2 - \mu + \sum_{n=1}^{\infty} \frac{\omega_n}{N^n} \tag{37}
$$

and the explicit forms for ω_n are deduced from the part of the Schwinger-Dyson equations corresponding to corrections to scaling. In other words the propagators of (28) are extended to

$$
\begin{aligned}
\langle\phi^i(0)\phi^j(x)\rangle &= \frac{\delta^{ij}A}{(x^2)^\alpha}\left[1 + A'(x^2)^\omega\right] \\
\langle\sigma(0)\sigma(x)\rangle &= \frac{B}{(x^2)^\beta}\left[1 + B'(x^2)^\omega\right] .
\end{aligned} \tag{38}
$$

In principle other corrections can appear here corresponding to other exponents such as that for the β-function of the nonlinear σ model but one tends to focus on one calculation at a time. The effect of the corrections is that to deduce ω_n within the Schwinger-Dyson formalism all Feynman diagrams with *one* correction insertion on a propagator contribute at each particular order in $1/N$. While the $O(1/N^2)$ expression for ω appeared in [15] the explicit evaluation of the contributing graphs has not been detailed. Thus we discuss one such diagram here as the approach can be readily adapted to the other graphs. It is given in Fig. 12. To see that it is $O(1/N^2)$ each closed loop of ϕ^i fields contributes a factor of N and each σ propagator is $O(1/N)$. This is due to the fact that the amplitude B is $O(1/N)$, [1,2]. As there are four of the former and five of the latter then this gives $O(1/N)$ overall which is one factor of $1/N$ more than the previous order graph of Fig. 10. Finally, another factor of $1/N$ derives from the actual Schwinger Dyson formalism used to determine ω_2. The double line on one σ propagator in Fig. 12 denotes the

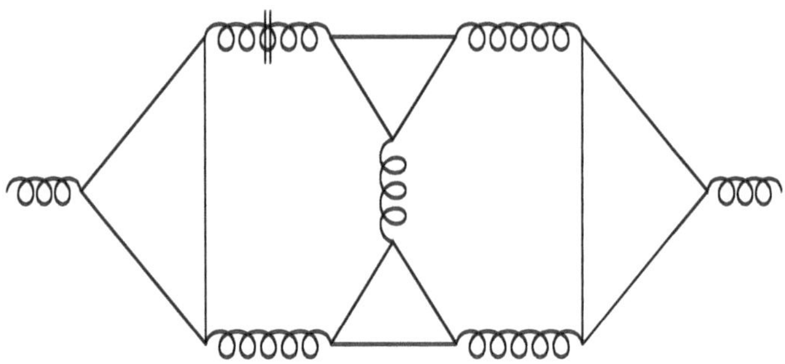

Fig. 12 Particular graph contributing to the ϕ^4 theory β-function at $O(1/N^2)$

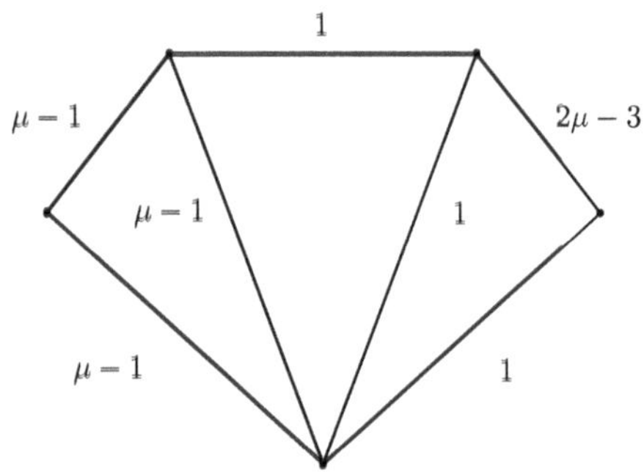

Fig. 13 Reduced integral of Fig. 12

B' correction. The presence of such a correction means that the graph is Δ-finite. Moreover, since we only want the value as a function of d rather than d and N we can replace the exponents of the lines by their canonical values. If one was computing ω_3 then the anomalous dimensions of each exponent would need to be retained at $O(1/N)$. The benefit of this restriction here is that of the ten vertices eight are unique. There are ten integrations to do over the vertices rather than the six of the loops as we are in coordinate space representation. Given this high degree of uniqueness the graph can be reduced rather quickly to one with fewer integrations. To do this one can use a variety of the rules we had earlier aside from uniqueness such as conformal transformation, unique triangle, insertion at an internal or external vertex. Ultimately one produces the graph of Fig. 13.

This graph cannot be reduced any further since there are no unique vertices or triangles. Though various vertices or triangles are one unit from uniqueness. Moreover, integration by parts cannot be used since at some point one produces an unregularized exponent, such as 0 or μ, or a zero in a denominator factor. In some sense this graph could be regarded as a master integral since it arises in several of the other graphs contributing to the σ Schwinger-Dyson equation. Moreover, it is worth noting that in strictly four dimensions the propagators of the graph would all have unit exponents. As an aside if an interested reader has been applying

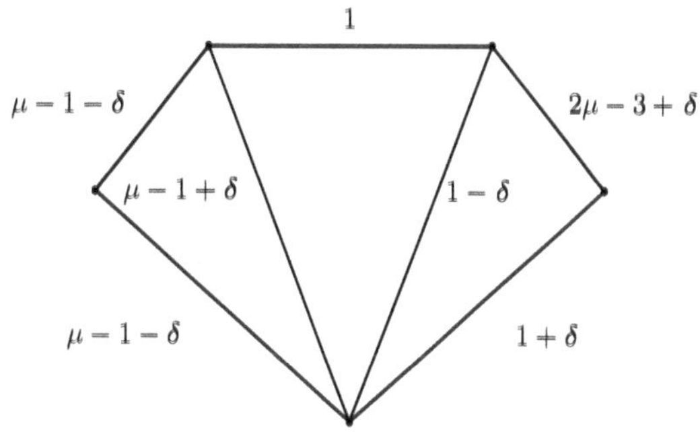

Fig. 14 Temporary regularization of previous graph to reduce it to two loop basic graphs

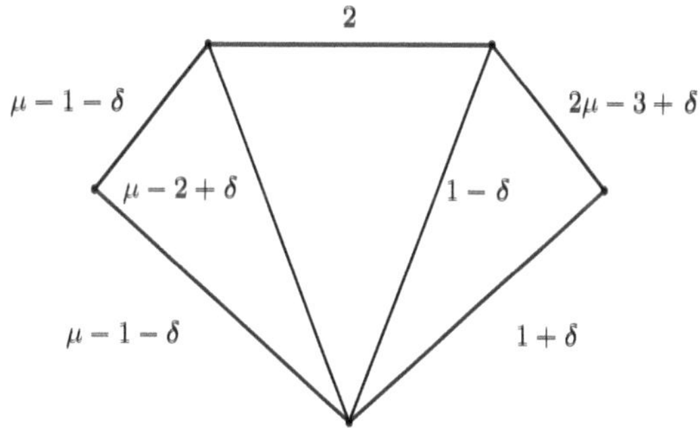

Fig. 15 First graph after integration by parts

the conformal techniques to reduce the diagram and obtains similar exponents but distributed differently around the diagram then it will be related to that of Fig. 13 by applying the transformations discussed for the master 2-loop self energy. We note here that if a conformal transformation is applied to the graph of Fig. 13 with the left internal point as the CL base, then that would introduce a new line from the top right internal vertex to the base. This illustrates comments made earlier.

To proceed further and reduce the graph to a known function of d requires an integration by parts but this requires modifying the integral first. Though before this can be achieved safely one has to introduce a temporary regularization to handle hidden singularities at a later stage of the computation. This technique has been applied by others, [4, 16]. For our case we have chosen the regularization of Fig. 14. How one chooses the temporary regularization is not unique. However, it is chosen here so that after application of the integration by parts rule of Fig. 7 the resulting four graphs have either unique vertices or triangles which are δ-dependent and which regularize any singularity after subsequent integration. For the integration by parts we use the top left internal vertex of Fig. 14 with the line joining the quartic vertex as the reference line of the rule of Fig. 7. This produces the four graphs of Figs. 15–18.

All but the third have at least one unique vertex while that has a unique triangle. In our earlier notation the first two graphs of Figs. 15 and 16 are

$$\nu(2, 1-\delta, 2\mu-3+\delta) I(\mu-1-\delta, \mu-1+\delta, \mu-1+\delta, \mu-1-\delta, 1) \quad (39)$$

Fig. 16 Second graph after integration by parts

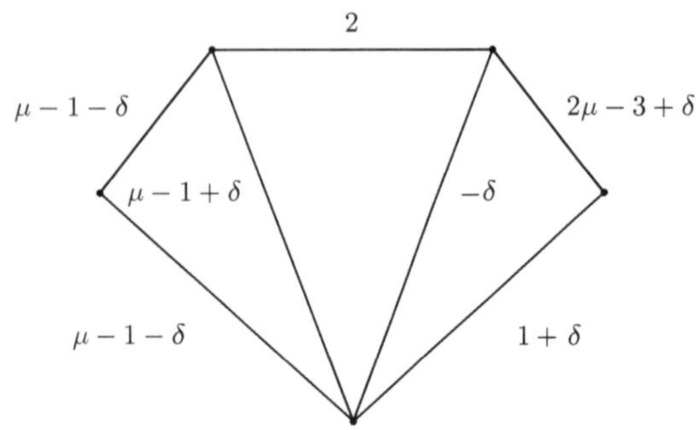

Fig. 17 Third graph after integration by parts

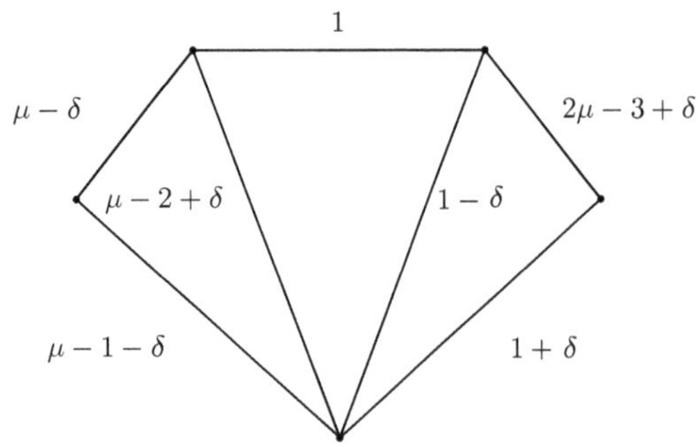

and

$$\nu(2, \mu - 1 - \delta, \mu - 1 + \delta) I(1 + \delta, 2\mu - 3, 1 + \delta, 2\mu - 3 - \delta, 1) \,. \tag{40}$$

As both of these are δ-finite and have no δ-singular coefficients, one can set δ to zero in each. The final evaluation is by a two loop reduction formula similar to those of Figs. 8 and 9. For the remaining two graphs of Figs. 17 and 18 one has to treat them together due to the singular propagator exponents as will be evident. After integrating the respective unique triangle and vertex they combine to produce

$$\begin{aligned} a^3(1) a(\mu - \delta) a(2\mu - 3 + \delta) \, [& I(\mu - 1, \mu - 1, 1 + \delta, \mu - 1 - \delta, \mu - 1 \\ & - \; I(\mu - 1, \mu - 1, 1 + \delta, \mu - 1 - 2\delta, \mu - 1 + \delta)] \,. \end{aligned} \tag{41}$$

As the external coefficient includes a factor of $1/\delta$ then the quantity inside the square brackets needs to be evaluated to $O(\delta)$. This is not possible exactly for both integrals. (It is for the first.) Instead since one only needs the $O(\delta)$ part itself one can achieve this by evaluating the integral

$$I(\mu - 1, \mu - 1, 1, \mu - 1 - \delta, \mu - 1 + \delta) \,. \tag{42}$$

Fig. 18 Fourth graph after integration by parts

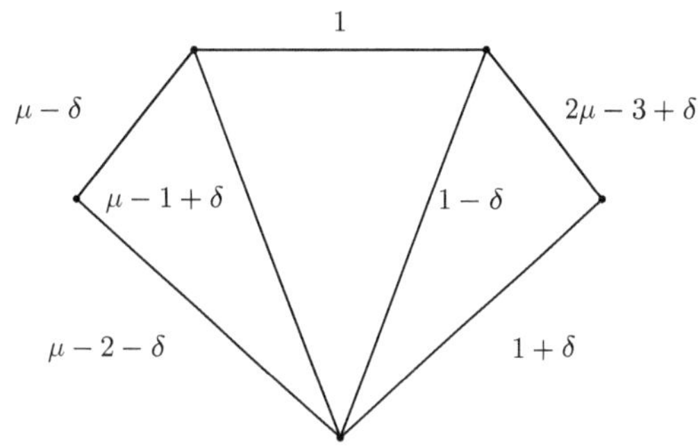

From the two 2-loop graphs we are interested in the $O(\delta)$ term of this integral clearly corresponds to the piece we require. Moreover, it can be evaluated exactly using $\downarrow$ as it then reduces to an integral to which one can apply a 2-loop recurrence relation similar to that of Fig. 9. The final expression for the graph of Fig. 12 is

$$\frac{a(\mu-1)a^2(2\mu-3)a(2\mu-2)}{2(\mu-3)(\mu-2)^9}\left[f_2 - f_1^2 - \frac{2f_1}{(\mu-2)} + 6f_3\right] \tag{43}$$

where

$$\begin{aligned} f_1 &= \psi(3-\mu) + \psi(2\mu-3) - \psi(\mu-1) - \psi(1) \\ f_2 &= \psi'(3-\mu) - \psi'(2\mu-3) + \psi'(\mu-1) - \psi'(1) \\ f_3 &= \psi'(\mu-1) - \psi'(1)\,. \end{aligned} \tag{44}$$

Setting $\mu = 2$ reproduces the established leading order value for the wheel of three spokes, [17], which provides a useful check. Finally, all the other contributing graphs are evaluated in a similar way and the full expression for ω_2, after using the Schwinger Dyson formalism, is given in [15].

5 Future Directions

We close the article by discussing several directions in which this approach could move. First, the extension of scalar field theories to non-abelian gauge theories has been considered in [18–22] for various applications where information is needed on the renormalization group functions of operators in deep inelastic scattering and the β-function. That approach is based on the observations of [23] using the number of quark flavours, N_f, as the expansion parameter. Rather than use the full QCD Lagrangian one exploits the critical point equivalence with the non-abelian Thirring model, [23],

$$L = i\bar{\psi}^i \not{D}\psi^i - \tfrac{1}{2}(A^a_\mu)^2 \tag{45}$$

where D_μ is the covariant derivative, T^a are the group generators and ψ^i is the quark field with $1 \leq i \leq N_f$. The spin-1 auxiliary field A^a_μ plays the role of the gluon in the higher dimensional theory. The triple and quartic gluon vertices of QCD are generated by the 3-point and 4-point functions of (45) with A^a_μ external legs respectively. Following the critical point analysis the propagators in a similar notation, but in momentum space, are

$$\langle \psi^i(-p)\psi^j(p) \rangle = \frac{\delta^{ij} A \not{p}}{(p^2)^{\mu-\alpha}}$$

$$\langle A^a_\mu(-p) A^b_\nu(p) \rangle = - \frac{\delta^{ab} B}{(p^2)^{\mu-\beta}} \left[\eta_{\mu\nu} - \xi \frac{p_\mu p_\nu}{p^2} \right] \tag{46}$$

where ξ is the gauge parameter with the Landau gauge corresponding to $\xi = 1$. From dimensional analysis the exponents are now, [18],

$$\alpha = \mu + \tfrac{1}{2}\eta \quad , \quad \beta = 1 - \eta - \chi \tag{47}$$

which means the basic vertex is one step from uniqueness. This complicates computations in that to proceed one has to break all contributing graphs into scalar integrals and treat them by transformations, subtractions or use integration by parts to reduce them to computable cases. While it has been possible to do this in certain instances, [19, 20, 22], it is not systematic.

Since the application of the method of [1, 2] to QCD an algorithm has been developed which allows one to exploit integration by parts. Known as the Laporta algorithm, [11], it creates all integration by parts relations between integrals of a particular topology and then algebraically solves them in such a way that all integrals are reduced to a basis set of master integrals. Once their values are known by other methods then the problem is complete. In the large N_f context once one moves to say $O(1/N_f^2)$ computations then graphs such as that of Fig. 12 need to be computed in QCD. Then the solid lines would represent quarks and the springs would correspond to gluons. However, taking the traces over the closed loops results in a huge number of irreducible numerator scalar products. While the propagators do not have integer powers, as is the case in perturbative calculations, there appears to be a similarity to the problem. In other words in principle a generalization of the Laporta algorithm should be able to produce a reduction of the irreducible graphs to a set of masters. The difficulty is that the presence of non-integer propagator powers means that the present Laporta algorithm would need to be modified in order to have a point, akin to a ground state, below which no more reductions could be possible. Though it is not clear under what conditions such a bottom point exists or whether for certain topologies or distribution of non-unit exponents it can be proved to be impossible. Indeed the latter point could be related to the issue of lack computability of a graph due to the presence of multiple-zeta values similar to (15). However, it seems that for the practical problem of deducing the QCD β-function at $O(1/N_f^2)$ such an extension to the Laporta algorithm is possibly the only feasible tactic at present.

Aside from this possible extension to the Laporta algorithm another interesting possibility is to what extent the conformal integration methods can be built into that algorithm to improve and speed reductions within a computer algebra programme for massless Feynman graphs. This may be important for higher loop topologies. For instance, earlier we derived recurrence relations for the two loop self energy topology based on the transformation deduced from the generalized uniqueness condition. While such relations are no doubt contained within integration by parts relations of the Laporta construction, that of Fig. 9 is particularly useful in that there is no increase in the power of any propagator. Therefore, it may be possible to construct similar relations using conformal transformations but for higher loop massless topologies. Indeed such transformations are not unrelated to the symmetry group of the topology as has already been studied in depth for the two loop self energy, [6, 7]. At the time of [6, 7] expanding a graph in terms of its group invariants was a promising approach which was complemented by later methods such as [8, 9]. However, it may be worth returning to a group theory analysis for topologies such as that represented in Fig. 13. This is because the high order expansion in terms of ϵ of this and other three and four loop topologies will soon be required for extending QCD to *five* and possibly higher loops. In this respect another direction of exploration may be to study the structure of the graph polynomials of a topology. The transformations of [2] have been derived from a graphical approach to understanding the structure of the two loop self energy graph. Understanding the effect such conformal transformations have on the graph polynomials of massless integrals may also give insight into the as yet undetermined group theory properties of higher order topologies.

Acknowledgements It is with pleasure that I thank the organizers of the meeting for permission to include this article in these proceedings. It is based on a talk presented at *Quantum Field Theory, Periods and Polylogarithms III*, Humboldt University, Berlin in June 2012 which was also in honour of Dr D.J. Broadhurst's 65th birthday. The AXODRAW package, [24], was used to draw the figures in the article.

References

1. Vasil'ev, A.N., Pismak, Y.M., Honkonen, J.R.: Simple method of calculating the critical indices in the $1/N$ expansion. Theor. Math. Phys. **46**, 104–113 (1981)
2. Vasil'ev, A.N., Pismak, Y.M., Honkonen, J.R.: $1/N$ expansion: calculation of the exponents η and ν in the order $1/N^2$ for arbitrary number of dimensions. Theor. Math. Phys. **47**, 465–475 (1981)
3. d'Eramo, M., Peliti, L., Parisi, G.: Theoretical predictions for critical exponents at the λ-point of Bose liquids. Lett. Nuovo Cim. **2**, 878–880 (1971)
4. Ussyukina, N.I.: Calculation of multiloop diagrams in high orders of perturbation theory. Phys. Lett. **B267**, 382–388 (1991)
5. Gracey, J.A.: On the evaluation of massless Feynman diagrams by the method of uniqueness JAG. Phys. Lett. **B277**, 469–473 (1992)

6. Broadhurst, D.J.: Exploiting the $1,440$-fold symmetry of the master two-loop diagram. Z. Phys. **C32**, 249–253 (1986)
7. Barfoot, D.T., Broadhurst, D.J.: $Z_2 \times S_6$ symmetry of the two-loop diagram. Z. Phys. **C41**, 81–85 (1988)
8. Broadhurst, D.J., Kreimer, D.: Knots and numbers in ϕ^4 theory to 7 loops and beyond. Int. J. Mod. Phys. **C6**, 519–524 (1995)
9. Bierenbaum, I., Weinzierl, S.: The massless two loop two point function. Eur. Phys. J. **C32**, 67–78 (2003)
10. Chetyrkin, K.G., Tkachov, F.V.: Integration by parts: the algorithm to calculate β-functions in 4 loops. Nucl. Phys. **B192**, 159–204 (1981)
11. Laporta, S.: High precision calculation of multiloop Feynman integrals by difference equations. Int. J. Mod. Phys. **A15**, 5087–5159 (2000)
12. Gracey, J.A.: Computation of critical exponent η at $O(1/N_f^2)$ in quantum electrodynamics in arbitrary dimensions. Nucl. Phys. **B414**, 614–648 (1994)
13. Vasil'ev, A.N., Pismak, Y.M., Honkonen, J.R.: $1/N$ expansion: calculation of the exponent η in the order $1/N^3$ by the conformal bootstrap method. Theor. Math. Phys. **50**, 127–134 (1982)
14. Zinn-Justin, J.: Quantum Field Theory and Critical Phenomena. International Series of Monographs on Physics. Clarendon Press, Oxford (2002)
15. Broadhurst, D.J., Gracey, J.A., Kreimer, D.: Beyond the triangle and uniqueness relations: non-zetas counterterms at large N from positive knots. Z. Phys. **C75**, 559–574 (1997)
16. Ussyukina, N.I., Davydychev, A.I.: New results for two-loop off-shell three-point diagrams. Phys. Lett. **B332**, 159–167 (1994)
17. Broadhurst, D.J.: Evaluation of a class of Feynman diagrams for all numbers of loops and dimensions. Phys. Lett. **B164**, 356–360 (1985)
18. Gracey, J.A.: Quark, gluon and ghost anomalous dimensions at $O(1/N_f)$ in quantum chromodynamics. Phys. Lett. **B318**, 177–183 (1993)
19. Gracey, J.A.: The QCD β-function at $O(1/N_f)$. Phys. Lett. **B373**, 178–184 (1996)
20. Ciuchini, M., Derkachov, S.E., Gracey, J.A., Manashov, A.N.: Computation of quark mass anomalous dimension at $O(1/N_f^2)$ in quantum chromodynamics. Nucl. Phys. **B579**, 56–100 (2000)
21. Gracey, J.A.: Anomalous dimension of non-singlet Wilson operators at $O(1/N_f)$ in deep inelastic scattering. Phys. Lett. **B322**, 141–146 (1994)
22. Bennett, J.F., Gracey, J.A.: Determination of the anomalous dimension of gluonic operators in deep inelastic scattering at $O(1/N_f)$. Nucl. Phys. **B517**, 241–268 (1998)
23. Hasenfratz, A., Hasenfratz, P.: The equivalence of the $SU(N)$ Yang-Mills theory with a purely fermionic model. Phys. Lett. **B297**, 166–170 (1992)
24. Vermaseren, J.A.M.: Axodraw. Comput. Phys. Commun. **83**, 45–58 (1994)

The Holonomic Toolkit

Manuel Kauers

Abstract This is an overview over standard techniques for holonomic functions, written for readers who are new to the subject. We state the definition for holonomy in a couple of different ways, including some concrete special cases as well as a more abstract and more general version. We give a collection of standard examples and state several fundamental properties of holonomic objects. Two techniques which are most useful in applications are explained in some more detail: closure properties, which can be used to prove identities among holonomic functions, and guessing, which can be used to generate plausible conjectures for equations satisfied by a given function.

1 What Is This All About?

This tutorial is an attempt to further advertize a concept which already is quite popular in some communities, but still not as popular in others. It is about the concept of holonomic functions and what computations can be done with them. The part of symbolic computation which is concerned with algorithms for transcendental functions faces a fundamental dilemma. On the one hand, problems arising from applications seem to induce a demand for algorithms that can answer questions about given analytic functions, or about given infinite sequences. On the other hand, general algorithms that take an "arbitrary" analytic functions or infinite sequences as input cannot exist, because the objects in question do in general not admit a finite representation on which an algorithm could operate. The dilemma is resolved by introducing classes of "nice" functions whose members admit a uniform finite description which can serve as data structure for algorithms.

M. Kauers (✉)
Research Institute for Symbolic Computation, Johannes Kepler University, A4040 Linz, Austria
e-mail: mkauers@risc.jku.at

C. Schneider and J. Blümlein (eds.), *Computer Algebra in Quantum Field Theory*, 119
Texts & Monographs in Symbolic Computation, DOI 10.1007/978-3-7091-1616-6_5,

A class being small means that strong assumptions are imposed on its elements. This makes the design of algorithms easier. A typical example for a small class is the set of all polynomial functions. They clearly admit a finite representation (for instance, the finite list of coefficients), and many questions about the elements of this class can be answered algorithmically. Implementations of these algorithms form the heart of every computer algebra system. The disadvantage of a small class is that quantities arising in applications are too often beyond the scope of the class.

Algorithms for bigger classes are more likely to be useful. An example for an extremely big class is the set of all functions y which admit a power series expansion $y = \sum_{n=0}^{\infty} a_n x^n$ whose coefficients are algebraic numbers and for which there exists an algorithm that computes for a given index n the corresponding coefficient a_n. Elements of this class clearly admit a finite representation (for instance, a piece of code implementing the algorithm for computing the nth coefficient), but hardly any interesting questions can be answered algorithmically for the functions in this class. It is for example impossible to decide algorithmically whether two representations actually represent the same function. So although the class contains virtually everything we may ever encounter in practical applications, it is not very useful.

The class of holonomic functions has proven to be a good compromise between these two extremes. On the one hand, it is small enough that algorithms could be designed for efficiently answering many important questions for a given element of the class. In particular, there are algorithms for proving identities among holonomic functions, for computing asymptotic expansions of them, and for evaluating them numerically to any desired accuracy. On the other hand, the class is big enough that it contains a lot of quantities that arise in applications. In particular, many Feynman integrals [7] and many generalized harmonic sums are holonomic. Every generalized polylogarithm and every hypergeometric term is holonomic. Every algebraic function and every quasi-polynomial is holonomic. According to Salvy [31], more than 60 % of the entries of Abramowitz/Stegun's table of mathematical functions [1] are holonomic, as well as some 25 % of the entries of Sloane's online encyclopedia of integer sequences (OEIS) [33].

The concept of holonomy was introduced in the 1970s by Bernstein [4] in the theory of D-modules (see Björk's book [5] for this part of the story). Its relevance to symbolic computation and the theory of special functions was first recognized by Zeilberger [42]. His 1990 article, which is still a good first reading for readers not familiar with the theory, has initiated a great amount of work both in combinatorics and in computer algebra. Stanley discusses the case of a single variable [34, 35] (see also Chap. 7 of [25]). Salvy and Zimmermann [32] and Mallinger [29] provide implementations for Maple and Mathematica, respectively. Algorithms for the case of several variables [14, 15, 40, 42] were implemented by Chyzak [14] for Maple and more recently by Koutschan [27, 28] for Mathematica. The applications in combinatorics are meanwhile too numerous to list a reasonable selection.

In this tutorial, it is not our aim to explain (or advertize) any particular software package. The goal is rather to give an overview over the various definitions of holonomy, the key properties of holonomic functions, and the most important algorithms for working with them. The text should provide any reader new to the topic with the necessary background for reading the manual of some software package without wondering how to make use of the functionality it provides.

2 What Is a Holonomic Function?

2.1 *Definitions and Basic Examples*

We give several variants of the definition of holonomy, discussing the most important special cases separately, before we describe the concept in more general (and more abstract) terms.

Definition 1. An infinite sequence $(a_n)_{n=0}^{\infty}$ of numbers is called *holonomic* (or *P-finite* or *P-recursive* or, rarely, *D-finite*) if there exists an integer $r \in \mathbb{N}$, independent of n, and univariate polynomials $p_0, \dots, p_r$, not all identically zero, such that for all $n \in \mathbb{N}$ we have $p_0(n)a_n + p_1(n)a_{n+1} + \dots + p_r(n)a_{n+r} = 0$.

Example 1. 1. The sequence $a_n = \frac{5n-3}{3n+5}$ is holonomic, because it satisfies the recurrence equation $(3n+5)(5n+2)a_n - (5n-3)(3n+8)a_{n+1} = 0$.

2. The sequence $a_n = n!$ is holonomic, because it satisfies the recurrence equation $a_{n+1} - (n+1)a_n = 0$.
3. The sequence $H_n = \sum_{k=1}^{n} \frac{1}{k}$ of harmonic numbers is holonomic, because it satisfies the recurrence equation $(n+1)H_n - (2n+3)H_{n+1} + (n+2)H_{n+2} = 0$.
4. The sequence $a_n = \sum_k \binom{n}{k}^2 \binom{n+k}{k}^2$ arising in Apery's proof [38] of the irrationality of $\zeta(3)$ is holonomic, because it satisfies the recurrence equation $(n+1)^3 a_n - (2n+3)(17n^2+51n+39)a_{n+1} + (n+2)^3 a_{n+2} = 0$.
5. The sequence $a_n = \int_0^1 \int_0^1 \frac{w^{-1-\varepsilon/2}(1-z)^{\varepsilon/2} z^{-\varepsilon/2}}{(z+w-wz)^{1-\varepsilon}} \bigl(1-w^{n+1}-(1-w)^{n+1}\bigr) dw\, dz$ coming from some Feynman diagram [27, p. 94f] is holonomic (regarding ε as a fixed parameter) because it satisfies the 3rd order recurrence equation $-(\varepsilon-n-3)(\varepsilon-n-2)(\varepsilon+2n+4)(\varepsilon+2n+6)a_{n+3} + (\varepsilon-n-2)(\varepsilon+2n+4)(\varepsilon^2+2\varepsilon n+5\varepsilon-6n^2-28n-34)a_{n+2} - (n+2)(\varepsilon^3-3\varepsilon^2 n-6\varepsilon^2-8\varepsilon n^2-30\varepsilon n-28\varepsilon+12n^3+64n^2+116n+72)a_{n+1} - 2(n+1)(n+2)^2(\varepsilon-2n-2)a_n = 0$.
6. For $n \in \mathbb{N}$, define $H_n(x)$ as the (uniquely determined) polynomial of degree n with the property $\int_{-\infty}^{\infty} H_n(x)H_k(x)e^{-x^2}dx = \sqrt{\pi}2^n n!\delta_{n,k}$ for all $n, k \in \mathbb{N}$, where $\delta_{n,k}$ is the Kronecker symbol. The $H_n(x)$ are called Hermite polynomials. Regarding them as a sequence with respect to n where x is some fixed parameter, the Hermite polynomials are holonomic, because they satisfy the recurrence equation $H_{n+2}(x) - 2xH_{n+1}(x) + (2+2n)H_n(x) = 0$.
7. The sequence $(a_n)_{n=0}^{\infty}$ defined recursively by $a_0 = 0$, $a_1 = 1$, $a_2 = 17$, and $a_{n+3} = (100+99n+97n^2)a_{n+2} + (96+95n+94n^2)a_{n+1} + (93+92n+91n^2)a_n$ has no particular significance, but it is nevertheless holonomic.

8. The sequences $a_n = \sqrt{n}$, $b_n = \mathrm{p}(n)$, $c_n = \zeta(n)$, $d_n = n^n$ where $\mathrm{p}(n)$ is the nth prime number and ζ denotes the Riemann's zeta function are **not** holonomic, i.e., none of them satisfies a linear recurrence equation with polynomial coefficients [19].

According to Definition 1, a sequence is holonomic if it can be viewed as a solution of some linear recurrence equation with polynomial coefficients. The definition for analytic functions is analogous:

Definition 2. An analytic function $y\colon U \to \mathbb{C}$ defined on some domain $U \subseteq \mathbb{C}$ (or more generally, any object y for which multiplication by polynomials, addition, and repeated differentiation is defined) is called *holonomic* (or *D-finite* or *P-finite*) if there exists an integer $r \in \mathbb{N}$ and univariate polynomials $p_0, \dots, p_r$, not all identically zero, such that $p_0(z)y(z) + p_1(z)y'(z) + \cdots + p_r(z)y^{(r)}(z) = 0$.

Example 2. 1. The function $y(z) = \frac{5z-3}{3z+5}$ is holonomic, because it satisfies the differential equation $(5z-3)(3z+5)y'(z) - 34y(z) = 0$.
2. The functions $\exp(z)$ and $\log(z)$ are holonomic, because they satisfy the differential equations $\exp'(z) - \exp(z) = 0$ and $z\log''(z) + \log'(z) = 0$, respectively.
3. The function $y(z) = 1/(1 + \sqrt{1-z^2})$ is holonomic, because it satisfies the differential equation $(z^3 - z)y''(z) + (4z^2 - 3)y'(z) + 2zy(z) = 0$.
4. The function $y(z) = \sum_{n=0}^{\infty} a_n z^n$ where a_n is the sequence from Example 1.4 is holonomic because it satisfies the differential equation $(z^2 - 34z + 1)z^2 y'''(z) + 3(2z^2 - 51z + 1)zy''(z) + (7z^2 - 112z + 1)y'(z) + (z-5)y(z) = 0$.
5. The function $y\colon [-1,1] \to \mathbb{R}$ uniquely determined by the conditions $y(0) = 0$, $y'(0) = 17$, $y''(z) = (100 + 99z + 98z^2)y'(z) + (97 + 96z + 95z^2)y(z)$ has no particular significance, but is nevertheless holonomic.
6. The functions $\exp(\exp(z))$, $1/(1 + \exp(z))$, $\zeta(z)$, $\Gamma(z)$, $W(z)$ (the Lambert W function [17]) are **not** holonomic, i.e., neither of them satisfies a linear differential equation with polynomial coefficients.

Note that as introduced in the two definitions above, the word "holonomic" is ambiguous. We need to distinguish between discrete variables and continuous variables. If a function depends on a discrete variable (typically named n, m, or k), then it is called holonomic if it satisfies a recurrence, and if it depends on a continuous variable (typically named x, t, or z), then it is called holonomic if it satisfies a differential equation. For example, the Gamma function is holonomic if we regard its argument as a discrete variable, but it is not holonomic if we regard its argument as a continuous variable. For a connection of the two notions, see Theorem 1 below.

The definition for the differential case extends as follows to functions in several variables.

Definition 3. An analytic function $y\colon U \to \mathbb{C}$, defined on some domain $U \subseteq \mathbb{C}^q$ ($q \in \mathbb{N}$ fixed) is called *holonomic* (or *D-finite* or *P-finite*) if for every variable z_i ($i = 1, \dots, q$) there exists an integer $r \in \mathbb{N}$ and polynomials $p_0, \dots, p_r$, possibly depending on all q variables $z_1, \ldots, z_q$ but not all identically zero, such that for all $z = (z_1, \ldots, z_q) \in U$ we have $p_0(z)y(z) + p_1(z)\frac{\partial}{\partial z_i}y(z) + \cdots + p_r(z)\frac{\partial^r}{\partial z_i^r}y(z) = 0$.

In other words, a multivariate analytic function is holonomic if it can be viewed as a solution of a system of q linear differential equations with polynomial coefficients. The polynomial coefficients may involve all the variables, but there is a restriction on the derivatives: the ith equation ($i = 1, \ldots, q$) may only contain differentiations with respect to the variable z_i. Again, in addition to analytic functions the definition extends more generally to any objects y for which multiplication by polynomials, addition, and repeated partial differentiation is defined.

For sequences with several indices, and more generally for functions depending on some discrete as well as some continuous variables, several different extensions of Definition 1 are in use. We give here two of them. Assigning the words "holonomic" and "D-finite" to these two properties seems to be in accordance with most of the recent literature. However, it should be observed that other authors use slightly different definitions.

Definition 4. Let $U \subseteq \mathbb{C}^q$ be a domain, and let

$$y = y(n_1, \ldots, n_p, z_1, \ldots, z_q) : \mathbb{N}^p \times U \to \mathbb{C}$$

be a function which is analytic in $z_1, \ldots, z_q$ for every fixed choice of $n_1, \ldots, n_p \in \mathbb{N}^p$.

1. y is called *D-finite* (or *P-finite*) if for every i ($i = 1, \ldots, p$) there exists a number $r \in \mathbb{N}$ and polynomials $u_0, \ldots, u_r$, possibly depending on $n_1, \ldots, n_p, z_1, \ldots, z_q$ and not all identically zero, such that for all $n = (n_1, \ldots, n_p) \in \mathbb{N}^p$ and all $z = (z_1, \ldots, z_q) \in U$ we have

$$\begin{aligned} &u_0(n, z)y(n_1, \ldots, n_{i-1}, n_i, n_{i+1}, \ldots, n_p, z) \\ &+ u_1(n, z)y(n_1, \ldots, n_{i-1}, n_i + 1, n_{i+1}, \ldots, n_p, z) \\ &+ u_2(n, z)y(n_1, \ldots, n_{i-1}, n_i + 2, n_{i+1}, \ldots, n_p, z) \\ &\quad\vdots \\ &+ u_r(n, z)y(n_1, \ldots, n_{i-1}, n_i + r, n_{i+1}, \ldots, n_p, z) = 0, \end{aligned}$$

and for every j ($j = 1, \ldots, q$) there exists a number $r \in \mathbb{N}$ and polynomials $u_0, \ldots, u_r$, possibly depending on $n_1, \ldots, n_p, z_1, \ldots, z_q$ and not all identically zero, such that for all $n = (n_1, \ldots, n_p) \in \mathbb{N}^p$ and all $z = (z_1, \ldots, z_q) \in U$ we have

$$u_0(n, z)y(n, z) + u_1(n, z)\frac{\partial}{\partial z_i}y(n, z) + \cdots + u_r(n, z)\frac{\partial^r}{\partial z_i^r}y(n, z) = 0.$$

2. y is called *holonomic* if the (formal) power series

$$\tilde{y}(x_1,\dots,x_p,z_1,\dots,z_q) := \sum_{n_1=1}^{\infty}\sum_{n_2=1}^{\infty}\cdots\sum_{n_p=1}^{\infty} y(n_1,\dots,n_p,z_1,\dots,z_q)x_1^{n_1}x_2^{n_2}\cdots x_p^{n_p}$$

is holonomic as function of $x_1,\dots,x_p,z_1,\dots,z_q$ in the sense of Definition 3.

Example 3. 1. The bivariate function $f(x,y) = \frac{1}{1-y-xy}$ is D-finite and holonomic because it satisfies the differential equations $(1-y-xy)\frac{\partial}{\partial x}f(x,y) - yf(x,y) = 0$ and $(1-y-xy)\frac{\partial}{\partial y}f(x,y) - (x+1)f(x,y) = 0$.

2. The bivariate sequence $a_{n,k} = \binom{n}{k}$ is D-finite, because it satisfies the recurrence equations $(n-k+1)a_{n+1,k} - (n+1)a_{n,k} = 0$ and $(n-k)a_{n,k+1} - (k+1)a_{n,k} = 0$, and it is holonomic because $f(x,y) = \sum_{n=0}^{\infty}\sum_{k=0}^{\infty}\binom{n}{k}x^k y^n = \frac{1}{1-y-xy}$ is holonomic.
3. Regarded as a function of one discrete variable n and a continuous variable x, the Hermite polynomials $H_n(x)$ are D-finite because they satisfy the equations $H_{n+2}(x) - 2xH_{n+1}(x) + 2(n+1)H_n(x) = 0$ and $H_n''(x) - 2xH_n'(x) + 2nH_n(x) = 0$. They are also holonomic because the formal power series $f(x,z) := \sum_{n=0}^{\infty} H_n(x)z^n$ satisfies the differential equations $z\frac{\partial^3}{\partial x^3}f(x,z) + (1-4xz)\frac{\partial^2}{\partial x^2}f(x,z) + (4x^2z - 4z - 2x)\frac{\partial}{\partial x}f(x,z) + 4xzf(x,z) = 0$ and $2z^3\frac{\partial^3}{\partial z^3}f(x,z) + (14z^2 - 2xz + 1)\frac{\partial^2}{\partial z^2}f(x,z) + (20z - 4x)\frac{\partial}{\partial z}f(x,z) + 4f(x,z) = 0$.
4. The integrand $\frac{w^{-1-\varepsilon/2}(1-z)^{\varepsilon/2}z^{-\varepsilon/2}}{(z+w-wz)^{1-\varepsilon}}(1-w^{n+1}-(1-w)^{n+1})$ of the Feynman integral in Example 1.5 is D-finite when w and z are regarded as continuous and n and ε are regarded as discrete variables.
5. The Kronecker symbol $\delta_{n,k}$, viewed as a bivariate sequence in n and m, is holonomic but not D-finite. The bivariate sequence $1/(n^2+k^2)$ is D-finite but not holonomic [40].
6. The bivariate sequence $S_1(n,k)$ of Stirling numbers of the first kind is not D-finite although it satisfies the recurrence equation $S_1(n+1,k+1) + nS_1(n,k+1) - S_1(n,k) = 0$. This recurrence equation does not suffice to establish D-finiteness because it involves shifts in both variables. It can be shown that $S_1(n,k)$ does not satisfy any recurrence equations containing only shifts in n or only shifts in k.

Although the two properties in Definition 4 are not equivalent, the difference does not play a big role in practice: multivariate functions arising in applications typically either have both properties or none of the two.

It is sometimes more transparent to work with operators acting on functions rather than with functional equations. In order to rephrase the previous definition using operators, consider the algebra

$$\mathbb{A} := \mathbb{C}(n_1,\dots,n_p,z_1,\dots,z_q)[S_1,\dots,S_p,D_1,\dots,D_q]$$

consisting of all the multivariate polynomials in the variables $S_1, \dots, S_p$, $D_1, \dots, D_q$ with coefficients that are rational functions (i.e., quotients of polynomials) in the variables $n_1, \dots, n_p, z_1, \dots, z_q$. We regard the elements of $\mathbb{A}$ as operators and let them act in the natural way on functions y: application of S_i corresponds to a shift $n_i \rightsquigarrow n_i + 1$, application of D_i causes a partial derivation $\frac{\partial}{\partial z_i}$, and application of some rational function u maps y to the function uy. We write $A \cdot y$ for the result obtained by applying an operator $A \in \mathbb{A}$ to the function y.

If A, B are operators and y is a function, then $(A + B) \cdot y = A \cdot y + B \cdot y$ (the $+$ on the left hand side being the addition in $\mathbb{A}$, and the $+$ on the right hand side being the pointwise addition of functions). For the product of two operators, we want to have $(AB) \cdot y = A \cdot (B \cdot y)$, i.e., multiplication of operators should be compatible with composition of application. This is not the case for the usual multiplication, but it works if we use a noncommutative multiplication which is such that for a rational function $u(n_1, \dots, n_p, z_1, \dots, z_q)$ we have

$$S_i\, u(n_1, \dots, n_p, z_1, \dots, z_q) = u(n_1, \dots, n_{i-1}, n_i + 1, n_{i+1}, \dots, n_p, z_1, \dots, z_q)\, S_i$$

for every i $(i = 1, \dots, p)$, and

$$\begin{aligned} D_j\, u(n_1, \dots, n_p, z_1, \dots, z_q) = u(n_1, \dots, n_p, z_1, \dots, z_q)\, D_j \\ + \frac{\partial}{\partial z_i} u(n_1, \dots, n_p, z_1, \dots, z_q) \end{aligned}$$

for every j $(j = 1, \dots, q)$. These rules for example imply $S_i n_i = (n_i + 1) S_i$ and $D_j z_j = z_j D_j + 1$ for $i = 1, \dots, p$ and $j = 1, \dots, q$. By furthermore requiring that $S_i S_j = S_j S_i$ and $S_i D_j = D_j S_i$ and $D_i D_j = D_j D_i$ for all i and j, the multiplication is uniquely determined. With this multiplication, we have for example

$$\begin{aligned} D^2\, a(z) = D(D\, a(z)) &= D(a(z) D + a'(z)) \\ &= D\, a(z) D + D a'(z) \\ &= (a(z) D + a'(z)) D + (a'(z) D + a''(z)) \\ &= a(z) D^2 + 2a'(z) D + a''(z). \end{aligned}$$

In terms of operators, part 1 of Definition 4 can be stated as follows: Let y be a function as in Definition 4, and let $\mathfrak{a} \subseteq \mathbb{A}$ be the set of all operators which map y to the zero function. Then y is called D-finite if

- For all $i = 1, \dots, p$ we have $\mathfrak{a} \cap \mathbb{C}(n_1, \dots, n_p, z_1, \dots, z_q)[S_i] \neq \{0\}$, and
- For all $j = 1, \dots, q$ we have $\mathfrak{a} \cap \mathbb{C}(n_1, \dots, n_p, z_1, \dots, z_q)[D_j] \neq \{0\}$.

The set $\mathfrak{a}$ is called the *annihilator* of y. It has the algebraic structure of an ideal of $\mathbb{A}$, i.e., it has the properties $A, B \in \mathfrak{a} \Rightarrow A + B \in \mathfrak{a}$ and $A \in \mathfrak{a}, B \in \mathbb{A} \Rightarrow BA \in \mathfrak{a}$.

Operator algebras can be used to abstract away the difference between shift and derivation, and to allow other operations as well. We will not use this most general form in the remainder of this tutorial, but only quote the definition of D-finiteness in this language. Let R be a commutative ring (for example, the set $\mathbb{C}(x_1, \dots, x_m)$ of all rational functions in m variables $x_1, \dots, x_m$ with coefficients in $\mathbb{C}$), and consider the algebra $\mathbb{A} = R[\partial_1, \dots, \partial_n]$ of multivariate polynomials in the indeterminates $\partial_1, \dots, \partial_n$ with coefficients in R. Let $\sigma_1, \dots, \sigma_n \colon R \to R$ be automorphisms (i.e., $\sigma_i(a + b) = \sigma_i(a) + \sigma_i(b)$ and $\sigma_i(ab) = \sigma_i(a)\sigma_i(b)$ for all $a, b \in R$) and for each i, let $\delta_i \colon R \to R$ be a so-called skew-derivation for σ_i. A skew-derivation is a map which satisfies $\delta_i(a + b) = \delta_i(a) + \delta_i(b)$ and the generalized Leibniz law $\delta_i(ab) = \delta_i(a)b + \sigma_i(a)\delta_i(b)$. Then consider the (noncommutative) multiplication on $\mathbb{A}$ defined through the properties $ab = ba$ for all $a, b \in R$, $\partial_i\partial_j = \partial_j\partial_i$ for all $i, j = 1, \dots, n$ and $\partial_i a = \sigma_i(a)\partial_i + \delta_i(a)$ for all $a \in R$ and all $i = 1, \dots, n$. Such an algebra $\mathbb{A}$ is called an Ore algebra. Details about arithmetic for such algebras are explained in a nice tutorial by Bronstein and Petkovšek [10].

Observe that the generators ∂_i of an Ore algebra can be used to represent shift operators (by choosing σ_i such that $\sigma_i(x) = x + 1$ for a variable x in R and δ_i the zero function) as well as derivations (by choosing σ_i the identity function and $\delta_i = \frac{\partial}{\partial x}$ for a variable x in R). In addition, further operations can be encoded, for example the q-shift (set $\sigma_i(x) := qx$ where x is a variable and q some fixed element of R).

We let the elements of an Ore algebra act ("operate") on the elements of some set F of "functions". To make this action precise, we need to assume that F is an R-module (i.e., there is an addition in F and a multiplication of elements in R by elements in F which is compatible with the addition), and that there are functions $d_1, \dots, d_n \colon F \to F$ ("partial pseudo-derivations") which satisfy certain compatibility conditions with the addition, the multiplication, and the σ_i and δ_i so as to ensure that the action of some $A \in \mathbb{A}$ on some $y \in F$, written $A \cdot y$, has the properties $(A + B) \cdot y = (A \cdot y) + (B \cdot y)$ and $(AB) \cdot y = A \cdot (B \cdot y)$ for all $A, B \in \mathbb{A}$, $a \cdot y = ay$ for all $a \in R \subseteq \mathbb{A}$, and $\partial_i \cdot y = d_i(y)$ for all i. As an example, if $\mathbb{A}$ is a ring of differential operators, a natural choice for F would be the set of all meromorphic functions, and if $\mathbb{A}$ is a ring of shift operators, a natural choice for F may be some vector space of sequences.

Definition 5. Let $\mathbb{A} = R[\partial_1, \dots, \partial_n]$ be an Ore algebra whose elements act on some set F as described above, and let $y \in F$. Let $\mathfrak{a} := \{ A \in \mathbb{A} : A \cdot y = 0 \}$ be the set of all operators which map y to the zero element of F. Then y is called ∂-finite if for all $i = 1, \dots, n$ we have $\mathfrak{a} \cap R[\partial_i] \neq \{0\}$.

Example 4. 1. Set $R = \mathbb{C}(n_1, \dots, n_p, z_1, \dots, z_q)$ and consider the Ore algebra $\mathbb{A} = R[\partial_1, \dots, \partial_p, \partial_{p+1}, \dots, \partial_{p+q}]$ defined by the automorphisms $\sigma_1, \dots, \sigma_{p+q} \colon R \to R$, and the skew-derivations $\delta_1, \dots, \delta_{p+q} \colon R \to R$ satisfying $\sigma_i(c) = c$ and $\delta_i(c) = 0$ for $i = 1, \dots, p + q$ and all $c \in \mathbb{C}$, and

$$\sigma_i(n_i) = n_i + 1, \quad \sigma_i(n_j) = n_j \quad (i \neq j), \qquad \sigma_i(z_j) = z_j \quad (j = 1, \dots, q)$$
$$\delta_i(n_j) = 0 \quad (j = 1, \dots, p), \qquad \delta_i(z_j) = 0 \quad (j = 1, \dots, q)$$

for $i = 1, \ldots, p$, and

$$\sigma_i(n_j) = n_j \quad (j = 1, \ldots, p), \qquad \sigma_i(z_j) = z_j \quad (j = 1, \ldots, q)$$
$$\delta_i(n_j) = 0 \quad (j = 1, \ldots, p), \qquad \delta_i(z_{i-p}) = 1, \quad \delta_i(z_{j-p}) = 0 \quad (i \neq j)$$

for $i = p+1, \ldots, p+q$. Then $\partial_1, \ldots, \partial_p$ act as shift operators for the variables $n_1, \ldots, n_p$, respectively, and $\partial_{p+1}, \ldots, \partial_{p+q}$ act as derivations for the variables $z_1, \ldots, z_q$, respectively.

For this choice of $\mathbb{A}$, Definition 5 reduces to part 1 of Definition 4.

2. Let $R = \mathbb{Q}(q, Q)$ and define $\sigma\colon R \to R$ by $\sigma(c) = c$ for all $c \in \mathbb{Q}(q)$ and $\sigma(Q) = qQ$, so that σ acts on Q like the shift $n \rightsquigarrow n+1$ acts on q^n. Consider the Ore algebra $\mathbb{A} = R[\partial]$ with $\delta = 0$ and this σ.

   Let F denote the vector space of all sequences over $\mathbb{Q}(q)$ and let $\mathbb{A}$ act on F by $\partial \cdot (a_n)_{n=0}^{\infty} := (a_{n+1})_{n=0}^{\infty}$ and $r(q, Q) \cdot (a_n)_{n=0}^{\infty} := (r(q, q^n)a_n)_{n=0}^{\infty}$ for $r(q, Q) \in \mathbb{Q}(q, Q)$ and $(a_n)_{n=0}^{\infty} \in F$.

   Consider the sequence $a_n := \prod_{k=1}^{n} \frac{1-q^k}{1-q}$, which is known as q-analog of the factorial in the literature [2, Chap. 10]. Because of

$$\big((1 - qQ) - (1 - Q)\partial\big) \cdot a_n = (1 - q^{n+1})a_n - (1 - q^n)a_{n+1} = 0$$

   it is ∂-finite with respect to the algebra $\mathbb{A}$.

Most of the algorithms and features explained below for the shift and/or differential case generalize to objects that are D-finite with respect to arbitrary Ore algebras $\mathbb{A}$. Even more, it has recently been observed [16] that for some of the properties a weaker assumption than D-finiteness is sufficient. However, the underlying ideas can best be explained for the univariate case, and for reasons of simplicity we will focus on this case.

2.2 *Fundamental Properties*

A key property of holonomic functions is that they can be described by a finite amount of data, and hence faithfully represented in a computer. This is almost obvious for univariate holonomic sequences: all the (infinitely many) terms of such a sequence are uniquely determined by the linear recurrence and a suitable (finite) number of initial values. If $(a_n)_{n=0}^{\infty}$ satisfies the recurrence

$$p_0(n)a_n + p_1(n)a_{n+1} + \cdots + p_{r-1}(n)a_{n+r-1} + p_r(n)a_{n+r} = 0$$

for all $n \in \mathbb{N}$, where $p_0, \ldots, p_r$ are certain polynomials, and p_r is not the zero polynomial, then the recurrence uniquely determines the value of a_{n+r} once we know the values of $a_n, \ldots, a_{n+r-1}$, unless n is a root of the polynomial p_r. In order to fix a particular solution of the recurrence, it is therefore enough to fix the values $a_0, \ldots, a_{r-1}$ as well as the values a_{n+r} for every positive integer root n of p_r. Note that p_r is a univariate polynomial, so it can have only finitely many roots.

The situation is not much different for holonomic functions in a continuous variable. In order to fix a particular solution of a given differential equation

$$q_0(z)y(z) + q_1(z)y'(z) + \cdots + q_{s-1}(z)y^{(s-1)}(z) + q_s(z)y^{(s)}(z) = 0,$$

where $q_0, \ldots, q_s$ are polynomials, it suffices to fix the initial conditions $y(0)$, $y'(0), \ldots, y^{(s-1)}(0)$, plus possibly some finitely many further values $y^{(n)}(0)$. For which indices n the value of $y^{(n)}(0)$ does not follow from the earlier values by the differential equation is not as obvious as in the case of a recurrence equation. One possibility is to make use of the following theorem, which associates to a given differential equation a recurrence equation from which the relevant indices n can then be read off as described before.

Theorem 1. *Let $(a_n)_{n=0}^{\infty}$ be a sequence and $y(z) = \sum_{n=0}^{\infty} a_n z^n$ the (formal) power series whose coefficient sequence is $(a_n)_{n=0}^{\infty}$. Then $(a_n)_{n=0}^{\infty}$ is holonomic in the sense of Definition 1 if and only if $y(z)$ is holonomic in the sense of Definition 2. A differential equation satisfied by $y(z)$ can be computed from a known recurrence equation for $(a_n)_{n=0}^{\infty}$ and vice versa.*

The theorem is based on the observation that multiplying a series by z^{-1} corresponds to a forward shift of the coefficient sequence, and a differentiation followed by a multiplication with z corresponds to a multiplication by n. Here is an example for obtaining a recurrence equation for $(a_n)_{n=0}^{\infty}$ from a given differential equation for the power series $y(z) = \sum_{n=0}^{\infty} a_n z^n$.

$$\begin{aligned}
&(z-2)y''(z) + 5zy'(z) - y(z) = 0\\
\Rightarrow\quad &(z-2)\sum_{n=0}^{\infty} a_n n(n-1)z^{n-2} + 5z\sum_{n=0}^{\infty} a_n n z^{n-1} - \sum_{n=0}^{\infty} a_n z^n = 0\\
\Rightarrow\quad &\sum_{n=0}^{\infty} a_n n(n-1)z^{n-1} - 2\sum_{n=0}^{\infty} a_n n(n-1)z^{n-2} + 5\sum_{n=0}^{\infty} a_n n z^n - \sum_{n=0}^{\infty} a_n z^n = 0\\
\Rightarrow\quad &\sum_{n=0}^{\infty} a_{n+1}(n+1)n z^n - 2\sum_{n=0}^{\infty} a_{n+2}(n+2)(n+1)z^n + 5\sum_{n=0}^{\infty} a_n n z^n - \sum_{n=0}^{\infty} a_n z^n = 0\\
\Rightarrow\quad &\sum_{n=0}^{\infty}\Big((n+1)n\,a_{n+1} - 2(n+2)(n+1)a_{n+2} + 5n\,a_n - a_n\Big)z^n = 0\\
\Rightarrow\quad &-2(n+2)(n+1)a_{n+2} + (n+1)n\,a_{n+1} + (5n-1)a_n = 0 \qquad (n \geq 0).
\end{aligned}$$

See *The Concrete Tetrahedron* [25, Theorem 7.1] for the general case. The reverse direction works similarly.

Theorem 1 does not generalize to multivariate D-finite functions, it does however hold (by definition) for holonomic functions in several variables. In fact, Theorem 1 is the motivation for defining multivariate holonomy as in Definition 4.

A second useful feature of holonomic functions is that their asymptotic behaviour can be described easily. We say that two sequences $(a_n)_{n=0}^{\infty}, (b_n)_{n=0}^{\infty}$ are asymptotically equivalent if a_n/b_n converges to 1 for $n \to \infty$. Similary, two functions $f(z), g(z)$ are called asymptotically equivalent at some point ζ if $f(z)/g(z)$ converges to 1 for $z \to \zeta$. The following theorem describes the possible asymptotic behaviour of holonomic sequences and functions. Unlike Theorem 1, it is not straightforward.

Theorem 2. *[18, 24, 41]*

1. *If $(a_n)_{n=0}^{\infty}$ is a holonomic sequence, then there exist constants $c_1, \dots, c_m$, polynomials $p_1, \dots, p_m$, natural numbers $r_1, \dots, r_m$, constants $\gamma_1, \dots, \gamma_m, \phi_1, \dots, \phi_m, \alpha_1, \dots, \alpha_m$ and natural numbers $\beta_1, \dots, \beta_m$ such that*

$$a_n \approx \sum_{k=1}^{m} c_k\, e^{p_k(n^{1/r_k})} n^{\gamma_k n} \phi_k^n n^{\alpha_k} \log(n)^{\beta_k} \qquad (n \to \infty).$$

2. *If $y(z)$ is a holonomic analytic function with a singularity at $\zeta \in \mathbb{C}$, then there exist constants $c_1, \dots, c_m$, polynomials $p_1, \dots, p_m$, natural numbers $r_1, \dots, r_m$, constants $\alpha_1, \dots, \alpha_m$, and natural numbers $\beta_1, \dots, \beta_m$ such that*

$$y(x) \sim \sum_{k=1}^{m} c_k\, e^{p_k((z-\zeta)^{-1/r_k})} (z-\zeta)^{\alpha_k} \log(z-\zeta)^{\beta_k} \qquad (z \to \zeta).$$

Typically, one of the terms in the sum dominates all the others, so we can take $m = 1$. As an example, for the sequence $(a_n)_{n=0}^{\infty}$ from Example 1.4 we have $a_n \sim c\,(12 + 17\sqrt{2})^n n^{-3/2}$ where

$$c \approx 0.220043767112643037850689759810486656678158042907.$$

All the data in the asymptotic expression can be calculated exactly from a given recurrence or differential equation, except for the multiplicative constants c_k. These can however calculated numerically to arbitrarily high precision. In typical examples, it is easy to compute at least a few dozen decimal digits for them.

It is also possible to compute numerically the values of an analytic holonomic function to arbitrary precision, as stated in part 2 of the following theorem. The statement about sequences in part 1 is trivial (all terms of the sequence can be computed using the recurrence), but part 2 is not because it also covers the case where the evaluation point is outside of the disk of convergence of the series. This is known as effective analytic continuation.

Theorem 3. *1. If a holonomic sequence $(a_n)_{n=0}^{\infty}$ is given in terms of recurrence equation and a suitable number of initial values $a_0, a_1, \dots, a_k$, then we can efficiently compute the nth term a_n of the sequence for every given index n.*

2. [13, 30, 36, 37] If a holonomic analytic function $y(z)$ is given in terms of a differential equation and a suitable number of initial values $y(0), y'(0), \dots, y^{(k)}(0)$, and if we are given some complex number ζ with rational real and imaginary part and a polygonal path from 0 to ζ whose vertices have rational real and imaginary path, and some positive rational number ε, then we can efficiently compute a number $\tilde{y}$ such that for the value $y(\zeta)$ of the analytic continuation of y along the given path to ζ we have $|y(\zeta) - \tilde{y}| < \varepsilon$.

3 What Are Closure Properties?

If p and q are polynomials, then also their sum $p + q$, their product pq, the composition $p \circ q$, the derivative p', and the indefinite integral $\int p$ are polynomials. We say that the class of polynomials is closed under these operations. Also the class of holonomic functions is closed under a number of operations.

Theorem 4. *[25, 35, 42]*

1. *If $(a_n)_{n=0}^{\infty}$ and $(b_n)_{n=0}^{\infty}$ are holonomic sequences, then so are $(a_n + b_n)_{n=0}^{\infty}$ and $(a_n b_n)_{n=0}^{\infty}$ and $\left(\sum_{k=0}^{n} a_k b_{n-k}\right)_{n=0}^{\infty}$.*
2. *If $a(z)$ and $b(z)$ are holonomic functions, then so are $a(z) + b(z)$ and $a(z)b(z)$.*
3. *If $(a_n)_{n=0}^{\infty}$ is a holonomic sequence and $\alpha, \beta \in \mathbb{Q}$ are nonnegative constants then $(a_{\lfloor \alpha n + \beta \rfloor})_{n=0}^{\infty}$ is a holonomic sequence.*
4. *If $a(z)$ is a holonomic function, then so are $a'(z)$ and $\int a(z)dz$.*
5. *If $a(z)$ is a holonomic function and $b(z)$ is an algebraic function, i.e., there is a nonzero bivariate polynomial $p(z, y)$ such that $p(z, b(z))$ is identically zero, then the composition $a(b(z))$ is holonomic.*

The theorem is most useful for recognizing a quantity given in terms of some expression as holonomic. For example, using the theorem, it is easy to see that

$$y(z) = \exp\bigl(1 - \sqrt{1 - z^2}\bigr) + \int \log(1 - z)^2 dz$$

is holonomic: the innermost functions $\exp(z)$ and $\log(z)$ are holonomic (Example 2.2), by part 3 of the theorem $\exp(1 - \sqrt{1 - z^2})$ and $\log(1 - z)$ are holonomic (the arguments are algebraic because they satisfy the equations $(1 - y)^2 - (1 - z^2) = 0$ and $y - (1 - z) = 0$, respectively), then by part 2 also $\log(1 - z)^2$ is holonomic, and then by part 4 also $\int \log(1 - z)^2 dz$ is holonomic. Finally, using once more part 2 it follows that $y(z)$ is holonomic.

By a similar reasoning, it is clear by inspection that

$$a_n = \sum_{k=1}^{n} \frac{1+2^k}{3+k^2} k! - (2n+5)! + \sum_{k=1}^{n} \frac{\sum_{i=1}^{k} \frac{\sum_{j=1}^{i} \frac{1}{j}}{i}}{k}$$

is holonomic.

By just looking at an expression, the closure properties in Theorem 4 are often sufficient to assert that some quantity is holonomic, which means that there does exist some differential equation or recurrence which has the object in question as solution. The equations can usually not be read off directly, but it is possible to compute them with computer algebra. For the two examples above, computer algebra packages for holonomic functions need virtually no time to find a differential equation for $y(z)$ of order 5 with polynomial coefficients of degree 14 and a recurrence for a_n of order 7 with polynomial coefficients of order 37.

The idea behind these algorithms is as follows. Consider for example two sequences $(a_n)_{n=0}^{\infty}$ and $(b_n)_{n=0}^{\infty}$ satisfying recurrence equations

$$a_{n+2} = u_1(n)a_{n+1} + u_0(n)a_n, \qquad b_{n+2} = v_1(n)b_{n+1} + v_0(n)b_n$$

for some known rational functions u_0, v_0, u_1, v_1. Let $(c_n)_{n=0}^{\infty}$ be the sum of these two sequences, i.e., $c_n = a_n + b_n$ for all $n \in \mathbb{N}$. Our goal is to compute a recurrence for $(c_n)_{n=0}^{\infty}$. Let us make an ansatz for a recurrence of order 4,

$$p_0(n)c_n + p_1(n)c_{n+1} + p_2(n)c_{n+2} + p_3(n)c_{n+3} + p_4(n)c_{n+4} = 0,$$

with undetermined polynomials $p_0, \ldots, p_4$. We will see in a moment that 4 is a good choice. By definition of the c_n, in order for the recurrence to hold, we must have

$$\begin{aligned} p_0(n)\big(a_n + b_n\big) + p_1(n)\big(a_{n+1} + b_{n+1}\big) + p_2(n)\big(a_{n+2} + b_{n+2}\big) \\ + p_3(n)\big(a_{n+3} + b_{n+3}\big) + p_4(n)\big(a_{n+4} + b_{n+4}\big) = 0. \end{aligned}$$

Using the known recurrences, we can reduce the higher order shifts to lower order shifts:

$$\begin{aligned} a_{n+2} &= u_1(n)a_{n+1} + u_0(n)a_n \\ a_{n+3} &= u_1(n+1)a_{n+2} + u_0(n+1)a_{n+1} \\ &= u_1(n+1)\big(u_1(n)a_{n+1} + u_0(n)a_n\big) + u_0(n+1)a_{n+1} \\ &= \big(u_1(n+1)u_1(n) + u_0(n+1)\big)a_{n+1} + u_1(n+1)u_0(n)a_n \\ a_{n+4} &= u_1(n+2)a_{n+3} + u_0(n+2)a_{n+2} \end{aligned}$$

$$\begin{aligned}
&= u_1(n+2)\big((u_1(n+1)u_1(n) + u_0(n+1))a_{n+1} + u_1(n+1)u_0(n)a_n\big)\\
&\quad + u_0(n+2)\big(u_1(n)a_{n+1} + u_0(n)a_n\big)\\
&= \big(u_0(n+2)u_1(n) + u_0(n+1)u_1(n+2) + u_1(n)u_1(n+1)u_1(n+2)\big)a_{n+1}\\
&\quad + u_0(n)\big(u_0(n+2) + u_1(n+1)u_1(n+2)\big)a_n,
\end{aligned}$$

and analogously for the shifted versions of b_n. After applying these substitutions, the ansatz for the recurrence for c_n takes the form

$$\begin{aligned}
&p_0(n)(a_n + b_n) + p_1(n)(a_{n+1} + b_{n+1})\\
&\quad + p_2(n)(\boxed{}\,a_{n+1} + \boxed{}\,a_n + \boxed{}\,b_{n+1} + \boxed{}\,b_n)\\
&\quad + p_3(n)(\boxed{}\,a_{n+1} + \boxed{}\,a_n + \boxed{}\,b_{n+1} + \boxed{}\,b_n)\\
&\quad + p_4(n)(\boxed{}\,a_{n+1} + \boxed{}\,a_n + \boxed{}\,b_{n+1} + \boxed{}\,b_n) = 0,
\end{aligned}$$

where the symbol $\boxed{}$ represents certain expressions involving the known rational functions u_0, u_1, v_0, v_1 as indicated above. Reordering the equation leads to

$$\begin{aligned}
&(p_0(n) + \boxed{}\,p_2(n) + \boxed{}\,p_3(n) + \boxed{}\,p_4(n))a_n\\
&+(p_1(n) + \boxed{}\,p_2(n) + \boxed{}\,p_3(n) + \boxed{}\,p_4(n))a_{n+1}\\
&+(p_0(n) + \boxed{}\,p_2(n) + \boxed{}\,p_3(n) + \boxed{}\,p_4(n))b_n\\
&+(p_1(n) + \boxed{}\,p_2(n) + \boxed{}\,p_3(n) + \boxed{}\,p_4(n))b_{n+1} = 0,
\end{aligned}$$

where we write again $\boxed{}$ to denote certain expressions of the u_0, u_1, v_0, v_1 which are a bit too messy to be spelled out here explicitly. This latter equation is certainly valid if we choose polynomials $p_0, \ldots, p_4$ that turn the four expressions in front of $a_n, a_{n+1}, b_n, b_{n+1}$ to zero. Such polynomials can be found by solving the linear system

$$\begin{pmatrix}
1 & 0 & \boxed{} & \boxed{} & \boxed{}\\
0 & 1 & \boxed{} & \boxed{} & \boxed{}\\
1 & 0 & \boxed{} & \boxed{} & \boxed{}\\
0 & 1 & \boxed{} & \boxed{} & \boxed{}
\end{pmatrix}
\begin{pmatrix} p_0\\ p_1\\ p_2\\ p_3\\ p_4 \end{pmatrix} = 0.$$

This is an underdetermined homogeneous linear system with four equations and five variables, so it must have a nontrivial solution vector, and the coordinates of this vector correspond do the coefficients of the recurrence we want to compute. Note that the system involves the variable n as parameter, and it has to be solved with this parameter kept symbolic.

The order 4 in the ansatz for the recurrence of $(c_n)_{n=0}^{\infty}$ was chosen such as to ensure that the resulting linear system has more variables than equations. In general, if $(a_n)_{n=0}^{\infty}$ satisfies a recurrence of order r and $(b_n)_{n=0}^{\infty}$ satisfies a recurrence of order s, the linear system obtained for constructing a recurrence for the sum will have at most $r + s$ equations, and therefore it must have a nontrivial solution as soon as we have at least $r + s + 1$ variables $p_0, \ldots, p_{r+s}$. Likewise, for the product sequence $(a_n b_n)_{n=0}^{\infty}$ a similar construction leads to a linear system with rs equations, which hence has a nontrivial solution once we supply at least $rs + 1$ variables $p_0, \ldots, p_{rs}$. The arguments for the other operations listed in Theorem 4 are similar.

Holonomic closure properties are not only interesting for finding appropriate holonomic descriptions of objects that are given in some other form. They can also be used for proving identities. If two holonomic objects A and B are given in some form, it may not be obvious at first glance whether they are actually equal. Using closure properties, we can compute a recurrence for $A - B$ (or, if A and B depend on a continuous variable, a recurrence for the Taylor coefficients of $A - B$ by way of Theorem 1). Then if the identity is valid for a certain finite number of initial values, it follows by induction that it is true.

Example 5. Consider the following identity for Hermite polynomials. We regard it as a (formal) power series with respect to t, where x and y are viewed as constant parameters. In the first term on the left the expression $H_n(x)H_n(y)\frac{1}{n!}$ is regarded as a sequence in the discrete variable n, with x and y as parameters. Apply the closure properties algorithms as indicated by the braces to obtain a linear recurrence for the coefficients in the series expansion of the whole left hand side.

$$\underbrace{\underbrace{\sum_{n=0}^{\infty} \underbrace{\underbrace{H_n(x)}_{\text{rec. of ord. 2}}\underbrace{H_n(y)}_{\text{rec. of ord. 2}}\underbrace{\frac{1}{n!}}_{\text{rec. of ord. 1}}}_{\text{rec. of order 4}} t^n}_{\text{differential equation of order 5}} - \underbrace{\underbrace{\underbrace{\frac{1}{\sqrt{1-4t^2}}}_{\text{alg.eq. of deg. 2}}}_{\text{diff.eq. of ord. 1}} \underbrace{\underbrace{\exp}_{\text{diff.eq. of ord. 1}} \underbrace{\left(\frac{4t(xy - t(x^2+y^2))}{1-4t^2}\right)}_{\text{alg.eq. of degree 1}}}_{\text{differential equation of order 1}}}_{\text{differential equation of order 1}}}_{\text{differential equation of order 5} \;\rightsquigarrow\; \text{recurrence equation of order 4}} = 0$$

If c_n denotes the coefficient of t^n in the series expansion of the left hand side, we obtain the recurrence

$$(n+4)c_{n+4} - 4xy\, c_{n+3} = 4(2n - 2x^2 - 2y^2 + 5)c_{n+2} - 16xy\, c_{n+1} + 16(n+1)c_n = 0$$

for all $n \geq 0$. Direct calculation confirms that $c_0 = c_1 = c_2 = c_3 = 0$, which together with the recurrence implies inductively that $c_n = 0$ for all $n \geq 0$. This proves the identity.

In general, once a recurrence of a holonomic sequence is known, it always suffices to check a certain finite number of initial values for being zero in order to decide whether the whole sequence is zero. The number of terms needed is the maximum of the order of the recurrence and the largest integer root of the leading coefficient (if there is any such root). Although integer roots in the leading coefficient are not common, their possible existence must always be taken into account. It is in general not sufficient to only estimate the order of a recurrence (which could be done very quickly) without actually computing it.

The example above is a typical application of the holonomic toolkit, except that it is usually not possible to prove an identity using only Theorems 1 and 4 and the mere definitions of the objects involved. A more realistic scenario is that we want to prove an identity, say of the form $A = BC$, which involves holonomic quantities A, B, C for which we calculated defining equations using other techniques, and then the algorithms behind Theorem 4 are only used to complete the proof by combining the partial results into a defining equation for the whole equation.

Closure properties are also available in several variables. The class of D-finite functions in several (discrete or continuous) variables is closed under addition and multiplication, under linear translates $n \rightsquigarrow \lfloor \alpha n + \beta \rfloor$ of discrete variables n (for fixed positive rational numbers α, β), and under compositions $z \rightsquigarrow y(z)$ of continuous variables z by some multivariate algebraic functions y that must not be constant with respect to z, may or may not depend on the other continuous variables, and must not depend on any of the discrete variables. The underlying ideas of the algorithms is the same as in the univariate case.

Also the class of holonomic functions in several variables enjoys these closure properties, as well as some further ones which in general do not apply to D-finite functions.

Theorem 5. *[42] Let* $a = a(n_1, \ldots, n_p, z_1, \ldots, z_q)$ *and* $b = b(n_1, \ldots, n_p, z_1, \ldots, q)$ *be holonomic with respect to the discrete variables* $n_1, \ldots, n_p$ *and the continuous variables* $z_1, \ldots, z_q$. *Then:*

1. *The sum* $a + b$ *and the product* ab *are holonomic,*
2. *If* b *is algebraic, not constant with respect to* z_1, *and independent of* $n_1, \ldots, n_p$, *then* $a(n_1, \ldots, n_p, b, z_2, \ldots, z_q)$ *is holonomic,*
3. $a(\lfloor \alpha n_1 + \beta \rfloor, n_2, \ldots, n_p, z_1, \ldots, z_q)$ *is holonomic for any fixed positive constants* $\alpha, \beta \in \mathbb{Q}$.
4. $a(0, n_2, \ldots, n_p, z_1, \ldots, z_q)$ *and* $a(n_1, \ldots, n_p, 0, z_2, \ldots, z_q)$ *are holonomic,*
5. $\sum_{k=0}^{n_1} a(k, n_2, \ldots, n_p, z_1, \ldots, z_q)$ *and* $\int_0^{z_1} a(n_1, \ldots, n_p, t, z_2, \ldots, z_q)dt$ *are holonomic (provided the integral converges),*
6. $\sum_{k=-\infty}^{\infty} a(k, n_2, \ldots, n_p, z_1, \ldots, z_q)$ *and* $\int_{-\infty}^{\infty} a(n_1, \ldots, n_p, t, z_2, \ldots, z_q)dt$ *are holonomic as functions in* $n_2, \ldots, n_p, z_1, \ldots, z_p$ *and* $n_1, \ldots, n_p, z_2, \ldots, z_q$, *respectively (provided these quantities are meaningful),*

This theorem is considerably more deep than Theorem 4, and the algorithms behind it are less straightforward than those sketched before for the univariate case. See the chapter on symbolic summation and integration in this volume for further information about these algorithms.

4 What Is Guessing?

We have seen that closure properties are useful for finding the holonomic representation of holonomic objects which are given in terms of holonomic functions for which defining equations are known (possibly recursively descending a nested expression). But when no such relation to known functions is available (yet), we cannot obtain defining equations in this way. We may in fact be faced with objects of which we do not know whether they are holonomic or not.

How can we check an arbitrary object for being holonomic? Of course, this question makes only sense relative to some choice of assumptions we are willing to make about how the object is "given", or more generally, what information about it we want to consider known. A very weak assumption which is almost always satisfied in practice is that we can calculate for every specific index n the nth term of the sequence (or the nth term of the power series) of interest. For example, suppose the first few terms of a sequence $(a_n)_{n=0}^{\infty}$ are known to be

5, 12, 21, 32, 45, 180, 797, 2,616, 6,837, 15,260, 30,405, 55,632, 95,261, 154,692.

How can we check whether this sequence is, say, a polynomial sequence? Strictly speaking, we cannot tell this at all without taking into account all the terms of the sequence. But from the available finite amount of data we can at least get an idea. By means of interpolation [39], we can easily compute for any tuple of $n+1$ numbers $x_0, \dots, x_n$ the (unique) polynomial p of degree at most n such that $p(0) = x_0$, $p(1) = x_1$, …, $p(n) = x_n$. For example, for the first two terms we find $p(n) = 7n + 5$, which however cannot be correct for all $n \geq 0$ because already $p(2) = 19 \neq 21$. The interpolating polynomial for the first three points is $p(n) = n^2 + 6n + 5$, which is correct (by construction) for $n = 0, 1, 2$, happens to be correct also for $n = 3$ and $n = 4$, although these values had not been used in the construction of p. However, also this polynomial cannot be correct for all $n \geq 0$ because we have $p(5) = 60 \neq 180$. Interpolation of the first six terms gives $p(n) = n^5 - 10n^4 + 35n^3 - 49n^2 + 30n + 5$ which turns out to match all the terms listed above. Of course, this does not prove that the polynomial is correct for all greater indices as well, but the more terms match, the more tempting it becomes to believe so. Interpolating polynomials based on a finite number of terms of some infinite sequence can therefore be considered as a guess for a possible description of the entire sequence, and the difference between the number of terms taken into account and the degree of the resulting interpolating polynomial can be considered as measuring the confidence of the guess (e.g., 0: no evidence, 1: somewhat reliable, 10: reasonably trustworthy, 100: almost certain).

In a similar fashion, it is also possible to come up with reliable guesses for recurrence equations possibly satisfied by some infinite sequence of which only a finite number of terms are known, or for differential equations possibly satisfied by a function of which the first few terms of the series expansion are known. To illustrate the technique, suppose we are given a sequence $(a_n)_{n=0}^{\infty}$ starting like

$$1,\ 2,\ 14,\ 106,\ 838,\ 6{,}802,\ 56{,}190,\ 470{,}010,\ 3{,}967{,}310,\ 33{,}747{,}490.$$

Let us search for a recurrence of order $r = 2$ with polynomial coefficients of degree $d = 1$, i.e., a recurrence of the form

$$(c_{0,0} + c_{0,1}n)a_n + (c_{1,0} + c_{1,1}n)a_{n+1} + (c_{2,0} + c_{2,1}n)a_{n+2} = 0$$

for constants $c_{i,j}$ yet to be determined. Since the recurrence is supposed to hold for $n = 0, \dots, 7$ (at least), we obtain the following system of linear constraints:

$$\begin{aligned}
n{=}0 &: (c_{0,0} + c_{0,1}0)1 + (c_{1,0} + c_{1,1}0)2 + (c_{2,0} + c_{2,1}0)14 = 0\\
n{=}1 &: (c_{0,0} + c_{0,1}1)2 + (c_{1,0} + c_{1,1}1)14 + (c_{2,0} + c_{2,1}1)106 = 0\\
n{=}2 &: (c_{0,0} + c_{0,1}2)14 + (c_{1,0} + c_{1,1}2)106 + (c_{2,0} + c_{2,1}2)838 = 0\\
&\ \ \vdots\\
n{=}7 &: (c_{0,0} + c_{0,1}7)470{,}010 + (c_{1,0} + c_{1,1}7)3{,}968{,}310 + (c_{2,0} + c_{2,1}7)33{,}747{,}490 = 0.
\end{aligned}$$

In other words, any choice of the $c_{i,j}$ which corresponds to a recurrence that holds for all $n \in \mathbb{N}$ must in particular correspond to recurrence that holds for $n = 0, \dots, 7$, and the choices for $c_{i,j}$ that correspond to a recurrence valid for $n = 0, \dots, 7$ are precisely the solutions of the following homogeneous linear system:

$$\begin{pmatrix}
1 & 0 & 2 & 0 & 14 & 0\\
2 & 2 & 14 & 14 & 106 & 106\\
14 & 28 & 106 & 212 & 838 & 1{,}676\\
106 & 318 & 838 & 2{,}514 & 6{,}802 & 20{,}406\\
838 & 3{,}352 & 6{,}802 & 27{,}208 & 56{,}190 & 224{,}760\\
6{,}802 & 34{,}010 & 56{,}190 & 280{,}950 & 470{,}010 & 2{,}350{,}050\\
56{,}190 & 337{,}140 & 470{,}010 & 2{,}820{,}060 & 3{,}968{,}310 & 23{,}809{,}860\\
470{,}010 & 3{,}290{,}070 & 3{,}968{,}310 & 27{,}778{,}170 & 33{,}747{,}490 & 236{,}232{,}430
\end{pmatrix}
\begin{pmatrix} c_{0,0}\\ c_{0,1}\\ c_{1,0}\\ c_{1,1}\\ c_{2,0}\\ c_{2,1} \end{pmatrix} = 0$$

This system has the solution $(0, 9, -14, -10, 2, 1)$, which means that the infinite sequence $(a_n)_{n=0}^{\infty}$ of which we were given the first 10 terms above satisfies the recurrence

$$9n\,a_n + (-14 - 10n)a_{n+1} + (2n + 1)a_{n+2} = 0,$$

at least for $n = 0, 1, \ldots, 7$. Note that the linear system had more equations than variables so that a priori we would not have expected that it has a nonzero solution at all. This makes it reasonable to guess that the recurrence we found is not just a match of the given data, but in fact a "true" recurrence, valid for all $n \in \mathbb{N}$. The reliability of such a guess can be estimated by the difference between number of variables and number of equations in the linear system (e.g. 0: already some indication, 10: convincing evidence, 100: strong evidence).

Guessing is a very popular technique in experimental mathematics, it is certainly a more widely used (and known?) part of the holonomic toolkit than the algorithms for closure properties. Several software packages provide efficient implementations of the algorithm sketched above, or of more sophisticated algorithms based on Hermite-Pade approximation [3]. Maple users can use gfun [32], Mathematica users can use the old package of Mallinger [29] or Kauers's package [23], which also supports multivariate guessing. For Axiom there is a package by Hebisch and Rubey [20]. Recent versions of these packages have no trouble processing hundreds or even thousands of terms.

Note the computational difference between the linear algebra problems for guessing and closure properties: For guessing, we solve large overdetermined systems with constant coefficients, whereas for closure properties we solve small underdetermined systems with polynomial coefficients.

Note also that if no equation can be found by guessing, then there definitely does not exist an equation of the specified order and degree. On the other hand, a guessed equation may be incorrect, although this very rarely happens in practice. The requirement that a dense overdetermined linear system should have a nontrivial solution acts as a strong filter against false guesses. In case of doubt, there are some other tests which can be applied to a guessing result to estimate how plausible it is [8].

4.1 *Trading Order for Degree*

The first step in the guessing procedure is to make a choice for the order r and the degree d of the equation to be searched. The possible choices are limited by the number N of available terms, because we want to end up with an overdetermined linear system. (An underdetermined system will always have nontrivial solutions, but these have no reason to have any significance for the infinite object from which the data sample originates.) An overdetermined system is obtained for r and d such that $(d + 2)(r + 1) \leqslant N$. The possible choices for r and d are thus the points (r, d) under a hyperbola determined by the number of available terms.

If for some point (r,d) below the hyperbola no equation is found, it may still be that there is an equation for some other point (r',d') (unless $r' \leq r$ and $d' \leq d$). An exhaustive search needs to go through all the integer points right below the hyperbola. These are only finitely many.

If an object is holonomic, it satisfies not only a single equation but infinitely many of them. First of all we can pass from any given equation to a higher degree one by simply multiplying it by n or z, respectively, and we can produce higher order equations by shifting or differentiating, respectively. This means that if there is an equation of order r and degree d, then there is also one of order r' and degree d' for every (r',d') with $r' \geq r$ and $d' \geq d$. In addition, in examples coming from applications, there usually exist further equations. A typical shape for the region of all points (r,d) for which there exists an equation of order r and degree d is shown in Fig. 2. As indicated by the curves in this figure, the equations which can be recovered with the smallest amount of data are those for which $r/d \approx 1$. In contrast, the minimal order operator tends to require the maximal number of terms. This is unfortunate because this operator is for many applications the most interesting one. Modern guessing packages [23] use an algorithm which guesses several nonminimal order recurrences (taking advantage of their small size) and construct from them a guess for the minimal order operator (which is most interesting to the user) [6, 9]. Very recent results [11, 12, 21] offer further improvements by giving precise a priori knowledge about the shape of the region where equations can be found.

Example 6. The technique explained above has been utilized for certain sequences arising in particle physics [6]. A certain quantity arising in this context (named $C^{(2)}_{2,q,C_F^3}$ in Table 5 of [6]) satisfies a recurrence of order 35 and degree 938. This is the minimal order recurrence for this sequence. In order to guess it directly, at least 33,841 terms of the sequence are needed. However, by first guessing a smaller recurrence with nonminimal order (in this case, order 51 and degree 92), it was sufficient to know 5,114 terms of the sequence. See Fig. 1 for an illustration.

4.2 Modular Techniques

A common problem in computer algebra is the intermediate growth of expressions during a calculation. In the context of holonomic functions, it is not unusual that the output of a calculation is much longer than the input, and yet the intermediate expressions can still be much longer than that, thereby causing severe time and memory problems. A classical technique in computer algebra for dealing with this problem is the use of homomorphic images (a.k.a. modular arithmetic) [22, 39]. The basic idea is that instead of solving a problem involving polynomials, one evaluates the polynomials at several points, then solves the resulting small problems, which no longer involve polynomials but only numbers, and afterwards combines the various solutions by interpolation to a solution of the original problem. To the same effect,

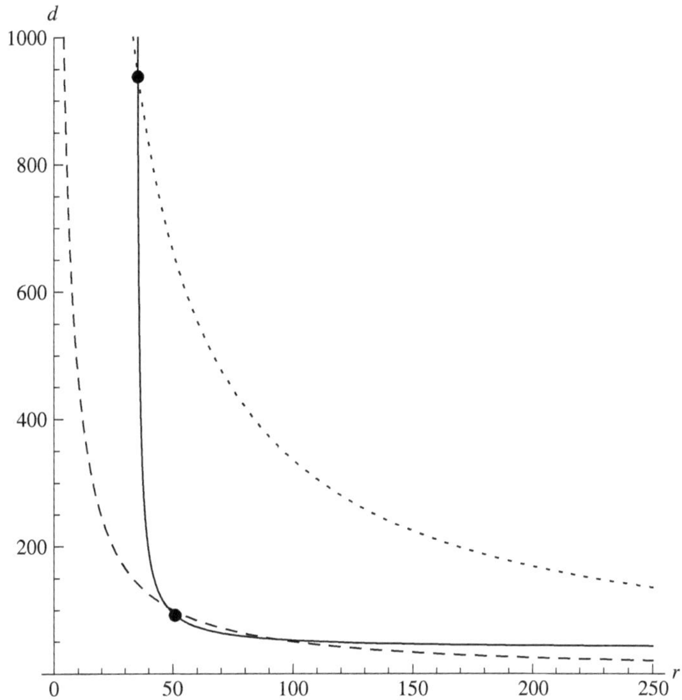

Fig. 1 For the sequence mentioned in Example 6, it turns out that there exists a recurrence of order r and degree d whenever (r, d) is above the *solid curve*. With 5,114 terms available, guessing can find recurrence equations of order r and degree d whenever (r, d) is below the *dashed curve*. With 33,841 terms, guessing can find recurrences of order r and degree d for all (r, d) below the *dotted curve*. The two *dots* mark the position of the minimal order recurrence and the recurrences which were actually guessed, and from which the minimal order recurrence was constructed

problems involving rational numbers are mapped to problems in finite fields, solved in these fields, and afterwards the modular solutions are combined to a rational solution using the Chinese remainder algorithm and rational reconstruction.

Implementations use this technique internally to speed up the computations and to save memory. The user does not see this, and does not need to care. Modular techniques are however also useful for the user, because in large problems the hard part of the computation is usually not the guessing itself but the generation of sufficiently many terms of the sequence or series. It is typically about one order of magnitude faster to compute the terms only modulo some fixed prime. Therefore, one should first compute the data only modulo some fixed prime p (for instance $p = 2^{31} - 1 = 2,147,483,647$), and then apply the guesser modulo this prime. If it does not find anything, then (with high probability) it would also not find any equation for the actual data, and there is no point in computing it.

On the other hand, if a modular equation for the modular data is found, then one can still go ahead and calculate the data modulo many other primes, reconstruct the non-modular data from the results, and apply the guesser to those in order to get the non-modular equation. This is possible, but there is a better way: calculate the data modulo several other primes, then for each prime guess an equation modulo the prime, and in the end reconstruct the non-modular equation from the modular ones. In practice, this strategy tends to require much fewer primes, and is therefore much more efficient.

Example 7. Consider the sequence $a_n = \sum_{k=0}^{n}(n+k)^3 3^{4k}\binom{n}{k}^2$. It satisfies a recurrence of order 2 and degree 4 which a guesser can recover from the first 20 terms of the sequence. The example is so small that both the computation of these twenty terms via the sum and the guessing can be done in virtually no time. The arithmetic effect described above can nevertheless be observed already here.

The term a_{20}, which is the largest in the sample, has 42 decimal digits, so if we work with primes of 11 decimal digits, we need four of them in order to reconstruct the values of $a_0, \dots, a_{20}$ from their images modulo the primes.

On the other hand, the largest coefficient in the recurrence has only 12 decimal digits. (It is a fraction with a 7-digit numerator and a 5-digit denominator.) Therefore we can already recover it if we know the coefficients of the recurrence modulo two different primes. See Fig. 2 for an illustration.

Example 8. For the sequence from Example 6, the largest among the first 5,114 terms is a fraction with 13,388 decimal digits in the numerator and 13,381 digits in the denominator. In contrast, the largest coefficient in the minimal order recurrence is a fraction with 1,187 decimal digits in the numerator and 7 digits in the denominator. The resulting speed-up is a factor of $(13,388+13,381)/(1,187+7) \approx 22.4$.

4.3 Boot Strapping

We have remarked that the generation of a sufficient amount of data is often more expensive than guessing an equation from the data. Of course, once we have an equation, it is very cheap to calculate as much data as we please—this is one of the fundamental properties of holonomic objects. But if we already know an equation, we don't need to guess one. To some extent, the situation has the character of a chicken/egg problem: in order to guess a recurrence most efficiently, the best thing would be if we could already use it for generating data. Sometimes the conflict can be resolved by guessing auxiliary equations: in a first step, use a naive way to compute a small number of terms, then use them to guess some equation which can be used to generate further terms, and iterate until you have enough terms to guess the equation of interest. We conclude with two examples for this strategy.

Example 9. [26] Consider the lattice $\mathbb{N}^4$. We are interested in walks starting at $(0,0,0,0)$ and going to (i,j,k,l) which may consist of any number of steps, where a single step can be of the form $(m,0,0,0)$ or $(0,m,0,0)$ or $(0,0,m,0)$

mod 100000000003	mod 100000000019	mod 100000000057	mod 100000000063
0	0	0	0
325	325	325	325
107896	107896	107896	107896
20619774	20619774	20619774	20619774
3180175360	3180175360	3180175360	3180175360
39281494338	39281494274	39281494122	39281494098
44150634536	44150625448	44150603864	44150600456
21005523377	21004395297	21001716107	21001293077
47849449427	47713597155	47390948009	47340003407
10040664517	94031491240	56009704700	50006264720
21985378201	66674685234	60311791810	64570282631
33735899035	69655913193	79966218922	18436302163
43166832072	5147734490	89883154498	50634811944
44406348808	99474064104	33709016642	55347756722
50094685503	23789791408	32316196105	19843824994
48089246095	80233657716	43056414134	27698338095
46630298409	55837852211	47226351528	21769511873
28184365337	25424314762	78343837230	83639568961
97965056397	76012039079	60755450198	76086130993
82688649229	13097788119	92702069386	96377809322
19000479750	73020719731	24797347614	67792308900

mod 10000000002200000000057	mod 10000000001200000003591
0	0
325	325
107896	107896
20619774	20619774
3180175360	3180175360
439281494350	439281494350
56844150636240	56844150636240
7050521005734892	7050521005734892
849076747874921728	849076747874921728
100057333113042384510	100057333113042384510
1595691831469856133143	1595691821669856129609
6025499912914500896416	6025498619314500429928
3987619360762795412890	3987472380362742409958
9655826781634081152247	9639354343828140989473
7039405589661276853185	5207872858000803723455
4174097428373312168941	1922634602838958137103
8692452790707403882121	8757614004539066331243
6892250317634951874859	1578404473178034386384
8887206360264581247194	8077821996665113985638
9184942883958236935737	6605404342457782542388
9037373502190121684807	9283417323516345218988

mod 1000000001420000000628800000085842000000204687
0
325
107896
20619774
3180175360
439281494350
56844150636240
7050521005734892
849076747874921728
100057333113042384510
11595691833669856133200
1326025500203314500903940
149983987652356362796267776
16808619659524675834176961324
1868910957450565998671929645600
206379043919577487080949672719000
22652994757462351298463229473971200
2473240118789349718278365703625872670
268742205192237274018023266433839996880
29076651329353997890401249406299092602900
3133800494309454721113304131063624227776000

mod 100000000003	mod 100000000019
0	0
325	325
107896	107896
20619774	20619774
3180175360	3180175360
39281494338	39281494274
44150634536	44150625448
21005523377	21004395297
47849449427	47713597155
10040664517	94031491240
21985378201	66674685234
33735899035	69655913193
43166832072	5147734490
44406348808	99474064104
50094685503	23789791408
48089246095	80233657716
46630298409	55837852211
28184365337	25424314762
97965056397	76012039079
82688649229	13097788119
19000479750	73020719731

↓ ↓

Guessed equation modulo 100000000003:

$$(n^4 + 12337892631n^3 + 24675785259n^2 + 12337892629n)a_{n+2} + (99999999839n^4 + 76585608333n^3 + 24849274151n^2 + 43356113667n)a_{n+1} + (6400n^4 + 62512855233n^3 + 50051350526n^2 + 12564160956n + 25025659263)a_n = 0$$

Guessed equation modulo 100000000019:

$$(n^4 + 17278992424n^3 + 34557984845n^2 + 17278992422n)a_{n+2} + (99999999855n^4 + 66245242769n^3 + 40899461546n^2 + 16817950742n)a_{n+1} + (6400n^4 + 85551511805n^3 + 42205976763n^2 + 27757443749n + 71102972391)a_n = 0$$

↓ ↓

$$(n^4 + \tfrac{260983}{104329}n^3 + \tfrac{208979}{104329}n^2 + \tfrac{52325}{104329}n)a_{n+2} - (164n^4 + \tfrac{68466146}{104329}n^3 + \tfrac{4539888}{6137}n^2 + \tfrac{17371256}{104329}n)a_{n+1} + (6400n^4 + \tfrac{3673408000}{104329}n^3 + \tfrac{7348870400}{104329}n^2 + \tfrac{6348339200}{104329}n + \tfrac{2005171200}{104329})a_n = 0$$

Fig. 2 Two ways of using modular arithmetic in guessing, illustrated with the data of Example 7

or $(0,0,0,m)$ for some positive integer m. If $a_{i,j,k,l}$ denotes the number of walks ending at the lattice point (i,j,k,l), we are interested in the sequence $a_{n,n,n,n}$ counting the walks that end on the diagonal. There is a simple way to compute these numbers, but this algorithm is costly. It would be somewhat hard to generate 1,000 terms with it.

Therefore we proceed as follows: compute the terms $a_{n,n,k,k}$ for $0 \leq n,k \leq 25$, say, and use bivariate guessing to guess recurrence equations for this sequence with respect to n and k. Using these recurrences, it is much easier to calculate $a_{n,n,n,n}$ for $n = 0,\ldots,1{,}000$. These terms can finally be used to guess the desired recurrence for $a_{n,n,n,n}$.

Example 10. (Rechnitzer, A.: Personal communication (2012)) For a certain application in combinatorial group theory, it was necessary to find a differential equation for the power series $[q^0]f(q,z)$, where $f(q,z)$ is a certain power series with respect to z whose coefficients are Laurent polynomials in q. The notation $[q^0]$ is meant to pick the constant term of each coefficient:

$$\begin{aligned} f(q,z) &= 1 + (q^{-1} + q)z + (q^{-2} + 4 + q^2)z^2 + \cdots \\ [q^0]f(q,z) &= 1 + 0z + 4z^2 + \cdots. \end{aligned}$$

The power series $f(q,z)$ was given in terms of a defining equation $p(q,z,f(q,z)) = 0$, where p is a polynomial in three variables which is too large to be printed here. Using Newton polygons [25, Chap. 6], it is possible to compute from p the first terms in the expansion of $f(q,z)$ with respect to z. But it is hard to generate enough terms to recover the differential equation for $[q^0]f(z)$.

Therefore we proceed as follows: compute the first 30 terms in the expansion of $f(q,z)$ for symbolic q, and use these to guess a general recurrence of the coefficients of $f(q,z)$ with symbolic q. Using this recurrence, generate many further terms of $f(q,z)$ with symbolic q, pick the constant terms and apply guessing to the result.

References

1. Abramowitz, M., Stegun, I.A.: Handbook of Mathematical Functions, 9th edn. Dover Publications, Inc., New York (1972)
2. Andrews, G.E., Askey, R., Roy, R.: Special Functions. Encyclopedia of Mathematics and Its Applications, vol. 71. Cambridge University Press, Cambridge/New York (1999)
3. Beckermann, B., Labahn, G.: Fraction-free computation of matrix rational interpolants and matrix gcds. SIAM J. Matrix Anal. Appl. **22**(1), 114–144 (2000)
4. Bernstein, J.N.: The analytic continuation of generalized functions with respect to a parameter. Funct. Anal. Appl. **6**(4), 26–40 (russian); 273–285 (english translation) (1972)
5. Bjork, J.E.: Rings of Differential Operators. North-Holland, Amsterdam/New York (1979)
6. Blümlein, J., Kauers, M., Klein, S., Schneider, C.: Determining the closed forms of the $o(a_s^3)$ anomalous dimension and Wilson coefficients from Mellin moments by means of computer algebra. Comput. Phys. Commun. **180**(11), 2143–2165 (2009)

7. Blümlein, J., Klein, S., Schneider, C., Stan, F.: A symbolic summation approach to Feynman integral calculus. J. Symb. Comput. **47**(10), 1267–1289 (2012)
8. Bostan, A., Kauers, M.: Automatic classification of restricted lattice walks. In: Proceedings of FPSAC'09, Hagenberg, pp. 201–215 (2009)
9. Bostan, A., Kauers, M., with an appendix by van Hoeij, M.: The complete generating function for Gessel walks is algebraic. Proc. AMS **138**(9), 3063–3078 (2010)
10. Bronstein, M., Petkovšek, M.: An introduction to pseudo-linear algebra. Theor. Comput. Sci. **157**(1), 3–33 (1996)
11. Chen, S., Kauers, M.: Order-degree curves for hypergeometric creative telescoping. In: Proceedings of ISSAC'12, Grenoble, pp. 122–129. ACM (2012). ISBN 978-1-4503-1269-1
12. Chen, S., Kauers, M.: Trading order for degree in creative telescoping. J. Symb. Comput. **47**(8), 968–995 (2012)
13. Chudnovsky, D.V., Chudnovsky, G.V.: Computer algebra in the service of mathematical physics and number theory. In: Chudnovsky, D.V., Jenks, R.D. (eds.) Computers in Mathematics. Lecture Notes in Pure and Applied Mathematics, vol. 125, pp. 109–232. Dekker/Stanford University (1986). ISBN 0-8247-8341-7
14. Chyzak, F.: An extension of Zeilberger's fast algorithm to general holonomic functions. Discret. Math. **217**, 115–134 (2000)
15. Chyzak, F., Salvy, B.: Non-commutative elimination in Ore algebras proves multivariate identities. J. Symb. Comput. **26**, 187–227 (1998)
16. Chyzak, F., Kauers, M., Salvy, B.: A non-holonomic systems approach to special function identities. In: May, J. (ed.) Proceedings of ISSAC'09, Seoul, pp. 111–118. ACM (2009). ISBN 978-1-4503-1269-1
17. Corless, R.M., Gonnet, G.H., Hare, D.E., Jeffrey, D.J., Knuth, D.E.: On the Lambert W function. Adv. Comput. Math. **5**, 329–359 (1996)
18. Flajolet, P., Sedgewick, R.: Analytic Combinatorics. Cambridge University Press, Cambridge/New York (2009)
19. Gerhold, S.: Combinatorial sequences: non-holonomicity and inequalities. Ph.D. thesis, RISC-Linz, Johannes Kepler Universität Linz (2005)
20. Hebisch, W., Rubey, M.: Extended Rate, more GFUN. J. Symb. Comput. **46**(8), 889–903 (2011)
21. Jaroschek, M., Kauers, M., Chen, S., Singer, M.F.: Desingularization explains order-degree curves for Ore operators. Technical report 1301.0917, ArXiv (2013)
22. Kauers, M.: Fast solvers for dense linear systems. Nucl. Phys. B (Proc. Suppl.) **183**, 245–250 (2008)
23. Kauers, M.: Guessing handbook. Technical report 09–07, RISC-Linz (2009)
24. Kauers, M.: A mathematica package for computing asymptotic expansions of solutions of p-finite recurrence equations. Technical report 11–04, RISC-Linz (2011)
25. Kauers, M., Paule, P.: The Concrete Tetrahedron. Springer, Wein/New York (2011)
26. Kauers, M., Zeilberger, D.: The computational challenge of enumerating high dimensional rook walks. Adv. Appl. Math. **47**(4), 813–819 (2011)
27. Koutschan, C.: Advanced applications of the holonomic systems approach. Ph.D. thesis, RISC-Linz, Johannes Kepler Universität Linz (2009)
28. Koutschan, C.: HolonomicFunctions (User's Guide). Technical report 10–01, RISC report series, University of Linz, Austria (2010). URL http://www.risc.uni-linz.ac.at/research/combinat/software/HolonomicFunctions/
29. Mallinger, C.: Algorithmic manipulations and transformations of univariate holonomic functions and sequences. Master's thesis, Johannes Kepler University, Linz (1996)
30. Mezzarobba, M., Salvy, B.: Effective bounds for P-recursive sequences. J. Symb. Comput. **45**(10), 1075–1096 (2010)
31. Salvy, B.: D-finiteness: algorithms and applications. In: Kauers, M. (ed.) Proceedings of ISSAC'05, Beijing, pp. 2–3. ACM (2005). ISBN 978-1-4503-1269-1. Invited talk
32. Salvy, B., Zimmermann, P.: Gfun: a Maple package for the manipulation of generating and holonomic functions in one variable. ACM Trans. Math. Softw. **20**(2), 163–177 (1994)

33. Sloane, N.J.: The on-line encyclopedia of integer sequences. http://www.oeis.org/
34. Stanley, R.P.: Differentiably finite power series. Eur. J. Comb. **1**, 175–188 (1980)
35. Stanley, R.P.: Enumerative Combinatorics, Volume 2. Cambridge Studies in Advanced Mathematics, vol. 62. Cambridge University Press, Cambridge (1999)
36. van der Hoeven, J.: Fast evaluation of holonomic functions. Theor. Comput. Sci. **210**(1), 199–216 (1999)
37. van der Hoeven, J.: Efficient accelero-summation of holonomic functions. J. Symb. Comput. **42**(4), 389–428 (2007)
38. van der Poorten, A.: A proof that Euler missed... – Apéry's proof of the irrationality for $\zeta(3)$. Math. Intell. **1**, 195–203 (1979)
39. von zur Gathen, J., Gerhard, J.: Modern Computer Algebra. Cambridge University Press, Cambridge/New York (1999)
40. Wilf, H.S., Zeilberger, D.: An algorithmic proof theory for hypergeometric (ordinary and q) multisum/integral identities. Invent. Math. **108**, 575–633 (1992)
41. Wimp, J., Zeilberger, D.: Resurrecting the asymptotics of linear recurrences. J. Math. Anal. Appl. **111**, 162–176 (1985)
42. Zeilberger, D.: A holonomic systems approach to special function identities. J. Comput. Appl. Math. **32**, 321–368 (1990)

Orthogonal Polynomials

Tom H. Koornwinder

Abstract This chapter gives a short introduction to orthogonal polynomials, both the general theory and some special classes. It ends with some remarks about the usage of computer algebra for this theory.

1 Introduction

This chapter gives a short introduction to orthogonal polynomials, both the general theory and some special classes. After the definition and first examples in Sect. 2, important but mainly elementary aspects of the general theory associated with the three-term recurrence relation are treated in Sect. 3. Sections 4, 6, and 7 discuss special classes of orthogonal polynomials, interrupted by Sect. 5 about Gauss quadrature. Section 8 collects some more advanced results in the general theory of orthogonal polynomials. Finally Sect. 9 discusses the role of computer algebra in the theory of (special) orthogonal polynomials.

Everything treated here is well-known from the literature. I mention a few books which can be recommended for more detailed study. A great classical introduction to orthogonal polynomials, both the general theory and the special polynomials, is Szegő [24]. A very readable textbook, in particular for the general theory, is Chihara [5]. As a textbook emphasizing the special theory I recommend Andrews, Askey and Roy [2]. Very good is also Ismail [10], but more focusing on the q-case. Two recent compendia of formulas for special orthogonal polynomials are Olver et al. [18, Chap. 18] and Koekoek, Lesky and Swarttouw [12, Chaps. 9 and 14].

T.H. Koornwinder (✉)
Korteweg-de Vries Institute, University of Amsterdam, P.O. Box 94248, 1090 GE Amsterdam, The Netherlands
e-mail: T.H.Koornwinder@uva.nl

C. Schneider and J. Blümlein (eds.), *Computer Algebra in Quantum Field Theory*, Texts & Monographs in Symbolic Computation, DOI 10.1007/978-3-7091-1616-6_6,

2 Definition of Orthogonal Polynomials and First Examples

Let $\mathscr{P}$ be the real vector space of all polynomials in one variable with real coefficients. Assume on $\mathscr{P}$ a (positive definite) inner product $\langle f, g\rangle$ ($f, g \in \mathscr{P}$). Orthogonalize the sequence of monomials $1, x, x^2, \ldots$ with respect to the inner product (Gram-Schmidt). This results into the sequence $p_0(x), p_1(x), p_2(x), \ldots$ of polynomials in x. So $p_0(x) = 1$ and, if $p_0(x), p_1(x), \ldots, p_{n-1}(x)$ are already produced and mutually orthogonal, then

$$p_n(x) := x^n - \sum_{k=0}^{n-1} \frac{\langle x^n, p_k\rangle}{\langle p_k, p_k\rangle}\, p_k(x).$$

Indeed, $p_n(x)$ is a linear combination of $1, x, x^2, \ldots, x^n$, and

$$\begin{aligned}\langle p_n, p_j\rangle &= \langle x^n, p_j\rangle - \sum_{k=0}^{n-1} \frac{\langle x^n, p_k\rangle}{\langle p_k, p_k\rangle} \langle p_k, p_j\rangle \\ &= \langle x^n, p_j\rangle - \frac{\langle x^n, p_j\rangle}{\langle p_j, p_j\rangle} \langle p_j, p_j\rangle = 0 \qquad (j = 0, 1, \ldots, n-1).\end{aligned}$$

Throughout we will use the constants h_n and k_n associated with the orthogonal system:

$$\langle p_n, p_n\rangle = h_n, \qquad p_n(x) = k_n x^n + \text{polynomial of lower degree}. \tag{1}$$

The p_n are unique up to a nonzero constant real factor. We may take them, for instance, *orthonormal* ($h_n = 1$; this determines p_n uniquely if also $k_n > 0$) or *monic* ($k_n = 1$).

In general we want

$$\langle x\, f, g\rangle = \langle f, x\, g\rangle.$$

This is true, for instance, if

$$\langle f, g\rangle := \int_a^b f(x)\, g(x)\, w(x)\, dx \qquad \text{or} \qquad \langle f, g\rangle := \sum_{j=0}^{\infty} f(x_j)\, g(x_j)\, w_j$$

for a *weight function* $w(x) \geq 0$ or for *weights* $w_j > 0$, respectively. These are special cases of an inner product

$$\langle f, g\rangle := \int_{\mathbb{R}} f(x)\, g(x)\, d\mu(x), \tag{2}$$

where μ is a (positive) *measure* on $\mathbb{R}$, namely the cases $d\mu(x) = w(x)\, dx$ on an interval I, and $\mu = \sum_{j=1}^{\infty} w_j\, \delta_{x_j}$, respectively.

A measure μ on $\mathbb{R}$ can also be thought as a *non-decreasing function* $\tilde{\mu}$ on $\mathbb{R}$. Then

$$\int_{\mathbb{R}} f(x)\,d\mu(x) = \int_{\mathbb{R}} f(x)\,d\tilde{\mu}(x) = \lim_{M\to\infty} \int_{-M}^{M} f(x)\,d\tilde{\mu}(x)$$

can be considered as a *Riemann-Stieltjes integral*. The measure μ has in x a *mass point* of *mass* $c > 0$ if the non-decreasing function $\tilde{\mu}$ has a jump c at x, i.e., if $\lim_{\delta\downarrow 0}\big(\tilde{\mu}(x+\delta) - \tilde{\mu}(x-\delta)\big) = c > 0$. The number of mass points is countable. More generally, the *support* of the measure μ consists of all $x \in \mathbb{R}$ such that $\tilde{\mu}(x+\delta) - \tilde{\mu}(x-\delta) > 0$ for all $\delta > 0$. This set $\operatorname{supp}(\mu)$ is always a closed subset of $\mathbb{R}$.

In the most general case let μ be a (positive) measure on $\mathbb{R}$ (of infinite support, i.e., not $\mu = \sum_{j=1}^{N} w_j\,\delta_{x_j}$) such that $\int_{\mathbb{R}} |x^n|\,d\mu(x) < \infty$ for all $n = 0, 1, 2, \ldots$. A system $\{p_0, p_1, p_2, \ldots\}$ obtained by orthogonalization of $\{1, x, x^2, \ldots\}$ with respect to the inner product (2) is called a system of *orthogonal polynomials* with respect to the orthogonality measure μ.

Here follow some first examples of explicit orthogonal polynomials.

- *Legendre polynomials* $P_n(x)$, orthogonal on $[-1, 1]$ with respect to the weight function 1. Normalized by $P_n(1) = 1$.
- *Hermite polynomials* $H_n(x)$, orthogonal on $(-\infty, \infty)$ with respect to the weight function e^{-x^2}. Normalized by $k_n = 2^n$.
- *Charlier polynomials* $c_n(x, a)$, orthogonal on the points $x = 0, 1, 2, \ldots$ with respect to the weights $a^x/x!$ $(a > 0)$. Normalized by $c_n(0; a) = 1$.

The h_n (see (1)) can be computed for these examples:

$$\tfrac{1}{2}\int_{-1}^{1} P_m(x)\,P_n(x)\,dx = \frac{1}{2n+1}\,\delta_{m,n}\,,$$

$$\pi^{-\frac{1}{2}}\int_{-\infty}^{\infty} H_m(x)\,H_n(x)\,e^{-x^2}\,dx = 2^n n!\,\delta_{m,n}\,,$$

$$e^{-a}\sum_{x=0}^{\infty} c_m(x,a)\,c_n(x,a)\,\frac{a^x}{x!} = a^{-n} n!\,\delta_{m,n}\,.$$

3 Three-Term Recurrence Relation and Some Consequences

3.1 Three-Term Recurrence Relation

The following theorem is fundamental for the general theory of orthogonal polynomials.

Theorem 1. *Orthogonal polynomials p_n satisfy*

$$\begin{aligned} xp_n(x) &= a_n p_{n+1}(x) + b_n p_n(x) + c_n p_{n-1}(x) \quad (n > 0), \\ xp_0(x) &= a_0 p_1(x) + b_0 p_0(x) \end{aligned} \tag{3}$$

with a_n, b_n, c_n real constants and $a_n c_{n+1} > 0$. Also $a_n = \dfrac{k_n}{k_{n+1}}$, $\dfrac{c_{n+1}}{h_{n+1}} = \dfrac{a_n}{h_n}$.

Moreover (Favard theorem), *if polynomials p_n of degree n ($n = 0, 1, 2, \ldots$) satisfy* (3) *with a_n, b_n, c_n real constants and $a_n c_{n+1} > 0$ then there exists a (positive) measure μ on $\mathbb{R}$ such that the polynomials p_n are orthogonal with respect to μ.*

The proof of the first part is easy. Indeed, $xp_n(x) = \sum_{k=0}^{n+1} \alpha_k p_k(x)$, and if $k \le n-2$ then $\langle xp_n, p_k \rangle = \langle p_n, xp_k \rangle = 0$, hence $\alpha_k = 0$. Furthermore,

$$c_{n+1} = \frac{\langle xp_{n+1}, p_n \rangle}{\langle p_n, p_n \rangle} = \frac{\langle xp_n, p_{n+1} \rangle}{h_n} = \frac{\langle xp_n, p_{n+1} \rangle}{\langle p_{n+1}, p_{n+1} \rangle} \frac{h_{n+1}}{h_n} = a_n \frac{h_{n+1}}{h_n}.$$

Hence $a_n c_{n+1} = a_n^2 h_{n+1}/h_n > 0$. Hence $c_{n+1}/h_{n+1} = a_n/h_n$.

The proof of the second part is much deeper (see Cihara [5, Chap. 2]).

Remarks

1. For orthonormal polynomials the recurrence relation (3) becomes

$$\begin{aligned} xp_n(x) &= a_n p_{n+1}(x) + b_n p_n(x) + a_{n-1} p_{n-1}(x) \quad (n > 0), \\ xp_0(x) &= a_0 p_1(x) + b_0 p_0(x), \end{aligned} \tag{4}$$

   and for monic orthogonal polynomials

$$\begin{aligned} xp_n(x) &= p_{n+1}(x) + b_n p_n(x) + c_n p_{n-1}(x) \quad (n > 0), \\ xp_0(x) &= p_1(x) + b_0 p_0(x), \end{aligned} \tag{5}$$

   with $c_n = h_n/h_{n-1} > 0$ in (5). If orthonormal polynomials p_n satisfy (4) then the corresponding monic polynomials $k_n^{-1} p_n$ satisfy (5) with $c_n = a_{n-1}^2$.
2. If the orthogonality measure is *even* ($\mu(E) = \mu(-E)$) then $p_n(-x) = (-1)^n p_n(x)$, hence $b_n = 0$, so $xp_n(x) = a_n p_{n+1}(x) + c_n p_{n-1}(x)$. Examples of orthogonal polynomials with even orthogonality measure are the Legendre and Hermite polynomials.
3. The recurrence relation (3) determines the polynomials p_n uniquely (up to a constant factor because of the choice of the constant p_0).
4. The orthogonality measure μ for a system of orthogonal polynomials may not be unique (up to a constant positive factor). See Example 1.

5. If μ is unique then $\mathscr{P}$ is dense in $L^2(\mu)$. See Shohat and Tamarkin [21, Theorem 2.14].
6. If there is an orthogonality measure μ with bounded support then μ is unique. See Chihara [5, Chap. 2, Theorem 5.6].

3.2 Moments

The *moment functional* $M: p \mapsto \langle p, 1\rangle : \mathscr{P} \to \mathbb{R}$ associated with an orthogonality measure μ is already determined by the rule $M(p_n) = \langle p_n, 1\rangle = 0$ for $n > 0$. Hence M is determined (up to a constant factor) by the system of orthogonal polynomials p_n, independent of the choice of the orthogonality measure, and hence M is also determined by (3). The same is true for the inner product $\langle f, g\rangle = \langle fg, 1\rangle$ on $\mathscr{P}$.

The moment functional M is also determined by the *moments* $\mu_n := \langle x^n, 1\rangle$ $(n = 0, 1, 2, \ldots)$. The condition $a_n c_{n+1} > 0$ is equivalent to *positive definiteness* of the moments, stated as

$$\Delta_n := \det(\mu_{i+j})_{i,j=0}^{n} > 0 \quad (n = 0, 1, 2, \ldots).$$

For given moments μ_n and corresponding orthogonal polynomials p_n a positive measure μ is an orthogonality measure for the p_n iff μ is a *solution of the (Hamburger) moment problem*

$$\int_{\mathbb{R}} x^n \, d\mu(x) = \mu_n \qquad (n = 0, 1, 2, \ldots). \tag{6}$$

Uniqueness of the orthogonality measure is equivalent to uniqueness of the moment problem.

Example 1 (non-unique orthogonality measure). The following goes back to Stieltjes [23, §56]. In the easily verified formula

$$\int_{-\infty}^{\infty} e^{-u^2}(1 + C \sin(2\pi u))\, du = \pi^{1/2}$$

make a transformation of integration variable $u = \log x - \frac{1}{2}(n+1)$ and take $-1 < C < 1$. Then

$$\pi^{-\frac{1}{2}} e^{-\frac{1}{4}} \int_0^{\infty} x^n (1 + C \sin(2\pi \log x))\, e^{-\log^2 x}\, dx = e^{\frac{1}{4}n(n+2)}. \tag{7}$$

Thus a one-parameter family of measures yields moments which are independent of C. The corresponding orthogonal polynomials p_n are a special case of the *Stieltjes-Wigert polynomials* [12, §14.27]: $p_n(x) = S_n(q^{\frac{1}{2}}x; q)$ with $q = e^{-\frac{1}{2}}$, see Christiansen [6, p. 223].

It is also elementary to show that

$$\left(\sum_{k=-\infty}^{\infty} e^{-\frac{1}{4}k^2}\right)^{-1} \sum_{k=-\infty}^{\infty} e^{-\frac{1}{2}kn} e^{-\frac{1}{4}(k+1)^2} = e^{\frac{1}{4}n(n+2)}, \tag{8}$$

which means that the same moments, up to a constant factor, as in (7) are obtained with the measure $\sum_{k=-\infty}^{\infty} e^{-\frac{1}{4}(k+1)^2} \delta_{\exp(-\frac{1}{2}k)}$.

3.3 Christoffel-Darboux Formula

Let $\mathscr{P}_n$ be the space of polynomials of degree $\le n$. Let $\{p_n\}$ be a system of orthogonal polynomials with respect to the measure μ. The *Christoffel-Darboux kernel* is defined by

$$K_n(x,y) := \sum_{j=0}^{n} \frac{p_j(x)p_j(y)}{h_j}. \tag{9}$$

Then

$$(\Pi_n f)(x) := \int_{\mathbb{R}} K_n(x,y)\, f(y)\, d\mu(y)$$

defines an orthogonal projection $\Pi_n\colon \mathscr{P} \to \mathscr{P}_n$. Indeed, if $f(y) = \sum_{k=0}^{\infty} \alpha_k\, p_k(y)$ (finite sum) then

$$(\Pi_n f)(x) = \sum_{j=0}^{n} p_j(x) \sum_{k=0}^{\infty} \frac{\alpha_k}{h_j} \int_{\mathbb{R}} p_j(y)\, p_k(y)\, d\mu(y) = \sum_{j=0}^{n} \alpha_j\, p_j(x).$$

The *Christoffel-Darboux formula* for $K_n(x,y)$ given by (9) is as follows.

$$\sum_{j=0}^{n} \frac{p_j(x)p_j(y)}{h_j} = \begin{cases} \dfrac{k_n}{h_n k_{n+1}} \dfrac{p_{n+1}(x)p_n(y) - p_n(x)p_{n+1}(y)}{x-y} & (x \ne y), \quad (10) \\[2ex] \dfrac{k_n}{h_n k_{n+1}} (p_{n+1}'(x)p_n(x) - p_n'(x)p_{n+1}(x)) & (x = y). \quad (11) \end{cases}$$

For the proof of (10) note that

$$x p_j(x) = a_j\, p_{j+1}(x) + b_j\, p_j(x) + c_j\, p_{j-1}(x),$$
$$y p_j(y) = a_j\, p_{j+1}(y) + b_j\, p_j(y) + c_j\, p_{j-1}(y).$$

Hence

$$(x-y)p_j(x)p_j(y)/h_j$$
$$= \frac{a_j}{h_j}(p_{j+1}(x)p_j(y) - p_j(x)p_{j+1}(y)) - \frac{c_j}{h_j}(p_j(x)p_{j-1}(y) - p_{j-1}(x)p_j(y)).$$

Use that $c_j/h_j = a_{j-1}/h_{j-1}$. Sum from $j = 0$ to n. Use that $a_n = k_n/k_{n+1}$. This yields (10). For (11) let $y \to x$ in (10).

3.4 Zeros of Orthogonal Polynomials

Theorem 2. *Let p_n be an orthogonal polynomial of degree n. Let μ have support within the closure of the interval (a,b). Then p_n has n distinct zeros on (a,b). Furthermore, the zeros of p_n and p_{n+1} alternate.*

Proof. For the proof of the first part suppose p_n has precisely $k < n$ sign changes on (a,b) at $x_1, x_2, \ldots, x_k$. Hence, after possibly multiplying p_n by -1, we have $p_n(x)(x - x_1)\ldots(x - x_k) \geq 0$ on $[a,b]$. Hence $\int_a^b p_n(x)$ $(x - x_1)\ldots(x - x_k)\,d\mu(x) > 0$. But by orthogonality we have $\int_a^b p_n(x)$ $(x - x_1)\ldots(x - x_k)\,d\mu(x) = 0$. Contradiction.

For the proof of the second part use (11): If $k_n, k_{n+1} > 0$ then

$$p'_{n+1}(x)p_n(x) - p'_n(x)p_{n+1}(x) = \frac{h_n k_{n+1}}{k_n}\sum_{j=0}^{n}\frac{p_j(x)^2}{h_j} > 0.$$

Hence, if y, z are two successive zeros of p_{n+1} then

$$p'_{n+1}(y)p_n(y) > 0, \quad p'_{n+1}(z)p_n(z) > 0.$$

Since $p'_{n+1}(y)$ and $p'_{n+1}(z)$ have opposite signs, $p_n(y)$ and $p_n(z)$ must have opposite signs. Hence p_n must have a zero in the interval (y,z). □

3.5 Kernel Polynomials

Recall the Christoffel-Darboux formula (10). Suppose the orthogonality measure μ has support within $(-\infty, b]$ and fix $y \geq b$. Then

$$\int_{-\infty}^{b} K_n(x,y)\,x^k\,(y-x)\,d\mu(x) = y^k(y-y) = 0 \qquad (k < n).$$

Hence $x \mapsto q_n(x) := K_n(x,y)$ is an orthogonal polynomial of degree n on $(-\infty, b]$ with respect to the measure $(y-x)\,d\mu(x)$. Hence

$$q_n(x) - q_{n-1}(x) = \frac{p_n(y)}{h_n}\, p_n(x),$$

$$p_n(y)p_{n+1}(x) - p_{n+1}(y)p_n(x) = \frac{h_n k_{n+1}}{k_n}\,(x-y)q_n(x).$$

The orthogonal polynomials q_n are called *kernel polynomials*. Of course, they depend on the choice of μ and of y.

4 Very Classical Orthogonal Polynomials

These are the Jacobi, Laguerre and Hermite polynomials. They are usually called *classical orthogonal polynomials*, but I prefer to call them *very classical* and to consider all polynomials in the (q-)Askey scheme (see Sects. 6 and 7) as classical.

We will need *hypergeometric series* [2, Chap. 2]:

$${}_rF_s\left(\begin{matrix} a_1,\dots,a_r \\ b_s,\dots,b_s \end{matrix}; z\right) := \sum_{k=0}^{\infty} \frac{(a_1)_k \dots (a_r)_k}{(b_1)_k \dots (b_s)_k} \frac{z^k}{k!}, \tag{12}$$

where $(a)_k := a(a+1)\dots(a+k-1)$ for $k = 1,2,\dots$ and $(a)_0 := 1$ is the *shifted factorial*. If one of the upper parameters in (12) equals a non-positive integer $-n$ then the series terminates after the term with $k = n$.

4.1 Jacobi Polynomials

Jacobi polynomials $P_n^{(\alpha,\beta)}$ [12, §9.8] are orthogonal on $(-1,1)$ with respect to the weight function $w(x) := (1-x)^\alpha(1+x)^\beta$ $(\alpha,\beta > -1)$ and they are normalized by $P_n^{(\alpha,\beta)}(1) = (\alpha+1)_n/n!$. They can be expressed as terminating Gauss hypergeometric series:

$$\begin{aligned} P_n^{(\alpha,\beta)}(x) &= \frac{(\alpha+1)_n}{n!}\, {}_2F_1\left(\begin{matrix} -n, n+\alpha+\beta+1 \\ \alpha+1 \end{matrix}; \tfrac12(1-x)\right) \\ &= \sum_{k=0}^{n} \frac{(n+\alpha+\beta+1)_k\,(\alpha+k+1)_{n-k}}{k!\,(n-k)!} \frac{(x-1)^k}{2^k}. \end{aligned} \tag{13}$$

They satisfy (because of the orthogonality property) the symmetry $P_n^{(\alpha,\beta)}(-x) = (-1)^n P_n^{(\beta,\alpha)}(x)$. Thus we conclude (much easier than by manipulation of the hypergeometric series) that

$$ {}_2F_1\left(\begin{matrix}-n, n+\alpha+\beta+1\\ \alpha+1\end{matrix}; z\right) = \frac{(-1)^n(\beta+1)_n}{(\alpha+1)_n}\, {}_2F_1\left(\begin{matrix}-n, n+\alpha+\beta+1\\ \beta+1\end{matrix}; 1-z\right). $$

For $p_n(x) := P_n^{(\alpha,\beta)}(x)$ there is the second order differential equation

$$ (1-x^2)\,p_n''(x) + \big(\beta-\alpha-(\alpha+\beta+2)x\big)\,p_n'(x) = -n(n+\alpha+\beta+1)\,p_n(x). \tag{14} $$

This can be split up by the *shift operator relations*

$$ \frac{d}{dx}\,P_n^{(\alpha,\beta)}(x) = \tfrac{1}{2}(n+\alpha+\beta+1)\,P_{n-1}^{(\alpha+1,\beta+1)}(x), \tag{15} $$

$$ \begin{aligned} &(1-x^2)\,\frac{d}{dx}\,P_{n-1}^{(\alpha+1,\beta+1)}(x) + \big(\beta-\alpha-(\alpha+\beta+2)x\big)\,P_{n-1}^{(\alpha+1,\beta+1)}(x)\\ &= (1-x)^{-\alpha}(1+x)^{-\beta}\,\frac{d}{dx}\Big((1-x)^{\alpha+1}(1+x)^{\beta+1}\,P_{n-1}^{(\alpha+1,\beta+1)}(x)\Big)\\ &= -2n\,P_n^{(\alpha,\beta)}(x). \end{aligned} \tag{16} $$

Note that the operator d/dx acting at the left-hand side of (15) raises the parameters and lowers the degree of the Jacobi polynomial, while the operator acting at the left-hand side of (16) lowers the parameters and raises the degree. Iteration of (16) gives the *Rodrigues formula*

$$ P_n^{(\alpha,\beta)}(x) = \frac{(-1)^n}{2^n n!}\,(1-x)^{-\alpha}(1+x)^{-\beta}\left(\frac{d}{dx}\right)^n\left((1-x)^{\alpha+n}(1+x)^{\beta+n}\right). $$

4.1.1 Special Cases

- *Gegenbauer* or *ultraspherical polynomials* ($\alpha = \beta = \lambda - \frac{1}{2}$):

$$ C_n^\lambda(x) := \frac{(2\lambda)_n}{(\lambda+\frac{1}{2})_n}\,P_n^{(\lambda-\frac{1}{2},\lambda-\frac{1}{2})}(x). $$

- *Legendre polynomials* ($\alpha = \beta = 0$): $P_n(x) := P_n^{(0,0)}(x)$.
- *Chebyshev polynomials* ($\alpha = \beta = \pm\frac{1}{2}$):

$$T_n(\cos\theta) := \cos(n\,\theta) = \frac{n!}{(\frac{1}{2})_n}\,P_n^{(-\frac{1}{2},-\frac{1}{2})}(\cos\theta),$$

$$U_n(\cos\theta) := \frac{\sin(n+1)\theta}{\sin\theta} = \frac{(2)_n}{(\frac{3}{2})_n}\,P_n^{(\frac{1}{2},\frac{1}{2})}(\cos\theta).$$

4.1.2 Quadratic Transformations

Since $P_{2n}^{(\alpha,\alpha)}(x)$ is an even polynomial of degree $2n$ in x, it is also a polynomial $p_n(2x^2-1)$ of degree n in $2x^2-1$. For $m \neq n$ we have

$$0 = \int_0^1 p_m(2y^2-1)\,p_n(2y^2-1)\,(1-y^2)^\alpha\,dy$$
$$= \text{const.}\int_{-1}^1 p_m(x)\,p_n(x)\,(1-x)^\alpha(1+x)^{-\frac{1}{2}}\,dx.$$

Hence

$$\frac{P_{2n}^{(\alpha,\alpha)}(x)}{P_{2n}^{(\alpha,\alpha)}(1)} = \frac{P_n^{(\alpha,-\frac{1}{2})}(2x^2-1)}{P_n^{(\alpha,-\frac{1}{2})}(1)}.$$

Similarly,

$$\frac{P_{2n+1}^{(\alpha,\alpha)}(x)}{P_{2n+1}^{(\alpha,\alpha)}(1)} = \frac{x\,P_n^{(\alpha,\frac{1}{2})}(2x^2-1)}{P_n^{(\alpha,\frac{1}{2})}(1)}.$$

Theorem 3. [5, Chap. 1, §8] *Let $\{p_n\}$ be a system of orthogonal polynomials with respect to an even weight function w on $\mathbb{R}$. Then there are systems $\{q_n\}$ and $\{r_n\}$ of orthogonal polynomials on $[0,\infty)$ with respect to weight functions $x \mapsto x^{-\frac{1}{2}}w(x^{\frac{1}{2}})$ and $x \mapsto x^{\frac{1}{2}}w(x^{\frac{1}{2}})$, respectively, such that $p_{2n}(x) = q_n(x^2)$ and $p_{2n+1}(x) = x\,r_n(x^2)$.*

4.2 Electrostatic Interpretation of Zeros

Let $p_n(x) :=$ const. $P_n^{(2p-1,2q-1)}(x) = (x-x_1)(x-x_2)\ldots(x-x_n)$ be monic Jacobi polynomials ($p,q>0$). By (14) we have

$$(1-x^2)p_n''(x) + 2(q-p-(p+q)x)p_n'(x) = -n(n+2p+2q-1)p_n(x).$$

Hence

$$0 = (1 - x_k^2) p_n''(x_k) + 2(q - p - (p+q)x_k) p_n'(x_k)$$

$$= \frac{1}{2} \frac{p_n''(x_k)}{p_n'(x_k)} + \frac{p}{x_k - 1} + \frac{q}{x_k + 1} = \sum_{j,\, j \neq k} \frac{1}{x_k - x_j} + \frac{p}{x_k - 1} + \frac{q}{x_k + 1}.$$

This can be reformulated as

$$(\nabla V)(x_1, \ldots, x_n) = 0,$$

where

$$V(y_1, \ldots, y_n) := -\sum_{i<j} \log(y_j - y_i) - p \sum_j \log(1 - y_j) - q \sum_j \log(1 + y_j)$$

is the logarithmic potential obtained from $n+2$ charges $q, 1, \ldots, 1, p$ at successive points $-1 < y_1 < \ldots < y_n < 1$. It achieves a minimum at the zeros of $P_n^{(2p-1,2q-1)}$. This result goes back to Stieltjes [22].

4.3 Laguerre Polynomials

Laguerre polynomials L_n^α [12, §9.12] are orthogonal on $[0, \infty)$ with respect to the weight function $w(x) := x^\alpha e^{-x}$ $(\alpha > -1)$. They are normalized by $L_n^\alpha(0) = (\alpha+1)_n / n!$. They can be expressed in terms of terminating confluent hypergeometric functions by

$$L_n^\alpha(x) = \frac{(\alpha+1)_n}{n!} \, {}_1F_1\left(\begin{matrix} -n \\ \alpha+1 \end{matrix} ; x \right) = \sum_{k=0}^{n} \frac{(\alpha+k+1)_{n-k}}{k!\,(n-k)!} (-x)^k. \tag{17}$$

For $p_n(x) := L_n^\alpha(x)$ there is the second order differential equation

$$x\, p_n''(x) + (\alpha + 1 - x)\, p_n'(x) = -n\, p_n(x).$$

This can be split up by the shift operator relations

$$\frac{d}{dx} L_n^\alpha(x) = -L_{n-1}^{\alpha+1}(x),$$

and

$$x\frac{d}{dx}L_{n-1}^{\alpha+1}(x)+(\alpha+1-x)L_{n-1}^{\alpha+1}(x)=x^{-\alpha}e^{x}\frac{d}{dx}\left(x^{\alpha+1}e^{-x}L_{n-1}^{\alpha+1}(x)\right)=n\,L_n^\alpha(x). \tag{18}$$

Iteration of (18) gives the Rodrigues formula

$$L_n^\alpha(x)=\frac{x^{-\alpha}e^x}{n!}\left(\frac{d}{dx}\right)^n\left(x^{n+\alpha}e^{-x}\right).$$

4.4 Hermite Polynomials

Hermite polynomials H_n [12, §9.15] are orthogonal with respect to the weight function $w(x) := e^{-x^2}$ on $\mathbb{R}$ and they are normalized by $H_n = 2^n x^n + \cdots$. They have the explicit expression

$$H_n(x)=n!\sum_{j=0}^{[n/2]}\frac{(-1)^j(2x)^{n-2j}}{j!\,(n-2j)!}. \tag{19}$$

There is the second order differential equation

$$H_n''(x)-2xH_n'(x)=-2nH_n(x).$$

This can be split up by the shift operator relations

$$H_n'(x)=2n\,H_{n-1}(x),\quad H_{n-1}'(x)-2xH_{n-1}(x)=e^{x^2}\frac{d}{dx}\left(e^{-x^2}H_{n-1}(x)\right)=-H_n(x). \tag{20}$$

Iteration of the last equality in (20) gives the Rodrigues formula

$$H_n(x)=(-1)^n e^{x^2}\left(\frac{d}{dx}\right)^n\left(e^{-x^2}\right).$$

4.5 General Method to Derive the Standard Formulas

The previous formulas can be derived by the following general method. Let (a,b) be an open interval and let $w, w_1 > 0$ be strictly positive C^1-functions on (a,b). Let $\{p_n\}$ and $\{q_n\}$ be systems of monic orthogonal polynomials on (a,b) with respect to the weight function w resp. w_1. Then under suitable boundary assumptions for w and w_1 we have

$$\int_a^b p_n'(x)\, q_{m-1}(x)\, w_1(x)\, dx = -\int_a^b p_n(x)\, w(x)^{-1} \frac{d}{dx}\big(w_1(x)\, q_{m-1}(x)\big)\, w(x)\, dx.$$

Suppose that for certain $a_n \neq 0$:

$$w(x)^{-1} \frac{d}{dx}\big(w_1(x)\, x^{n-1}\big) = -a_n\, x^n + \text{polynomial of degree} < n.$$

Then we easily derive a pair of first order differentiation formulas connecting $\{p_n\}$ and $\{q_n\}$, an eigenvalue equation for p_n involving a second order differential operator, and a formula connecting the quadratic norms for p_n and q_{n-1}:

$$p_n'(x) = n\, q_{n=1}(x), \quad w(x)^{-1} \frac{d}{dx}\left(w_1(x)\, q_{n-1}(x)\right) = -a_n\, p_n(x),$$

$$w(x)^{-1} \frac{d}{dx}\big(w_1(x)\, p_n'(x)\big) = -n a_n\, p_n(x),$$

$$n \int_a^b q_{n-1}(x)^2\, w_1(x)\, dx = a_n \int_a^b p_n(x)^2\, w(x)\, dx.$$

In particular, if we work with monic Jacobi polynomials $p_n^{(\alpha,\beta)}$, then $(a,b) = (-1,1)$, $w(x) = (1-x)^\alpha(1+x)^\beta$, $p_n(x) = p_n^{(\alpha,\beta)}(x)$, $w_1(x) = (1-x)^{\alpha+1}(1+x)^{\beta+1}$, $q_m(x) = p_m^{(\alpha+1,\beta+1)}(x)$. Then $a_n = (n+\alpha+\beta+1)$. Hence

$$\frac{d}{dx}\, p_n^{(\alpha,\beta)}(x) = n\, p_{n-1}^{(\alpha+1,\beta+1)}(x), \tag{21}$$

$$\left((1-x^2)\frac{d}{dx} + (\beta - \alpha - (\alpha+\beta+2)x)\right) p_{n-1}^{(\alpha+1,\beta+1)}(x) = -(n+\alpha+\beta+1)\, p_n^{(\alpha,\beta)}(x). \tag{22}$$

For $x = 1$ (22) yields

$$p_n^{(\alpha,\beta)}(1) = \frac{2(\alpha+1)}{n+\alpha+\beta+1}\, p_{n-1}^{(\alpha+1,\beta+1)}(1).$$

Iteration gives

$$p_n^{(\alpha,\beta)}(1) = \frac{2^n(\alpha+1)_n}{(n+\alpha+\beta+1)_n}. \tag{23}$$

So for $p_n = \text{const.}\, p_n^{(\alpha,\beta)} = k_n x^n + \cdots$ we know $p_n(1)/k_n$, independent of the normalization.

The hypergeometric series representation of Jacobi polynomials is next obtained from (21) by Taylor expansion:

$$\frac{p_n^{(\alpha,\beta)}(x)}{p_n^{(\alpha,\beta)}(1)} = \sum_{k=0}^{n} \frac{(x-1)^k}{k!} \left(\frac{d}{dx}\right)^k p_n^{(\alpha,\beta)}(x)\Big|_{x=1}$$

$$= \sum_{k=0}^{n} \frac{(x-1)^k}{k!} \frac{n!}{(n-k)!} \frac{p_{n-k}^{(\alpha+k,\beta+k)}(1)}{p_n^{(\alpha,\beta)}(1)} = {}_2F_1\left(\begin{matrix}-n, n+\alpha+\beta+1\\ \alpha+1\end{matrix}; \tfrac{1}{2}(1-x)\right).$$

The quadratic norm h_n can be obtained by iteration of

$$\int_{-1}^{1} p_n^{(\alpha,\beta)}(x)^2 \,(1-x)^\alpha (1+x)^\beta \,dx$$
$$= \frac{n}{n+\alpha+\beta+1} \int_{-1}^{1} p_{n-1}^{(\alpha+1,\beta+1)}(x)^2 \,(1-x)^{\alpha+1}(1+x)^{\beta+1}\,dx.$$

So for $p_n = \text{const.}\, p_n^{(\alpha,\beta)} = k_n x^n + \cdots$ we know h_n/k_n^2, independent of the normalization.

4.6 *Characterization Theorems*

Up to a constant factors and up to transformations $x \to ax + b$ of the argument the very classical orthogonal polynomials (Jacobi, Laguerre and Hermite) are uniquely determined as orthogonal polynomials p_n satisfying any of the following three criteria. (In fact there are more ways to characterize these polynomials, see Al-Salam [1].)

- (*Bochner theorem*) The p_n are eigenfunctions of a second order differential operator.
- The polynomials p'_{n+1} are again orthogonal polynomials.
- The polynomials are orthogonal with respect to a positive C^∞ weight function w on an open interval I and there is a polynomial X such that the *Rodrigues formula* holds on I:

$$p_n(x) = \text{const.}\, w(x)^{-1} \left(\frac{d}{dx}\right)^n \big(X(x)^n w(x)\big).$$

4.7 Limit Results

The very classical orthogonal polynomials are connected to each other by limit relations. We give these limits below for the monic versions $p_n^{(\alpha,\beta)}, \ell_n^\alpha, h_n$ of these polynomials, and on each line we give also the corresponding limit for the weight functions:

$$\alpha^{n/2} p_n^{(\alpha,\alpha)}(x/\alpha^{1/2}) \to h_n(x), \quad (1-x^2/\alpha)^\alpha \to e^{-x^2}, \quad \alpha \to \infty, \tag{24}$$

$$(-\beta/2)^n \, p_n^{(\alpha,\beta)}(1-2x/\beta) \to \ell_n^\alpha(x), \quad x^\alpha(1-x/\beta)^\beta \to x^\alpha e^{-x}, \quad \beta \to \infty, \tag{25}$$

$$\frac{\ell_n^\alpha((2\alpha)^{1/2}x+\alpha)}{(2\alpha)^{n/2}} \to h_n(x), \quad (1+(2/\alpha)^{1/2}x)^\alpha e^{-(2\alpha)^{1/2}x} \to e^{-x^2}, \quad \alpha \to \infty. \tag{26}$$

The limits of the orthogonal polynomials in (24) and (25) immediately follow from (13), (17) and (19). For various ways to prove (26) see [17, §2].

5 Gauss Quadrature

Let be given n real points $x_1 < x_2 < \ldots < x_n$. Put $p_n(x) := (x-x_1)\ldots(x-x_n)$. For $k = 1,\ldots,n$ let l_k be the unique polynomial of degree $< n$ such that $l_k(x_j) = \delta_{k,j}$ $(j = 1,\ldots,n)$. This polynomial, called the *Lagrange interpolation polynomial*, equals

$$l_k(x) = \frac{\prod_{j;\, j\neq k}(x-x_j)}{\prod_{j;\, j\neq k}(x_k-x_j)} = \frac{p_n(x)}{(x-x_k)\, p_n'(x_k)}.$$

For all polynomials r of degree $< n$ we have

$$r(x) = \sum_{k=1}^{n} r(x_k)\, l_k(x).$$

Theorem 4 (Gauss quadrature). *Let p_n be an orthogonal polynomial with respect to μ and let the l_k be the Lagrange interpolation polynomials associated with the zeros $x_1,\ldots,x_n$ of p_n. Put*

$$\lambda_k := \int_{\mathbb{R}} l_k(x)\, d\mu(x).$$

Then

$$\lambda_k = \int_{\mathbb{R}} l_k(x)^2 \, d\mu(x) > 0$$

and for all polynomials of degree $\le 2n-1$ *we have*

$$\int_{\mathbb{R}} f(x)\, d\mu(x) = \sum_{k=1}^{n} \lambda_k \, f(x_k). \tag{27}$$

Proof. Let f be a polynomial of degree $\le 2n-1$. Then for certain polynomials q and r of degree $\le n-1$ we have $f(x) = q(x)p_n(x) + r(x)$. Hence $f(x_k) = r(x_k)$ and

$$\int_{\mathbb{R}} f(x)\, d\mu(z) = \int_{\mathbb{R}} r(x)\, d\mu(x) = \sum_{k=1}^{n} r(x_k) \int_{\mathbb{R}} l_k(x)\, d\mu(x)$$

$$= \sum_{k=1}^{n} \lambda_k r(x_k) = \sum_{k=1}^{n} \lambda_k f(x_k).$$

Also

$$\lambda_k = \sum_{j=1}^{n} \lambda_j \, l_k(x_j)^2 = \int_{\mathbb{R}} l_k(x)^2 \, d\mu(x) > 0.$$

From (27) we see in particular that, for $i, j \le n-1$,

$$h_j \delta_{i,j} = \int_{\mathbb{R}} p_i(x) p_j(x)\, d\mu(x) = \sum_{k=1}^{n} \lambda_k \, p_i(x_k)\, p_j(x_k).$$

Thus the finite system $p_0, p_1, \ldots, p_{n-1}$ forms a set of orthogonal polynomials on the finite set $\{x_1, \ldots, x_n\}$ of the n zeros of p_n with respect to the weights λ_k and with quadratic norms h_j. All information about this system is already contained in the finite system of recurrence relations

$$x p_j(x) = a_j p_{j+1}(x) + b_j p_j(x) + c_j p_{j-1}(x) \quad (j = 0, 1, \ldots, n-1)$$

with $a_j c_{j+1} > 0$ $(j = 0, 1, \ldots, n-2)$. In particular, the λ_k are obtained up to a constant factor by solving the system

$$\sum_{k=1}^{n} \lambda_k p_j(x_k) = 0 \quad (j = 1, \ldots, n-1).$$

6 Askey Scheme

As an example of a finite system of orthogonal polynomials as described at the end of the previous section, consider orthogonal polynomials $p_0, p_1, \ldots, p_N$ on the zeros $0, 1, \ldots, N$ of the polynomial $p_{N+1}(x) := x(x-1)\ldots(x-N)$ with respect to nice explicit weights w_x $(x = 0, 1, \ldots, N)$ like:

- $w_x := \binom{n}{x} p^x (1-p)^{N-x}$ $(0 < p < 1)$. Then the p_n are the *Krawtchouk polynomials*

$$K_n(x; p, N) := {}_2F_1\left(\begin{matrix}-n, -x \\ -N\end{matrix}; \frac{1}{p}\right) = \sum_{k=0}^{n} \frac{(-n)_k (-x)_k}{(-N)_k\, k!} \frac{1}{p^k}.$$

- $w_x := \dfrac{(\alpha+1)_x}{x!} \dfrac{(\beta+1)_{N-x}}{(N-x)!}$ $(\alpha, \beta > -1)$. Then the p_n are the *Hahn polynomials*

$$Q_n(x; \alpha, \beta, N) := {}_3F_2\left(\begin{matrix}-n, n+\alpha+\beta+1, -x \\ \alpha+1, -N\end{matrix}; 1\right).$$

Hahn polynomials are discrete versions of Jacobi polynomials:

$$Q_n(Nx; \alpha, \beta, N) = {}_3F_2\left(\begin{matrix}-n, n+\alpha+\beta+1, -Nx \\ \alpha+1, -N\end{matrix}; 1\right) \to$$
$$ {}_2F_1\left(\begin{matrix}-n, n+\alpha+\beta+1 \\ \alpha+1\end{matrix}; x\right) = \text{const.}\ P_n^{(\alpha,\beta)}(1-2x)$$

and

$$N^{-1} \sum_{x \in \{0, \frac{1}{N}, \frac{2}{N}, \ldots, 1\}} Q_m(Nx; \alpha, \beta, N) Q_n(Nx; \alpha, \beta, N)\, w_{Nx} \to$$
$$\text{const.} \int_0^1 P_m^{(\alpha,\beta)}(1-2x) P_n^{(\alpha,\beta)}(1-2x)\, x^\alpha (1-x)^\beta\, dx.$$

Jacobi and Krawtchouk polynomials are different ways of looking at the matrix elements of the irreducible representations of SU(2), see [16]. The $3j$ coefficients or Clebsch-Gordan coefficients for SU(2) can be expressed as Hahn polynomials, see for instance [15].

While we saw that the Jacobi, Laguerre and Hermite polynomials are eigenfunctions of a second order differential operator,

$$A(x) p_n''(x) + B(x) p_n'(x) + C(x) p_n(x) = \lambda_n p_n(x), \tag{28}$$

the Hahn and Krawtchouk polynomials are examples of orthogonal polynomials p_n on $\{0, 1, \ldots, N\}$ which are eigenfunctions of a second order difference operator,

$$A(x)\,p_n(x-1) + B(x)\,p_n(x) + C(x)\,p_n(x+1) = \lambda_n\, p_n(x). \tag{29}$$

If we also allow orthogonal polynomials on the infinite set $\{0, 1, 2, \ldots\}$ then *Meixner polynomials* $M_n(x; \beta, c)$ and *Charlier polynomials* $C_n(x; a)$ appear. Here

$$M_n(x;\beta,c) := {}_2F_1\left(\begin{matrix}-n,-x\\ \beta\end{matrix};1-\frac{1}{c}\right), \quad w_x := \frac{(\beta_x)}{x!}\,c^x,$$

$$C_n(x;a) := {}_2F_0(-n,-x;;-a^{-1}), \quad w_x := a^x/x!\,.$$

If we also include orthogonal polynomials which are eigenfunctions of a second order operator as follows,

$$A(x)\,p_n(x+i) + B(x)\,p_n(x) + C(x)\,p_n(x-i) = \lambda_n\, p_n(x), \tag{30}$$

then we have collected all families of orthogonal polynomials which belong to the *Hahn class.*

Similarly, with an eigenvalue equation of the form

$$A(x)\,p_n(q(x+1)) + B(x)\,p_n(q(x)) + C(x)\,p_n(q(x-1)) = \lambda_n\, p_n(q(x)), \tag{31}$$

where q is a fixed polynomial of second degree, we obtain the orthogonal polynomials on a *quadratic lattice*. All orthogonal polynomials satisfying an equation of the form (28)–(31) have been classified. There are only 13 families, depending on at most four parameters, and all expressible as hypergeometric functions, ${}_4F_3$ in the most complicated case. They can be arranged hierarchically according to limit transitions denoted by arrows. This is the famous *Askey scheme*, see for instance [17, Fig. 1].

7 The q-Case

On top of the Askey-scheme is lying the q-Askey scheme [12, beginning of Chap. 14], from which there are also arrows to the Askey scheme as $q \to 1$. We take always $0 < q < 1$ and let $q \uparrow 1$ for the limit to the classical case. Some typical examples of q-analogues of classical concepts are (see Gasper and Rahman [9]):

- q-number: $[a]_q := \dfrac{1-q^a}{1-q} \to a$
- q-shifted factorial: $(a;q)_n := \prod_{k=0}^{n-1}(1-aq^k)$ (also for $n = \infty$), $\dfrac{(q^a;q)_k}{(1-q)^a} \to$ $(a)_k$.

- q-hypergeometric series:

$$ {}_{s+1}\phi_s\left(\begin{matrix} a_1,\ldots,a_{s+1} \\ b_1,\ldots,b_s \end{matrix};q,z\right) := \sum_{k=0}^{\infty} \frac{(a_1;q)_k \ldots (a_{s+1};q)_k}{(b_1;q)_k \ldots (b_s;q)_k} \frac{z^k}{(q;q)_k}, $$

$$ {}_{s+1}\phi_s\left(\begin{matrix} q^{a_1},\ldots,q^{a_{s+1}} \\ q^{b_1},\ldots,q^{b_s} \end{matrix};q,z\right) \to {}_{s+1}F_s\left(\begin{matrix} a_1,\ldots,a_{s+1} \\ b_1,\ldots,b_s \end{matrix};z\right). $$

- q-derivative: $(D_q f)(x) := \dfrac{f(x)-f(qx)}{(1-q)x} \to f'(x)$.
- q-integral: $\displaystyle\int_0^1 f(x)\,d_qx := (1-q)\sum_{k=0}^{\infty} f(q^k)\,q^k \to \int_0^1 f(x)\,dx$.

The q-case allows more symmetry which may be broken when taking limits for q to 1. In the elliptic case [9, Chap. 11] lying above the q-case there is even more symmetry.

On the highest level in the q-Askey scheme are the *Askey-Wilson polynomials* [3]. They are given by

$$ p_n(\cos\theta;a,b,c,d \mid q) := \frac{(ab;q)_n(ac;q)_n(ad;q)_n}{a^n} \times {}_4\phi_3\left(\begin{matrix} q^{-n}, q^{n-1}abcd, ae^{i\theta}, ae^{-i\theta} \\ ab, ac, ad \end{matrix};q,q\right), $$

and they are symmetric in the parameters a,b,c,d. For suitable restrictions on the parameters they are orthogonal with respect to an explicit weight function on $(-1,1)$. In the special case $a=-c=\beta^{\frac{1}{2}}$, $b=-d=(q\beta)^{\frac{1}{2}}$ we get the *continuous q-ultraspherical polynomials* [12, §14.10.1]. They satisfy the orthogonality relation

$$ \int_0^{\pi} C_m(\cos\theta;\beta \mid q)\,C_n(\cos\theta;\beta \mid q)\left|\frac{(e^{2i\theta};q)_\infty}{(\beta e^{2i\theta};q)_\infty}\right|^2 d\theta = 0 \qquad (m \neq n), $$

and they have the generating function

$$ \left|\frac{(\beta e^{i\theta}t;q)_\infty}{(e^{i\theta}t;q)_\infty}\right|^2 = \sum_{n=0}^{\infty} C_n(x;\beta \mid q)t^n. $$

For $q \uparrow 1$ they tend to ultraspherical (or Gegenbauer) polynomials: $C_n(x;q^\lambda \mid q) \to C_n^\lambda(x)$. The Gegenbauer polynomials have the generating function

$$ (1-2xt+t^2)^{-\lambda} = \sum_{n=0}^{\infty} C_n^\lambda(x)t^n. $$

8 Some Deeper Properties of General Orthogonal Polynomials

8.1 True Interval of Orthogonality

Consider a system of orthogonal polynomials $\{p_n\}$. Let p_n have zeros $x_{n,1} < x_{n,2} < \ldots < x_{n,n}$. Then

$$\begin{aligned} x_{l,l} > x_{l+1,l} > \ldots > x_{n,l} \downarrow \xi_l \geq -\infty, \\ x_{j,1} < x_{j+1,2} < \ldots < x_{n,n-j+1} \uparrow \eta_j \leq \infty. \end{aligned} \tag{32}$$

The closure I of the interval (ξ_1, η_1) is called the *true interval of orthogonality* of the system $\{p_n\}$. It has the following properties (see [21, p. 112]).

1. I is the smallest closed interval containing all zeros $x_{n,i}$.
2. There is an orthogonality measure μ for the $p_n(x)$ such that I is the smallest closed interval containing the support of μ.
3. If μ is any orthogonality measure for the $p_n(x)$ and J is a closed interval containing the support of μ then $I \subset J$.

8.2 Criteria for Bounded Support of Orthogonality Measure

Recall the three-term recurrence relation (5) for a system of monic orthogonal polynomials $\{p_n\}$. Let ξ_1, η_1 be as in (32). The following theorem gives criteria for the support of an orthogonality measure in terms of the behaviour of the coefficients b_n, c_n in (5) as $n \to \infty$.

Theorem 5.

1. ([5, p. 109]) *If* $\{b_n\}$ *is bounded and* $\{c_n\}$ *is unbounded then* $(\xi_1, \eta_1) = (-\infty, \infty)$.
2. ([5, Theorem 2.2]) *If* $\{b_n\}$ *and* $\{c_n\}$ *are bounded then* $[\xi_1, \eta_1]$ *is bounded.*
3. ([5, Theorem 4.5 and p. 121]) *If* $b_n \to b$ *and* $c_n \to c$ $(b, c$ *finite) then* $\operatorname{supp}(\mu)$ *is bounded with at most countably many points outside* $[b - 2\sqrt{c}, b + 2\sqrt{c}]$, *and* $b \pm 2\sqrt{c}$ *are limit points of* $\operatorname{supp}(\mu)$.

Example 2. Monic Jacobi polynomials $p_n^{(\alpha,\beta)}(x)$:

$$b_n = \frac{\beta^2 - \alpha^2}{(2n+\alpha+\beta)(2n+\alpha+\beta+2)} \to 0,$$

$$c_n = \frac{4n(n+\alpha)(n+\beta)(n+\alpha+\beta)}{(2n+\alpha+\beta-1)(2n+\alpha+\beta)^2(2n+\alpha+\beta+1)} \to \frac{1}{4}.$$

Hence $[b - 2\sqrt{c}, b + 2\sqrt{c}] = [-1, 1]$.

8.3 Criteria for Uniqueness of Orthogonality Measure

Put

$$\rho(z) := \Big(\sum_{n=0}^{\infty} |p_n(z)|^2\Big)^{-1} \quad (z \in \mathbb{C}).$$

Then $0 \le \rho(z) < \infty$. Note that $\rho(z) = 0$ iff $\sum_{n=0}^{\infty} |p_n(z)|^2$ diverges and that $\rho(z) > 0$ iff $\sum_{n=0}^{\infty} |p_n(z)|^2$ converges.

Theorem 6. ([21, pp. 49–51]) *The orthogonality measure is not unique iff $\rho(z) > 0$ for all $z \in \mathbb{C}$. Equivalently, the orthogonality measure is unique iff $\rho(z) = 0$ for some $z \in \mathbb{C}$.*

In the case of a unique orthogonality measure μ, we have $\rho(z) = 0$ for $z \in \mathbb{C}\backslash\mathbb{R}$ and $\rho(x) = \mu(\{x\})$ (the mass at x) for $x \in \mathbb{R}$, which implies that $\rho(x) \neq 0$ iff x is a mass point of μ.

In case of non-uniqueness, for each $x \in \mathbb{R}$ the largest possible mass of a measure μ at x is $\rho(x)$ and there is a measure realizing this mass at x.

Recall the moments $\mu_n := \langle x^n, 1\rangle = \int_{\mathbb{R}} x^n \, d\mu(x)$, which are uniquely determined (up to a constant factor) by the system $\{p_n\}$, and also recall the three-term recurrence relation (4) for a system of orthonormal polynomials $\{p_n\}$.

Theorem 7 (Carleman). ([21, Theorem 1.10 and pp. 47, 59]) *There is a unique orthogonality measure for the p_n if one of the following two conditions is satisfied.*

$$\text{(i)} \ \sum_{n=1}^{\infty} \mu_{2n}^{-1/(2n)} = \infty, \qquad \text{(ii)} \ \sum_{n=1}^{\infty} a_n^{-1} = \infty.$$

Example 3 (Hermite).

$$\mu_{2n} = \int_{-\infty}^{\infty} x^{2n} e^{-x^2}\, dx = \Gamma(n+\tfrac{1}{2}) \quad \text{and}$$

$$\log \Gamma(n+\tfrac{1}{2}) = n \log(n+\tfrac{1}{2}) + O(n) \quad \text{as } n \to \infty,$$

so $\mu_{2n}^{-1/(2n)} \sim (n+\frac{1}{2})^{-\frac{1}{2}}$. Hence $\sum_{n=1}^{\infty} \mu_{2n}^{-1/(2n)} = \infty$, i.e., the orthogonality measure for the Hermite polynomials is unique.

Example 4 (Laguerre). Monic Laguerre polynomials p_n satisfy

$$x p_n(x) = p_{n+1}(x) + (2n+\alpha+1) p_n(x) + n(n+\alpha) p_{n-1}(x).$$

Since $\displaystyle\sum_{n=0}^{\infty} \frac{1}{(n(n+\alpha))^{1/2}} = \infty$, the orthogonality measure is unique. Also note that

$$\frac{L_n^\alpha(0)^2}{h_n} = \frac{\Gamma(n+\alpha+1)}{\Gamma(n+1)\Gamma(\alpha+1)} \sim n^\alpha.$$

Since $\sum_{n=1}^\infty n^\alpha = \infty$ ($\alpha > -1$) we conclude once more that the orthogonality measure is unique.

Example 5 (Stieltjes-Wigert). Consider the moments μ_n given by the right-hand side of (7). Then

$$\sum_{n=1}^\infty \mu_{2n}^{-1/(2n)} = \sum_{n=1}^\infty e^{-\frac{1}{2}(n+1)} < \infty.$$

Since the corresponding moment problem is undetermined, the above inequality agrees with Theorem 7(i). Furthermore, from [12, (14.27.4)] with $q = e^{-\frac{1}{2}}$ we see that the corresponding orthonormal polynomials $p_n(x) = \text{const.}\, S_n(q^{\frac{1}{2}}x;q)$ with $q = e^{-\frac{1}{2}}$ have $a_{n-1}^2 = e^{2n}(1-e^{-\frac{1}{2}n})$, by which $\sum_{n=1}^\infty a_n^{-1} < \infty$, in agreement with Theorem 7(ii).

8.4 Orthogonal Polynomials and Continued Fractions

Let monic orthogonal polynomials p_n be recursively defined by

$$p_0(x) = 1,\ p_1(x) = x - b_0,$$
$$xp_n(x) = p_{n+1}(x) + b_n p_n(x) + c_n p_{n-1}(x)(n \geq 1,\ c_n > 0).$$

Then define monic orthogonal polynomials $p_n^{(1)}$ by

$$p_0^{(1)}(x) = 1,\ p_1^{(1)}(x) = x - b_1,$$
$$xp_n^{(1)}(x) = p_{n+1}^{(1)}(x) + b_{n+1}p_n^{(1)}(x) + c_{n+1}p_{n-1}^{(1)}(x)\ (n \geq 1).$$

They are called *first associated orthogonal polynomials* or *numerator polynomials*.

Define

$$F_1(x) := \frac{1}{x-b_0},\quad F_2(x) := \frac{1}{x-b_0-\frac{c_1}{x-b_1}},\quad F_3(x) := \frac{1}{x-b_0-\frac{c_1}{x-b_1-\frac{c_2}{x-b_2}}},$$

and recursively obtain $F_{n+1}(x)$ from $F_n(x)$ by replacing b_{n-1} by $b_{n-1} + \dfrac{c_n}{x-b_n}$. This is a *continued fraction*, which can be notated as

$$F_n(z) = \frac{1}{z - b_0 - |} \frac{|c_1}{z - b_1 - |} \cdots \frac{|c_{n-2}}{z - b_{n-2} - |} \frac{|c_{n-1}}{z - b_{n-1}}.$$

Theorem 8 (essentially due to Stieltjes). ([5, Chap. 3, §4])

$$F_n(z) = \frac{p_{n-1}^{(1)}(z)}{p_n(z)} \quad \text{and} \quad p_{n-1}^{(1)}(y) = \frac{1}{\mu_0} \int_{\mathbb{R}} \frac{p_n(y) - p_n(x)}{y - x}\, d\mu(x).$$

Theorem 9 (Markov). ([5, Chap. 3, (4.8)]) *Suppose that there is a (unique) orthogonality measure μ of bounded support for the p_n. Let $[\xi_1, \eta_1]$ be the true interval of orthogonality. Then*

$$\lim_{n\to\infty} F_n(z) = \frac{1}{\mu_0} \int_{\xi_1}^{\eta_1} \frac{d\mu(x)}{z - x},$$

uniformly on compact subsets of $\mathbb{C} \backslash [\xi_1, \eta_1]$.

8.5 Measures in Case of Non-uniqueness

Take p_n and $p_n^{(1)}$ orthonormal:

$$\begin{aligned}
&p_0(x) = 1, \quad p_1(x) = (x - b_0)/a_0,\\
&x p_n(x) = a_n p_{n+1}(x) + b_n p_n(x) + a_{n-1} p_{n-1}(x) \quad (n \ge 1),\\
&p_0^{(1)}(x) = 1, \quad p_1^{(1)}(x) = (x - b_1)/a_1,\\
&x p_n^{(1)}(x) = a_{n+1} p_{n+1}^{(1)}(x) + b_{n+1} p_n^{(1)}(x) + a_n p_{n-1}^{(1)}(x) \quad (n \ge 1),
\end{aligned}$$

where $a_n > 0$. Let $\mu_0 = 1, \ \mu_1, \mu_2, \ldots$ be the moments for the p_n. Suppose that the orthogonality measure for the p_n is not unique. Then the possible orthogonality measures are precisely the positive measures μ solving the moment problem (6). The set of these solutions is convex and weakly compact.

We will need the following entire analytic functions.

$$A(z) := z \sum_{n=0}^{\infty} p_n^{(1)}(0)\, p_n^{(1)}(z), \quad B(z) := -1 + z \sum_{n=1}^{\infty} p_{n-1}^{(1)}(0)\, p_n(z),$$

$$C(z) := 1 + z \sum_{n=1}^{\infty} p_n(0)\, p_{n-1}^{(1)}(z), \quad D(z) = z \sum_{n=0}^{\infty} p_n(0)\, p_n(z).$$

By a *Pick function* we mean a holomorphic function ϕ mapping the open upper half plane into the closed upper half plane. Let **P** denote the set of all Pick functions.

In the theorem below we will associate with a Pick function ϕ a certain measure μ_ϕ. There μ_t for $t \in \mathbb{R}$ will mean the measure μ_ϕ with ϕ the constant Pick function $z \mapsto t$, and μ_∞ will mean the measure μ_ϕ with ϕ the constant function $z \mapsto \infty$ (not a Pick function).

Theorem 10 (Nevanlinna, M. Riesz). ([21, Theorem 2.12]) *Suppose the moment problem* (6) *is undetermined. The identity*

$$\int_{\mathbb{R}} \frac{d\mu_\phi(t)}{t-z} = -\frac{A(z)\phi(z)-C(z)}{B(z)\phi(z)-D(z)} \quad (\Im z > 0)$$

gives a one-to-one correspondence $\phi \to \mu_\phi$ between $\mathbf{P} \cup \{\infty\}$ and the set of measures solving the moment problem (6).

Furthermore the measures μ_t $(t \in \mathbb{R} \cup \{\infty\})$ are precisely the extremal elements of the convex set, and also precisely the measures μ solving (6) *for which the polynomials are dense in $L^2(\mu)$. All measures μ_t are discrete. The mass points of μ_t are the zeros of the entire function $tB - D$ (or of B if $t = \infty$).*

Example 6 (Stieltjes-Wigert). The measure which gives in (8) a solution for the moment problem associated with special Stieltjes-Wigert polynomials, has a support which is almost discrete, but not completely, since 0 is a limit point of the support. Therefore (see the above theorem) this measure cannot be extremal. As observed by Christiansen at the end of [6], finding explicit extremal measures for this case seems to be completely out of reach. Since the measure in (8) is not extremal, the polynomials will not be dense in the corresponding L^2 space. Christiansen and Koelink [7, Theorem 3.5] give an explicit orthogonal system in this L^2 space which complements the orthogonal system of Stieltjes-Wigert polynomials to a complete orthogonal system.

9 Orthogonal Polynomials in Connection with Computer Algebra

Undoubtedly, computer algebra is nowadays a powerful tool which many mathematicians and physicists use in daily practice for their research, often using wide spectrum computer algebra programs like Mathematica or Maple, to which further specialized packages are possibly added. This is certainly also the case for research in orthogonal polynomials, in particular when it concerns special families. Jacobi, Laguerre and Hermite polynomials can be immediately called in Mathematica and Maple, while other polynomials in the (q-)Askey scheme can be defined by their (q-)hypergeometric series interpretation. Even more general special orthogonal polynomials can be generated by their three-term recurrence relation.

Typical kinds of computations being done are:

1. Checking a symbolic computation on computer which was first done by hand.
2. Doing a symbolic computation first on computer and then find a hopefully elegant derivation which can be written up.

3. Doing a symbolic computation on computer and then write in the paper something like: "By using Mathematica we found ...".
4. Checking general theorems, with (hopefully correct) proofs available, for special examples by computer algebra.
5. Formulating general conjectures in interaction with output of symbolic computation for special examples.
6. Trying to find a simple evaluation of a parameter dependent expression by extrapolating from outputs for special cases of the parameters.
7. Building large collections of formulas, to be made available on the internet, which are fully derived by computer algebra, and which can be made adaptive for the user.
8. Applying full force computer algebra, often using special purpose programs, for obtaining massive output which is a priori hopeless to get by hand or to be rewritten into an elegant expression.

While item 8 is common practice in high energy physics, I have little to say about this from my own experience. Concerning item 3 there may be a danger that we become lazy, and no longer look for an elegant analytic proof when the result was already obtained by computer algebra. In particular, many formulas for terminating hypergeometric series can be derived much quicker when we recognize them as orthogonal polynomials and use some orthogonality argument.

As an example of item 1, part of the formulas in the NIST handbook [18] was indeed checked by computer algebra. Concerning item 7, it is certainly a challenge for computer algebraists how much of a formula database for special functions can be produced purely by computer algebra. Current examples are CAOP [4] (maintained by Wolfram Koepf, Kassel) and DDMF [8] (maintained by Frédéric Chyzak et al. at INRIA).

The most spectacular success of computer algebra for special functions has been the Zeilberger algorithm, now already more than 20 years old. It is treated in several books: Petkovšek et al. [19], Koepf [13], and Kauers and Paule [11]. In particular, [13] contains quite a lot of examples of application of this algorithm to special orthogonal polynomials, including the discrete and the q-case.

Various applications of computer algebra to special orthogonal polynomials can be found in other chapters of the present volume.

A very desirable application of computer algebra would be to recognize from a given three-term recurrence relation with explicit, possibly still parameter dependent coefficients, whether it comes from a system of orthogonal polynomials in the (q-)Askey scheme, and if so, which system precisely. A very heuristic algorithm was implemented in the procedure Rec2ortho [20] (started by Swarttouw and maintained by the author). It is only up to the level of 2 parameters in the Askey scheme. On the other hand Koepf and Schmersau [14] give an algorithm how to go back and forth between an explicit eigenvalue equation (28) or (29) and a corresponding three-term recurrence relation with explicit coefficients.

References

1. Al-Salam, W.A.: Characterization theorems for orthogonal polynomials. In: Nevai, P. (ed.) Orthogonal Polynomials: Theory and Practice. NATO ASI Series C, vol. 294, pp. 1–24. Kluwer, Dordrecht (1990)
2. Andrews, G.E., Askey, R., Roy, R.: Special Functions. Encyclopedia of Mathematics and Its Applications, vol. 71. Cambridge University Press, Cambridge (1999)
3. Askey, R., Wilson, J.: Some basic hypergeometric orthogonal polynomials that generalize Jacobi polynomials. Mem. Am. Math. Soc. **54**(319), iv+55 (1985)
4. CAOP – Computer Algebra & Orthogonal Polynomials. http://www.caop.org/ (2012)
5. Chihara, T.S.: An introduction to orthogonal polynomials. Gordon and Breach, New York (1978); reprinted, Dover (2011)
6. Christiansen, J.S.: The moment problem associated with the Stieltjes-Wigert polynomials. J. Math. Anal. Appl. **277**, 218–245 (2003)
7. Christiansen, J.S., Koelink, E.: Self-adjoint difference operators and classical solutions to the Stieltjes-Wigert moment problem. J. Approx. Theory **140**, 1–26 (2006)
8. DDMF – Dynamic Dictionary of Mathematical Functions. http://ddmf.msr-inria.inria.fr/, version 1.9.1 (2013)
9. Gasper, G., Rahman, M.: Basic Hypergeometric Series. Encyclopedia of Mathematics and Its Applications, vol. 96, 2nd edn. Cambridge University Press, Cambridge (2004)
10. Ismail, M.E.H.: Classical and Quantum Orthogonal Polynomials in One Variable. Encyclopedia of Mathematics and Its Applications, vol. 98, 1st paperback edn. Cambridge University Press, Cambridge (2009)
11. Kauers, M., Paule, P.: The Concrete Tetrahedron. Springer, Wien/New York (2011)
12. Koekoek, R., Lesky, P.A., Swarttouw, R.F.: Hypergeometric Orthogonal Polynomials and Their q-Analogues. Springer, Berlin/London (2010)
13. Koepf, W.: Hypergeometric Summation. Vieweg, Braunschweig (1998)
14. Koepf, W., Schmersau, D.: Recurrence equations and their classical orthogonal polynomial solutions. Appl. Math. Comput. **128**, 303–327 (2002)
15. Koornwinder, T.H.: Clebsch-Gordan coefficients for SU(2) and Hahn polynomials. Nieuw Arch. Wisk. (3) **29**, 140–155 (1981)
16. Koornwinder, T.H.: Krawtchouk polynomials, a unification of two different group theoretic interpretations. SIAM J. Math. Anal. **13**, 1011–1023 (1982)
17. Koornwinder, T.H.: The Askey scheme as a four-manifold with corners. Ramanujan J. **20**, 409–439 (2009)
18. Olver, F.W.J. et al.: NIST Handbook of Mathematical Functions. Cambridge University Press, Cambridge (2010); DLMF, http://dlmf.nist.gov
19. Petkovšek, M., Wilf, H.S., Zeilberger, D.: $A = B$. A.K. Peters. http://www.math.upenn.edu/~wilf/Downld.html (1996)
20. Rec2ortho – Recurrence relation to orthogonal polynomials. http://staff.science.uva.nl/~thk/art/software/rec2ortho/ (2005)
21. Shohat, J.A., Tamarkin, J.D.: The Problem of Moments. American Mathematical Society, Providence (1943)
22. Stieltjes, T.J.: Sur les polynômes de Jacobi. C. R. Acad. Sci. Paris **100**, 620–622 (1885); Also in Stieltjes, T.J. In: van Dijk, G. (ed.) Collected Papers, vol. I. Springer (1993)
23. Stieltjes, T.J.: Recherches sur les fractions continues. Ann. Fac. Sci. Toulouse **8**, 1–122; **9**, 1–47 (1894, 1895); Also in Stieltjes, T.J. In: van Dijk, G. (ed.) Collected Papers, vol. II. Springer (1993)
24. Szegő, G.: Orthogonal Polynomials, 4th edn. American Mathematical Society, Providence (1975)

Creative Telescoping for Holonomic Functions

Christoph Koutschan

Abstract The aim of this article is twofold: on the one hand it is intended to serve as a gentle introduction to the topic of creative telescoping, from a practical point of view; for this purpose its application to several problems is exemplified. On the other hand, this chapter has the flavour of a survey article: the developments in this area during the last two decades are sketched and a selection of references is compiled in order to highlight the impact of creative telescoping in numerous contexts.

1 Introduction

The method of *creative telescoping* is a widely used paradigm in computer algebra, in order to treat symbolic sums and integrals in an algorithmic way. Its modus operandi is to derive, from an implicit description of the summand resp. integrand, e.g., in terms of recurrences or differential equations, an implicit description for the sum resp. integral. The latter can be used for proving an identity or for finding a closed form for the expression in question. Algorithms that use this idea are nowadays implemented in all major computer algebra systems. Meanwhile, they have been successfully applied to many problems from various areas of mathematics and physics, see Sect. 7 for a selection of such applications.

The key idea of creative telescoping is rather simple and works for summation problems as well as for integrals. For example, consider the problem of evaluating a sum of the form $F(n) = \sum_{k=a}^{b} f(k,n)$ for $a, b \in \mathbb{Z}$ and some bivariate sequence f. If one succeeds to find another bivariate sequence g and univariate sequences c_0 and c_1 such that the equation

C. Koutschan (✉)
Johann Radon Institute for Computational and Applied Mathematics (RICAM), Austrian Academy of Sciences (ÖAW), Linz, Austria
e-mail: christoph.koutschan@ricam.oeaw.ac.at

C. Schneider and J. Blümlein (eds.), *Computer Algebra in Quantum Field Theory*, Texts & Monographs in Symbolic Computation, DOI 10.1007/978-3-7091-1616-6_7,

$$c_1(n)f(k,n+1)+c_0(n)f(k,n)=g(k+1,n)-g(k,n) \tag{1}$$

holds, then a recurrence for the sum F is obtained by summing (1) with respect to k from a to b, and then telescoping the right-hand side:

$$c_1(n)F(n+1)+c_0(n)F(n)=g(b+1,n)-g(a,n).$$

For this reasoning to be nontrivial, one stipulates that the sequence g is given as a closed-form expression in terms of the input (this will be made precise later). Note that on the left-hand side of (1) one can have a longer linear combination of $f(k,n),\ldots,f(k,n+d)$, giving rise to a higher-order recurrence for F. This procedure works similarly for integrals, see Sect. 4 for a detailed exposition. In order to guarantee that a creative telescoping equation, like (1), exists, one requires that the summand f satisfies sufficiently many equations. This requirement leads to the concepts of *holonomic functions* and *∂-finite functions*; they will be introduced in Sect. 3.

The class of holonomic functions is quite rich and thus the method of creative telescoping applies to a wide variety of summation and integration problems. Just to give the reader an impression of this diversity, we list a random selection of identities that can be proven by the methods described in this article (where $P_n^{(a,b)}(x)$ denotes the Jacobi polynomials, $L_n^a(x)$ the Laguerre polynomials, $J_n(x)$ the Bessel function of the first kind, $H_n(x)$ the Hermite polynomials, $C_n^{(\lambda)}(x)$ the Gegenbauer polynomials, $\Gamma(n)$ the Gamma function, and $y_n(x)$ the spherical Bessel function of the second kind):

$$\sum_{k=0}^{n}\binom{n}{k}^2\binom{k+n}{k}^2=\sum_{k=0}^{n}\binom{n}{k}\binom{k+n}{k}\sum_{j=0}^{k}\binom{k}{j}^3,$$

$$\int_0^\infty \frac{1}{(x^4+2ax^2+1)^{m+1}}\,\mathrm{d}x=\frac{\pi P_m^{(m+\frac{1}{2},-m-\frac{1}{2})}(a)}{2^{m+\frac{3}{2}}(a+1)^{m+\frac{1}{2}}},$$

$$\int_0^\infty e^{-t}t^{\frac{a}{2}+n}J_a\left(2\sqrt{tx}\right)\mathrm{d}t=e^{-x}x^{a/2}n!L_n^a(x),$$

$$\sum_{n=0}^{\infty}\frac{(-t)^n y_{n-1}(z)}{n!}=\frac{1}{z}\sin\left(\sqrt{z^2+2tz}\right),$$

$$\int_{-\infty}^{\infty}\sum_{m=0}^{\infty}\sum_{n=0}^{\infty}\frac{H_m(x)H_n(x)r^m s^n e^{-x^2}}{m!n!}\,\mathrm{d}x=\sqrt{\pi}e^{2rs},$$

$$\int_{-1}^{1}\left(1-x^2\right)^{\nu-\frac{1}{2}}e^{iax}C_n^{(\nu)}(x)\,\mathrm{d}x=\frac{\pi 2^{1-\nu}i^n\Gamma(n+2\nu)a^{-\nu}J_{n+\nu}(a)}{n!\Gamma(\nu)}.$$

Further examples are discussed in Sect. 6 where we also demonstrate the usage of our Mathematica package HolonomicFunctions:

In[1]:= **<< HolonomicFunctions.m**

HolonomicFunctions package by Christoph Koutschan, RISC-Linz, Version 1.6 (12.04.2012)

For further reading, we recommend the following textbooks: the classic source for hypergeometric summation is the wonderful book [80], although Zeilberger's algorithm made it already into the second edition of Concrete Mathematics [45], as well as into its recent "algorithmic supplement" [52]. A book that is completely dedicated to hypergeometric summation is [57]. We also would like to point the reader to the excellent survey articles [24, 33, 59, 77, 100] and to the theses [32, 61] for more detailed introductions to the topic of creative telescoping in the context of holonomic functions.

2 History and Developments

The notion *creative telescoping* was first coined by van der Poorten in his essay [92] on Apéry's proof of the irrationality of $\zeta(3)$. But certainly, the underlying principle was known and used long before as an ad hoc trick to tackle sums and integrals. The most famous example is the practice of *differentiating under the integral sign*, that was made popular by Feynman in his enjoyable book "Surely You're Joking, Mr. Feynman!" [40], see also [4]. It was Zeilberger who equipped creative telescoping with a concrete well-defined meaning and connected it to an algorithmic method [99].

The seminal paper that initiated all the developments presented here is Zeilberger's 1990 *holonomic systems approach* paper [98]. It sketches an algorithmic proof theory for identities among a large class of elementary and special functions, involving summation quantifiers and integrals. The main theorems are based on the theory of D-modules [13, 38], as well as the creative telescoping algorithm which uses a general, but inefficient, elimination procedure. Therefore, it was not really suited to be applied to real problems, except from some toy examples, and was later called "the slow algorithm" by Zeilberger, see Sect. 5.1. But very quickly, one realized the big potential that lied in these ideas. Takayama designed a method that is still based on elimination, but in a more sophisticated way using modules [90], see Sect. 5.2. In the same year—we're still in 1990—more efficient creative telescoping algorithms for special cases were formulated: Zeilberger's celebrated "fast algorithm" for hypergeometric single sums [97] and its differential analogue, the Almkvist-Zeilberger algorithm for the integration of hyperexponential functions [4]. The theory on which these two algorithms are built was developed by Wilf and Zeilberger [94] and was named *WZ theory* after its inventors, who were awarded the Leroy P. Steele Prize in 1998 for this seminal work.

In the following years the main focus of research in this field concentrated on hypergeometric summation. Certain extensions [55] and optimizations [84] of Zeilberger's algorithm and its q-analogue [75] were published. The problem of dealing with multiple sums was studied in more detail [10, 29, 93], also for q-hypergeometric terms [85]. Based on estimates on the order of the output recurrence and the largest integer root of its leading coefficient, Yen derived an a priori bound for the number of instances one has to check in order to get a rigorous proof of a (q-) hypergeometric summation identity [95,96]; although these bounds are too large for real applications, this in principle allows to prove such identities by just verifying them on a finite set of special cases, without executing Zeilberger's algorithm explicitly. This bound was later improved drastically in [47]. Sharp bounds for the order of the telescoper that is computed by Zeilberger's algorithm and its q-analogue were derived in [72]. Abramov considered the question for which inputs the algorithm succeeds [2, 3].

In the late 1990s a return to the original ideas of Zeilberger started, namely to consider general holonomic functions instead of only (q-) hypergeometric/hyperexponential expressions. This development was initiated by Chyzak and Salvy [32, 35] and culminated in a generalization of Zeilberger's algorithm to holonomic functions [34] that is now known as Chyzak's algorithm, see Sect. 5.3. This work was picked up in [61] where several nontrivial applications of creative telescoping were presented. A fast but heuristic approach to the computation of creative telescoping relations for general holonomic functions was then given in [63], see Sect. 5.4.

During the last few years, a new interest in creative telescoping algorithms arose. The main motivation was to understand the complexity of such algorithms, a question that had been neglected during the two preceding decades. This research finally also led to new algorithmic ideas. A first attempt to study the complexity of creative telescoping was made in [18], but this investigation was restricted to bivariate rational functions as inputs. The problem of predicting the order and the degree of the coefficients of the output was largely solved in [27] for the hyperexponential case and in [26] for the hypergeometric case. Both articles also discuss the trading of order for degree, i.e., the option of computing an equation with lower coefficient degree at the cost of a larger order and vice versa; this trade-off can be used to reduce the complexity of the algorithms. The question of existence criteria for creative telescoping relations for mixed hypergeometric terms was answered in [30]. Concerning new creative telescoping algorithms, the use of residues for the computation of telescopers has been investigated in [28] for rational functions and in [31] for algebraic functions. Further innovations include an algorithm for hyperexponential functions based on Hermite reduction [21] and a new algorithm for rational functions [22] using the Griffiths-Dwork method.

Since our focus is on creative telescoping for holonomic functions, we mention only briefly some other settings in which this method can be realized. The first algorithm for a class of non-holonomic sequences was given in [71], where Abel-type sums were considered. An algorithm for summation of expressions involving

Stirling numbers and similar non-holonomic bivariate sequences was invented in [50]. Closure properties and creative telescoping for general non-holonomic functions were presented in [36]. In the setting of difference fields, Schneider developed a sophisticated symbolic summation theory [86] whose core again is creative telescoping. For more information on this topic we refer to the chapter [87] in this volume. Similarly, see [82] for creative telescoping in differential fields.

We have already mentioned that algorithms based on creative telescoping are part of many computer algebra systems. For example, Zeilberger's fast algorithm [97] for hypergeometric summation has been implemented in Maple [59, 80], shortly after its invention. In current Maple versions it is available by the command SumTools[Hypergeometric][Zeilberger]. Other implementations of Zeilberger's algorithm are in Mathematica [76], in Reduce [56], and in Macsyma [25]. Its differential analogue, the Almkvist-Zeilberger algorithm [4], can be called by DEtools[Zeilberger] in Maple. For the q-analogue, Zeilberger's algorithm for q-hypergeometric summation, there exist implementations in Mathematica [75, 83] and in Maple [16], see also the command QDifferenceEquations[Zeilberger] there. Packages for multiple sums have been written in Mathematica, namely MultiSum [93] for hypergeometric summands and its q-version qMultiSum [85] for q-hypergeometric multi-sums. Multiple integrals can be treated with the Maple package MultInt [91]. Finally, there are two software packages for creative telescoping of general holonomic functions, which are not restricted to (q-) hypergeometric/hyperexponential inputs, i.e., expressions satisfying first-order equations: Mgfun [32] for Maple and HolonomicFunctions [64] for Mathematica.

3 Holonomic and ∂-Finite Functions

In order to state, in an algebraic language, the concepts that are introduced in this section, and for writing mixed difference-differential equations in a concise way, the following operator notation is employed: let D_x denote the partial derivative operator with respect to x (x is then called a *continuous variable*) and S_n the forward shift operator with respect to n (n is then called a *discrete variable*); they act on a function f by

$$D_x f = \frac{\partial f}{\partial x} \quad \text{and} \quad S_n f = f\big|_{n \to n+1}.$$

They allow us to write linear homogeneous difference-differential equations in terms of operators, e.g.,

$$\frac{\partial}{\partial x} f(k,n+1,x,y) + n\frac{\partial}{\partial y} f(k,n,x,y) + xf(k+1,n,x,y) - f(k,n,x,y) = 0$$

turns into

$$(D_x S_n + n D_y + x S_k - 1) f(k,n,x,y) = 0,$$

in other words, such equations are represented by polynomials in the operator symbols D_x, S_n, etc., with coefficients in some field $\mathbb{F}$ which we assume to be of characteristic 0. Note that the polynomial ring $\mathbb{F}\langle D_x, S_n, \dots\rangle$ is not necessarily commutative, a fact that is indicated by the angle brackets. Its multiplication is subject to the rules

$$D_x \cdot a(x) = a(x) \cdot D_x + a'(x) \quad \text{and} \quad S_n \cdot a(n) = a(n+1) \cdot S_n.$$

Typically, $\mathbb{F}$ is a rational function field in the variables x, n, etc. over $\mathbb{Q}$ or over some other field $\mathbb{K}$. Such non-commutative rings of operators were introduced in [73] and are called *Ore algebras*. We use the symbol ∂ to denote an arbitrary operator symbol from an Ore algebra, so that ∂_w may stand for S_w or D_w, for example. Thus, a generic Ore algebra can be written as $\mathbb{O} = \mathbb{F}\langle \boldsymbol{\partial}_{\boldsymbol{w}}\rangle$ with, e.g., $\mathbb{F} = \mathbb{Q}(\boldsymbol{w})$, where $\boldsymbol{w} = w_1, \dots, w_\ell$ and $\boldsymbol{\partial}_{\boldsymbol{w}} = \partial_{w_1}, \dots, \partial_{w_\ell}$. We define the *annihilator* (w.r.t. some Ore algebra $\mathbb{O}$) of a function f:

$$\operatorname{Ann}_{\mathbb{O}}(f) := \{P \in \mathbb{O} \mid P(f) = 0\}.$$

It can easily be seen that $\operatorname{Ann}_{\mathbb{O}}(f)$ is a left ideal in $\mathbb{O}$. Every left ideal $I \subseteq \operatorname{Ann}_{\mathbb{O}}(f)$ is called an *annihilating ideal* for f. In the holonomic systems approach, functions are represented by annihilating ideals (plus initial values) as a data structure. When working with left ideals, we use *left Gröbner bases* [23, 49] which are an important tool for executing certain operations algorithmically (e.g., for deciding the ideal membership problem).

Definition 1. Let $\mathbb{O} = \mathbb{F}\langle \boldsymbol{\partial}_{\boldsymbol{w}}\rangle$ be an Ore algebra. A function f is called *∂-finite* or D-finite w.r.t. $\mathbb{O}$ if $\mathbb{O}/\operatorname{Ann}_{\mathbb{O}}(f)$ is a finite-dimensional $\mathbb{F}$-vector space. Its dimension is called the *rank* of f w.r.t. $\mathbb{O}$.

Example 1. Consider the family of Laguerre polynomials $L_n^a(x)$ as an example of a ∂-finite function w.r.t. $\mathbb{O} = \mathbb{Q}(n,a,x)\langle S_n, S_a, D_x\rangle$. The left ideal $I = \operatorname{Ann}_{\mathbb{O}}(L_n^a(x))$ is generated by the following three operators that can be easily obtained with the HolonomicFunctions package:

In[2]:= **Annihilator[LaguerreL[*n*, *a*, *x*], {S[*n*], S[*a*], Der[*x*]}]**

Out[2]= $\{S_a + D_x - 1, (n+1)S_n - xD_x + (-a-n+x-1), xD_x^2 + (a-x+1)D_x + n\}$

These operators represent well-known identities for Laguerre polynomials. Moreover, they are a left Gröbner basis of I with respect to the degree-lexicographic order. Thus, from the leading monomials S_a, S_n, and D_x^2, one can easily read off that the dimension of the $\mathbb{Q}(n,a,x)$-vector space $\mathbb{O}/I$ is two, in other words: $L_n^a(x)$ is ∂-finite w.r.t. $\mathbb{O}$ of rank 2.

Without proof we state the following theorem about *closure properties* of ∂-finite functions; its proof can be found in [61, Chap. 2.3]. We remark that all of them are algorithmically executable, and the algorithms work with the above mentioned data structure.

Theorem 1. *Let $\mathbb{O}$ be an Ore algebra and let f and g be ∂-finite w.r.t. $\mathbb{O}$ of rank r and s, respectively. Then*

(a) $f + g$ is ∂-finite of rank $\leq r + s$.
(b) $f \cdot g$ is ∂-finite of rank $\leq r \cdot s$.
(c) Pf is ∂-finite of rank $\leq r$ for any $P \in \mathbb{O}$.
(d) $f|_{x\to A(x,y,\ldots)}$ is ∂-finite of rank $\leq r \cdot d$ if $x, y, \ldots$ are continuous variables and if A satisfies a polynomial equation of degree d.
(e) $f|_{n\to A(m,n,\ldots)}$ is ∂-finite of rank $\leq r$ if A is an integer-linear expression in the discrete variables $m, n, \ldots$.

If we want to consider integration and summation problems, then the function in question needs to be *holonomic*, a concept that is closely related to ∂-finiteness. The precise definition is a bit technical and therefore skipped here; the interested reader can find it, e.g., in [38, 61, 98]. The closure properties for ∂-finite functions are also valid for holonomic functions. Additionally, the following theorem establishes the closure of holonomic functions with respect to sums and integrals; for its proof, we once again refer to [61, 98].

Theorem 2. *Let the function f be holonomic w.r.t. D_x (resp. S_n). Then also $\int_a^b f\,\mathrm{d}x$ (resp. $\sum_{n=a}^b f$) is holonomic.*

All holonomic functions that appear in this chapter are also ∂-finite and vice versa; therefore we will not continue to care about this subtle distinction, but only talk about holonomic functions from now on. A more elaborate introduction to holonomic and ∂-finite functions is given in the chapter [51] of this book.

4 Creative Telescoping for Holonomic Functions

In order to treat a sum of the form $F(\boldsymbol{w}) = \sum_{k=a}^b f(k, \boldsymbol{w})$ with creative telescoping, one has to find an operator P which annihilates f, i.e., $Pf = 0$, and which is of the form

$$P = T(\boldsymbol{w}, \boldsymbol{\partial}_{\boldsymbol{w}}) + (S_k - 1) \cdot C(k, \boldsymbol{w}, S_k, \boldsymbol{\partial}_{\boldsymbol{w}}) \tag{2}$$

where $\boldsymbol{\partial}_{\boldsymbol{w}}$ stands for some operators that act on the variables $\boldsymbol{w} = w_1, \ldots, w_\ell$. The operator T is called the *telescoper*, and we will refer to C as the *certificate* or *delta part*. Written as an equation, (2) turns into $-Tf(k, \boldsymbol{w}) = g(k+1, \boldsymbol{w}) - g(k, \boldsymbol{w})$ with $g(k, \boldsymbol{w}) = Cf(k, \boldsymbol{w})$, compare also with (1). With such an operator P we can immediately derive a relation for $F(\boldsymbol{w})$:

$$
\begin{aligned}
0 &= \sum_{k=a}^{b} P(k, \boldsymbol{w}, S_k, \boldsymbol{\partial}_{\boldsymbol{w}}) f(k, \boldsymbol{w}) \\
&= \sum_{k=a}^{b} T(\boldsymbol{w}, \boldsymbol{\partial}_{\boldsymbol{w}}) f(k, \boldsymbol{w}) \;+\; \sum_{k=a}^{b} \big((S_k - 1) C(k, \boldsymbol{w}, S_k, \boldsymbol{\partial}_{\boldsymbol{w}})\big) f(k, \boldsymbol{w}) \\
&= T(\boldsymbol{w}, \boldsymbol{\partial}_{\boldsymbol{w}}) \underbrace{\sum_{k=a}^{b} f(k, \boldsymbol{w})}_{F(\boldsymbol{w})} \;+\; \underbrace{\Big[C(k, \boldsymbol{w}, S_k, \boldsymbol{\partial}_{\boldsymbol{w}}) f(k, \boldsymbol{w})\Big]_{k=a}^{k=b+1}}_{\text{inhomogeneous part}}. \qquad (3)
\end{aligned}
$$

If the *inhomogeneous part* evaluates to zero then T is an annihilating operator for the sum, otherwise we get an inhomogeneous relation. In the latter case, one can homogenize it by multiplying an annihilating operator for the inhomogeneous part to T from the left. Note that in general, the summation bounds a and b may depend on $\boldsymbol{w}$ in which case some correction terms need to be added which are created when the operator T is pulled in front of the sum.

In terms of closure properties for holonomic functions, see Theorem 2, this reads as follows: the summand $f(k, \boldsymbol{w})$ is given by an annihilating ideal and the operator P must be a member of this ideal. The goal is to compute an annihilating ideal for the function $F(\boldsymbol{w})$ that is sufficiently large (to testify its holonomicity). We have seen that every operator P with the above properties yields an annihilating operator for F, so one continues to compute such creative telescoping operators until the left ideal generated by them is large enough.

Multiple sums can be done by iteratively applying the above procedure. Alternatively, one can use creative telescoping operators of the form

$$T(\boldsymbol{w}, \boldsymbol{\partial}_{\boldsymbol{w}}) + (S_{k_1} - 1) \cdot C_1(\boldsymbol{k}, \boldsymbol{w}, \boldsymbol{S}_{\boldsymbol{k}}, \boldsymbol{\partial}_{\boldsymbol{w}}) + \cdots + (S_{k_j} - 1) \cdot C_j(\boldsymbol{k}, \boldsymbol{w}, \boldsymbol{S}_{\boldsymbol{k}}, \boldsymbol{\partial}_{\boldsymbol{w}}) \quad (4)$$

where $\boldsymbol{k} = k_1, \ldots, k_j$ are the summation variables.

Similarly one derives annihilating operators for an integral $I(\boldsymbol{w}) = \int_a^b f(x, \boldsymbol{w})\,\mathrm{d}x$. In this case we look for creative telescoping operators that annihilate f and that are of the form

$$P = T(\boldsymbol{w}, \boldsymbol{\partial}_{\boldsymbol{w}}) + D_x \cdot C(x, \boldsymbol{w}, D_x, \boldsymbol{\partial}_{\boldsymbol{w}}). \qquad (5)$$

Again, it is straightforward to deduce a relation for the integral

$$
\begin{aligned}
0 &= \int_a^b P(x, \boldsymbol{w}, D_x, \boldsymbol{\partial}_{\boldsymbol{w}}) f(x, \boldsymbol{w})\,\mathrm{d}x \\
&= \int_a^b T(\boldsymbol{w}, \boldsymbol{\partial}_{\boldsymbol{w}}) f(x, \boldsymbol{w})\,\mathrm{d}x \;+\; \int_a^b \big(D_x C(x, \boldsymbol{w}, D_x, \boldsymbol{\partial}_{\boldsymbol{w}})\big) f(x, \boldsymbol{w})\,\mathrm{d}x \\
&= T(\boldsymbol{w}, \boldsymbol{\partial}_{\boldsymbol{w}}) \underbrace{\int_a^b f(x, \boldsymbol{w})\,\mathrm{d}x}_{I(\boldsymbol{w})} \;+\; \underbrace{\Big[C(x, \boldsymbol{w}, D_x, \boldsymbol{\partial}_{\boldsymbol{w}}) f(x, \boldsymbol{w})\Big]_{x=a}^{x=b}}_{\text{inhomogeneous part}} \qquad (6)
\end{aligned}
$$

which may be homogeneous or inhomogeneous, as before. Analogously to the summation case, multiple integrals can be treated iteratively or by creative telescoping operators of the form

$$T(\boldsymbol{w}, \boldsymbol{\partial}_{\boldsymbol{w}}) + D_{x_1} \cdot C_1(\boldsymbol{x}, \boldsymbol{w}, \boldsymbol{D}_{\boldsymbol{x}}, \boldsymbol{\partial}_{\boldsymbol{w}}) + \cdots + D_{x_j} \cdot C_j(\boldsymbol{x}, \boldsymbol{w}, \boldsymbol{D}_{\boldsymbol{x}}, \boldsymbol{\partial}_{\boldsymbol{w}}). \tag{7}$$

where now $\boldsymbol{x} = x_1, \ldots, x_j$ are the integration variables.

In practice it happens very often that the inhomogeneous part vanishes. The reason for that is because many sums and integrals run over *natural boundaries*. This concept is often used, e.g., in Takayama's algorithm, to argue a priori that there will be no inhomogeneous parts after telescoping. For that purpose, we define that $\sum_{k=a}^{b} f$ resp. $\int_a^b f \, \mathrm{d}x$ has natural boundaries if for any arbitrary operator $P \in \mathbb{O}$ for a suitable Ore algebra $\mathbb{O}$ the expression $\big[Pf\big]_{k=a}^{k=b+1}$ resp. $\big[Pf\big]_{x=a}^{x=b}$ evaluates to zero. Typical examples for natural boundaries are sums with finite support, or integrals over the whole real line that involve something like $\exp(-x^2)$. Likewise contour integrals along a closed path do have natural boundaries.

5 Algorithms for Computing Creative Telescoping Relations

In this section some algorithms for computing creative telescoping relations are described briefly; for a detailed exposition see [61]. We focus on algorithms that are applicable to general holonomic functions and omit those which are designed for special cases of holonomic functions—like rational, hypergeometric, or hyperexponential functions—and refer to Sect. 2 and the references given there. In the following, the summation and integration variables are denoted by $\boldsymbol{v} = v_1, \ldots, v_j$ whereas $\boldsymbol{w} = w_1, \ldots, w_\ell$ are the surviving parameters. So the most general case to consider is a holonomic function $f(\boldsymbol{v}, \boldsymbol{w})$ which has to be summed and integrated several times, thus some of the $\boldsymbol{v}$ may be discrete variables and the others continuous ones. The task is to find operators in the (given) annihilating ideal of f which can be written in the form

$$T(\boldsymbol{w}, \boldsymbol{\partial}_{\boldsymbol{w}}) + \Delta_{v_1} \cdot C_1(\boldsymbol{v}, \boldsymbol{w}, \boldsymbol{\partial}_{\boldsymbol{v}}, \boldsymbol{\partial}_{\boldsymbol{w}}) + \cdots + \Delta_{v_j} \cdot C_j(\boldsymbol{v}, \boldsymbol{w}, \boldsymbol{\partial}_{\boldsymbol{v}}, \boldsymbol{\partial}_{\boldsymbol{w}}) \tag{8}$$

where $\Delta_v = S_v - 1$ if v is a discrete variable and $\Delta_v = D_v$ if v is a continuous variable; compare also with (4) and (7).

5.1 *Zeilberger's Slow Algorithm*

In [98] Zeilberger suggested to approach holonomic sums or integrals by finding operators whose coefficients are completely free of the summation and integration variables $\boldsymbol{v}$. Once such an operator is found, it is immediate to rewrite it into the form (8) using division with remainder, since the corresponding operators ∂_v now commute with all remaining variables w and with all other operators ∂_w. The theory of holonomic D-modules answers the question whether this elimination is possible at all in an affirmative way. The same argument justifies the termination of all other algorithms described in this section. Operators that are free of some variables can be found, e.g., by a Gröbner basis computation in $\mathbb{K}(w)[v]\langle\partial_v, \partial_w\rangle$ or by ansatz and coefficient comparison. In any case, this algorithm searches for creative telescoping operators that are not as general as possible—also the certificates are free of v in contrast to what is indicated in (8)—and therefore is very slow in practice and often does not find the minimal telescoper.

5.2 Takayama's Algorithm

In order to avoid the overhead that results in a complete elimination of the v, Takayama came up with an algorithm that he termed an "infinite dimensional analog of Gröbner basis" [90]. He formulated it only in the differential setting and in a quite theoretical fashion. Chyzak and Salvy [35] later presented optimizations that are relevant in practice and extended it to the more general setting of Ore operators. Compared to Zeilberger's slow algorithm, Takayama's algorithm is faster and delivers better results, i.e., larger annihilating ideals.

The idea in a nutshell is the following: while in Zeilberger's slow algorithm first the v were eliminated and then the certificates were divided out, the order is now reversed. In Takayama's algorithm one first reduces modulo the right ideals $\partial_{v_1}\mathbb{O}, \dots, \partial_{v_J}\mathbb{O}$ and then performs the elimination of the v. The consequence is that the certificates $C_1, \dots, C_J$ are not computed at all because everything that would contribute to them is thrown away in the first step. Hence one has to assume a priori that the inhomogeneous parts vanish, e.g., in the case of natural boundaries.

There is one technical complication in this approach: one starts with a left ideal and then divides out some right ideals. After that there is no ideal structure any more and therefore, one is not allowed to multiply by either of the variables v from the left. In order to solve this problem one enlarges, at the very beginning, the set of generators of the input annihilating ideal by some of their left multiples by v-powers and, at the end, computes a Gröbner basis w.r.t. to POT ordering (position over term) in the module that is generated by the power products of v.

5.3 Chyzak's Algorithm

Chyzak presented his algorithm [34] as an extension of Zeilberger's algorithm to general holonomic functions. Like the latter, Chyzak's algorithm can only find creative telescoping operators for single sums or single integrals. Hence the goal is to find operators of the form

$$T(\mathbf{w}, \boldsymbol{\partial}_{\mathbf{w}}) + \Delta_v \cdot C(v, \mathbf{w}, \partial_v, \boldsymbol{\partial}_{\mathbf{w}}) \tag{9}$$

in the annihilating ideal $I \subseteq \mathbb{K}(v, \mathbf{w})\langle \partial_v, \boldsymbol{\partial}_{\mathbf{w}} \rangle$ of the summand or integrand $f(v, \mathbf{w})$. The idea of the algorithm is to make an ansatz with undetermined coefficients for T and C. Since we may assume that C is in normal form w.r.t. I, its ansatz is as follows:

$$C(v, \mathbf{w}, \partial_v, \boldsymbol{\partial}_{\mathbf{w}}) = c_1(v, \mathbf{w})U_1 + \cdots + c_r(v, \mathbf{w})U_r \tag{10}$$

where $U_1, \ldots, U_r$ are the monomials which cannot be reduced by I. Given a Gröbner basis for I, these are exactly the monomials under its staircase and r is the rank of f. The ansatz for T is of the form

$$T(\mathbf{w}, \boldsymbol{\partial}_{\mathbf{w}}) = t_1(\mathbf{w})\boldsymbol{\partial}_{\mathbf{w}}^{\boldsymbol{\alpha}_1} + \cdots + t_s(\mathbf{w})\boldsymbol{\partial}_{\mathbf{w}}^{\boldsymbol{\alpha}_s} \tag{11}$$

where $\boldsymbol{\alpha}_i \in \mathbb{N}^\ell$ for $1 \leq i \leq s$. The ansatz $T + \Delta_v \cdot C$ is reduced with the Gröbner basis of I which leads to a system of equations for the unknown rational functions $c_1, \ldots, c_r, t_1, \ldots, t_s$. In the summation (resp. integration) case, this is a parametrized linear first-order system of difference (resp. differential) equations in the unknown functions $c_1 \ldots, c_r$ and with parameters $t_1, \ldots, t_s$. One has to find rational function solutions of this system and for the parameters, a problem for which several algorithms exist. Finally, Chyzak's algorithm proceeds by increasing the support of T in (11) until the ansatz yields a solution; doing this in a certain systematic way guarantees that the computed telescopers form a Gröbner basis in $\mathbb{K}(\mathbf{w})\langle \boldsymbol{\partial}_{\mathbf{w}} \rangle$.

5.4 A Heuristic Approach

In [63] a variant of Chyzak's algorithm was developed that is based on a refined ansatz for the unknown rational functions $c_1, \ldots, c_r$. The motivation comes from the fact that the bottleneck in Chyzak's algorithm is to solve the coupled first-order system. The key observation is that good candidates for the denominators of the c_i can be obtained from the leading coefficients of the input Gröbner basis. Thus the ansatz (10) is refined in the following way:

$$c_i(v, \mathbf{w}) = \frac{c_{i,0}(\mathbf{w}) + c_{i,1}(\mathbf{w})v + \cdots + c_{i,e_i}(\mathbf{w})v^{e_i}}{d_i(v, \mathbf{w})}, \quad 1 \leq i \leq r,$$

where the d_i are explicit polynomials and the e_i are degree bounds for the numerator; both quantities are determined heuristically. In many examples this approach is faster than Chyzak's algorithm, but due to its heuristics it may not always succeed. Note also that this approach can be generalized to multiple sums and integrals, see Sect. 6.5.

6 Demonstration of the HolonomicFunctions Package

6.1 Differential Equations for Bivariate Hypergeometric Functions

The most studied concept in the area of special functions are hypergeometric functions, whose most prominent representative is the Gauss hypergeometric function ${}_2F_1$. We consider here the Appell hypergeometric function F_1 defined by

$$F_1(\alpha,\beta,\beta',\gamma;x,y) = \sum_{m=0}^{\infty}\sum_{n=0}^{\infty} \frac{x^m y^n (\beta)_m (\beta')_n (\alpha)_{m+n}}{m!n!(\gamma)_{m+n}} \tag{12}$$

for $|x| < 1$ and $|y| < 1$. Classical mathematical tables like [44] list systems of differential equations for such functions, e.g., entry 9.181 for the Appell functions. The nature of this example is that no closed form is desired, but a system of partial differential equations. These equations are now derived completely automatically from (12) using Takayama's algorithm.

The input for Takayama's algorithm is an annihilating ideal for the summand which is obtained by the command Annihilator. We need to introduce the shift operators S_m and S_n for the summation variables and the partial derivatives D_x and D_y since we are interested in PDEs w.r.t. x and y. The computation of the annihilating ideal is direct since the summand is hypergeometric in all discrete variables and hyperexponential in all continuous variables:

```
In[3]:= ann = Annihilator[Pochhammer[α, m + n] * Pochhammer[β, m] *
    Pochhammer[b, n] / (Pochhammer[γ, m + n] * m! * n!) * ^m * y^n,
    {S[m], S[n], Der[x], Der[y]}]
```

Out[3]= $\{yD_y - n, xD_x - m,$
$(mn+m+n^2+n\gamma+n+\gamma)S_n-(bmy+bny+by\alpha+mny+n^2y+ny\alpha),$
$(m^2+mn+m\gamma+m+n+\gamma)S_m-(m^2x+mnx+mx\alpha+mx\beta+nx\beta+x\alpha\beta)\}$

Next the double summation is performed and a Gröbner basis for the left ideal containing partial differential equations satisfied by the series F_1 is computed:

```
In[4]:= pde = Takayama[ann, {m, n}]
```

Out[4]= $\{(xy^2 - xy - y^3 + y^2)D_y^2 + (bx^2 - bx)D_x + (bxy - by^2 + xy\alpha - xy\beta +$
$xy + x\beta - x\gamma - y^2\alpha - y^2 + y\gamma)D_y + (bx\alpha - by\alpha),$
$(x - y)D_xD_y - bD_x + \beta D_y,$
$(x^3 - x^2y - x^2 + xy)D_x^2 + (bxy - by + x^2\alpha + x^2\beta + x^2 - xy\alpha - xy\beta -$
$xy - x\gamma + y\gamma)D_x + (y\beta - y^2\beta)D_y + (x\alpha\beta - y\alpha\beta)\}$

Observe that the two equations given in [44, 9.181] do not appear in the above result. To verify that they are nevertheless correct, one has to show that they are members of the derived annihilating ideal. This is achieved by reducing them with the Gröbner basis and check whether the remainder is zero:

In[5]:= **OreReduce[$(x(y-1))$**(Der[x]Der[y]) + $(y(y-1))$**Der[y]^2 + (bx)**Der[x] + $(y(\alpha+b+1)-\gamma)$**Der[y] + αb, pde]**

Out[5]= 0

On the other hand, the desired equations can be produced automatically by observing that the first is free of β' and the second does not involve β. The command FindRelation finds operators in a given annihilating ideal that satisfy certain properties, to be specified by options:

In[6]:= **FindRelation[pde, Eliminate → β]**

Out[6]= $\{(xy-x)D_xD_y + (y^2-y)D_y^2 + bxD_x + (by+y\alpha+y-\gamma)D_y + b\alpha\}$

This is precisely the form in which the first partial differential equation appears in [44] and an analogous computation yields the second one.

6.2 An Integral Involving Chebyshev Polynomials

It has been pointed out that creative telescoping does not deliver closed-form solutions. The next example demonstrates how it can be used to prove an identity, in this case the evaluation of a definite integral which appears in [44, 7.349]:

$$\int_{-1}^{1} \left(1-x^2\right)^{-1/2} T_n\left(1-x^2y\right) \mathrm{d}x = \frac{\pi}{2}\left(P_{n-1}(1-y)+P_n(1-y)\right). \tag{13}$$

Here $T_n(x)$ denotes the Chebyshev polynomials of the first kind defined by

$$T_n(x) = \cos(n \arccos x)$$

and the evaluation is given in terms of Legendre polynomials $P_n(x)$ defined by

$$P_n(x) = \frac{1}{2^n n!}\frac{\mathrm{d}^n}{\mathrm{d}x^n}\left(x^2-1\right)^n.$$

This relatively simple example is chosen not only to demonstrate Chyzak's algorithm but also to enlighten the concept of closure properties.

The starting point is the computation of an annihilating ideal for the integrand $f(n,x,y) = (1-x^2)^{-1/2}\,T_n(1-x^2y)$ in (13) which, in this instance, we will discuss in some more detail. For this purpose, recall the three-term recurrence

$$T_{n+2}(z) - 2zT_{n+1}(z) + T_n(z) = 0 \tag{14}$$

and the second-order differential equation

$$(z^2 - 1)T_n''(z) + zT_n'(z) - n^2 T_n(z) = 0 \tag{15}$$

for the Chebyshev polynomials which are both classic and well-known. The HolonomicFunctions package has these relations stored in a kind of database. Clearly, the integrand f also satisfies the recurrence (14) if z is replaced by $1-x^2y$. The same substitution is performed in (15) and considering $T_n(1-x^2y)$ as a function in y yields

$$\frac{(1-x^2y)^2-1}{x^4}\frac{\partial^2}{\partial y^2}T_n(1-x^2y) + \frac{1-x^2y}{-x^2}\frac{\partial}{\partial y}T_n(1-x^2y) - n^2 T_n(1-x^2y) = 0.$$

Multiplying with x^2 produces another annihilating operator

$$(x^2y^2 - 2y)D_y^2 + (x^2y - 1)D_y - n^2x^2$$

for the integrand f. Note that the square root term can be ignored since it is free of y. Finally, observe that

$$\begin{aligned}
\frac{df}{dx} &= \frac{-2xy}{\sqrt{1-x^2}}T_n'(1-x^2y) + \frac{x}{(1-x^2)^{3/2}}T_n(1-x^2y)\\
\frac{df}{dy} &= \frac{-x^2}{\sqrt{1-x^2}}T_n'(1-x^2y)
\end{aligned}$$

giving rise to the operator

$$xD_x - 2yD_y - \frac{x^2}{1-x^2}$$

which also annihilates f. The above ad hoc derivation of annihilating operators for a compound expression can be turned into an algorithmic method, and this is implemented in the Annihilator command:

In[7]:= **Annihilator[ChebyshevT[n, 1 − x^2y]/Sqrt[1 − x^2], {S[n], Der[x], Der[y]}]**

Out[7]= $\{(x^3 - x)D_x + (2y - 2x^2y)D_y + x^2,$
$nS_n + (x^2y^2 - 2y)D_y + (nx^2y - n),$
$(x^2y^2 - 2y)D_y^2 + (x^2y - 1)D_y - n^2x^2\}$

The above operators form a left Gröbner basis, and therefore differ slightly from the ones that were derived by hand; but the latter can be obtained as simple linear combinations of the previous ones.

Now we are ready to perform creative telescoping: we apply Chyzak's algorithm to find operators of the form $T_i + D_x C_i$ in the annihilating ideal. Our implementation returns two such operators, with the property that $\{T_1, T_2\}$ is a Gröbner basis:

In[8]:= **{{T1, T2}, {C1, C2}} = CreativeTelescoping[%, Der[x]]**

Out[8]= $$\Big\{\{(2n^2+2n)S_n + (2ny^2-4ny+y^2-2y)D_y + (2n^2y-2n^2+ny-2n),$$
$$(y^2-2y)D_y^2 + (y-2)D_y - n^2\},$$
$$\Big\{\frac{y\,(x^4y-x^2y-2x^2+2)}{x}D_y + y\,(nx^3-nx),\ \frac{x^2-1}{x}D_y\Big\}\Big\}$$

With the help of Mathematica, it is easily verified that the inhomogeneous part, see (6), vanishes:

In[9]:= **Limit[ApplyOreOperator[C1, ChebyshevT[n, 1 − x^2 y]/Sqrt[1 − x^2]], x → 1]**

Out[9]= 0

(Similar checks have to be done for the lower bound and for C_2.) It follows that T_1 and T_2 generate an annihilating ideal for the integral. For the convenience of the user, all the previous steps can be performed at once by typing a single command:

In[10]:= **Annihilator[Integrate[ChebyshevT[n, 1−x^2 y]/Sqrt[1−x^2], {x, −1, 1}], {S[n], Der[y]}]**

Out[10]= $$\{(2n^2+2n)S_n + (2ny^2-4ny+y^2-2y)D_y + (2n^2y-2n^2+ny-2n),$$
$$(y^2-2y)D_y^2 + (y-2)D_y - n^2\}$$

The next step is to compute an annihilating ideal for the right-hand side of (13). Instead of applying the Annihilator command to the expression itself which would produce an annihilating ideal of rank 4 by assertion (a) of Theorem 1, the fact that the sum of the two Legendre polynomials can be written as $Q(P_{n-1}(1-y))$ with $Q = S_n + 1$ is employed. This observation produces an annihilating ideal of rank 2, see part (c) of Theorem 1:

In[11]:= **rhs=Annihilator[ApplyOreOperator[S[n]+1, LegendreP[n−1, 1−y]], {S[n], Der[y]}]**

Out[11]= $$\{(2n^2+2n)S_n + (2ny^2-4ny+y^2-2y)D_y + (2n^2y-2n^2+ny-2n),$$
$$(y^2-2y)D_y^2 + (y-2)D_y - n^2\}$$

Finally, one realizes that the annihilating ideals for both sides of the identity coincide. The proof is completed by comparing two initial values, e.g., for $n = 0$ and $n = 1$. This has to be done by hand (of course, with the help of the computer algebra system), but is not part of the functionality of the HolonomicFunctions package.

6.3 A q-Holonomic Summation Problem from Knot Theory

The colored Jones function is a powerful knot invariant; it is a q-holonomic sequence of Laurent polynomials [42]. Its recurrence equation is of interest since it seems to be closely related with the A-polynomial of a knot. The recurrence for the colored Jones function $J_{7_4,n}(q)$ of the knot 7_4 was derived in [41] using creative telescoping, starting from the sum representation

$$J_{7_4,n}(q) = \sum_{k=0}^{n-1} (-1)^k \big(c_k(q)\big)^2 q^{-kn-\frac{k(k+3)}{2}} (q^{n-1};q^{-1})_k (q^{n+1};q)_k \tag{16}$$

where $(x;q)_n$ denotes the q-Pochhammer symbol defined as $\prod_{j=0}^{n-1}(1-xq^j)$ and where the sequence $c_k(q)$ satisfies a second-order recurrence:

$$c_{k+2}(q)+(q^{k+3}+q^{k+4}-q^{2k+5}+q^{3k+7})c_{k+1}(q)+(q^{2k+6}-q^{3k+7})c_k(q) = 0. \tag{17}$$

Note that the summand in (16) is not q-hypergeometric and therefore the q-version of Zeilberger's algorithm cannot be applied.

Again, we start by constructing an annihilating ideal for the summand. The one for the sequence $c_k(q)$ is given by its definition (17), we just have to add the trivial relation w.r.t. n and convert everything to operator form (note the usage of q-shift operators):

In[12]:= **annc = ToOrePolynomial[{QS[qn, q^n] − 1, QS[qk, q^k]^2 + ($q^{k+3}(1+q-q^{k+2}+q^{2k+4})$)**QS[qk, q^k] + $q^{2k+6}(1-q^{k+1})$}]**

Out[12]= $\{S_{\mathrm{qn},q}-1,\ S_{\mathrm{qk},q}^2+\left(q^7\mathrm{qk}^3-q^5\mathrm{qk}^2+q^4\mathrm{qk}+q^3\mathrm{qk}\right)S_{\mathrm{qk},q}+\left(q^6\mathrm{qk}^2-q^7\mathrm{qk}^3\right)\}$

Next, the closure property "multiplication", see Theorem 1 (b), is applied (the result is about 2 pages long and therefore not displayed here):

In[13]:= **annSmnd = DFiniteTimes[annc, annc, Annihilator[$(-1)^k$ q^$(-kn-k(k+3)/2)$ QPochhammer[q^{n-1}, $1/q$, k] QPochhammer[q^{n+1}, q, k], {QS[qk, q^k], QS[qn, q^n]}]];**

The stage is now prepared for calling Chyzak's algorithm which delivers a pair (T, C) consisting of the telescoper and certificate:

In[14]:= **{T, C} = CreativeTelescoping[annSmnd, QS[qk, q^k] − 1]**

This computation takes about 2 min and the result is again too large to be printed here. We remark that the inhomogeneous part does not vanish so that we obtain an inhomogeneous recurrence for the function $J_{7_4,n}(q)$. The result is in accordance with the AJ conjecture and the previously known A-polynomial of the knot 7_4.

6.4 A Double Integral Related to Feynman Diagrams

We study the double integral

$$\int_0^1 \int_0^1 \frac{w^{-1-\varepsilon/2}(1-z)^{\varepsilon/2} z^{-\varepsilon/2}}{(z+w-wz)^{1-\varepsilon}} \left(1 - w^{n+1} - (1-w)^{n+1}\right) \mathrm{d}w\, \mathrm{d}z \qquad (18)$$

than can be found in [54, (J.17)]. The task is to compute a recurrence in n where ε is just a parameter. We are aware of the fact that (18) is not a hard challenge for physicists, and we use it only as a proof of concept here. We are going to apply Chyzak's algorithm iteratively.

For computing an annihilating ideal for the inner integral, we simply use the command Annihilator that takes care of the inhomogeneous part automatically:

In[15]:= $\boldsymbol{f = w\hat{}(-1-\varepsilon/2)\,(1-z)\hat{}(\varepsilon/2)\,z\hat{}(-\varepsilon/2)/(w+z-w\,z)\hat{}(1-\varepsilon)}$
$\boldsymbol{(1 - w\hat{}(n+1) - (1-w)\hat{}(n+1));}$

In[16]:= **ann = Annihilator[Integrate[f, {w, 0, 1}], {S[n], Der[z]}];**

This result is quite large so that we do not want to display it here. But it can be used again as input to Chyzak's algorithm, in order to treat the outer integral.

In[17]:= **{{T}, {C}} = CreativeTelescoping[ann, Der[z], S[n]];**

It is a little bit tricky to handle the inhomogeneous part of the outer integral since it involves an integral itself:

$$\left[C \int_0^1 f\, \mathrm{d}w\right]_{z=0}^{z=1} = \int_0^1 \left[C f\right]_{z=0}^{z=1} \mathrm{d}w. \qquad (19)$$

It turns out that the right-hand side of (19) is preferable to show that the inhomogeneous part evaluates to zero. Therefore the operator T annihilates the double integral, and this is the desired recurrence in n (which is of order 3):

In[18]:= **Factor[T]**

Out[18]= $-(\varepsilon-n-3)(\varepsilon-n-2)(\varepsilon+2n+4)(\varepsilon+2n+6)S_n^3 + (\varepsilon-n-2)$
$(\varepsilon+2n+4)(\varepsilon^2+2\varepsilon n+5\varepsilon-6n^2-28n-34)S_n^2 - (n+2)(\varepsilon^3-3\varepsilon^2 n$
$-6\varepsilon^2-8\varepsilon n^2-30\varepsilon n-28\varepsilon+12n^3+64n^2+116n+72)S_n$
$-2(n+1)(n+2)^2(\varepsilon-2n-2)$

6.5 A Hypergeometric Double Sum

We finally turn to a binomial double sum which was investigated in [8]:

$$\sum_i \sum_j \binom{i+j}{i}^2 \binom{4n-2i-2j}{2n-2i} = (2n+1)\binom{2n}{n}^2. \qquad (20)$$

We apply the heuristic approach from Sect. 5.4 to it. The corresponding command in the HolonomicFunctions package is FindCreativeTelescoping:

In[19]:= **FindCreativeTelescoping[Binomial[i + j, i]^2 × Binomial[4n − 2i − 2j, 2n − 2i], {S[i] − 1, S[j] − 1}, S[n]]**

Out[19]= $$\left\{\{1\},\left\{\left\{\frac{-2i^2j+i^2n-i^2-2ij^2+3ijn-2ij+3in}{(j+1)(i+j-2n)},\frac{-2i^2j-2ij^2+3ijn-2ij+j^2n-j^2+3jn}{(i+1)(i+j-2n)}\right\}\right\}\right\}$$

The output consists of the telescoper and the two certificates. At first glance it may seem contradictory that the telescoper is 1, but there are contributions from the certificates that make the recurrence for the double sum inhomogeneous. So we don't claim that the operator 1 annihilates the double sum, which would imply that it is zero.

7 Selected Applications of Creative Telescoping

In this section we want to give an extensive, but certainly not complete, collection of examples which show the beneficial use of creative telescoping in diverse areas of mathematics and physics.

Zeilberger's algorithm for hypergeometric sums is a meanwhile so classic tool that it is impossible to list all papers where it has been used to prove some binomial sum identity. We therefore restrict ourselves to publications where this algorithm plays a more or less central role. In [39] it was used to prove Ramanujan's famous formula for π, and in [46] for some formulas of similar type. The whole paper [89] is dedicated to binomial identities that arise in combinatorics and how to prove them algorithmically. Two proofs of the notorious binomial double sum identity (20) are given in [8] where, due to the lack of multi-summation software packages at that time, the problem was reduced in a tricky way to a single sum identity. A "triumph of computer algebra" is celebrated in [81] where the computation of factorial moments and probability generating functions for heap ordered trees is based on Zeilberger's algorithm. In [5] it is used to derive formulas for hypergeometric series acceleration, among them a pretty formula for $\zeta(3)$ that allowed to evaluate this constant to a large number of digits. In the article [60], Zeilberger's algorithm is combined with asymptotic estimates in order to give automated proofs of non-terminating series identities of Saalschütz type. Applications in the context of orthogonal polynomials are given in [58]. A fast way of computing Catalan's constant is derived in [103] by means of creative telescoping. While the recurrence that plays a crucial role in Apéry's proof of the irrationality of $\zeta(3)$ is nowadays a popular example for demonstrating these techniques, they were not available to Apéry when he came up with his proof. A new, elementary proof, still using Zeilberger's algorithm, is given in [104]. We conclude this paragraph by mentioning [6] where a binomial identity that arose in the study of a certain integral is investigated.

We turn to applications of creative telescoping that go beyond Zeilberger's algorithm. As an application of its q-analogue we cite [74] where computer proofs for the Rogers-Ramanujan identities are constructed. Multi-summation techniques for q-hypergeometric terms were used in [12] to prove a partition theorem of Göllnitz. Computer proofs for summation identities involving Stirling numbers are given in [53]. In [17] creative telescoping was used to obtain bounds on the order and degree of differential equations satisfied by algebraic functions. Chyzak's algorithm was applied to the generating function of 3-dimensional rook paths [19] in order to derive an explicit formula. Creative telescoping proofs for a selection of special function identities, mostly involving integrals, are presented in [66]. Another application to the evaluation of integrals is [7].

In [101] Zeilberger proposed an approach how to evaluate determinants of matrices with holonomic entries with the method of creative telescoping. This approach applies to determinants of the form $\det_{1\leq i,j\leq n}(a_{i,j})$ whose entries are bivariate holonomic sequences, not depending on the dimension n. The so-called "holonomic ansatz" celebrated its greatest success so far when it was employed to prove the qTSPP conjecture [68], a long-standing prominent problem in enumerative combinatorics, which previously had been reduced to a certain determinant evaluation of the above type. This conjecture is the q-analogue of what is known as Stembridge's theorem about the enumeration of totally symmetric plane partitions. Based on creative telescoping, this theorem was re-proved twice, both times using the formulation as a determinant evaluation: the first time by applying symbolic summation techniques to a decomposition of the matrix [9], the second time following the holonomic ansatz [62]. Some extensions of the holonomic ansatz were presented in [67] and were applied to solve several conjectures about determinants. An analogous method for the evaluation of Pfaffians was developed in [48].

In the field of quantum topology and knot theory, a prominent object of interest is the so-called *colored Jones function* of a knot. This function is actually an infinite sequence of Laurent polynomials and in [42] it has been shown that this sequence is always q-holonomic, by establishing an explicit multisum representation with proper q-hypergeometric summand. The corresponding minimal-order recurrence is called the *non-commutative A-polynomial* of the knot. Creative telescoping was used to compute it for a family of twist knots [43] and for a few double twist knots [41].

We are turning to applications in the area of numerical analysis. A widely used method for computer simulations of real-world phenomena described by partial differential equations is the *finite element method* (FEM). A short motivation of using symbolic summation techniques in this area is given in [79], and a concrete application where hypergeometric summation algorithms deliver certain recurrence equations which allow for a fast evaluation of the basis functions, is described in [11]. Further examples, where creative telescoping is used for verifying identities arising in the context of FEM or for finding identities that help to speed up the numerical simulations, can be found in [14, 15, 69].

Last but not least we want to point out that creative telescoping has extensively supported computations in physics. We will not detail on the very fruitful interaction of summation methods in difference fields with the computation of Feynman integrals in particle physics [1], but refer to the survey [87], also contained in this volume, and the references therein. The estimation of the entropy of a certain process [70] was supported by computer algebra. In the study of generalized two-Qubit Hilbert-Schmidt separability probabilities [88] creative telescoping was employed to simplify a complicated expression involving generalized hypergeometric functions. The authors of [20] underline the particular importance that creative telescoping may play in the evaluation of the n-fold integrals $\chi^{(n)}$ of the magnetic susceptibility of the Ising model. Also relativistic Coulomb integrals have been treated with the holonomic systems approach [78]. Likewise it was used in the proof of a third-order integrability criterion for homogeneous potentials of degree -1 [37]. One branch of statistical physics deals with random walks on lattices; some results in this area [65, 102] were obtained by creative telescoping.

References

1. Ablinger, J., Blümlein, J., Klein, S., Schneider, C.: Modern summation methods and the computation of 2- and 3-loop Feynman diagrams. Nucl. Phys. B Proc. Suppl. **205–206**(0), 110–115 (2010)
2. Abramov, S.A.: When does Zeilberger's algorithm succeed? Adv. Appl. Math. **30**, 424–441 (2003)
3. Abramov, S.A., Le, H.Q.: A criterion for the applicability of Zeilberger's algorithm to rational functions. Discrete Math. **259**(1–3), 1–17 (2002)
4. Almkvist, G., Zeilberger, D.: The method of differentiating under the integral sign. J. Symb. Comput. **10**(6), 571–591 (1990)
5. Amdeberhan, T., Zeilberger, D.: Hypergeometric series acceleration via the WZ method. Electron. J. Comb. **4**(2), R3 (1997)
6. Amdeberhan, T., de Angelis, V., Lin, M., Moll, V.H., Sury, B.: A pretty binomial identity. Elem. Math. **67**(1), 18–25 (2012)
7. Amdeberhan, T., Koutschan, C., Moll, V.H., Rowland, E.S.: The iterated integrals of $\ln(1 + x^n)$. Int. J. Number Theory **8**(1), 71–94 (2012)
8. Andrews, G.E., Paule, P.: Some questions concerning computer-generated proofs of a binomial double-sum identity. J. Symb. Comput. **16**, 147–153 (1993)
9. Andrews, G.E., Paule, P., Schneider, C.: Plane partitions VI. Stembridge's TSPP theorem. Adv. Appl. Math. **34**, 709–739 (2005)
10. Apagodu, M., Zeilberger, D.: Multi-variable Zeilberger and Almkvist-Zeilberger algorithms and the sharpening of Wilf-Zeilberger theory. Adv. Appl. Math. **37**(2), 139–152 (2006)
11. Bećirović, A., Paule, P., Pillwein, V., Riese, A., Schneider, C., Schöberl, J.: Hypergeometric summation algorithms for high order finite elements. Computing **78**(3), 235–249 (2006)
12. Berkovich, A., Riese, A.: A computer proof of a polynomial identity implying a partition theorem of Göllnitz. Adv. Appl. Math. **28**, 1–16 (2002)
13. Bernstein, J.N.: The analytic continuation of generalized functions with respect to a parameter. Funct. Anal. Appl. **6**(4), 273–285 (1972)
14. Beuchler, S., Pillwein, V.: Sparse shape functions for tetrahedral p-FEM using integrated Jacobi polynomials. Computing **80**(4), 345–375 (2007)

15. Beuchler, S., Pillwein, V., Zaglmayr, S.: Sparsity optimized high order finite element functions for H(curl) on tetrahedra. Adv. Appl. Math. **50**(5), 749–769 (2013)
16. Böing, H., Koepf, W.: Algorithms for q-hypergeometric summation in computer algebra. J. Symb. Comput. **28**, 777–799 (1999)
17. Bostan, A., Chyzak, F., Lecerf, G., Salvy, B., Schost, É.: Differential equations for algebraic functions. In: Proceedings of the International Symposium on Symbolic and Algebraic Computation (ISSAC), Waterloo. ACM, New York (2007)
18. Bostan, A., Chen, S., Chyzak, F., Li, Z.: Complexity of creative telescoping for bivariate rational functions. In: Proceedings of the International Symposium on Symbolic and Algebraic Computation (ISSAC), Munich, pp. 203–210. ACM, New York (2010)
19. Bostan, A., Chyzak, F., van Hoeij, M., Pech, L.: Explicit formula for the generating series of diagonal 3D rook paths. Séminaire Lotharingien de Combinatoire **66**, B66a (2011)
20. Bostan, A., Boukraa, S., Christol, G., Hassani, S., Maillard, J.-M.: Ising n-fold integrals as diagonals of rational functions and integrality of series expansions: integrality versus modularity. Technical report 1211.6031, arXiv (2012)
21. Bostan, A., Chen, S., Chyzak, F., Li, Z., Xin, G.: Hermite reduction and creative telescoping for hyperexponential functions. In: Proceedings of the International Symposium on Symbolic and Algebraic Computation (ISSAC), Boston, pp. 77–84. ACM, New York (2013)
22. Bostan, A., Lairez, P., Salvy, B.: Creative telescoping for rational functions using the Griffiths-Dwork method. In: Proceedings of the International Symposium on Symbolic and Algebraic Computation (ISSAC), Boston, pp. 93-100. ACM, New York (2013)
23. Buchberger, B.: Ein Algorithmus zum Auffinden der Basiselemente des Restklassenrings nach einem nulldimensionalen Polynomideal. PhD thesis, University of Innsbruck, Innsbruck (1965)
24. Cartier, P.: Démonstration "automatique" d'identités et fonctions hypergéometriques [d'après D. Zeilberger]. Astérisque **206**, 41–91 (1991). Séminaire Bourbaki, 44ème année, 1991–1992, n°746
25. Caruso, F.: A Macsyma implementation of Zeilberger's fast algorithm. In: Strehl, V. (ed.) Séminaire Lotharingien de Combinatoire, S43c, pp. 1–8. Institut Girard Desargues, Université Claude Bernard Lyon-I, Villeurbanne (2000)
26. Chen, S., Kauers, M.: Order-degree curves for hypergeometric creative telescoping. In: Proceedings of the International Symposium on Symbolic and Algebraic Computation (ISSAC), Grenoble, pp. 122–129 (2012)
27. Chen, S., Kauers, M.: Trading order for degree in creative telescoping. J. Symb. Comput. **47**(8), 968–995 (2012)
28. Chen, S., Singer, M.F.: Residues and telescopers for bivariate rational functions. Adv. Appl. Math. **49**(2), 111–133 (2012)
29. Chen, W.Y.C., Hou, Q.-H., Mu, Y.-P.: A telescoping method for double summations. J. Comput. Appl. Math. **196**(2), 553–566 (2006)
30. Chen, S., Chyzak, F., Feng, R., Fu, G., Li, Z.: On the existence of telescopers for mixed hypergeometric terms. Technical report 1211.2430, arXiv (2012)
31. Chen, S., Kauers, M., Singer, M.F.: Telescopers for rational and algebraic functions via residues. In: Proceedings of the International Symposium on Symbolic and Algebraic Computation (ISSAC), Grenoble, pp. 130–137. ACM, New York (2012)
32. Chyzak, F.: Fonctions holonomes en calcul formel. PhD thesis, École polytechnique (1998)
33. Chyzak, F.: Gröbner bases, symbolic summation and symbolic integration. In: Buchberger, B., Winkler, F. (eds.) Gröbner Bases and Applications. Volume 251 of London Mathematical Society Lecture Notes Series, pp. 32–60. Cambridge University Press, Cambridge (1998). Proceedings of the Conference 33 Years of Gröbner Bases
34. Chyzak, F.: An extension of Zeilberger's fast algorithm to general holonomic functions. Discrete Math. **217**(1–3), 115–134 (2000)
35. Chyzak, F., Salvy, B.: Non-commutative elimination in Ore algebras proves multivariate identities. J. Symb. Comput. **26**, 187–227 (1998)

36. Chyzak, F., Kauers, M., Salvy, B.: A non-holonomic systems approach to special function identities. In: Proceedings of the International Symposium on Symbolic and Algebraic Computation (ISSAC), Seoul, pp. 111–118. ACM, New York (2009)
37. Combot, T., Koutschan, C.: Third order integrability conditions for homogeneous potentials of degree −1. J. Math. Phys. **53**(8), 082704 (2012)
38. Coutinho, S.C.: A Primer of Algebraic D-Modules. Volume 33 of London Mathematical Society Student Texts. Cambridge University Press, Cambridge/New York (1995)
39. Ekhad, S.B., Zeilberger, D.: A WZ proof of Ramanujan's formula for π. In: Rassias, J.M. (ed.) Geometry, Analysis, and Mechanics, pp. 107–108. World Scientific, Singapore (1994)
40. Feynman, R.P., Leighton, R. (eds.): Surely You're Joking, Mr. Feynman!: Adventures of a Curious Character. W. W. Norton, New York (1985)
41. Garoufalidis, S., Koutschan, C.: Irreducibility of q-difference operators and the knot 7_4. Algebr. Geom. Topol. (2013, To appear). Preprint on arXiv:1211.6020
42. Garoufalidis, S., Lê, T.T.Q.: The colored Jones function is q-holonomic. Geom. Topol. **9**, 1253–1293 (2005). (electronic)
43. Garoufalidis, S., Sun, X.: The non-commutative A-polynomial of twist knots. J. Knot Theory Ramif. **19**(12), 1571–1595 (2010)
44. Gradshteyn, I.S., Ryzhik, J.M., Jeffrey, A., Zwillinger, D. (eds.): Table of Integrals, Series, and Products, 7th edn. Academic/Elsevier, Amsterdam (2007)
45. Graham, R.L., Knuth, D.E., Patashnik, O: Concrete Mathematics, 2nd edn. Addison-Wesley, Reading (1994)
46. Guillera, J.: Generators of some Ramanujan formulas. Ramanujan J. **11**, 41–48 (2006)
47. Guo, Q.H., Hou, Q.-H., Sun, L.H.: Proving hypergeometric identities by numerical verifications. J. Symb. Comput. **43**(12), 895–907 (2008)
48. Ishikawa, M., Koutschan, C.: Zeilberger's holonomic ansatz for Pfaffians. In: Proceedings of the International Symposium on Symbolic and Algebraic Computation (ISSAC), Grenoble, pp. 227–233. ACM, (2012)
49. Kandri-Rody, A., Weispfenning, V.: Non-commutative Gröbner bases in algebras of solvable type. J. Symb. Comput. **9**(1), 1–26 (1990)
50. Kauers, M.: Summation algorithms for Stirling number identities. J. Symb. Comput. **42**(10), 948–970 (2007)
51. Kauers, M.: The holonomic toolkit. In: Blümlein, J., Schneider, C. (eds.) Computer Algebra in Quantum Field Theory: Integration, Summation and Special Functions. Springer, Wien (2013)
52. Kauers, M., Paule, P.: The Concrete Tetrahedron. Text and Monographs in Symbolic Computation, 1st edn. Springer, Wien (2011)
53. Kauers, M., Schneider, C.: Automated proofs for some Stirling number identities. Electron. J. Comb. **15**(1), 1–7 (2008). R2
54. Klein, S.: Heavy flavor coefficient functions in deep-inelastic scattering at $O(a_s^2)$ and large virtualities. Diplomarbeit, Universität Potsdam (2006)
55. Koepf, W.: Algorithms for m-fold hypergeometric summation. J. Symb. Comput. **20**, 399–417 (1995)
56. Koepf, W.: REDUCE package for the indefinite and definite summation. SIGSAM Bull. **29**(1), 14–30 (1995)
57. Koepf, W.: Hypergeometric Summation. An Algorithmic Approach to Summation and Special Function Identities. Advanced Lectures in Mathematics. Vieweg Verlag, Braunschweig/Wiesbaden (1998)
58. Koepf, W., Schmersau, D.: Representations of orthogonal polynomials. J. Comput. Appl. Math. **90**, 57–94 (1998)
59. Koornwinder, T.H.: On Zeilberger's algorithm and its q-analogue. J. Comput. Appl. Math. **48**(1–2), 91–111 (1993)
60. Koornwinder, T.H.: Identities of nonterminating series by Zeilberger's algorithm. J. Comput. Appl. Math. **99**(1–2), 449–461 (1998)

61. Koutschan, C.: Advanced applications of the holonomic systems approach. PhD thesis, Research Institute for Symbolic Computation (RISC), Johannes Kepler University, Linz (2009)
62. Koutschan, C.: Eliminating human insight: an algorithmic proof of Stembridge's TSPP theorem. In: Amdeberhan, T., Medina, L.A., Moll, V.H. (eds.) Gems in Experimental Mathematics. Volume 517 of Contemporary Mathematics, pp. 219–230. American Mathematical Society, Providence (2010)
63. Koutschan, C.: A fast approach to creative telescoping. Math. Comput. Sci. **4**(2–3), 259–266 (2010)
64. Koutschan, C.: HolonomicFunctions (user's guide). Technical report 10-01, RISC Report Series, Johannes Kepler University, Linz (2010). http://www.risc.jku.at/research/combinat/software/HolonomicFunctions/
65. Koutschan, C.: Lattice Green's functions of the higher-dimensional face-centered cubic lattices. J. Phys. A Math. Theor. **46**(12), 125005 (2013)
66. Koutschan, C., Moll, V.H.: The integrals in Gradshteyn and Ryzhik. Part 18: some automatic proofs. SCIENTIA Ser. A Math. Sci. **20**, 93–111 (2011)
67. Koutschan, C., Thanatipanonda, T.: Advanced computer algebra for determinants. Ann. Comb. (2013, To appear). Preprint on arXiv:1112.0647
68. Koutschan, C., Kauers, M., Zeilberger, D.: Proof of George Andrews's and David Robbins's q-TSPP conjecture. Proc. Natl. Acad. Sci. **108**(6), 2196–2199 (2011)
69. Koutschan, C., Lehrenfeld, C., Schöberl, J.: Computer algebra meets finite elements: an efficient implementation for Maxwell's equations. In: Langer, U., Paule, P. (eds.) Numerical and Symbolic Scientific Computing: Progress and Prospects. Volume 1 of Texts and Monographs in Symbolic Computation, pp. 105–121. Springer, Wien (2012)
70. Lyons, R., Paule, P., Riese, A.: A computer proof of a series evaluation in terms of harmonic numbers. Appl. Algebra Eng. Commun. Comput. **13**, 327–333 (2002)
71. Majewicz, J.E.: WZ-style certification and Sister Celine's technique for Abel-type sums. J. Differ. Equ. Appl. **2**(1) 55–65 (1996)
72. Mohammed, M. Zeilberger, D.: Sharp upper bounds for the orders of the recurrences output by the Zeilberger and q-Zeilberger algorithms. J. Symb. Comput. **39**(2), 201–207 (2005)
73. Ore, Ø.: Theory of non-commutative polynomials. Ann. Math. **34**(3), 480–508 (1933)
74. Paule, P.: Short and easy computer proofs of the Rogers-Ramanujan identities and of identities of similar type. Electron. J. Comb. **1**, 1–9 (1994)
75. Paule, P., Riese, A.: A Mathematica q-analogue of Zeilberger's algorithm based on an algebraically motivated approach to q-hypergeometric telescoping. In: Ismail, M.E.H., Masson, D.R., Rahman, M. (eds.) Special Functions, q-Series and Related Topics. Volume 14 of Fields Institute Communications, pp. 179–210. American Mathematical Society, Providence (1997)
76. Paule, P., Schorn, M.: A Mathematica version of Zeilberger's algorithm for proving binomial coefficient identities. J. Symb. Comput. **20**(5/6), 673–698 (1995). http://www.risc.jku.at/research/combinat/software/PauleSchorn/
77. Paule, P., Strehl, V.: Symbolic summation—some recent developments. In: Fleischer, J. et al. (eds.) Computer Algebra in Science and Engineering—Algorithms, Systems, and Applications, pp. 138–162. World Scientific, Singapore (1995)
78. Paule, P., Suslov, S.: Relativistic Coulomb integrals and Zeilbergers holonomic systems approach I. In: Blümlein, J., Schneider, C. (eds.) Computer Algebra in Quantum Field Theory: Integration, Summation and Special Functions. Springer, Wien (2013)
79. Paule, P., Pillwein, V., Schneider, C., Schöberl, J.: Hypergeometric summation techniques for high order finite elements. In: PAMM, Weinheim, vol. 6, pp. 689–690. Wiley InterScience (2006)
80. Petkovšek, M., Wilf, H.S., Zeilberger, D.: $A = B$. A. K. Peters, Ltd., Wellesley (1996)
81. Prodinger, H.: Descendants in heap ordered trees or a triumph of computer algebra. Electron. J. Comb. **3**(1), R29 (1996)
82. Raab, C.G.: Definite integration in differential fields. PhD thesis, Johannes Kepler Universität Linz (2012)

83. Riese, A.: A Mathematica q-analogue of Zeilberger's algorithm for proving q-hypergeometric identities. Master's thesis, RISC, Johannes Kepler University Linz (1995)
84. Riese, A.: Fine-tuning Zeilberger's algorithm—the methods of automatic filtering and creative substituting. In: Garvan, F.G., Ismail, M.E.H. (eds.) Symbolic Computation, Number Theory, Special Functions, Physics and Combinatorics. Volume 4 of Developments in Mathematics, pp. 243–254. Kluwer, Dordrecht (2001)
85. Riese, A.: qMultiSum—A package for proving q-hypergeometric multiple summation identities. J. Symb. Comput. **35**, 349–376 (2003)
86. Schneider, C.: Symbolic summation in difference fields. PhD thesis, RISC, Johannes Kepler University, Nov 2001
87. Schneider, C.: Simplifying multiple sums in difference fields. In: Blümlein, J., Schneider, C. (eds.) Computer Algebra in Quantum Field Theory: Integration, Summation and Special Functions. Springer, Wien (2013)
88. Slater, P.B.: A concise formula for generalized two-qubit Hilbert-Schmidt separability probabilities. Technical report 1301.6617, arXiv (2013)
89. Strehl, V.: Binomial identities—combinatorial and algorithmic aspects. Discrete Math. **136** (1–3), 309–346 (1994)
90. Takayama, N.: An algorithm of constructing the integral of a module—an infinite dimensional analog of Gröbner basis. In: Proceedings of the International Symposium on Symbolic and Algebraic Computation (ISSAC), Tokyo, pp. 206–211. ACM, New York (1990)
91. Tefera, A.: MultInt, a MAPLE package for multiple integration by the WZ method. J. Symb. Comput. **34**(5), 329–353 (2002)
92. van der Poorten, A.: A proof that Euler missed—Apéry's proof of the irrationality of $\zeta(3)$. An informal report. Math. Intell. **1**, 195–203 (1979)
93. Wegschaider, K.: Computer generated proofs of binomial multi-sum identities. Master's thesis, RISC, Johannes Kepler University Linz, May 1997. http://www.risc.jku.at/research/combinat/software/MultiSum/
94. Wilf, H.S., Zeilberger, D.: An algorithmic proof theory for hypergeometric (ordinary and "q") multisum/integral identities. Invent. Math. **108**(1), 575–633 (1992)
95. Yen, L.: A two-line algorithm for proving terminating hypergeometric identities. J. Math. Anal. Appl. **198**(3), 856–878 (1996)
96. Yen, L.: A two-line algorithm for proving q-hypergeometric identities. J. Math. Anal. Appl. **213**(1), 1–14 (1997)
97. Zeiberger, D.: A fast algorithm for proving terminating hypergeometric identities. Discrete Math. **80**(2), 207–211 (1990)
98. Zeilberger, D.: A holonomic systems approach to special functions identities. J. Comput. Appl. Math. **32**(3), 321–368 (1990)
99. Zeilberger, D.: The method of creative telescoping. J. Symb. Comput. **11**, 195–204 (1991)
100. Zeilberger, D.: Three recitations on holonomic systems and hypergeometric series. J. Symb. Comput. **20**(5–6), 699–724 (1995)
101. Zeilberger, D.: The holonomic ansatz II. Automatic discovery(!) and proof(!!) of holonomic determinant evaluations. Ann. Comb. **11**(2), 241–247 (2007)
102. Zhang, X.-K., Wan, J., Lu, J.-J., Xu, X.-P.: Recurrence and Pólya number of general one-dimensional random walks. Commun. Theor. Phys. **56**(2), 293 (2011)
103. Zudilin, W.: An Apéry-like difference equation for Catalan's constant. Electron. J. Comb. **10**(1), #R14 (2003)
104. Zudilin, W.: Apéry's theorem. Thirty years after [an elementary proof of Apéry's theorem]. Int. J. Math. Comput. Sci. **4**(1), 9–19 (2009)

Renormalization and Mellin Transforms

Dirk Kreimer and Erik Panzer

Abstract We study renormalization in a kinetic scheme (realized by subtraction at fixed external parameters as implemented in the BPHZ and MOM schemes) using the Hopf algebraic framework, first summarizing and recovering known results in this setting. Then we give a direct combinatorial description of renormalized amplitudes in terms of Mellin transform coefficients, featuring the universal property of rooted trees H_R. In particular, a special class of automorphisms of H_R emerges from the action of changing Mellin transforms on the Hochschild cohomology of perturbation series.

Furthermore, we show how the Hopf algebra of polynomials carries a refined renormalization group property, implying its coarser form on the level of correlation functions. Application to scalar quantum field theory reveals the scaling behaviour of individual Feynman graphs.

1 Introduction

As was shown in [6, 16, 24], we may decompose Feynman integrals into functions of a single *scale* parameter s only (further forking into logarithmic divergent parts multiplied by suitable powers of s) and scale-independent functions of the other kinematic variables, called *angles*. Furthermore, the Hopf algebra H_R of rooted trees suffices to encode the full structure of subdivergences in quantum field theory by [8, 9, 16].

We can therefore study such generic Feynman rules in a purely algebraic framework as pioneered in [9, 18]. Renormalizing short-distance singularities by subtraction at a reference scale μ (which we refer to as *kinetic scheme*) leads to

D. Kreimer (✉) · E. Panzer
Departments of Mathematics and Physics, Humboldt University,
Unter den Linden 6, 10099 Berlin, Germany
e-mail: kreimer@physik.hu-berlin.de; panzer@mathematik.hu-berlin.de

C. Schneider and J. Blümlein (eds.), *Computer Algebra in Quantum Field Theory*, Texts & Monographs in Symbolic Computation, DOI 10.1007/978-3-7091-1616-6_8,

amplitudes of a distinguished algebraic kind: Theorem 4 proves them to implement the universal property of H_R, delivering an explicit combinatorial evaluation in terms of Mellin transform coefficients.

Further investigating the role of Hochschild cohomology, in Sect. 6 we define a class of automorphisms of H_R which transform the perturbation series in a way equivalent to changing the Feynman rules. This clarifies how exact one-cocycles describe variations.

In Sects. 4 and 5 we advertise to think about the renormalization group property as a Hopf algebra morphism to polynomials, determining higher logarithms in (28). We show how it implies the renormalization group on correlation functions and extend the *propagator-coupling-duality* of [5] which yields the functional equation (34).

After analysing the differences to the minimal subtraction scheme in Sect. 7, we show explicitly how our general results manifest themselves in scalar field theory.

2 Connected Hopf Algebras

The fundamental mathematical structure behind perturbative renormalization is the Hopf algebra as discovered in [16]. We briefly summarize the results on Hopf algebras we need and recommend [21, 22] for detailed introductions with a focus on renormalization.

All vector spaces live over a field $\mathbb{K}$ of zero characteristic (in examples $\mathbb{K} = \mathbb{R}$), $\operatorname{Hom}(\cdot,\cdot)$ denotes $\mathbb{K}$-linear maps and $\operatorname{lin} M$ the linear span. Every algebra $(\mathscr{A}, m, u)$ shall be unital, associative and commutative, any bialgebras $(H, m, u, \Delta, \varepsilon)$ in addition also counital and coassociative. They split into the scalars and the *augmentation ideal* $\ker\varepsilon$ as $H = \mathbb{K}\mathbb{1} \oplus \ker\varepsilon = \operatorname{im} u \oplus \ker\varepsilon$, inducing the projection $P := \operatorname{id} - u\circ\varepsilon\colon\ H \twoheadrightarrow \ker\varepsilon$. We use Sweedler's notation $\Delta(x) = \sum_x x_1 \otimes x_2$ and $\tilde{\Delta}(x) = \sum_x x' \otimes x''$ to abbreviate the *reduced coproduct* $\tilde{\Delta} := \Delta - \mathbb{1}\otimes\operatorname{id} - \operatorname{id}\otimes\mathbb{1}$.

We assume a *connected grading* $H = \bigoplus_{n\geq 0} H_n$ ($H_0 = \mathbb{K}\cdot\mathbb{1}$) and write $|x| := n$ for homogeneous $0 \neq x \in H_n$, defining the *grading operator* $Y \in \operatorname{End}(H)$ by $Yx = |x|\cdot x$. Exponentiation yields a one-parameter group $\mathbb{K} \ni t \mapsto \theta_t$ of Hopf algebra automorphisms

$$\theta_t := \exp(tY) = \sum_{n\in\mathbb{N}_0} \frac{(tY)^n}{n!}, \quad \forall n \in \mathbb{N}_0: \quad H_n \ni x \mapsto \theta_t(x) = e^{t|x|}x = e^{nt}x. \tag{1}$$

An algebra $(\mathscr{A}, m_{\mathscr{A}}, u_{\mathscr{A}})$ induces the associative *convolution product* on $\operatorname{Hom}(H, \mathscr{A})$ by

$$\operatorname{Hom}(H,\mathscr{A}) \ni \phi, \psi \mapsto \phi \star \psi := m_{\mathscr{A}} \circ (\phi\otimes\psi)\circ\Delta \in \operatorname{Hom}(H,\mathscr{A}),$$

with unit given by $e := u_{\mathscr{A}} \circ \varepsilon$. As outcome of the connectedness of H we stress

1. The *characters* $G^H_{\mathscr{A}} := \{\phi \in \operatorname{Hom}(H, \mathscr{A})\colon\ \phi \circ u = u_{\mathscr{A}}$ and $\phi \circ m = m_{\mathscr{A}} \circ (\phi \otimes \phi)\}$ (morphisms of unital algebras) form a group under $\star$.
2. Hence $\operatorname{id} \in G^H_H$ has a unique inverse $S := \operatorname{id}^{\star -1}$, called *antipode*, turning H into a Hopf algebra. For all $\phi \in G^H_{\mathscr{A}}$ we have $\phi^{\star -1} = \phi \circ S$.
3. The bijection $\exp_\star\colon \mathfrak{g}^H_{\mathscr{A}} \to G^H_{\mathscr{A}}$ with inverse $\log_\star\colon G^H_{\mathscr{A}} \to \mathfrak{g}^H_{\mathscr{A}}$ between $G^H_{\mathscr{A}}$ and the *infinitesimal characters* $\mathfrak{g}^H_{\mathscr{A}} := \{\phi \in \operatorname{Hom}(H, \mathscr{A})\colon\ \phi \circ m = \phi \otimes e + e \otimes \phi\}$ is given by the pointwise finite series

$$\exp_\star(\phi) := \sum_{n \in \mathbb{N}_0} \frac{\phi^{\star n}}{n!} \quad \text{and} \quad \log_\star(\phi) := \sum_{n \in \mathbb{N}} \frac{(-1)^{n+1}}{n} (\phi - e)^{\star n}. \tag{2}$$

2.1 Hochschild Cohomology

The Hochschild cochain complex [1,8,22] we associate to H contains the functionals $H' = \operatorname{Hom}(H, \mathbb{K})$ as zero-cochains. One-cocycles $L \in \mathrm{HZ}^1_\varepsilon(H) \subset \operatorname{End}(H)$ are linear maps such that $\Delta \circ L = (\operatorname{id} \otimes L) \circ \Delta + L \otimes \mathbb{1}$ and the differential

$$\delta : H' \to \mathrm{HZ}^1_\varepsilon(H), \alpha \mapsto \delta\alpha := (\operatorname{id} \otimes \alpha) \circ \Delta - u \circ \alpha \in \mathrm{HB}^1_\varepsilon(H) := \delta\left(H'\right) \tag{3}$$

determines the first cohomology group by $\mathrm{HH}^1_\varepsilon(H) := \mathrm{HZ}^1_\varepsilon(H)/\mathrm{HB}^1_\varepsilon(H)$.

Lemma 1. *Cocycles* $L \in HZ^1_\varepsilon(H)$ *fulfil* $\operatorname{im} L \subseteq \ker \varepsilon$ *and* $L(\mathbb{1}) \in \operatorname{Prim}(H) := \ker \tilde{\Delta}$ *is primitive. The map* $HH^1_\varepsilon(H) \to \operatorname{Prim}(H)$, $[L] \mapsto L(\mathbb{1})$ *is well-defined since* $\delta\alpha(\mathbb{1}) = 0$ *for all* $\alpha \in H'$.

2.2 Rooted Trees

The Hopf algebra H_R of rooted trees serves as the domain of Feynman rules. As an algebra, $H_R = S(\operatorname{lin} \mathscr{T}) = \mathbb{K}[\mathscr{T}]$ is free commutative[1] generated by the *rooted trees* $\mathscr{T}$ and spanned by their disjoint unions (products) called *rooted forests* $\mathscr{F}$:

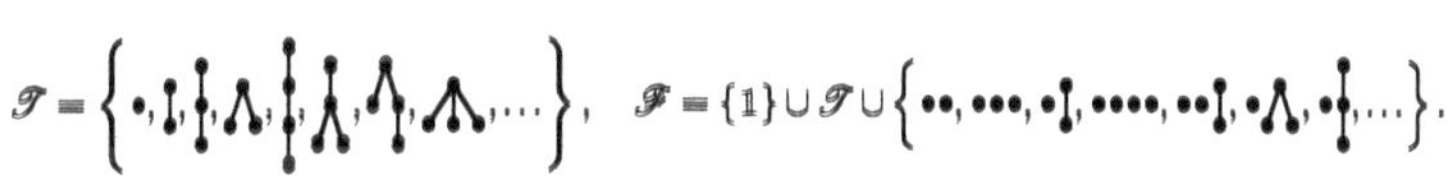

Every $w \in \mathscr{F}$ is just the monomial $w = \prod_{t \in \pi_0(w)} t$ of its multiset of tree components $\pi_0(w)$, while $\mathbb{1}$ denotes the empty forest. The number $|w| := |V(w)|$ of nodes $V(w)$ induces the grading $H_{R,n} = \operatorname{lin} \mathscr{F}_n$ where $\mathscr{F}_n := \{w \in \mathscr{F}\colon\ |w| = n\}$.

[1] We consider *unordered* trees 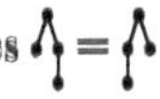 and forests 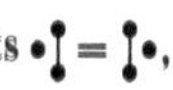, sometimes called *non-planar*.

Definition 1. The (linear) *grafting operator* $B_+ \in \operatorname{End}(H_R)$ attaches all trees of a forest to a new root, so for example $B_+(\mathbb{1}) = \bullet$, $B_+(\bullet) = \begin{smallmatrix}\bullet\\ |\\ \bullet\end{smallmatrix}$ and $B_+(\bullet\bullet) = \begin{smallmatrix} & \bullet & \\ \diagup & & \diagdown \\ \bullet & & \bullet\end{smallmatrix}$.

Clearly, B_+ is homogenous of degree one with respect to the grading and restricts to a bijection $B_+ \colon \mathscr{F} \to \mathscr{T}$. The coproduct Δ is defined to make B_+ a cocycle by requiring

$$\Delta \circ B_+ = B_+ \otimes \mathbb{1} + (\operatorname{id} \otimes B_+) \circ \Delta. \tag{4}$$

Lemma 2. *In cohomology,* $0 \neq [B_+] \in HH^1_\varepsilon(H_R)$ *is non-trivial by* $B_+(\mathbb{1}) = \bullet \neq 0$.

It characterizes H_R through the well-known (Theorem 2 of [8]) *universal property* of

Theorem 1. *To an algebra* $\mathscr{A}$ *and* $L \in \operatorname{End}(\mathscr{A})$ *there exists a unique morphism* ${}^L\rho \colon H_R \to \mathscr{A}$ *of unital algebras such that*

$$ {}^L\rho \circ B_+ = L \circ {}^L\rho, \quad \textit{equivalently} \quad \begin{array}{ccc} H_R & \xrightarrow{\ {}^L\rho\ } & \mathscr{A} \\ {\scriptstyle B_+}\downarrow & & \downarrow{\scriptstyle L} \\ H_R & \xrightarrow[\ {}^L\rho\ ]{} & \mathscr{A} \end{array} \quad \textit{commutes.} \tag{5}$$

In case of a bialgebra $\mathscr{A}$ *and a cocycle* $L \in HZ^1_\varepsilon(\mathscr{A})$, ${}^L\rho$ *is a morphism of bialgebras and even of Hopf algebras when* $\mathscr{A}$ *is Hopf.*

This morphism ${}^L\rho$ simply replaces B_+, m_{H_R} and $\mathbb{1}$ by L, $m_{\mathscr{A}}$ and $\mathbb{1}_{\mathscr{A}}$ as in

$${}^L\rho\left(\begin{smallmatrix} & \bullet & \\ \diagup & & \diagdown \\ \bullet & & \bullet\end{smallmatrix} - 3\bullet\right) = {}^L\rho\left\{B_+\left([B_+(\mathbb{1})]^2\right) - 3B_+(\mathbb{1})\right\} = L\left([L(\mathbb{1}_{\mathscr{A}})]^2\right) - 3L(\mathbb{1}_{\mathscr{A}}).$$

Example 1. The cocycle $\int_0 \in \mathrm{HZ}^1_\varepsilon(\mathbb{K}[x])$ of Sect. 4 induces the character

$$\varphi := {}^{\int_0}\rho \in G^{H_R}_{\mathbb{K}[x]} \quad \text{fulfilling} \quad \varphi(w) = \frac{x^{|w|}}{w!} \quad \text{for any forest} \quad w \in \mathscr{F}, \quad \text{using} \tag{6}$$

Definition 2. The *tree factorial* $(\cdot)! \in G^{H_R}_{\mathbb{K}}$ is equivalently determined by requesting

$$[B_+(w)]! = w! \cdot |B_+(w)| \quad \text{or} \quad w! \overset{}{\underset{2}{=}} \prod_{v \in V(w)} |w_v| \quad \text{for all} \quad w \in \mathscr{F}. \tag{7}$$

[2] By w_v we denote the subtree of w rooted at the node $v \in V(w)$.

3 The Generic Model

As explained in the introduction we consider Feynman rules as characters $\phi \in G_{\mathscr{A}}^{H_R}$, mapping a rooted tree to a function of the parameter s (by Proposition 1 it lies in the algebra $\mathscr{A} = \mathbb{K}[z^{-1}, z]][s^{-z}]$). Since B_+ mimics the insertion of a subdivergence into a fixed graph γ (restricting to a single insertion place by a result from [24]), applying ϕ yields a subintegral and therefore

Definition 3. The generic Feynman rules ${}_z\phi$ are given through Theorem 1 by

$$ {}_z\phi_s \circ B_+ = \int_0^\infty \frac{f(\frac{\zeta}{s})\zeta^{-z}}{s}\, {}_z\phi_\zeta \,\mathrm{d}\zeta = \int_0^\infty f(\zeta)(s\zeta)^{-z}\, {}_z\phi_{s\zeta}\,\mathrm{d}\zeta. \tag{8} $$

The integration kernel f is specified by γ after *Wick rotation* to Euclidean space, with the asymptotic behaviour $f(\zeta) \sim \zeta^{-1}$ for $\zeta \to \infty$ generating the (logarithmic) divergences of these integrals (we do not address infrared problems and exclude any poles in f). The regulator ζ^{-z} ensures convergence when $0 < \Re(z) < 1$, with results depending analytically on z. We can perform all the integrals using this *Mellin transform*

$$ F(z) := \int_0^\infty f(\zeta)\zeta^{-z}\,\mathrm{d}\zeta = \sum_{n=-1}^{\infty} c_n z^n, \quad \text{by} \tag{9} $$

Proposition 1. *For any forest $w \in \mathscr{F}$ we have (called* BPHZ model *in [4])*

$$ {}_z\phi_s(w) = s^{-z|w|} \prod_{v\in V(w)} F\left(z\,|w_v|\right). \tag{10} $$

Proof. As both sides of (10) are clearly multiplicative, it is enough to prove the claim inductively for trees. Let it be valid for some forest $w \in \mathscr{F}$, then for $t = B_+(w)$ observe

$$ {}_z\phi_s \circ B_+(w) = \int_0^\infty (s\zeta)^{-z} f(\zeta)\, {}_z\phi_{s\zeta}(w)\,\mathrm{d}\zeta = \int_0^\infty (s\zeta)^{-z} f(\zeta)(s\zeta)^{-z|w|} \prod_{v\in V(w)} F\left(z\,|w_v|\right)\mathrm{d}\zeta $$

$$ = s^{-z|B_+(w)|}\left[\prod_{v\in V(w)} F\left(z\,|w_v|\right)\right] F\left(z\,|B_+(w)|\right) = s^{-z|t|} \prod_{v\in V(t)} F\left(z\,|t_v|\right). $$

Example 2. Using (10), we can directly write down the Feynman rules like

${}_z\phi_s(\bullet) = s^{-z}F(z)$, ${}_z\phi_s($$) = s^{-2z}F(z)F(2z)$ and ${}_z\phi_s($$) = s^{-3z}[F(z)]^2F(3z)$.

Many examples (choices of F) are discussed in [4], the particular case of the one-loop propagator graph γ of Yukawa theory is in [5] and for scalar Yukawa theory in six dimensions one has $F(z) = \frac{1}{z(1-z)(2-z)(3-z)}$ as in [22]. Already noted in [17], the highest order pole of ${}_z\phi_s(w)$ is independent of s and just the tree factorial

$$ {}_z\phi_s(w) \in s^{-z|w|} \prod_{v \in V(w)} \left\{ \tfrac{c_{-1}}{z|w_v|} + \mathbb{K}[[z]] \right\} \underset{(7)}{=} \frac{1}{w!} \left(\tfrac{c_{-1}}{z}\right)^{|w|} + z^{1-|w|}\mathbb{K}[\ln s][[z]]. \tag{11} $$

3.1 Renormalization

Algebraically, renormalization of a character $\phi \in G^H_{\mathscr{A}}$ equals a *Birkhoff decomposition* [9, 21, 22] into the *renormalized* character $\phi_R := \phi_+ \in G^H_{\mathscr{A}}$ and the *counterterms* $Z := \phi_- \in G^H_{\mathscr{A}}$ defined by the conditions

$$ \phi = \phi_-^{\star -1} \star \phi_+ \quad \text{and} \quad \phi_\pm(\ker\varepsilon) \subseteq \mathscr{A}_\pm, \tag{12} $$

with respect to a splitting $\mathscr{A} = \mathscr{A}_+ \oplus \mathscr{A}_-$ determined by the *renormalization scheme* (the projection $R: \mathscr{A} \twoheadrightarrow \mathscr{A}_-$). Turning to *minimal subtraction* in Sect. 7 we now focus on

Definition 4. On the target algebra $\mathscr{A}$ of regularized Feynman rules depending on a single external variable s, define the *kinetic scheme* by evaluation at $s = \mu$:

$$ \operatorname{End}(\mathscr{A}) \ni R_\mu := \operatorname{ev}_\mu = \left(\mathscr{A} \ni f \mapsto f|_{s=\mu}\right). \tag{13} $$

This scheme exploits that subtraction improves the decay at infinity: Let $f(\zeta) \sim \frac{1}{\zeta}$, meaning $f(\zeta) = \frac{1}{\zeta} + \tilde{f}(\zeta)$ for some $\tilde{f}(\zeta) \in \mathscr{O}\left(\zeta^{-1-\varepsilon}\right)$ with $\varepsilon > 0$. Then ${}_z\phi_s(\bullet)$ is logarithmically divergent (would it not be for the regulator ζ^{-z}), but subtraction

$$ {}_z\phi_s(\bullet) - {}_z\phi_\mu(\bullet) = \int_0^\infty \left[\frac{f(\frac{\zeta}{s})}{s} - \frac{f(\frac{\zeta}{\mu})}{\mu} \right] \zeta^{-z} = \int_0^\infty \left[\frac{\tilde{f}(\frac{\zeta}{s})}{s} - \frac{\tilde{f}(\frac{\zeta}{\mu})}{\mu} \right] \zeta^{-z} \tag{14} $$

yields a convergent integral even for $z = 0$. As R_μ is a character of $\mathscr{A}$, the Birkhoff recursion simplifies to $Z = R_\mu \circ {}_z\phi \circ S = {}_z\phi_\mu \circ S$ and ${}_z\phi_R = {}_z\phi_\mu^{\star -1} \star {}_z\phi_s$.

Example 3. We find ${}_z\phi_{R,s}(\bullet) = (s^{-z} - \mu^{-z})\, F(z)$ and $S\left(\begin{smallmatrix}\bullet\\|\\\bullet\end{smallmatrix}\right) = -\begin{smallmatrix}\bullet\\|\\\bullet\end{smallmatrix} + \bullet\bullet$ results in

$$ {}_z\phi_{R,s}\left(\begin{smallmatrix}\bullet\\|\\\bullet\end{smallmatrix}\right) = \left(s^{-2z} - \mu^{-2z}\right) F(z)F(2z) - (s^{-z} - \mu^{-z})\, \mu^{-z} F^2(z). \tag{15} $$

The goal of renormalization is to assure the *finiteness* of the *physical limit*

$$ {}_0\phi_R := \lim_{z \to 0} {}_z\phi_R, \tag{16} $$

and indeed we find the finite ${}_0\phi_{R,s}(\bullet) = -c_{-1}\ln\frac{s}{\mu}$. In the case of (15) check

$$
\begin{aligned}
{}_0\phi_{R,s}\left(\begin{smallmatrix}\bullet\\ |\\ \bullet\end{smallmatrix}\right) &= \lim_{z\to 0}\left\{-\left[-z\ln\tfrac{s}{\mu} + \tfrac{z^2}{2}\left(\ln^2 s + 2\ln s\ln\mu - 3\ln^2\mu\right)\right]\cdot\left[\tfrac{c_{-1}^2}{z^2} + 2\tfrac{c_{-1}c_0}{z}\right]\right. \\
&\quad \left. + \left[-2z\ln\tfrac{s}{\mu} + 2z^2\left(\ln^2 s - \ln^2\mu\right)\right]\cdot\left[\tfrac{c_{-1}^2}{2z^2} + \tfrac{3c_0c_{-1}}{2z}\right]\right\} = \frac{c_{-1}^2}{2}\ln^2\tfrac{s}{\mu} - c_{-1}c_0\ln\tfrac{s}{\mu},
\end{aligned}
\tag{17}
$$

where all poles in z perfectly cancel. Note that ${}_0\phi_{R,s}$ maps a forest w to a polynomial in $\mathbb{K}[\ln\frac{s}{\mu}]$ of degree $\leq |w|$ without constant term (except for ${}_0\phi(\mathbb{1}) = 1$), due to the subtraction at $s = \mu$. We now prove these properties in general, extending work in [18].

3.2 Subdivergences

Inductively, the Birkhoff decomposition is constructed as $\phi_+(x) = (\mathrm{id} - R_\mu)\bar{\phi}(x)$ where the *Bogoliubov character* $\bar{\phi}(x)$ (*$\bar{R}$-operation*) serves to renormalize the *subdivergences*. It is defined by

$$
\bar{\phi}(x) := \phi(x) + \sum_x \phi_-(x')\phi(x'') = \phi(x) + [\phi_- \star \phi - \phi_- - \phi](x) = \phi_+(x) - \phi_-(x).
$$

Theorem 2. *For an endomorphism $L \in \mathrm{End}(\mathscr{A})$ consider the Feynman rules $\phi := {}^L\rho$ induced by (5). Given a renormalization scheme $R \in \mathrm{End}(\mathscr{A})$ such that*

$$
L \circ m_{\mathscr{A}} \circ (\phi_- \otimes \mathrm{id}) = m_{\mathscr{A}} \circ (\phi_- \otimes L), \tag{18}
$$

that is to say, L is linear over the counterterms, we have

$$
\bar{\phi} \circ B_+ = L \circ \phi_+. \tag{19}
$$

Proof. This is a straightforward consequence of the cocycle property of B_+:

$$
\begin{aligned}
\bar{\phi} \circ B_+ &= (\phi_- \star \phi - \phi_-) \circ B_+ \\
&= m_{\mathscr{A}} \circ (\phi_- \otimes \phi) \circ [(\mathrm{id} \otimes B_+) \circ \Delta + B_+ \otimes \mathbb{1}] - \phi_- \circ B_+ \\
&= \phi_- \star (\phi \circ B_+) = \phi_- \star (L \circ \phi) \underset{(18)}{=} L \circ (\phi_- \star \phi) = L \circ \phi_+.
\end{aligned}
$$

As the counterterms Z of our model are independent of s, they can be moved out of the integrals in (8) and (18) is fulfilled indeed. This is a general feature of quantum field theories: The counterterms to not depend on any external variables.[3]

The significance of (19) lies in the expression of the renormalized $\phi_{R,0}(t)$ for a tree $t = B_+(w)$ only in terms of the renormalized value ${}_z\phi_R(w)$. This allows for inductive proofs of properties of ${}_z\phi_R$ and also ${}_0\phi_R$, without having to consider the unrenormalized Feynman rules or their counterterms at all.

3.3 Finiteness

Proposition 2. *The physical limit ${}_0\phi_{R,s}$ exists and maps H_R into polynomials $\mathbb{K}[\ln \frac{s}{\mu}]$.*

Proof. We proceed inductively from ${}_0\phi_{R,s}(\mathbb{1}) = 1$ and, as ${}_0\phi_R$ is a character, only need to consider trees $t = B_+(w)$ in the induction step. Hence for this $w \in \mathcal{F}$ we already know that ${}_0\phi_{R,\zeta}(w) \in \mathcal{O}\left(\ln^N \zeta\right)$ for some $N \in \mathbb{N}_0$ such that dominated convergence yields

$$
\begin{aligned}
{}_0\phi_{R,s}(t) &\underset{(19)}{=} \lim_{z\to 0}(\mathrm{id} - R_\mu)\left[s \mapsto \int_0^\infty \frac{f(\zeta/s)}{s}\zeta^{-z}\, {}_z\phi_{R,\zeta}(w)\,\mathrm{d}\zeta\right] \\
&= \lim_{z\to 0}\int_0^\infty \left[\tfrac{f(\zeta/s)}{s} - \tfrac{f(\zeta/\mu)}{\mu}\right]\zeta^{-z}\, {}_z\phi_{R,\zeta}(w)\,\mathrm{d}\zeta = \int_0^\infty \left[\tfrac{f(\zeta/s)}{s} - \tfrac{f(\zeta/\mu)}{\mu}\right] {}_0\phi_{R,\zeta}(w)\,\mathrm{d}\zeta,
\end{aligned}
$$

recalling the term in square brackets to be from $\mathcal{O}\left(\zeta^{-1-\varepsilon}\right)$ as in (14). This proves the cancellation of all z-poles in ${}_z\phi_{R,s}(t)$ and we identify ${}_0\phi_{R,s}(t)$ with the $\propto z^0$ term, which is a polynomial in $\ln s$ and $\ln \mu$ of degree $|t|$ by inspection of (10): Each such logarithm comes with a factor z (expanding s^{-z}) which needs to cancel with a pole $\frac{c_{-1}}{z|t_v|}$ from some $F(z\,|t_v|)$ in order to contribute to the $\propto z^0$ term. Finally the substitution $\zeta \mapsto \zeta\mu$ gives

$$
{}_0\phi_{R,s}(t) = \int_0^\infty \left[\frac{f(\zeta\frac{\mu}{s})}{\frac{s}{\mu}} - f(\zeta)\right] {}_0\phi_{R,\mu\zeta}(w)\,\mathrm{d}\zeta, \tag{20}
$$

hence by induction ${}_0\phi_{R,\zeta\mu}$ only depends on ζ and ${}_0\phi_{R,s}$ is a function of $\frac{s}{\mu}$ only.

Using (20), the physical limit of the renormalized Feynman rules can be obtained inductively by convergent integrations after performing the subtraction at $s = \mu$ on the integrand, in particular without the need of any regulator. Therefore ${}_0\phi_R$ is independent of the choice of regularization prescription, so employing a *cutoff*

[3]Even if the divergence of a Feynman graph does depend on external momenta as happens for higher degrees of divergence, the Hopf algebra is defined such that the counterterms are evaluations on certain *external structures*, given by distributions in [9]. So in any case, ϕ_- maps to scalars.

regulator or *dimensional regularization* yields the same renormalized result in the physical limit.

4 The Hopf Algebra of Polynomials

We summarize relevant properties of the polynomials, focusing on their Hochschild cohomology (the relevance of $\int_0$ was already mentioned in [8]). First observe

Lemma 3. *Requiring* $\Delta(x) = x \otimes \mathbb{1} + \mathbb{1} \otimes x$ *induces a unique Hopf algebra structure on the polynomials* $\mathbb{K}[x]$. *It is graded by degree, connected, commutative and cocommutative with* $\Delta(x^n) = \sum_{i=0}^{n} \binom{n}{i} x^i \otimes x^{n-i}$ *and the primitive elements are* $\mathrm{Prim}(\mathbb{K}[x]) = \mathbb{K} \cdot x$.

The *integration operator* $\int_0 : x^n \mapsto \frac{1}{n+1} x^{n+1}$ furnishes a cocycle $\int_0 \in \mathrm{HZ}^1_\varepsilon(\mathbb{K}[x])$ as

$$\Delta \int_0 \left(\frac{x^n}{n!}\right) = \Delta\left(\frac{x^{n+1}}{(n+1)!}\right) = \sum_{k=0}^{n+1} \frac{x^k}{k!} \otimes \frac{x^{n+1-k}}{(n+1-k)!}$$
$$= \frac{x^{n+1}}{(n+1)!} \otimes \mathbb{1} + \sum_{k=0}^{n} \frac{x^k}{k!} \otimes \int_0 \left(\frac{x^{n-k}}{(n-k)!}\right) = \left[\int_0 \otimes \mathbb{1} + \left(\mathrm{id} \otimes \int_0\right) \circ \Delta\right]\left(\frac{x^n}{n!}\right),$$

and is not a coboundary since $\int_0 1 = x \neq 0$. In fact it generates the cohomology by

Theorem 3. $HH^1_\varepsilon(\mathbb{K}[x]) = \mathbb{K} \cdot [\int_0]$ *is one-dimensional as the 1-cocycles of* $\mathbb{K}[x]$ *are*

$$HZ^1_\varepsilon(\mathbb{K}[x]) = \mathbb{K} \cdot \int_0 \oplus \delta\left(\mathbb{K}[x]'\right) = \mathbb{K} \cdot \int_0 \oplus HB^1_\varepsilon(\mathbb{K}[x]). \tag{21}$$

Proof. For an arbitrary cocycle $L \in \mathrm{HZ}^1_\varepsilon(\mathbb{K}[x])$, Lemma 1 ensures $L(1) = x a_{-1}$ where $a_{-1} := \partial_0 L(1)$. Hence $\tilde{L} := L - a_{-1} \int_0 \in \mathrm{HZ}^1_\varepsilon$ fulfils $\tilde{L}(1) = 0$, so $L_0 := \tilde{L} \circ \int_0 \in \mathrm{HZ}^1_\varepsilon$ by

$$\Delta \circ L_0 = (\mathrm{id} \otimes \tilde{L}) \circ \Delta \circ \int_0 + (\tilde{L} \otimes 1) \circ \int_0 = (\mathrm{id} \otimes L_0) \circ \Delta + L_0 \otimes 1 + \tilde{L}(1) \cdot \int_0 .$$

Repeating the argument inductively yields $a_n := \partial_0 L_n(1) = \partial_0 \circ L \circ \int_0^{n+1}(1) \in \mathbb{K}$ and $L_{n+1} := (L_n - a_n \int_0) \circ \int_0 \in \mathrm{HZ}^1_\varepsilon$, so for any $n \in \mathbb{N}_0$ we may read off from

$$L \circ \int_0^n (1) = a_{-1} \int_0^{n+1} (1) + \ldots + a_{n-2} \int_0^2 (1) + L_{n-1}(1)$$
$$= a_{-1} \int_0 \left(\int_0^n 1\right) + \sum_{j=0}^{n-1} a_j \int_0^{n-j} (1)$$

that indeed $L = a_{-1} \int_0 + \delta\alpha$ for the functional $\alpha := \partial_0 \circ L \circ \int_0$ with $\alpha(\frac{x^n}{n!}) = a_n$.

Lemma 4. *Up to subtraction $P = \delta\varepsilon = \mathrm{id} - \mathrm{ev}_0 : \mathbb{K}[x] \twoheadrightarrow \ker \varepsilon = x\mathbb{K}[x]$ of the constant part, direct computation exhibits $\delta\alpha$ for any $\alpha \in \mathbb{K}[x]'$ as the differential operator*

$$\delta\alpha = P \circ \sum_{n \in \mathbb{N}_0} \alpha\left(\tfrac{x^n}{n!}\right) \partial^n \in \mathrm{End}(\mathbb{K}[x]). \tag{22}$$

Lemma 5. *As characters $\phi \in G_{\mathbb{K}}^{\mathbb{K}[x]}$ of $\mathbb{K}[x]$ are fixed by $\lambda := \phi(x)$, they form the group $G_{\mathbb{K}}^{\mathbb{K}[x]} = \{\mathrm{ev}_\lambda : \lambda \in \mathbb{K}\}$ of evaluations (the counit $\varepsilon = \mathrm{ev}_0$ equals the neutral element)*

$$\mathbb{K}[x] \ni p(x) \mapsto \mathrm{ev}_\lambda(p) := p(\lambda) \quad \textit{with the product} \quad \mathrm{ev}_a \star \mathrm{ev}_b = \mathrm{ev}_{a+b}. \tag{23}$$

Proof. Note $[\mathrm{ev}_a \star \mathrm{ev}_b](x^n) = [\mathrm{ev}_a(1) \cdot \mathrm{ev}_b(x) + \mathrm{ev}_a(x) \cdot \mathrm{ev}_b(1)]^n = (b + a)^n$.

Lemma 6. *The isomorphism $(\mathbb{K}, +) \ni a \mapsto \mathrm{ev}_a \in G_{\mathbb{K}}^{\mathbb{K}[x]}$ of groups is generated by the functional $\partial_0 = \mathrm{ev}_0 \circ \partial \in \mathfrak{g}_{\mathbb{K}}^{\mathbb{K}[x]}$, meaning $\log_\star \mathrm{ev}_a = a\partial_0$ and $\mathrm{ev}_a = \exp_\star(a\partial_0)$.*

Proof. Expanding the exponential series reveals $\exp_\star(a\partial_0)(x^n) = a^n$ as a direct consequence of $\partial_0^{\star k} = \varepsilon \circ \partial^{\star k} = \varepsilon \circ \partial^k$:

$$\partial_0^{\star k}\left(\frac{x^n}{n!}\right) = \sum_{l_1+\ldots+l_k=n} \left(\partial_0 \tfrac{x^{l_1}}{l_1!}\right) \cdots \left(\partial_0 \tfrac{x^{l_k}}{l_k!}\right) = \sum_{l_1+\ldots+l_k=n} \delta_{1,l_1} \cdots \delta_{1,l_k} = \delta_{k,n} = \partial^k\Big|_0 \left(\frac{x^n}{n!}\right).$$

4.1 Feynman Rules Induced by Cocycles

Let ${}_0\phi : H_R \to \mathbb{K}[x]$ denote the polynomials that evaluate to the renormalized Feynman rules ${}_0\phi_{R,s} = \mathrm{ev}_\ell \circ {}_0\phi$ at $\ell = \ln \frac{s}{\mu}$. We state

Theorem 4. *The renormalized Feynman rules ${}_0\phi = {}^L\rho$ arise out of the universal property of Theorem 1, where the coefficients c_n of (9) determine the cocycle*

$$L := -c_{-1} \int_0 + \delta\eta \in HZ_\varepsilon^1(\mathbb{K}[x]) \quad \textit{with} \quad \eta(x^n) := n!\,(-1)^n c_n \quad \textit{for any } n \in \mathbb{N}_0. \tag{24}$$

Proof. We may set $\mu = 1$ and produce logarithms of subdivergences by differentiation, exploiting analyticity of $zF(z)$ and $\frac{s^{-z}-1}{z}$ at $z = 0$ we obtain

$$\lim_{z\to 0}(\mathrm{id}-R_1)\left[s\mapsto\int_0^\infty f(\zeta)(s\zeta)^{-z}\ln^n(s\zeta)\,\mathrm{d}\zeta\right] = \left(-\frac{\partial}{\partial z}\right)^n_{z=0}(\mathrm{id}-R_1)\int_0^\infty f(\zeta)(s\zeta)^{-z}\,\mathrm{d}\zeta$$

$$= \left(-\frac{\partial}{\partial z}\right)^n_{z=0}\left[\frac{s^{-z}-1}{z}\cdot zF(z)\right] = (-1)^n\sum_{k=0}^{n}\binom{n}{k}k!\frac{(-\ln s)^{k+1}}{(k+1)!}(n-k)!\,c_{n-k-1}$$

$$= \mathrm{ev}_{\ln s}\left[-c_{-1}\frac{x^{n+1}}{n+1}+\sum_{i=1}^{n}\binom{n}{i}x^i(-1)^{n-i}c_{n-i}(n-i)!\right] = \mathrm{ev}_{\ln s}\circ L\left(x^n\right). \qquad (*)$$

By linearity we can replace $\ln^n(s\zeta)$ in the integrand by any polynomial to prove Theorem 4 inductively: As $_0\phi$ and $^L\rho$ are algebra morphisms, it suffices to consider a tree $t = B_+(w)$ for a forest $w \in \mathscr{F}$ already fulfilling $_0\phi(w) = {}^L\rho(w)$ in the induction step

$$_0\phi_{R,s}(t) \underset{(20)}{=} \lim_{z\to 0}(\mathrm{id}-R_1)\left[s\mapsto\int_0^\infty f(\zeta)(s\zeta)^{-z}\,\mathrm{ev}_{\ln s\zeta}\circ{}_0\phi(w)\,\mathrm{d}\zeta\right]$$

$$\underset{(*)}{=} \mathrm{ev}_\ell\circ L\left[{}_0\phi(w)\right] = \mathrm{ev}_\ell\circ L\circ{}^L\rho(w) \underset{1}{=} \mathrm{ev}_\ell\circ{}^L\rho\circ B_+(w) = \mathrm{ev}_\ell\circ{}^L\rho(t),$$

where the convergence of (20) allows to reintroduce ζ^{-z} into the integrand.

Corollary 1. *As L is a cocycle, by Theorem 1 the physical limit $_0\phi : H_R \to \mathbb{K}[x]$ of the renormalized Feynman rules* (8) *is a morphism of Hopf algebras.*

This key property naturally yields the renormalization group as we shall see in the sequel. For now observe the simple and explicit combinatorial recursion Example 4, expressing $_0\phi$ in terms of the Mellin transform coefficients without any need for series expansions in z, as shown in

Example 4. Using (24) we rederive $_0\phi\,(\bullet) = {}^L\rho\circ B_+(\mathbb{1}) = L(1) = -c_{-1}\,x$ and also

$$_0\phi\,(\vdots) = {}^L\rho\circ B_+(\bullet) = L\circ{}^L\rho(\bullet) = \left[-c_{-1}\int_0 + \delta\eta\right](-c_{-1}x) = c_{-1}^2\frac{x^2}{2} - c_{-1}c_0\,x,$$

$$_0\phi\,(\wedge) = {}^L\rho\circ B_+(\bullet\bullet) = L\circ{}^L\rho(\bullet\bullet) = \left[-c_{-1}\int_0 + \delta\eta\right]\left\{(-c_{-1}\,x)^2\right\}$$

$$= -c_{-1}^3\frac{x^3}{3} + c_{-1}^2\left[\eta(1)\,x^2 + 2\eta(x)\,x\right] = -c_{-1}^3\frac{x^3}{3} + c_{-1}^2c_0\,x^2 - 2c_{-1}^2c_1\,x.$$

Defining $\tilde{F}(z) := F(z) - \frac{c_{-1}}{z} = \sum_{n\in\mathbb{N}_0} c_n z^n$, (22) uncovers $\delta\eta = P\circ\tilde{F}(-\partial_x)$ and under the convention $\partial_x^{-1} := \int_0$ we may thus write $L = P\circ F(-\partial_x)$.

Corollary 2. *As in η only $-c_{-1}\int_0$ increases the degree in x, the highest order (called* leading log*) of ${}_0\phi$ is the tree factorial (note the analogy to* (11)*): For any forest $w \in \mathscr{F}$,*

$$ {}_0\phi(w) \in {}^{[-c_{-1}\int_0]}\rho(w) + \mathscr{O}\left(x^{|w|-1}\right) \underset{(6)}{=} \frac{\left(-c_{-1}x\right)^{|w|}}{w!} + \mathbb{K}[x]_{<|w|}. \tag{25} $$

4.2 Feynman Rules as Hopf Algebra Morphisms

As ${}_0\phi : H_R \to \mathbb{K}[x]$ is a morphism of Hopf algebras, the induced map $G_{\mathbb{K}}^{\mathbb{K}[x]} \to G_{\mathbb{K}}^{H_R}$ given by $\mathrm{ev}_a \mapsto {}_0\phi_a := \mathrm{ev}_a \circ {}_0\phi$ becomes a morphism of groups. In particular note

Corollary 3. *Using* (23) *we obtain the* renormalization group equation *(as in [17])*

$$ {}_0\phi_a \star {}_0\phi_b = {}_0\phi_{a+b}, \quad \textit{for any} \quad a, b \in \mathbb{K}. \tag{26} $$

Before we obtain the generator of this one-parameter group in Corollary 4, note how this result gives non-trivial relations between individual trees (graphs) like

$$ \begin{aligned} {}_0\phi_a \star {}_0\phi_b \left(\overset{\bullet}{\underset{\bullet}{|}}\right) &= {}_0\phi_a \left(\overset{\bullet}{\underset{\bullet}{|}}\right) + {}_0\phi_a\left(\bullet\right) {}_0\phi_b\left(\bullet\right) + {}_0\phi_b \left(\overset{\bullet}{\underset{\bullet}{|}}\right) \\ &\underset{(17)}{=} c_{-1}^2 \frac{a^2+b^2}{2} - c_{-1}c_0\left(a+b\right) + c_{-1}^2 ab \underset{(17)}{=} {}_0\phi_{a+b}\left(\overset{\bullet}{\underset{\bullet}{|}}\right). \end{aligned} $$

Proposition 3. *Let H be any connected bialgebra and $\phi \colon H \to \mathbb{K}[x]$ a morphism of bialgebras.*[4] *Then $\log_\star \phi$ is given by the linear term in x through*

$$ \log_\star \phi = x \cdot \partial_0 \circ \phi. \tag{27} $$

Proof. Letting $\phi : C \to H$ and $\psi : H \to \mathscr{A}$ denote morphisms of coalgebras and algebras, exploiting $(\psi \circ \phi - u_{\mathscr{A}} \circ \varepsilon_C)^{\star n} = \psi \circ (\phi - u_H \circ \varepsilon_H)^{\star n} = (\psi - u_{\mathscr{A}} \circ \varepsilon_H)^{\star n} \circ \phi$ in (2) proves $(\log_\star \psi) \circ \phi = \log_\star(\psi \circ \phi) = \psi \circ \log_\star \phi$. Now set $\psi = \mathrm{ev}_a$ and use Lemma 6.

Example 5. In the leading-log case (6) we read off $\partial_0 \circ \varphi = Z_\bullet \in \mathfrak{g}_{\mathbb{K}}^{H_R}$ where $Z_\bullet(w) := \delta_{w,\bullet}$. Comparing $\varphi = \exp_\star(xZ_\bullet)$ with (6) shows $|w|! = w! \cdot Z_\bullet^{\star|w|}(w)$, hence[5]

---

[4]This already implies ϕ to be a morphism of Hopf algebras.

[5]This combinatorial relation among tree factorials, noted in [17], thus drops out of $\Delta\varphi = (\varphi \otimes \varphi) \circ \Delta$.

$$\frac{|w|}{w!} = \frac{1}{(|w|-1)!}\sum_{w} Z_{\bullet}(w_1) Z_{\bullet}^{\star |w|-1}(w_2) = \sum_{w\colon w_1=\bullet} \frac{1}{|w_2|!} Z_{\bullet}^{\star |w_2|}(w_2) = \sum_{w\colon w_1=\bullet} \frac{1}{w_2!}.$$

Corollary 4. *The character* ${}_0\phi$ *is fully determined by the* anomalous dimension

$$H_R' \supset \mathfrak{g}_{\mathbb{K}}^{H_R} \ni \gamma := -\partial_0 \circ {}_0\phi \quad \textit{such that} \quad {}_0\phi = \exp_\star(-x\cdot\gamma) = \sum_{n\in\mathbb{N}_0} \frac{\gamma^{\star n}}{n!}(-x)^n. \tag{28}$$

An analogous phenomenon happens with the counterterms in the minimal subtraction scheme: The first order poles $\propto z^{-1}$ alone already determine the full counterterm via the *scattering formula* proved in [10]. However, (28) is much simpler as illustrated in

Example 6. Reading off $\gamma(\bullet) = c_{-1}$, $\gamma\left(\overset{\bullet}{\underset{\bullet}{|}}\right) = c_{-1}c_0$ and $\gamma\left(\overset{\bullet}{\bullet\wedge\bullet}\right) = 2c_{-1}^2 c_1$ from the Example 4 above, Corollary 4 determines the higher powers of x through

$$ {}_0\phi\left(\overset{\bullet}{\underset{\bullet}{|}}\right) \underset{(2)}{=} \left[e - x\gamma + x^2\frac{\gamma\star\gamma}{2}\right]\left(\overset{\bullet}{\underset{\bullet}{|}}\right) = 0 - x\gamma\left(\overset{\bullet}{\underset{\bullet}{|}}\right) + x^2\frac{\gamma^2(\bullet)}{2} = -c_{-1}c_0\,x + c_{-1}^2\frac{x^2}{2},$$

$$ {}_0\phi\left(\overset{\bullet}{\bullet\wedge\bullet}\right) = 0 - x\gamma\left(\overset{\bullet}{\bullet\wedge\bullet}\right) + x^2\frac{\gamma\otimes\gamma}{2}\left(2\bullet\otimes\overset{\bullet}{\underset{\bullet}{|}} + \bullet\bullet\otimes\bullet\right) - x^3\frac{\gamma\otimes\gamma\otimes\gamma}{6}\left(2\bullet\otimes\bullet\otimes\bullet\right)$$

$$= -\gamma^3(\bullet)\frac{x^3}{3} + x^2\gamma(\bullet)\gamma\left(\overset{\bullet}{\underset{\bullet}{|}}\right) - 2c_{-1}^2c_1\,x = -c_{-1}^3\frac{x^3}{3} + c_{-1}^2c_0\,x^2 - 2c_{-1}^2c_1\,x.$$

Note how the fragment $\bullet\bullet\otimes\bullet$ of $\Delta\left(\overset{\bullet}{\bullet\wedge\bullet}\right)$ does not contribute to the quadratic terms $\frac{x^2}{2}\gamma\star\gamma$, as γ vanishes on products. We will exploit this in (33) of Sect. 5.1 and close with a method of calculating γ emerging from

Lemma 7. *From* $\gamma\circ B_+ = -\partial_0\circ L\circ{}_0\phi = \mathrm{ev}_0\circ[zF(z)]_{-\partial_x}\circ\exp_\star(-x\gamma)$ *we obtain the inductive formula* $\gamma\circ B_+ = \sum_{n\in\mathbb{N}_0} c_{n-1}\gamma^{\star n}$.

Example 7. We can recursively calculate $\gamma(\bullet) = c_{-1}\varepsilon(\mathbb{1}) = c_{-1}$, similarly also

$$\gamma\left(\overset{\bullet}{\underset{\bullet}{|}}\right) = c_{-1}\varepsilon(\bullet) + c_0\gamma(\bullet) = c_{-1}c_0,$$

$$\gamma\left(\overset{\bullet}{\underset{\bullet}{\overset{|}{\underset{|}{\bullet}}}}\right) = c_{-1}\varepsilon\left(\overset{\bullet}{\underset{\bullet}{|}}\right) + c_0\gamma\left(\overset{\bullet}{\underset{\bullet}{|}}\right) + c_1\gamma\star\gamma\left(\overset{\bullet}{\underset{\bullet}{|}}\right) = c_{-1}c_0^2 + c_1\left[\gamma(\bullet)\right]^2 = c_{-1}c_0^2 + c_{-1}^2c_1,$$

$$\gamma\left(\overset{\bullet}{\bullet\wedge\bullet}\right) = c_{-1}\varepsilon(\bullet\bullet) + c_0\gamma(\bullet\bullet) + c_1\gamma\star\gamma(\bullet\bullet) = 2c_1\left[\gamma(\bullet)\right]^2 = 2c_{-1}^2c_1 \quad \text{and so on.}$$

5 Dyson-Schwinger Equations and Correlation Functions

We now study the implications for the *correlation functions* (31) as formal power series in the *coupling constant* g. For simplicity we restrict to a single equation and refer to [24] for systems. With detailed treatments in [1, 11], for our purposes suffices

Definition 5. To a parameter $\kappa \in \mathbb{K}$ and a family of cocycles $B_\cdot : \mathbb{N} \to \mathrm{HZ}^1_\varepsilon(H_R)$ we associate the *combinatorial Dyson-Schwinger equation*[6]

$$X(g) = \mathbb{1} + \sum_{n\in\mathbb{N}} g^n B_n \left(X^{1+n\kappa}(g)\right). \tag{29}$$

Lemma 8. *As* perturbation series $X(g) = \sum_{n\in\mathbb{N}_0} x_n g^n \in H_R[[g]]$, *Eq.* (29) *has a unique solution. It begins with* $x_0 = \mathbb{1}$ *while* x_{n+1} *is determined recursively from* $x_0, \ldots, x_n$. *These coefficients generate a Hopf subalgebra, explicitly we find*[7]

$$\Delta X(g) = \sum_{n\in\mathbb{N}_0} [X(g)]^{1+n\kappa} \otimes g^n x_n \in (H_R \otimes H_R)[[g]]. \tag{30}$$

Example 8. In [5,22], $X(g) = \mathbb{1} - gB_+\left(\frac{1}{X(g)}\right)$ features $\kappa = -2$, summing all trees

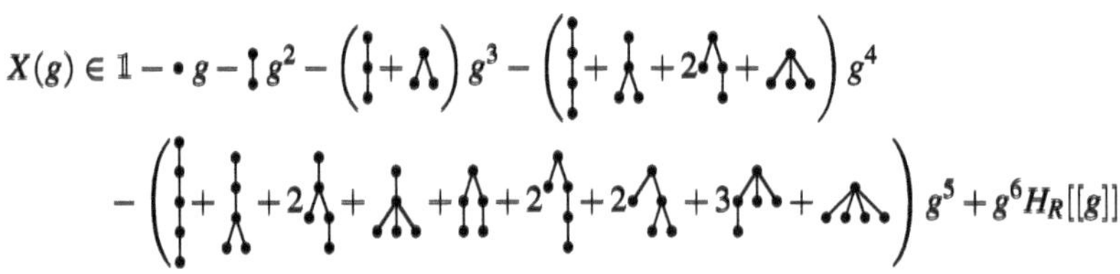

with a combinatorial factor.[8] Physically these correspond to (Yukawa) propagators

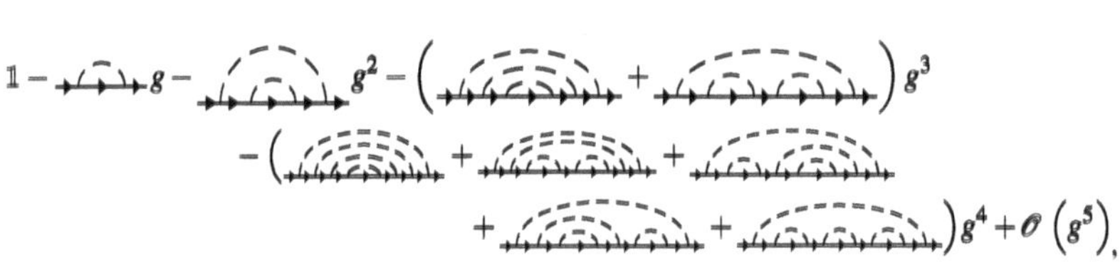

arising from insertions of the one-loop graph  into itself.

[6]As $x_0 = \mathbb{1}$, for arbitrary p the series $[X(g)]^p := \sum_{n\in\mathbb{N}_0} \binom{p}{n} [X(g) - \mathbb{1}]^n \in H_R[[g]]$ is well defined.

[7]A proof of (30) may be found in [11] and [12, 13] study systems of Dyson-Schwinger equations.

[8]Counting the number of corresponding ordered trees.

Definition 6. The *correlation function* $G(g)$ evaluates the renormalized Feynman rules ${}_0\phi : H_R \to \mathbb{K}[\ell]$ on the perturbation series $X(g)$, yielding the formal power series

$$G(g) := {}_0\phi \circ X(g) = \sum_{n \in \mathbb{N}_0} {}_0\phi(x_n) g^n \in (\mathbb{K}[\ell])\,[[g]]. \tag{31}$$

We call $\tilde{\gamma}(g) := \gamma \circ X(g) = -\partial_\ell|_0 G(g) \in \mathbb{K}[[g]]$ the *physical anomalous dimension.*

Example 9. The Feynman rules φ from (6) result in the convergent series $G(g) = \sqrt{1 - 2g\ell}$ and $\tilde{\gamma}(g) \underset{\S}{=} -Z_\bullet \circ X(g) = g$ for the propagator of Example 8. Perturbatively,

$$G(g) = 1 - \frac{(g\ell)}{\bullet!} - \frac{(g\ell)^2}{\overset{\bullet}{\underset{\bullet}{|}}!} - \frac{(g\ell)^3}{\overset{\bullet}{\underset{\bullet}{|}}\!\!\underset{\bullet}{|}!} - \frac{(g\ell)^3}{\Lambda!} - \ldots = 1 - g\ell - \frac{1}{2}(g\ell)^2 - \frac{1}{2}(g\ell)^3 + \mathcal{O}\left((g\ell)^4\right)$$

5.1 Propagator Coupling Duality

The Hopf subalgebra of the perturbation series allows to calculate convolutions in

Lemma 9. *Let* $\psi \in \mathfrak{g}_{\mathscr{A}}^{H_R}$ *denote an infinitesimal character,* $\Psi \in G_{\mathscr{A}}^{H_R}$ *a character and* $\lambda \in \mathrm{Hom}(H_R, \mathscr{A})$ *a linear map. Then (in suggestive notation)*

$$\begin{aligned}(\Psi \star \lambda) \circ X(g) &= [\Psi \circ X(g)] \cdot \lambda \circ X\left(g\,[\Psi \circ X(g)]^\kappa\right) \qquad (32)\\ &:= [\Psi \circ X(g)] \cdot \sum_{n \in \mathbb{N}_0} \lambda(x_n) \cdot \left(g\,[\Psi \circ X(g)]^\kappa\right)^n \in \mathscr{A}[[g]]\end{aligned}$$

$$(\psi \star \lambda) \circ X(g) = [\psi \circ X(g)] \cdot \left(\mathrm{id} + \kappa g \partial_g\right)[\lambda \circ X(g)] \in \mathscr{A}[[g]]. \tag{33}$$

Proof. These are immediate consequences of Lemma 8, for (33) consider

$$\psi\left([X(g)]^{1+n\kappa}\right) \cdot g^n = \sum_{i \in \mathbb{N}_0} \binom{1+n\kappa}{i} \psi\left([X(g) - \mathbb{1}]^i\right) g^n = \psi\left(X(g) - \mathbb{1}\right) \cdot (1 + n\kappa) g^n.$$

Example 10. Continuing 8 we deduce $Z_\bullet^{\star 2}(X(g)) = -g(1 - 2g\partial_g)(-g) = -g^2$ and

$$Z_\bullet^{\star n+1}(X(g)) \underset{(33)}{=} -g^{n+1}(2n-1)(2n-3)\cdots(1) = -g^{n+1}\frac{(2n)!}{2^n n!},$$

proving $\varphi(x_{n+1}) = -2^{-n} C_n \ell^{n+1}$ with the *Catalan numbers* C_n already noted in [20]. From Example 9 we find their generating function $2g \sum_{n\in\mathbb{N}_0} g^n C_n = 1 - \sqrt{1-4g}$.

Corollary 5. *As* ${}_0\phi$ *is a morphism of Hopf algebras, for any* $a, b \in \mathbb{K}$ *we can factor*

$$G_{a+b}(g) = ({}_0\phi_a \star {}_0\phi_b) \circ X(g) \underset{(32)}{=} G_a(g) \cdot G_b\left[g G_a^{\kappa}(g)\right] = G_b(g) \cdot G_a\left[g G_b^{\kappa}(g)\right]. \tag{34}$$

These functional equations of formal power series make sense for the non-perturbative correlation functions as well. Relating the scale- with the coupling-dependence, this integrated form of the renormalization group equation becomes infinitesimally

Corollary 6. *From* $-\frac{\mathrm{d}}{\mathrm{d}^x}\, {}_0\phi = \gamma \star {}_0\phi = {}_0\phi \star \gamma$ *or differentiating* (34) *by b at zero note*

$$G_\ell(g) \cdot \tilde{\gamma}\left[g G_\ell^{\kappa}(g)\right] \underset{(32)}{=} -\partial_\ell G_\ell(g) \underset{(33)}{=} \tilde{\gamma}(g) \cdot \left(1 + \kappa g \partial_g\right) G_\ell(g). \tag{35}$$

The first of these equations generalizes the *propagator coupling duality* in [5, 20]. For any fixed coupling g, it expresses the correlation function as the solution of the o.d.e.

$$-\frac{\mathrm{d}}{\mathrm{d}\ell} \ln G_\ell(g) = \tilde{\gamma}\left[g e^{\kappa \ln G_\ell(g)}\right] \quad \text{with} \quad \ln G_0(g) = 0, \tag{36}$$

determining $G_\ell(g)$ completely from $\tilde{\gamma}(g)$ in a non-perturbative manner as in (39).

Example 11. The *leading*-log expansion takes only the highest power of ℓ in each g-order. Equally, $\tilde{\gamma}(g) = c g^n$ for constants $c \in \mathbb{K}$, $n \in \mathbb{N}$ and (36) integrates to

$$G_{\text{leading-log}}(g) = \left[1 + c n \kappa \ell g^n\right]^{-\frac{1}{n\kappa}}. \tag{37}$$

As a special case we recover Example 9 for $n = c = 1$ and $\kappa = -2$.

Example 12. In the linear case $\kappa = 0$, (34) states $G_{a+b}(g) = G_a(g) \cdot G_b(g)$ in accordance with the *scaling solution* $G_\ell(g) = e^{-\ell \tilde{\gamma}(g)}$ of (36), well-known from [19].

Example 13. For vertex insertions as in [2] we have $\kappa = 1$, so $G_{a+b}(g) = G_b(g) \cdot G_a\left[\tilde{G}_b(g)\right]$ expresses the running of the coupling constant $\tilde{G} := g \cdot G$: A change in scale by b is (up to a multiplicative constant) equivalent to replacing the coupling g by $\tilde{G}_b(g)$.

5.2 The Physicist's Renormalization Group

To cast (34) and (35) into the common forms of (7.3.15) and (7.3.21) in [7], we introduce the *β-function* $\beta(g) := -\kappa g\tilde{\gamma}(g)$ and the *running coupling* $g(\mu)$ as the solution of

$$\mu\frac{\mathrm{d}}{\mathrm{d}\mu}g(\mu) = \beta(g(\mu)), \quad \text{hence} \quad \mu\frac{\mathrm{d}}{\mathrm{d}\mu}G\left(g(\mu), \ln\tfrac{s}{\mu}\right) \underset{(35)}{=} \tilde{\gamma}(g(\mu))G\left(g(\mu), \ln\tfrac{s}{\mu}\right). \tag{38}$$

Integration relates the correlation functions for different renormalization points μ in

$$\begin{aligned} G\left(g(\mu_2), \ln\tfrac{s}{\mu_2}\right) &= G\left(g(\mu_1), \ln\tfrac{s}{\mu_1}\right)\cdot \exp\left[\int_{\mu_1}^{\mu_2}\tilde{\gamma}(g(\mu))\tfrac{\mathrm{d}\mu}{\mu}\right] \\ &\underset{(38)}{=} G\left(g(\mu_1), \ln\tfrac{s}{\mu_1}\right)\cdot\left[\frac{g(\mu_2)}{g(\mu_1)}\right]^{-\frac{1}{\kappa}}. \end{aligned}$$

Setting $\mu_1 = s$ we may thus write $G_\ell(g)$ explicitly in terms of $\tilde{\gamma}(g)$ as

$$G_\ell(g) = \left[\frac{g}{g(s)}\right]^{-\frac{1}{\kappa}}, \quad \text{with } g(s) \text{ subject to} \quad \ell = \ln\frac{s}{\mu} = \int_g^{g(s)}\frac{\mathrm{d}g'}{\beta(g')}. \tag{39}$$

5.3 Relation to Mellin Transforms

We finally exploit the analytic input from Theorem 4 to the perturbation series in

$$G_\ell(g) \underset{(29)}{=} 1+\sum_{n\in\mathbb{N}} g^n{}_0\phi\circ B_n\left(X(g)^{1+n\kappa}\right) \underset{4}{=} 1+\sum_{n\in\mathbb{N}} g^n\left[-c_{-1}^{(n)}\int_0 + P\circ\widetilde{F}_n(\partial_{-\ell})\right]G_\ell(g)^{1+n\kappa},$$

with Mellin transforms $F_n(z) = \frac{1}{z}c_{-1}^{(n)} + \tilde{F}_n(z)$ corresponding to the insertions[9] B_n.

Corollary 7. *The power series $G_\ell(g) \in \mathbb{K}[\ell][[g]]$ is fully determined by*

$$G_\ell(0) = 1 \quad \textit{and} \quad \partial_{-\ell}G_\ell(g) \underset{(9)}{=} \sum_{n\in\mathbb{N}} g^n\left[zF_n(z)\right]_{z=-\partial_\ell}\left(G_\ell(g)^{1+n\kappa}\right). \tag{40}$$

Restricting to a single cocycle $F_k(z) = F(z)\delta_{k,n}$, choosing $F(z) = \frac{c_{-1}}{z}$ reproduces (37) from $\partial_{-\ell}G_\ell(g) = g^n c_{-1}G_\ell(g)^{1+n\kappa}$. More generally, for any rational $F(z) = \frac{p(z)}{q(z)} \in \mathbb{K}(z)$ with $q(0) = 0$, (40) collapses to a finite order ode $q(-\partial_\ell)G_\ell(g) =$

[9] For this generality we need decorated rooted trees as commented on in Sect. 6.1

$g^n p(-\partial_\ell) G_\ell(g)^{1+n\kappa}$ that makes perfect sense non-perturbatively (extending the algebraic $\partial_\ell \in \operatorname{End}(\mathbb{K}[\ell])$ to the analytic differential operator).

Example 14. For $F(z) = \frac{1}{z(1-z)}$, the propagator ($\kappa = -2$ as in Example 8) fulfils

$$\frac{g}{G_\ell(g)} = \partial_{-\ell}\left(1 - \partial_{-\ell}\right) G_\ell(g) \underset{(35)}{=} \tilde{\gamma}(g)\left(1 - 2g\partial_g\right)\left[1 - \tilde{\gamma}(g)\left(1 - 2g\partial_g\right)\right] G_\ell(g).$$

At $\ell = 0$ this evaluates to $\tilde{\gamma}(g) - \tilde{\gamma}(g)(1 - 2g\partial_g)\tilde{\gamma}(g) = g$, which is studied in [5, 24].

6 Automorphisms of H_R

Applying the universal property to H_R itself, adding coboundaries to B_+ leads to

Definition 7. For any $\alpha \in H_R'$, Theorem 1 defines the Hopf algebra morphism

$${}^{\alpha}\chi := {}^{B_+ + \delta\alpha}\rho\colon\ H_R \to H_R \quad \text{such that} \quad {}^{\alpha}\chi \circ B_+ = [B_+ + \delta\alpha] \circ {}^{\alpha}\chi. \tag{41}$$

Example 15. The action on the simplest trees yields

$${}^{\alpha}\chi(\bullet) = {}^{\alpha}\chi \circ B_+(\mathbb{1}) = B_+(\mathbb{1}) + (\delta\alpha)(\mathbb{1}) = B_+(\mathbb{1}) = \bullet,$$

$${}^{\alpha}\chi\left(\begin{smallmatrix}\bullet\\ \vert\\ \bullet\end{smallmatrix}\right) = {}^{\alpha}\chi \circ B_+(\bullet) = (B_+ + \delta\alpha)\,{}^{\alpha}\chi(\bullet) = \begin{smallmatrix}\bullet\\ \vert\\ \bullet\end{smallmatrix} + \delta\alpha(\bullet) = \begin{smallmatrix}\bullet\\ \vert\\ \bullet\end{smallmatrix} + \alpha(\mathbb{1})\bullet,$$

$${}^{\alpha}\chi\left(\begin{smallmatrix}\bullet\\ \vert\\ \bullet\\ \vert\\ \bullet\end{smallmatrix}\right) = \begin{smallmatrix}\bullet\\ \vert\\ \bullet\\ \vert\\ \bullet\end{smallmatrix} + 2\alpha(\mathbb{1})\begin{smallmatrix}\bullet\\ \vert\\ \bullet\end{smallmatrix} + \left\{[\alpha(\mathbb{1})]^2 + \alpha(\bullet)\right\}\bullet \quad \text{and} \quad {}^{\alpha}\chi\left(\Lambda\right) = \Lambda + 2\alpha(\bullet)\bullet + \alpha(\mathbb{1})\bullet\bullet.$$

These morphisms capture the change of ${}^{L}\rho$ under a variation of L by a coboundary:

Theorem 5. *Let H denote a bialgebra, $L \in HZ_\varepsilon^1(H)$ a 1-cocycle and further $\alpha \in H'$ a functional. Then for ${}^{L}\rho, {}^{L+\delta\alpha}\rho\colon\ H_R \to H$ given through Theorem 1 and ${}^{\alpha\circ{}^{L}\rho}\chi\colon\ H_R \to H_R$ from Definition 7, we have*

$${}^{L+\delta\alpha}\rho = {}^{L}\rho \circ {}^{[\alpha\circ{}^{L}\rho]}\chi, \quad \textit{equivalently} \quad \begin{array}{ccc} H_R & \xrightarrow{\ {}^{L+\delta\alpha}\rho\ } & H \\ {\scriptstyle {}^{\alpha\circ{}^{L}\rho}\chi}\big\downarrow & \nearrow{\scriptstyle {}^{L}\rho} & \\ H_R & & \end{array} \quad \textit{commutes.} \tag{42}$$

Proof. As both sides of (42) are algebra morphisms, it suffices to prove it inductively for trees: Let it be true for a forest $w \in \mathscr{F}$, then it holds as well for the tree $B_+(w)$ by

$$
\begin{aligned}
{}^{L}\rho \circ {}^{[\alpha\circ {}^{L}\rho]}\chi \circ B_+(w) &\underset{(5)}{=} {}^{L}\rho \circ \left[B_+ + \delta\left(\alpha \circ {}^{L}\rho\right)\right] \circ {}^{[\alpha\circ {}^{L}\rho]}\chi(w) \\
&= \left\{L \circ {}^{L}\rho + (\delta\alpha) \circ {}^{L}\rho\right\} \circ {}^{[\alpha\circ {}^{L}\rho]}\chi(w) = \{L + \delta\alpha\} \circ \underbrace{{}^{L}\rho \circ {}^{[\alpha\circ {}^{L}\rho]}\chi(w)}_{{}^{L+\delta\alpha}\rho(w)} \underset{(5)}{=} {}^{L+\delta\alpha}\rho \circ B_+(w).
\end{aligned}
$$

We used $(\delta\alpha) \circ {}^{L}\rho = {}^{L}\rho \circ \delta\left(\alpha \circ {}^{L}\rho\right)$, following from ${}^{L}\rho$ being a morphism of bialgebras.

Hence the action of a coboundary $\delta\alpha$ on the universal morphisms induced by L is given by ${}^{\alpha\circ {}^{L}\rho}\chi$. This turns out to be an automorphism of H_R as shown in

Theorem 6. *The map* ${}^{\cdot}\chi\colon\ H_R' \to \mathrm{End}_{Hopf}(H_R)$, *taking values in the space of Hopf algebra endomorphisms of* H_R, *fulfils the following properties:*

1. *For any* $w \in \mathscr{F}$ *and* $\alpha \in H_R'$, ${}^{\alpha}\chi(w)$ *differs from* w *only by lower order forests:*

$$
{}^{\alpha}\chi(w) \in w + H_R^{|w|-1} = w + \bigoplus_{n=0}^{|w|-1} H_{R,n}. \tag{43}
$$

2. ${}^{\cdot}\chi$ *maps* H_R' *into the Hopf algebra automorphisms* $\mathrm{Aut}_{Hopf}(H_R)$. *Its image is closed under composition, as for any* $\alpha, \beta \in H_R'$ *we have* ${}^{\alpha}\chi \circ {}^{\beta}\chi = {}^{\gamma}\chi$ *taking*

$$
\gamma = \alpha + \beta \circ {}^{\alpha}\chi^{-1}. \tag{44}
$$

3. *The maps* $\delta\colon H_R' \to HZ_\varepsilon^1(H_R)$ *and* ${}^{\cdot}\chi\colon H_R' \to \mathrm{Aut}_{Hopf}(H_R)$ *are injective, thus the subgroup* $\mathrm{im}\,{}^{\cdot}\chi = \left\{{}^{\alpha}\chi\colon\ \alpha \in H_R'\right\} \subset \mathrm{Aut}_{Hopf}(H_R)$ *induces a group structure on* H_R' *with neutral element* 0 *and group law* $\triangleright$ *given by*

$$
\alpha \triangleright \beta := {}^{\cdot}\chi^{-1}\left({}^{\alpha}\chi \circ {}^{\beta}\chi\right) \underset{(44)}{=} \alpha + \beta \circ {}^{\alpha}\chi^{-1} \quad \text{and} \quad \alpha^{\triangleright -1} = -\alpha \circ {}^{\alpha}\chi. \tag{45}
$$

Proof. Statement (43) is an immediate consequence of $\delta\alpha(H_R^n) \subseteq H_R^n$: Starting from ${}^{\alpha}\chi\,(\bullet) = \bullet$, suppose inductively (43) to hold for forests $w, w' \in \mathscr{F}$. Then it obviously also holds for $w \cdot w'$ as well and even so for $B_+(w)$ through

$$
{}^{\alpha}\chi \circ B_+(w) = [B_+ + \delta\alpha] \circ {}^{\alpha}\chi(w) \subseteq [B_+ + \delta\alpha]\left(w + H_R^{|w|-1}\right) \subseteq B_+(w) + H_R^{|w|}.
$$

This already implies bijectivity of ${}^{\alpha}\chi$, but applying (42) to $L = B_+ + \delta\alpha$ and ${}^{\tilde{\alpha}}\chi$ for $\tilde{\alpha} := -\alpha \circ {}^{\alpha}\chi$ shows $\mathrm{id} = {}^{\alpha}\chi \circ {}^{\tilde{\alpha}}\chi$ directly. We deduce bijectivity of all ${}^{\alpha}\chi$ and thus ${}^{\alpha}\chi \in \mathrm{Aut}_{\mathrm{Hopf}}(H_R)$ with the inverse ${}^{\alpha}\chi^{-1} = {}^{\tilde{\alpha}}\chi$. Now (44) follows from

$$
{}^{[\alpha+\beta\circ{}^{\alpha}\chi^{-1}]}\chi = {}^{[B_+ + \delta\alpha] + \delta(\beta\circ{}^{\alpha}\chi^{-1})}\rho \underset{(42)}{=} {}^{[B_+ + \delta\alpha]}\rho \circ {}^{\left[\beta\circ{}^{\alpha}\chi^{-1}\circ\left({}^{B_+ + \delta\alpha}\right)\rho\right]}\chi = {}^{\alpha}\chi \circ {}^{\beta}\chi.
$$

Finally consider $\alpha, \beta \in H_R'$ with ${}^{\alpha}\chi = {}^{\beta}\chi$, then $0 = ({}^{\alpha}\chi - {}^{\beta}\chi) \circ B_+ = \delta \circ (\alpha - \beta) \circ {}^{\alpha}\chi$ reduces the injectivity of ${}^{\cdot}\chi$ to that of δ. But if $\delta\alpha = 0$, for all $n \in \mathbb{N}_0$

$$0 = \delta\alpha\left(\bullet^{n+1}\right) = \sum_{i=0}^{n} \binom{n+1}{i} \alpha\left(\bullet^{i}\right) \bullet^{n+1-i} \quad \text{implies} \quad \alpha\left(\bullet^{n}\right) = 0.$$

Given an arbitrary forest $w \in \mathscr{F}$ and $n \in \mathbb{N}$, the expression

$$0 = \delta\alpha\left(\bullet^{n} w\right) = w \underbrace{\alpha\left(\bullet^{n}\right)}_{0} + \sum_{w} \sum_{i=0}^{n} \binom{n}{i} \bullet^{i} w' \alpha\left(\bullet^{n-i} w''\right)$$
$$+ \sum_{i=1}^{n} \binom{n}{i} \left[\bullet^{i} w \underbrace{\alpha\left(\bullet^{n-i}\right)}_{0} + \bullet^{i} \alpha\left(w \bullet^{n-i}\right) \right]$$

simplifies upon projection onto $\mathbb{K}\bullet$ to $\alpha\left(w\bullet^{n-1}\right) = -\frac{1}{n}\sum_{w:\, w'=\bullet} \alpha\left(\bullet^{n} w''\right)$. Iterating this formula exhibits $\alpha(w)$ as a scalar multiple of $\alpha\left(\bullet^{|w|}\right) = 0$ and proves $\alpha = 0$.

6.1 Decorated Rooted Trees

Our observations generalize straight forwardly to the Hopf algebra $H_R(\mathscr{D})$ of rooted trees with decorations drawn from a set $\mathscr{D}$. In this case, the universal property assigns to each $\mathscr{D}$-indexed family $L_\cdot\colon\ \mathscr{D} \to \mathrm{End}(\mathscr{A})$ the unique algebra morphism

$${}^{L_\cdot}\rho\colon H_R(\mathscr{D}) \to \mathscr{A} \quad \text{such that} \quad {}^{L_\cdot}\rho \circ B_+^d = L_d \circ {}^{L_\cdot}\rho \quad \text{for any } d \in \mathscr{D}.$$

For cocycles $\operatorname{im} L_\cdot \subseteq HZ_\varepsilon^1(\mathscr{A})$ this is a morphism of bialgebras and even of Hopf algebras (should $\mathscr{A}$ be Hopf). For a family $\alpha_\cdot\colon\ \mathscr{D} \to H_R'(\mathscr{D})$ of functionals, setting $L_d^{\alpha_\cdot} := B_+^d + \delta\alpha_d$ yields an automorphism ${}^{\alpha_\cdot}\chi := {}^{L^{\alpha_\cdot}}\rho$ of the Hopf algebra $H_R(\mathscr{D})$. Theorems 5 and 6 generalize in the obvious way.

In view of the Feynman rules, decorations d denote different graphs into which B_+^d inserts a subdivergence. Hence we gain a family of Mellin transforms $F_\cdot$ and Theorem 4 generalizes straightforwardly as ${}_0\phi \circ B_+^d = P \circ F_d(-\partial_\ell) \circ {}_0\phi$.

6.2 Subleading Corrections Under Variations of Mellin Transforms

As an application of (42) consider a change of the Mellin transform F to a different F' that keeps c_{-1} fixed but alters the other coefficients c_n. With $\alpha := \eta' - \eta$,

$${}_0\phi' = {}^{L'}\rho = {}^{L+\delta\alpha}\rho = {}^{L}\rho \circ {}^{[\alpha \circ {}^L\rho]}\chi = {}_0\phi \circ {}^{[\alpha \circ {}_0\phi]}\chi$$

translates the new renormalized Feynman rules ${}_0\phi'$ into the original ${}_0\phi$.

Fixing $c_{-1} = -1$, this in particular relates ${}_0\phi$ to $\varphi = \int_0 \rho$ using Example 15 together with $\eta \circ \varphi(w) = (-1)^{|w|} \frac{|w|!}{w!} c_{|w|}$ as

$$
\begin{aligned}
{}_0\phi\,(\bullet) &= x = \varphi\,(\bullet) = \varphi \circ {}^{\eta\circ\varphi}\chi\,(\bullet),\\
{}_0\phi\left(\begin{smallmatrix}\bullet\\|\\\bullet\end{smallmatrix}\right) &= \frac{x^2}{2} + c_0 x = \varphi\left\{\begin{smallmatrix}\bullet\\|\\\bullet\end{smallmatrix} + \eta(1)\bullet\right\} = \varphi \circ {}^{\eta\circ\varphi}\chi\left(\begin{smallmatrix}\bullet\\|\\\bullet\end{smallmatrix}\right),\\
{}_0\phi\left(\begin{smallmatrix}\bullet\\|\\\bullet\\|\\\bullet\end{smallmatrix}\right) &= \frac{x^3}{6} + x^2 c_0 + x(c_0^2 - c_1) = \varphi\left\{\begin{smallmatrix}\bullet\\|\\\bullet\\|\\\bullet\end{smallmatrix} + 2c_0 \begin{smallmatrix}\bullet\\|\\\bullet\end{smallmatrix} + \left[c_0^2 - c_1\right]\bullet\right\} = \varphi \circ {}^{\eta\circ\varphi}\chi\left(\begin{smallmatrix}\bullet\\|\\\bullet\\|\\\bullet\end{smallmatrix}\right)
\end{aligned}
$$

and

$$
{}_0\phi\left(\begin{smallmatrix}&\bullet&\\/&&\backslash\\\bullet&&\bullet\end{smallmatrix}\right) = \frac{x^3}{3} + c_0 \cdot x^2 - 2c_1 \cdot x = \varphi\left\{\begin{smallmatrix}&\bullet&\\/&&\backslash\\\bullet&&\bullet\end{smallmatrix} + c_0 \bullet\bullet - 2c_1 \bullet\right\} = \varphi \circ {}^{\eta\circ\varphi}\chi\left(\begin{smallmatrix}&\bullet&\\/&&\backslash\\\bullet&&\bullet\end{smallmatrix}\right).
$$

Corollary 8. *The new correlation function ${}_0\phi \circ X = \varphi \circ \tilde{X}$ equals the original φ applied to a modified perturbation series $\tilde{X}(g)$, fulfilling a Dyson-Schwinger equation differing by coboundaries. By* (43) *the leading logs coincide and explicitly*

$$
\tilde{X}(g) := {}^{\eta\circ\varphi}\chi \circ X(g) = \mathbb{1} + \sum_{n\in\mathbb{N}} g^n \left(B_n + \delta\eta_n\right)\left(\tilde{X}(g)^{1+n\kappa}\right).
$$

7 Locality, Finiteness and Minimal Subtraction

Consider the regularized but unrenormalized Feynman rules ${}_z\phi$. Now setting $\mathscr{A} := \mathbb{K}[z^{-1}, z]]$ and $\phi := {}_z\phi_1 \in G_{\mathscr{A}}^{H_R}$, (10) fixes the scale dependence ${}_z\phi_s = \phi \circ \theta_{-\ln sz}$.

Proposition 4. *For any character $\phi \in G_{\mathscr{A}}^{H_R}$, the following are equivalent:*

1. *$\phi^{\star-1} \star (\phi \circ Y) = \phi \circ (S \star Y)$ maps into $\frac{1}{z}\mathbb{K}[[z]]$, so $\lim_{z\to 0} \phi^{\star-1} \star (z\phi \circ Y)$ exists.*
2. *For every $n \in \mathbb{N}_0$, $\phi^{\star-1} \star (\phi \circ Y^n) = \phi \circ (S \star Y^n)$ maps into $z^{-n}\mathbb{K}[[z]]$.*
3. *For any $\ell \in \mathbb{K}$, $\phi^{\star-1} \star (\phi \circ \theta_{\ell z}) = \phi \circ (S \star \theta_{\ell z})$ maps into $\mathbb{K}[[z]]$.*

Proof. We refer to the accounts in [5, 10, 21], however only 1. $\Rightarrow$ 2. is non-trivial and

$$
\phi \circ \left(S \star Y^{n+1}\right) = \phi \circ (S \star Y^n) \circ Y + [\phi \circ (S \star Y)] \star [\phi \circ (S \star Y^n)]
$$

yields an inductive proof. It exploits $(S \circ Y) \star \mathrm{id} = -S \star Y$ in the formula (α arbitrary)

$$
S \star (\alpha \circ Y) - (S \star \alpha) \circ Y = -(S \circ Y) \star \alpha = -\left[(S \circ Y) \star \mathrm{id}\right] \star S \star \alpha = S \star Y \star S \star \alpha.
$$

Note that condition 3. is equivalent to the finiteness Proposition 2 of the physical limit ${}_0\phi$ as

$$ {}_z\phi_{R,s} = {}_z\phi_\mu^{\star-1} \star {}_z\phi_s = \phi \circ \left[(S \circ \theta_{-z\ln\mu}) \star \theta_{-z\ln s} \right] = \phi \circ (S \star \theta_{-z\ln\frac{s}{\mu}}) \circ \theta_{-z\ln\mu}. $$

Corollary 9. *The anomalous dimension can be obtained from the $\frac{1}{z}$-pole coefficients*

$$ \gamma = -\partial_0 \circ {}_0\phi = -\partial_0 \circ \lim_{z\to 0} \phi \circ (S \star \theta_{-zx}) = \operatorname{Res}\left[\phi \circ (S \star Y)\right]. \tag{46} $$

The minimal subtraction scheme R_{MS} projects onto the pole parts such that $\mathscr{A} = \mathscr{A}_- \oplus \mathscr{A}_+$ where $\mathscr{A}_- := z^{-1}\mathbb{K}[z^{-1}]$ and $\mathscr{A}_+ := \mathbb{K}[[z]]$. Though it renders finiteness trivial, its counterterms might depend on the scale s and violate locality. So from [10] we need

Definition 8. A Feynman rule $\phi \in G_{\mathscr{A}}^{H_R}$ is called *local* iff in the minimal subtraction scheme, the counterterm $(\phi \circ \theta_{\ell z})_-$ is independent of $\ell \in \mathbb{K}$.

Proposition 5. *Locality of $\phi \in G_{\mathscr{A}}^{H_R}$ is equivalent to the conditions of Proposition 4.*

Proof. In case of Proposition 4, $\phi \circ \theta_{\ell z} = (\phi_-)^{\star-1} \star \left\{\phi_+ \star [\phi^{\star-1} \star (\phi \circ \theta_{\ell z})]\right\}$ is a Birkhoff decomposition by condition 3. such that $(\phi \circ \theta_{\ell z})_- = \phi_-$ from uniqueness. Conversely, for local ϕ,

$$ 0 = R_{\text{MS}} \circ (\phi \circ \theta_{\ell z})_+ = R_{\text{MS}} \circ [(\phi \circ \theta_{\ell z})_- \star (\phi \circ \theta_{\ell z})] = R_{\text{MS}} \circ [\phi_- \star (\phi \circ \theta_{\ell z})] $$

implies $\mathbb{K}[[z]] = \ker R_{\text{MS}} \supseteq \operatorname{im} \phi_- \star (\phi \circ \theta_{\ell z})$ and convolution with $\phi_+^{\star-1} = \phi^{\star-1} \star \phi_-^{\star-1} : H_R \to \mathbb{K}[[z]]$ yields condition 3. of Proposition 4.

So we showed algebraically that the problems of finiteness in the kinetic scheme and locality in minimal subtraction are precisely the same. These schemes are related by

Lemma 10. *If ${}_z\phi_{MS,s}$ denotes the R_{MS}-renormalized Feynman rule, then its scale dependence is given by ${}_0\phi$ through ${}_z\phi_{MS} = \left(R_\mu \circ {}_z\phi_{MS}\right) \star {}_z\phi_R$ (as already exploited in [4]).*

Proof. Locality of the minimal subtraction counterterms ϕ_- implies $R_\mu \circ \phi_- = \phi_-$, hence

$$ \left(R_\mu \circ {}_z\phi_{\text{MS}}\right) \star {}_z\phi_R = \left[R_\mu \circ (\phi_- \star {}_z\phi)\right] \star \left(R_\mu \circ {}_z\phi\right)^{\star-1} \star {}_z\phi = \left(R_\mu \circ \phi_-\right) \star {}_z\phi = {}_z\phi_{\text{MS}}. $$

The physical limit $\operatorname{ev}_{\ln s} \circ {}_0\phi_{\text{MS}} = \lim_{z\to 0} {}_z\phi_{\text{MS},s}$ yields polynomials ${}_0\phi_{\text{MS}}$ and Lemma 10 becomes

Corollary 10. *The characters ${}_0\phi_{MS}, {}_0\phi : H_R \to \mathbb{K}[x]$ fulfil the relations*

$$ {}_0\phi_{MS} = \left(\varepsilon \circ {}_0\phi_{MS}\right) \star {}_0\phi, \quad \textit{equivalently} \quad \Delta \circ {}_0\phi_{MS} = \left({}_0\phi_{MS} \otimes {}_0\phi\right) \circ \Delta. \tag{47} $$

In particular, the constant parts $\varepsilon \circ {}_0\phi_{\text{MS}} = \text{ev}_0 \circ {}_0\phi_{\text{MS}} \in G_{\mathbb{K}}^{H_R}$ determine ${}_0\phi_{\text{MS}}$ completely as the scale dependence is governed by ${}_0\phi$. Using ${}_0\phi = \exp_\star(-x\gamma)$, the *β-functional* ${}_0\phi_{\text{MS}} \equiv \exp_\star(x\beta) \star (\varepsilon \circ {}_0\phi_{\text{MS}})$ from [10] relates to γ by conjugation:

$$\beta \star \left(\varepsilon \circ {}_0\phi_{\text{MS}}\right) = -\left(\varepsilon \circ {}_0\phi_{\text{MS}}\right) \star \gamma.$$

Corollary 11. *Applying* (32) *to* (47) *expresses the correlation function of the R_{MS}-scheme to the kinetic scheme by a redefinition of the coupling constant:*

$$G_{\text{MS},\ell}(g) = G_{\text{MS},0}(g) \cdot G_\ell\Big(g \cdot [G_{\text{MS},0}(g)]^\kappa\Big).$$

8 Feynman Graphs and Logarithmic Divergences

In a typical renormalizable scalar quantum field theory, the vertex function is logarithmically divergent and may be renormalized by a simple subtraction as studied above. Referring to [6] for quadratic divergences, we now restrict to logarithmically divergent graphs with only logarithmic subdivergences, in D dimensions of space-time.

Following the notation established in [3], the renormalized amplitude of a graph Γ in the Hopf algebra H of Feynman graphs is given by the *forest formula*[10]

$$\Phi_+(\Gamma) = \int \Omega_\Gamma \sum_{F \in \mathscr{F}(\Gamma)} \frac{(-1)^{|F|}}{\psi_F^{D/2}} \ln \frac{\frac{\varphi}{\psi}_{\Gamma/F} + \sum_{\Gamma \neq \gamma \in F} \frac{\tilde{\varphi}}{\psi}_{\gamma/F}}{\frac{\tilde{\varphi}}{\psi}_{\Gamma/F} + \sum_{\Gamma \neq \gamma \in F} \frac{\tilde{\varphi}}{\psi}_{\gamma/F}}. \tag{48}$$

The forests $\mathscr{F}(\Gamma)$ account for subdivergences, the first and second *Symanzik polynomials* $\psi_\Gamma, \varphi_\Gamma$ depend on the edge variables α_e (*Schwinger parameters*) and we integrate over $\mathbb{RP}_{\geqslant 0}^{|E(\Gamma)|-1}$ in projective space with canonical volume form Ω_Γ.

Apart from a scale s, φ_Γ depends on dimensionless *angle variables* $\Theta = \left\{\frac{m^2}{s}\right\} \cup \left\{\frac{p_i \cdot p_j}{s}\right\}$ built from the mass m and external momenta p_i. We abbreviate $\frac{\varphi}{\psi}_\Gamma := \frac{\varphi_\Gamma}{\psi_\Gamma}$ and denote evaluation at the renormalization point $(\tilde{s}, \tilde{\Theta})$ of the kinetic scheme by a tilde or $\cdot|_R := \cdot|_{(s,\Theta)\mapsto(\tilde{s},\tilde{\Theta})}$.

Definition 9. Holding the angles Θ fixed, the *period functional* $\mathscr{P} \in H'$ is given by

$$\mathscr{P}(\Gamma) := -\frac{\partial}{\partial \ln s} \Phi_+(\Gamma)\bigg|_R \quad \text{for any} \quad \Gamma \in H. \tag{49}$$

[10] We prefer to work in the *parametric* representation as introduced in [14, Sect. 6-2-3].

Corollary 12. *For any graph $\Gamma \in H$, the value $\mathscr{P}(\Gamma)$ is a period in the sense of [15] (provided that $\tilde{s}$ and all $\theta \in \tilde{\Theta}$ are rational) by the formula*

$$\mathscr{P}(\Gamma) \underset{(48)}{=} \int \Omega_\Gamma \sum_{F \in \mathscr{F}(\Gamma)} \frac{(-1)^{1+|F|}}{\psi_F^{D/2}} \frac{\frac{\tilde{\varphi}}{\psi}_{\Gamma/F}}{\frac{\tilde{\varphi}}{\psi}_{\Gamma/F} + \sum_{\Gamma \neq \gamma \in F} \frac{\tilde{\varphi}}{\psi}_{\gamma/F}}. \tag{50}$$

For primitive (subdivergence free) graphs, [23] gives equivalent definitions of this period in momentum and position space. The product rule, (49) and $\Phi_+|_R = \varepsilon$ show

Corollary 13. *The period is an infinitesimal character $\mathscr{P} \in \mathfrak{g}_{\mathbb{K}}^H$ (it vanishes on any graph that is not connected).*

8.1 Renormalization Group

Proposition 6. *Holding the angles Θ fixed, differentiation by the scale results in*[11]

$$-\frac{\partial}{\partial \ln s} \Phi_+ = \mathscr{P} \star \Phi_+. \tag{51}$$

Proof. Adding $0 = \mathscr{P}(\Gamma) - \mathscr{P}(\Gamma)$ and collecting the contributions of $\frac{\tilde{\varphi}}{\psi}_{\gamma/F}$ in $(*)$ we find

$$\begin{aligned}
&-\frac{\partial}{\partial \ln s} \Phi_+(\Gamma) \\
&\underset{(48)}{=} \int \Omega_\Gamma \left\{ \frac{1}{\psi_\Gamma^{D/2}} + \sum_{\{\Gamma\} \neq F \in \mathscr{F}(\Gamma)} \frac{(-1)^{1+|F|}}{\psi_F^{D/2}} \frac{\frac{\varphi}{\psi}_{\Gamma/F}}{\frac{\varphi}{\psi}_{\Gamma/F} + \sum_{\Gamma \neq \delta \in F} \frac{\tilde{\varphi}}{\psi}_{\delta/F}} \right\} \\
&\underset{(50)}{=} \mathscr{P}(\Gamma) + \int \Omega_\Gamma \sum_{\{\Gamma\} \neq F \in \mathscr{F}(\Gamma)} \frac{(-1)^{1+|F|}}{\psi_F^{D/2}} \frac{\left(\frac{\varphi}{\psi}_{\Gamma/F} - \frac{\tilde{\varphi}}{\psi}_{\Gamma/F}\right) \sum_{\Gamma \neq \gamma \in F} \frac{\tilde{\varphi}}{\psi}_{\gamma/F}}{\left[\frac{\varphi}{\psi}_{\Gamma/F} + \sum_{\Gamma \neq \delta \in F} \frac{\tilde{\varphi}}{\psi}_{\delta/F}\right] \cdot \left[\frac{\tilde{\varphi}}{\psi}_{\Gamma/F} + \sum_{\Gamma \neq \delta \in F} \frac{\tilde{\varphi}}{\psi}_{\delta/F}\right]} \\
&\underset{(*)}{=} \mathscr{P}(\Gamma) + \int \Omega_\Gamma \sum_{\substack{\gamma \prec \Gamma \\ |\pi_0(\gamma)|=1}} \sum_{\gamma \in F \in \mathscr{F}(\Gamma)} \frac{(-1)^{1+|F|}}{\psi_F^{D/2}} \frac{\left(\frac{\varphi}{\psi}_{\Gamma/F} - \frac{\tilde{\varphi}}{\psi}_{\Gamma/F}\right) \frac{\tilde{\varphi}}{\psi}_{\gamma/F}}{\left[\frac{\varphi}{\psi}_{\Gamma/F} + \sum_{\Gamma \neq \delta \in F} \frac{\tilde{\varphi}}{\psi}_{\delta/F}\right] \cdot \sum_{\delta \in F} \frac{\tilde{\varphi}}{\psi}_{\delta/F}}.
\end{aligned}$$

[11]This simple form circumvents the decomposition into one-scale graphs utilized in [6] and therefore holds in the original renormalization Hopf algebra H.

With $\gamma \prec \Gamma$ denoting a subdivergence $\gamma \neq \Gamma$, the forests $F \in \mathscr{F}(\Gamma)$ containing γ correspond bijectively to the forests of γ and Γ/γ by

$$\mathscr{F}_\gamma(\Gamma) := \{F \in \mathscr{F}(\Gamma)\colon\ \gamma \in F\} \ni F \mapsto \left(F|_\gamma,\ F/\gamma\right) \in \mathscr{F}(\gamma) \times \mathscr{F}(\Gamma/\gamma), \quad \text{using}$$

$$F|_\gamma := \{\delta \in F\colon\ \delta \preceq \gamma\} \quad \text{and} \quad F/\gamma := \{\delta/\gamma\colon\ \delta \in F \quad \text{and} \quad \delta \npreceq \gamma\}.$$

This is an immediate consequence of the definition of a forest, as for $F \in \mathscr{F}_\gamma(\Gamma)$, each $\delta \in F$ is either disjoint to γ or strictly containing γ (in both cases it is mapped to $\delta/\gamma \in F/\gamma$) or itself a subdivergence of γ. Thus integrating $\int_0^\infty \frac{A-\tilde{A}}{(A+tB)(\tilde{A}+tB)} \mathrm{d}t = B^{-1} \ln \frac{A}{\tilde{A}}$ in

$$= \mathscr{P}(\Gamma) + \int \sum_{\substack{\gamma \prec \Gamma \\ |\pi_0(\gamma)|=1}} \Omega_\gamma \wedge \Omega_{\Gamma/\gamma} \sum_{\substack{F_\gamma \in \mathscr{F}(\gamma) \\ F \in \mathscr{F}(\Gamma/\gamma)}} \frac{(-1)^{1+|F_\gamma|+|F|}}{\psi_{F_\gamma}^{D/2} \cdot \psi_F^{D/2}}$$

$$\times \int_0^\infty \frac{\mathrm{d}t_\gamma}{t_\gamma} \frac{\left(\frac{\varphi}{\psi}_{\Gamma/F} - \frac{\tilde{\varphi}}{\psi}_{\Gamma/F}\right) \cdot t_\gamma \cdot \frac{\tilde{\varphi}}{\psi}_{\gamma/F_\gamma}}{\left[\frac{\varphi}{\psi}_{\Gamma/F} + \sum_{\Gamma \neq \delta \in F} \frac{\tilde{\varphi}}{\psi}_{\delta/F} + t_\gamma \cdot \sum_{\delta \in F_\gamma} \frac{\tilde{\varphi}}{\psi}_{\delta/F_\gamma}\right] \cdot \left[\sum_{\delta \in F} \frac{\tilde{\varphi}}{\psi}_{\delta/F} + t_\gamma \cdot \sum_{\delta \in F_\gamma} \frac{\tilde{\varphi}}{\psi}_{\delta/F_\gamma}\right]}$$

$$= \mathscr{P}(\Gamma) + \int \sum_{\substack{\gamma \prec \Gamma \\ |\pi_0(\gamma)|=1}} \Omega_\gamma \wedge \Omega_{\Gamma/\gamma} \sum_{\substack{F_\gamma \in \mathscr{F}(\gamma) \\ F \in \mathscr{F}(\Gamma/\gamma)}} \frac{(-1)^{1+|F_\gamma|+|F|}}{\psi_{F_\gamma}^{D/2} \cdot \psi_F^{D/2}} \cdot \frac{\frac{\tilde{\varphi}}{\psi}_{\gamma/F_\gamma}}{\sum_{\delta \in F_\gamma} \frac{\tilde{\varphi}}{\psi}_{\delta/F_\gamma}} \cdot \ln \frac{\frac{\varphi}{\psi}_{(\Gamma/\gamma)/F} + \sum_{\delta \in F \setminus \{\Gamma/\gamma\}} \frac{\tilde{\varphi}}{\psi}_{\delta/F}}{\sum_{\delta \in F} \frac{\tilde{\varphi}}{\psi}_{\delta/F}}$$

reduces to the projective $\int \Omega_\gamma$ in the edge variables of the subgraph γ, making use of

$$|F| = \left|F|_\gamma\right| + |F/\gamma|, \quad \frac{\varphi}{\psi}_{\delta/F} = \begin{cases} \frac{\varphi}{\psi}_{(\delta/\gamma)/(F/\gamma)}, & \text{if } \gamma \npreceq \delta \in F \\ \frac{\varphi}{\psi}_{\delta/F|_\gamma}, & \text{if } \gamma \succeq \delta \in F \end{cases}$$

and

$$\psi_F = \psi_{F|_\gamma} \cdot \psi_{F/\gamma}.$$

The apparent factorization into $\mathscr{P}(\gamma)$ and $\Phi_+(\Gamma/\gamma)$ shows that we obtain convergent integrals for each $\gamma \prec \Gamma$ individually and may therefore separate into

$$= \mathscr{P}(\Gamma) + \sum_{\substack{\gamma \prec \Gamma \\ |\pi_0(\gamma)|=1}} \int \Omega_\gamma \sum_{F_\gamma \in \mathscr{F}(\gamma)} \frac{(-1)^{1+|F_\gamma|}}{\psi_{F_\gamma}^{D/2}} \cdot \frac{\frac{\tilde{\varphi}}{\psi}_{\gamma/F_\gamma}}{\sum_{\delta \in F_\gamma} \frac{\tilde{\varphi}}{\psi}_{\delta/F_\gamma}}$$

$$\times \int \Omega_{\Gamma/\gamma} \sum_{F \in \mathscr{F}(\Gamma/\gamma)} \frac{(-1)^{|F|}}{\psi_F^{D/2}} \cdot \ln \frac{\frac{\varphi}{\psi}_{(\Gamma/\gamma)/F} + \sum_{\delta \in F \setminus \{\Gamma/\gamma\}} \frac{\tilde{\varphi}}{\psi}_{\delta/F}}{\sum_{\delta \in F} \frac{\tilde{\varphi}}{\psi}_{\delta/F}} = \mathscr{P} \star \Phi_+(\Gamma).$$

Note that the terms $\gamma \otimes \Gamma/\gamma$ of $\Delta(\Gamma)$ with $|\pi_0(\gamma)| > 1$ do not contribute here by Corollary 13.

Together with Corollary 13 and the connected graduation of H, this shows

$$\Phi_+ = \sum_{n\in\mathbb{N}_0} \frac{(-\ell)^n}{n!} \left[\left(-\frac{\partial}{\partial \ln s} \right)^n \Phi_+ \right]_{s=\tilde{s}} \underset{(51)}{=} \sum_{n\in\mathbb{N}_0} \frac{(-\ell \cdot \mathscr{P})^{\star n}}{n!} \star \Phi_+|_{s=\tilde{s}},$$

where we set $\ell := \ln \frac{s}{\tilde{s}}$ and the series is pointwise finite. Hence note

Corollary 14. *The renormalized Feynman rules $\Phi_+ = \Phi_+|_{\Theta=\tilde{\Theta}} \star \Phi_+|_{s=\tilde{s}}$ factorize ([6] gives a different decomposition) into the angle-dependent part $\Phi_+|_{s=\tilde{s}}$ and the scale-dependence $\Phi_+|_{\Theta=\tilde{\Theta}}$ given as the Hopf algebra morphism*

$$\Phi_+|_{\Theta=\tilde{\Theta}} = \exp_\star(-\ell\mathscr{P}) : \quad H \to \mathbb{K}[\ell]. \tag{52}$$

Example 16. For primitive $\Gamma \in \mathrm{Prim}(H)$, $\Phi_+(\Gamma) = -\ell \cdot \mathscr{P}(\Gamma) + \Phi_+|_{s=\tilde{s}}(\Gamma)$ disentangles the scale- and angle-dependence. Subdivergences evoke higher powers of ℓ with angle-dependent factors. *Dunce's cap* of ϕ^4-theory gives $\mathscr{P}$() $= \mathscr{P}$ () $= 1$ such that

Φ_+ (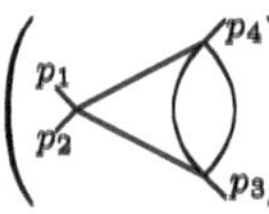) $= \frac{\ell^2}{2} - \ell - \ell\,\Phi_+|_{s=\tilde{s}}$ (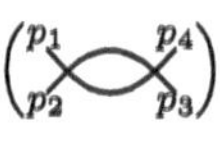) $+ \Phi_+|_{s=\tilde{s}}$ (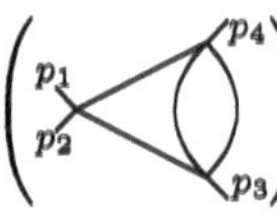).

8.2 Dimensional Regularization

The *dimensional regularization* of [7] assigns a Laurent series ${}_z\Phi(\Gamma)$ in $z \in \mathbb{C}$ to each Feynman graph $\Gamma \in H$, which for large $\Re z$ is given by the convergent parametric integral

$${}_z\Phi(\Gamma) = \left[\prod_{e\in E(\Gamma)} \int_0^\infty \alpha_e \right] \frac{e^{-\frac{\varphi}{\psi}_\Gamma}}{\psi_\Gamma^{D/2-z}}. \tag{53}$$

As $\frac{\varphi}{\psi}_\Gamma$ is linear in the scale s and homogeneous of degree one in the edge variables, simultaneously rescaling of all α_e yields (for logarithmically divergent graphs)

Corollary 15. *The scale dependence ${}_z\Phi = {}_z\Phi|_{s=\tilde{s}} \circ \theta_{-z\ell}$ of (53) is induced from the grading Y of H given by the loop number.*

Thus the finiteness of the physical limit $\Phi_+|_{\Theta=\tilde{\Theta}} = \lim_{z\to 0} {}_z\Phi|_R \circ (S \star \theta_{-z\ell})$ results by Proposition 4 in the local character ${}_z\Phi|_R \in G^H_{\mathscr{A}}$, evaluated at the renormalization point $(\tilde{s}, \tilde{\Theta})$.

Corollary 16. *In dimensional regularization, the period* (50) *is the* $\frac{1}{z}$*-pole coefficient*

$$\mathscr{P} \underset{(46)}{=} \mathrm{Res} \circ {}_z\Phi|_R \circ (S \star Y). \tag{54}$$

8.3 Dilatations

For $\lambda > 0$, consider the *dilatation operator* Λ_λ scaling masses $m \mapsto \lambda \cdot m$ and momenta $p_i \mapsto \lambda \cdot p_i$. It fixes all angles Θ, multiplies the scale s with λ^2 and therefore acts as

$$\Phi_+ \circ \Lambda_\lambda = \exp_\star\left(-\mathscr{P} \ln \frac{s}{\tilde{s}}\right) \star \Phi_+|_{s=\tilde{s}} \circ (s \mapsto s \cdot \lambda^2) = \exp_\star(-2\mathscr{P} \ln \lambda) \star \Phi_+.$$

In other words, the dilatations $\mathbb{R}_{>0} \ni \lambda \mapsto \Lambda_\lambda \mapsto \exp_\star(-2\mathscr{P} \ln \lambda) \star \cdot$ are represented on the group $G^H_{\mathscr{A}}$ of characters by a left convolution. As the unrenormalized logarithmically divergent graphs are dimensionless and naively invariant under Λ_λ, $\mathscr{P}$ precisely measures how renormalization breaks this symmetry, giving rise to *anomalous dimensions*.

9 Conclusion

We stress that the physical limit of the renormalized Feynman rules results in a morphism ${}_0\phi : H_R \to \mathbb{K}[x]$ of Hopf algebras in case of the kinetic scheme. This compatibility with the coproduct allows to obtain ${}_0\phi$ from the linear terms γ only. As we just exemplified, these relations are statements about individual Feynman graphs unraveling scale- and angle-dependence in a simple way. Again we recommend [6] for further reading.

Secondly we revealed how Hochschild cohomology governs not only the perturbation series through Dyson-Schwinger equations, but also determines the Feynman rules. Addition of exact one-cocycles captures variations of Feynman rules and the anomalous dimension γ can efficiently be calculated in terms of Mellin transform coefficients.

Note how this feature is lost upon substitution of the kinetic scheme by minimal subtraction: We do not obtain a Hopf algebra morphism anymore due to the constant terms, which are also more difficult to obtain in terms of the Mellin transforms F.

Finally we want to emphasize the remarks in Sect. 5 towards a non-perturbative framework. Though this relation between $F(z)$ and the anomalous dimension $\tilde{\gamma}(g)$ is still under investigation and so far only fully understood in special cases, these already give interesting results [5, 24].

Acknowledgements Alexander von Humboldt Chair in Mathematical Physics, supported by the Alexander von Humboldt Foundation and the BMBF.

References

1. Bergbauer, C., Kreimer, D.: Hopf algebras in renormalization theory: locality and Dyson-Schwinger equations from Hochschild cohomology. IRMA Lect. Math. Theor. Phys. **10**, 133–164 (2006)
2. Bierenbaum, I., Kreimer, D., Weinzierl, S.: The next-to-ladder approximation for Dyson-Schwinger equations. Phys. Lett. **B646**, 129–133 (2007). doi:10.1016/j.physletb.2007.01.018
3. Bloch, S., Kreimer, D.: Mixed Hodge structures and renormalization in physics. Commun. Number Theory Phys. **2**, 637–718 (2008)
4. Broadhurst, D.J., Kreimer, D.: Renormalization automated by Hopf algebra. J. Symb. Comput. **27**, 581 (1999). doi:10.1006/jsco.1999.0283
5. Broadhurst, D.J., Kreimer, D.: Exact solutions of Dyson-Schwinger equations for iterated one-loop integrals and propagator-coupling duality. Nucl. Phys. **B600**, 403–422 (2001). doi:10.1016/S0550-3213(01)00071-2
6. Brown, F., Kreimer, D.: Angles, scales and parametric renormalization. Lett. Math. Phys. **103**(9), 933–1007 (2013)
7. Collins, J.C.: Renormalization. Cambridge Monographs on Mathematical Physics. Cambridge University Press, Cambridge/New York (1984)
8. Connes, A., Kreimer, D.: Hopf algebras, renormalization and noncommutative geometry. Commun. Math. Phys. **199**, 203–242 (1998). doi:10.1007/s002200050499
9. Connes, A., Kreimer, D.: Renormalization in quantum field theory and the Riemann-Hilbert problem I: the Hopf algebra structure of graphs and the main theorem. Commun. Math. Phys. **210**, 249–273 (2000). doi:10.1007/s002200050779
10. Connes, A., Kreimer, D.: Renormalization in quantum field theory and the Riemann-Hilbert problem II: the β-function, diffeomorphisms and the renormalization group. Commun. Math. Phys. **216**, 215–241 (2001). doi:10.1007/PL00005547
11. Foissy, L.: Faà di Bruno subalgebras of the Hopf algebra of planar trees from combinatorial Dyson-Schwinger equations. Adv. Math. **218**(1), 136–162 (2008). doi:10.1016/j.aim.2007.12.003. http://www.sciencedirect.com/science/article/pii/S0001870807003301
12. Foissy, L.: Classification of systems of Dyson-Schwinger equations in the Hopf algebra of decorated rooted trees. Adv. Math. **224**(5), 2094–2150 (2010). doi:10.1016/j.aim.2010.01.024. http://www.sciencedirect.com/science/article/pii/S0001870810000435
13. Foissy, L.: General Dyson-Schwinger equations and systems. ArXiv e-prints (2011)
14. Itzykson, C., Zuber, J.B.: Quantum Field Theory. Dover, New York (2005)
15. Kontsevich, M., Zagier, D.: Periods. In: Mathematics Unlimited – 2001 and Beyond. Springer, Berlin (2001)
16. Kreimer, D.: On the Hopf algebra structure of perturbative quantum field theories. Adv. Theor. Math. Phys. **2**, 303–334 (1997)
17. Kreimer, D.: Chen's iterated integral represents the operator product expansion. Adv. Theor. Math. Phys. **3**, 3 (2000)
18. Kreimer, D.: Factorization in quantum field theory: an exercise in hopf algebras and local singularities. In: Cartier, P., Moussa, P., Julia, B., Vanhove, P. (eds.) On Conformal Field

Theories, Discrete Groups and Renormalization, Frontiers in Number Theory, Physics, and Geometry, vol. 2, pp. 715–736. Springer, Berlin/Heidelberg (2007). http://dx.doi.org/10.1007/978-3-540-30308-4_14
19. Kreimer, D.: Étude for linear Dyson-Schwinger equations. In: Albeverio, S., Marcolli, M., Paycha, S., Plazas, J. (eds.) Traces in Number Theory, Geometry and Quantum Fields. Aspects of Mathematics E, vol. 38, pp. 155–160. Vieweg, Wiesbaden (2008). http://preprints.ihes.fr/2006/P/P-06-23.pdf
20. Kreimer, D., Yeats, K.: An étude in non-linear Dyson-Schwinger equations. Nucl. Phys. Proc. Suppl. **160**, 116–121 (2006). doi:10.1016/j.nuclphysbps.2006.09.036. Proceedings of the 8th DESY workshop on elementary particle theory
21. Manchon, D.: Hopf Algebras in Renormalisation. pp. 365–427. Elsevier/North-Holland (2008). doi:10.1016/S1570-7954(07)05007-3. http://www.sciencedirect.com/science/article/pii/S1570795407050073
22. Panzer, E.: Hopf-algebraic renormalization of kreimer's toy model. Master's thesis, Humboldt-Universität zu Berlin (2011)
23. Schnetz, O.: Quantum periods: a census of ϕ^4-transcendentals. Commun. Number Theory Phys. **4**, 1–48 (2010)
24. Yeats, K.A.: Rearranging Dyson-Schwinger equations. Mem. Am. Math. Soc. **211**(995), 1–82 (2011)

Relativistic Coulomb Integrals and Zeilberger's Holonomic Systems Approach. I

Peter Paule and Sergei K. Suslov

Dedicated to Doron Zeilberger on the occasion of his 60th birthday

Science is what we understand well enough to explain to a computer. Art is everything else we do.

Donald E. Knuth [21]

Abstract With the help of computer algebra we study the diagonal matrix elements $\langle Or^p \rangle$, where $O = \{1, \beta, i\boldsymbol{\alpha}\mathbf{n}\beta\}$ are the standard Dirac matrix operators and the angular brackets denote the quantum-mechanical average for the relativistic Coulomb problem. Using Zeilberger's extension of Gosper's algorithm and a variant to it, three-term recurrence relations for each of these expectation values are derived together with some transformation formulas for the corresponding generalized hypergeometric series. In addition, the virial recurrence relations for these integrals are also found and proved algorithmically.

Electronic supplementary material The online version of this article (doi: 10.1007/978-3-7091-1616-6_9) contains supplementary material, which is available to authorized users.

P. Paule
Research Institute for Symbolic Computation (RISC), Johannes Kepler University Linz, Linz, Austria
e-mail: Peter.Paule@risc.jku.at

S.K. Suslov (✉)
School of Mathematical and Statistics Sciences, Mathematical, Computational and Modeling Sciences Center, Arizona State University, Tempe, AZ, USA
e-mail: sks@asu.edu

C. Schneider and J. Blümlein (eds.), *Computer Algebra in Quantum Field Theory*, Texts & Monographs in Symbolic Computation, DOI 10.1007/978-3-7091-1616-6_9,

1 Introduction

This work has been initiated by the following email regarding Doron Zeilberger's Z60 conference:

http://www.math.rutgers.edu/events/Z60/

```
Email to Peter Paule from Sergei Suslov [27 Feb 2010]
Subject: Uranium 91+ ion
"...I understand that you are coming to Doron's conference in May
and write to you with an unusual suggestion...
I am attaching two of my recent papers inspired by recent success
in checking Quantum Electrodynamics in strong fields [see Refs. [35] and
[37] in this chapter].
It is a very complicated problem theoretically, and fantastically,
enormously complicated (at the level of science fiction!)
experimentally, which has been solved - after 20 years of hard
work by theorists from Russia (Shabaev + 20 coauthors/students)
and experimentalists from Germany.
Experimentally they took a uranium 92 atom, got rid of all but one
electrons, and measured the energy shifts due to the quantization
of the electromagnetic radiation field!
Mathematically, among other things, the precise structure of the
energy levels of the U 91 + ion requires the evaluation of
certain relativistic Coulomb integrals, done, in a final form, in
my attached papers ...
Here is the problem:
These integrals have numerous recurrence relations found by
physicists on the basis of virial theorems. They are also sums of
3 (linearly dependent) 3F2 series.
Now you can imagine what a mess it is if one tries to derive those
relations at the level of hypergeometric series (3 times 3 = 9
functions usually!).
It looks as a perfect job for the G-Z algorithm in a realistic
(important) classical problem of relativistic quantum mechanics.
It looks as a good birthday present to Doron, if one could have
done that. I feel we can do that together.
Looking forward to your answer on my crazy suggestion, BW, Sergei"
```

The first named author's computer algebra response reported at the Z60 conference is presented in this joint paper.

2 Relativistic Coulomb Integrals

Recent experimental and theoretical progress has renewed interest in quantum electrodynamics of atomic hydrogenlike systems (see, for example, [3, 10, 11, 13, 14, 17, 31, 33, 34] and the references therein). In the last decade, the two-time Green's function method of deriving formal expressions for the energy shift of a bound-state level of high-Z few-electron systems was developed [31] and numerical calculations of QED effects in heavy ions were performed with an excellent agreement to current experimental data [10, 11, 33]. These advances motivate a detailed study of the expectation values of the Dirac matrix operators multiplied by the powers of the radius between the bound-state relativistic Coulomb wave functions. Special cases appear in calculations of the magnetic dipole hyperfine splitting, the electric quadrupole hyperfine splitting, the anomalous Zeeman effect, and the relativistic recoil corrections in hydrogenlike ions (see, for example, [1, 30, 32, 35] and the references therein). These expectation values can be used in calculations with hydrogenlike wave functions when a high precision is required. For applications of the off-diagonal matrix elements, see [22–24, 28, 29], and [32].

Two different forms of the radial wave functions F and G are available (see, for example, [18] and [38]). Given a set of parameters $a, \alpha_1, \alpha_2, \beta, \beta_1, \beta_2$, and γ, depending on physical constants ε, κ, μ, and ν, consider

$$\begin{pmatrix} F(r) \\ G(r) \end{pmatrix} = a^2\beta^{3/2}\sqrt{\frac{n!}{\gamma\, \Gamma(n+2\nu)}}(2a\beta r)^{\nu-1}e^{-a\beta r}\begin{pmatrix} \alpha_1 & \alpha_2 \\ \beta_1 & \beta_2 \end{pmatrix}\begin{pmatrix} L_{n-1}^{2\nu}(2a\beta r) \\ L_n^{2\nu}(2a\beta r) \end{pmatrix} \tag{1}$$

where, using the notation from [20], $L_n^{\lambda}(x)$ stands for the corresponding Laguerre polynomial of order n. Throughout this paper,

$$\begin{aligned} &\kappa = \pm(j+1/2), \qquad \nu = \sqrt{\kappa^2-\mu^2}, \\ &\mu = \alpha Z = Ze^2/\hbar c, \qquad a = \sqrt{1-\varepsilon^2}, \\ &\varepsilon = E/mc^2, \qquad \beta = mc/\hbar, \end{aligned} \tag{2}$$

and $\gamma = \mu(\kappa-\nu)(\varepsilon\kappa-\nu)$, with the total angular momentum $j = 1/2, 3/2, 5/2$, etc. (see [4, 5, 8, 25, 35], and [38] regarding the relativistic Coulomb problem). The following identities

$$\varepsilon\mu = a(\nu+n), \quad \varepsilon\mu + a\nu = a(n+2\nu), \quad \varepsilon\mu - a\nu = an, \tag{3}$$
$$\varepsilon^2\kappa^2 - \nu^2 = a^2 n(n+2\nu) = \mu^2 - a^2\kappa^2$$

are useful in the calculation of the matrix elements.

The relativistic Coulomb integrals of the radial functions,

$$A_p = \int_0^\infty r^{p+2}\left(F^2(r)+G^2(r)\right)\, dr, \tag{4}$$

$$B_p = \int_0^\infty r^{p+2}\left(F^2(r)-G^2(r)\right)\, dr, \tag{5}$$

$$C_p = \int_0^\infty r^{p+2}F(r)\,G(r)\, dr, \tag{6}$$

have been evaluated in Refs. [35] and [37] for all admissible integer powers p, in terms of linear combinations of special generalized hypergeometric ${}_3F_2$ series related to the Chebyshev polynomials of a discrete variable [18, 19].

Note. We concentrate on the radial integrals since, for problems involving spherical symmetry, one can reduce all expectation values to radial integrals by use of the properties of angular momentum.

Throughout the paper we use the following abbreviated form of the standard notation of the generalized hypergeometric series ${}_3F_2$; see, e.g., [20]:

$$ {}_3F_2\left(\begin{matrix} a_1, a_2, a_3 \\ b_1, b_2 \end{matrix}\right) := {}_3F_2\left(\begin{matrix} a_1, a_2, a_3 \\ b_1, b_2 \end{matrix}; 1\right) = \sum_{k=0}^{\infty} \frac{(a_1)_k (a_2)_k (a_3)_k}{(b_1)_k (b_2)_k k!}, \tag{7} $$

where $(a)_k := a(a+1)\dots(a+k-1)$ denotes the Pochhammer symbol.

Analogs of the traditional hypergeometric representations for the integrals are as follows [35]:

$$ \begin{aligned} 2\mu\,(2a\beta)^p \frac{\Gamma(2\nu+1)}{\Gamma(2\nu+p+1)} A_p &= 2p\varepsilon a n\; {}_3F_2\left(\begin{matrix} 1-n, -p, p+1 \\ 2\nu+1, \ 2 \end{matrix}\right) \\ &+ (\mu + a\kappa)\; {}_3F_2\left(\begin{matrix} 1-n, -p, p+1 \\ 2\nu+1, \ 1 \end{matrix}\right) + (\mu - a\kappa)\; {}_3F_2\left(\begin{matrix} -n, -p, p+1 \\ 2\nu+1, \ 1 \end{matrix}\right), \end{aligned} \tag{8} $$

$$ \begin{aligned} 2\mu\,(2a\beta)^p \frac{\Gamma(2\nu+1)}{\Gamma(2\nu+p+1)} B_p &= 2pan\; {}_3F_2\left(\begin{matrix} 1-n, -p, p+1 \\ 2\nu+1, \ 2 \end{matrix}\right) \\ &+\varepsilon (\mu + a\kappa)\; {}_3F_2\left(\begin{matrix} 1-n, -p, p+1 \\ 2\nu+1, \ 1 \end{matrix}\right) + \varepsilon (\mu - a\kappa)\; {}_3F_2\left(\begin{matrix} -n, -p, p+1 \\ 2\nu+1, \ 1 \end{matrix}\right), \end{aligned} \tag{9} $$

$$ \begin{aligned} &4\mu\,(2a\beta)^p \frac{\Gamma(2\nu+1)}{\Gamma(2\nu+p+1)} C_p \\ &= a(\mu + a\kappa)\; {}_3F_2\left(\begin{matrix} 1-n, -p, p+1 \\ 2\nu+1, \ 1 \end{matrix}\right) - a(\mu - a\kappa)\; {}_3F_2\left(\begin{matrix} -n, -p, p+1 \\ 2\nu+1, \ 1 \end{matrix}\right). \end{aligned} \tag{10} $$

The averages of r^p for the relativistic hydrogen atom, namely the integrals A_p, were evaluated in the late 1930s by Davis [6] as a sum of certain three ${}_3F_2$ functions.[1] But it has been realized only recently that these series are, in fact, linearly dependent and related to the Chebyshev polynomials of a discrete variable [35]. The most compact forms in terms of only two linearly independent generalized hypergeometric series are given in Ref. [37].

[1]He finishes his article by saying: "In conclusion I wish to thank Professors H. Bateman, P. S. Epstein, W. V. Houston, and J. R. Openheimer for their helpful suggestions."

In addition, the integrals themselves are linearly dependent:

$$(2\kappa + \varepsilon\,(p+1))\,A_p - (2\varepsilon\kappa + p + 1)\,B_p = 4\mu C_p; \tag{11}$$

see, for example, [1,28,29], and [35]. Thus, eliminating, say C_p, one can deal with A_p and B_p only.

The integrals (4)–(6) satisfy numerous recurrence relations in p, which provide an effective way of their evaluation for small p. A set of useful recurrence relations between the relativistic matrix elements was derived by Shabaev [29] (see also [1,7,28,32,35], and [39]) on the basis of a hypervirial theorem:

$$2\kappa A_p - (p+1)\,B_p = 4\mu C_p + 4\beta\varepsilon C_{p+1}, \tag{12}$$

$$2\kappa B_p - (p+1)\,A_p = 4\beta C_{p+1}, \tag{13}$$

$$\mu B_p - (p+1)\,C_p = \beta\left(A_{p+1} - \varepsilon B_{p+1}\right). \tag{14}$$

From these relations one can derive (see [1,29], and [32]) the linear relation (11) and the following computationally convenient recurrence formulas (15)–(18), stated in our notation as

$$\begin{aligned} A_{p+1} &= -(p+1)\,\frac{4\nu^2\varepsilon + 2\kappa\,(p+2) + \varepsilon\,(p+1)\,(2\kappa\varepsilon + p + 2)}{4\,(1-\varepsilon^2)\,(p+2)\,\beta\mu}\,A_p \\ &\quad + \frac{4\mu^2\,(p+2) + (p+1)\,(2\kappa\varepsilon + p + 1)\,(2\kappa\varepsilon + p + 2)}{4\,(1-\varepsilon^2)\,(p+2)\,\beta\mu}\,B_p, \end{aligned} \tag{15}$$

$$\begin{aligned} B_{p+1} &= -(p+1)\,\frac{4\nu^2 + 2\kappa\varepsilon\,(2p+3) + \varepsilon^2\,(p+1)\,(p+2)}{4\,(1-\varepsilon^2)\,(p+2)\,\beta\mu}\,A_p \\ &\quad + \frac{4\mu^2\varepsilon\,(p+2) + (p+1)\,(2\kappa\varepsilon + p + 1)\,(2\kappa + \varepsilon\,(p+2))}{4\,(1-\varepsilon^2)\,(p+2)\,\beta\mu}\,B_p \end{aligned} \tag{16}$$

and

$$\begin{aligned} A_{p-1} &= \beta\,\frac{4\mu^2\varepsilon\,(p+1) + p\,(2\kappa\varepsilon + p)\,(2\kappa + \varepsilon\,(p+1))}{\mu\,(4\nu^2 - p^2)\,p}\,A_p \\ &\quad - \beta\,\frac{4\mu^2\,(p+1) + p\,(2\kappa\varepsilon + p)\,(2\kappa\varepsilon + p + 1)}{\mu\,(4\nu^2 - p^2)\,p}\,B_p, \end{aligned} \tag{17}$$

$$\begin{aligned} B_{p-1} &= \beta\,\frac{4\nu^2 + 2\kappa\varepsilon\,(2p+1) + \varepsilon^2 p\,(p+1)}{\mu\,(4\nu^2 - p^2)}\,A_p \\ &\quad - \beta\,\frac{4\nu^2\varepsilon + 2\kappa\,(p+1) + \varepsilon p\,(2\kappa\varepsilon + p + 1)}{\mu\,(4\nu^2 - p^2)}\,B_p, \end{aligned} \tag{18}$$

respectively.

Note. (i) These recurrences are complemented by the symmetries of the integrals A_p, B_p, and C_p under the reflections $p \to -p-1$ and $p \to -p-3$ found in [35]; see also [2]. (ii) These relations were also derived in [36] by a different method using relativistic versions of the Kramers–Pasternack three-term recurrence relations.

3 Computer Algebra and Software

The general algorithmic background of the computer algebra applications in this paper is Zeilberger's path-breaking holonomic systems paper [40]. The examples given in the following sections restrict to applications: (i) of Zeilberger's extension [42] of Gosper's algorithm [9], also called Zeilberger's "fast algorithm" [21,41], and (ii) of a variant of it which has been described in the unpublished manuscript (Paule, 2001, Contiguous relations and creative telescoping, unpublished manuscript, 33p). Both of these algorithms have been implemented in the Fast Zeilberger package `zb.m` which is written in *Mathematica* and whose functionality is illustrated below. A very general framework of Zeilberger's creative telescoping (i), and also of its variant (ii), is provided by Schneider's extension of Karr's summation in difference fields [12]; see, for instance, [26, 27] and the references therein.

The Fast Zeilberger Package can be obtained freely from the site

http://www.risc.jku.at/research/combinat/software/

after sending a password request to the first named author. Put the package `zb.m` in some directory, e.g., `/home/mydirectory`, open a *Mathematica* session, and read in the package by

```
In[1]:= SetDirectory["/home/ppaule/RISC_Comb_Software_Sep05.dir/fastZeil"];
In[2]:= <<zb.m
    Fast Zeilberger Package by Peter Paule and Markus Schorn
    (enhanced by Axel Riese) - © RISC Linz - V 3.53 (02/22/05)
```

A Mathematica notebook containing a full account of the Mathematica sessions described below, together with some additional material, is available at:

http://hahn.la.asu.edu/~suslov/curres/index.htm

4 Unmixed Three-Term Recurrence Relations

The following relations purely in the A_p and B_p, respectively, have been established in [37]:

$$A_{p+1} = \frac{\mu P(p)}{a^2\beta\left(4\mu^2(p+1)+p(2\varepsilon\kappa+p)(2\varepsilon\kappa+p+1)\right)(p+2)} A_p - \frac{\left(4\nu^2-p^2\right)\left(4\mu^2(p+2)+(p+1)(2\varepsilon\kappa+p+1)(2\varepsilon\kappa+p+2)\right)p}{(2a\beta)^2\left(4\mu^2(p+1)+p(2\varepsilon\kappa+p)(2\varepsilon\kappa+p+1)\right)(p+2)} A_{p-1}, \tag{19}$$

$$B_{p+1} = \frac{\varepsilon\mu\, Q(p)}{a^2\beta\left(4\nu^2 + 2\varepsilon\kappa(2p+1) + \varepsilon^2 p(p+1)\right)(p+2)} B_p - \frac{\left(4\nu^2 - p^2\right)\left(4\nu^2 + 2\varepsilon\kappa(2p+3) + \varepsilon^2(p+1)(p+2)\right)(p+1)}{(2a\beta)^2\left(4\nu^2 + 2\varepsilon\kappa(2p+1) + \varepsilon^2 p(p+1)\right)(p+2)} B_{p-1}, \tag{20}$$

where

$$P(p) = 2\varepsilon p(p+2)(2\varepsilon\kappa + p)(2\varepsilon\kappa + p + 1) + \varepsilon\left[4\left(\varepsilon^2\kappa^2 - \nu^2\right) - p\left(4\varepsilon^2\kappa^2 + p(p+1)\right)\right] + (2p+1)\left[4\varepsilon^2\kappa + 2(p+2)\left(2\varepsilon\mu^2 - \kappa\right)\right], \tag{21}$$

$$Q(p) = (2p+3)\left[4\nu^2 + 2\varepsilon\kappa(2p+1) + p(p+1)\right] - a^2(2p+1)(p+1)(p+2). \tag{22}$$

In comparison with other papers (e.g., [1, 2, 28, 29, 35, 36], and the references therein), this approach provides an alternative way of the recursive evaluation of the special values A_p and B_p, when one deals separately with one of these integrals only. The corresponding initial data $A_0 = 1$ and $B_{-1} = a^2\beta/\mu$ can be found in [35].

Note. The derivation in [37] resembles the reduction (uncoupling) of the first order system of differential equations for relativistic radial Coulomb wave functions F and G to the second order differential equations; see, for example, [18] and [38].

With Zeilberger's definite extension [41, 42] of Gosper's algorithm [9] for indefinite hypergeometric summation, the derivation of such recurrences is fully automatic if the input is given as a terminating hypergeometric series (and provided that the input is of computationally feasible size). We illustrate this by a mechanical derivation of the following simple three-term recurrence relation for the integral C_p, not found in [37]:

$$C_{p+1} = \mu(2p+1)\frac{2\kappa + \varepsilon\left[p(p+1) - 4\kappa^2\right]}{a^2\beta\left(p^2 - 4\kappa^2\right)(p+1)} C_p + p\,\frac{\left(p^2 - 4\nu^2\right)\left[(p+1)^2 - 4\kappa^2\right]}{(2a\beta)^2\left(p^2 - 4\kappa^2\right)(p+1)} C_{p-1}. \tag{23}$$

As input for C_p we take the hypergeometric sum representation from (10). We start our *Mathematica* session by reading in the RISC "Fast Zeilberger" package:

```
In[1]:= <<zb.m

In[2]:= (a_)k_ := Pochhammer[a,k];
```

$$\texttt{F1[k\_]} := \frac{(1-n)_k\,(-p)_k\,(p+1)_k}{(2\,\nu+1)_k\,(1)_k\,k!};\quad \texttt{F2[k\_]} := \frac{(-n)_k\,(-p)_k\,(p+1)_k}{(2\,\nu+1)_k\,(1)_k\,k!};$$

```
In[3]:= FullSimplify[{F1[k]/F1[k], F2[k]/F1[k]}]
```

$$\text{Out[3]}=\left\{1,\ -\frac{n}{k-n}\right\}$$

$$\text{In[4]}:=\ f[k\_]:=\left(4\ \mu\ (2\ a\ \beta)^{p}\ \frac{\text{Gamma}[2\ \nu+1]}{\text{Gamma}[2\ \nu+p+1]}\right)^{-1}\ \text{F1}[k]\left(a\ (\mu+a\ \kappa)-a\ (\mu-a\ \kappa)*\frac{n}{n-k}\right);$$

```
In[5]:= SuslovRec=Zb[f[k] , k, 0, Infinity, p, 2] // Simplify
```

$$\text{Out[5]}=\{4\ a\ \beta\ (-(3+2\ p)\ (-(2+3\ p+p^2)\ \mu^2\ (n+\nu)-2\ a\ n\ \kappa\ \mu\ (n+2\ \nu)+$$
$$a^2\ \kappa^2\ (n+\nu)\ (2+4\ n^2+3\ p+p^2+8\ n\ \nu))\ \text{SUM}[1+p]+$$
$$a\ (2+p)\ \beta\ (-(1+p)^2\ \mu^2+a^2\ \kappa^2\ (4\ n^2+(1+p)^2+8\ n\ \nu))\ \text{SUM}[p+2]==$$
$$(1+p)\ (1+2\ p+p^2-4\ \nu^2)\ (-(2+p)^2\ \mu^2+a^2\ \kappa^2\ (4\ n^2+(2+p)^2+8\ n\ \nu))\ \text{SUM}[p]\}$$

Here $C_p =$ `SUM[p]`. Utilizing two of the identities (2)–(3) brings `Out[5]` into the form (23). In order to *prove* the correctness of `Out[5]`, just type

```
In[6]:= Prove[]
```

and the program generates automatically a pretty print version of a proof in a separate window or file, respectively.
The computerized derivations and proofs of (19)–(20) are analogous; one finds the details in the corresponding *Mathematica* notebooks on the article's website.[2]

5 Related Transformations of Generalized Hypergeometric Series

Several relations between two pairs of the generalized hypergeometric series under consideration are given in [35] and [37]:

$$
\begin{aligned}
{}_3F_2\left(\begin{matrix}1-n,\ -p,\ p+1\\ 2\nu+1,\ \ 1\end{matrix}\right) \qquad (24)\\
=\frac{(2\nu+n)\,(2\nu+p+1)\,(2\nu+p+2)\,(2n+p+1)}{4\nu\,(2\nu+1)\,(\nu+n)\,(p+1)}\\
\times{}_3F_2\left(\begin{matrix}1-n,\ p+2,\ -p-1\\ 2\nu+2,\ \ 1\end{matrix}\right)\\
-\frac{n\,(4\nu+2n+p+1)}{2\,(\nu+n)\,(p+1)}\,{}_3F_2\left(\begin{matrix}-n,\ p+2,\ -p-1\\ 2\nu,\ \ 1\end{matrix}\right)
\end{aligned}
$$

and

$$
\begin{aligned}
{}_3F_2\left(\begin{matrix}-n,\ -p,\ p+1\\ 2\nu+1,\ \ 1\end{matrix}\right) \qquad (25)\\
=\frac{n\,(4\nu+2n-p-1)\,(2\nu+p+1)\,(2\nu+p+2)}{4\nu\,(2\nu+1)\,(\nu+n)\,(p+1)}
\end{aligned}
$$

[2] See RelativisticCoulombIntegralsISupplementaryI.nb

$$\times {}_3F_2\left(\begin{matrix} 1-n, & p+2, & -p-1 \\ & 2\nu+2, & 1 \end{matrix}\right)$$

$$-\frac{(2\nu+n)(2n-p-1)}{2(\nu+n)(p+1)}\, {}_3F_2\left(\begin{matrix} -n, & p+2, & -p-1 \\ & 2\nu, & 1 \end{matrix}\right).$$

In addition,

$$\frac{p(p+1)}{2\nu+n}\, {}_3F_2\left(\begin{matrix} 1-n, & p+1, & -p \\ & 2\nu+1, & 2 \end{matrix}\right) \tag{26}$$

$$= \frac{(p-2\nu)(2\nu+p+1)}{2(2\nu+1)(\nu+n)}\, {}_3F_2\left(\begin{matrix} 1-n, & p+1, & -p \\ & 2\nu+2, & 1 \end{matrix}\right)$$

$$+\frac{\nu}{\nu+n}\, {}_3F_2\left(\begin{matrix} -n, & p+1, & -p \\ & 2\nu, & 1 \end{matrix}\right),$$

and

$$\frac{p(p+1)}{n+2\nu}\, {}_3F_2\left(\begin{matrix} 1-n, & -p, & p+1 \\ & 2\nu+1, & 2 \end{matrix}\right)$$

$$= {}_3F_2\left(\begin{matrix} -n, & -p, & p+1 \\ & 2\nu+1, & 1 \end{matrix}\right) - {}_3F_2\left(\begin{matrix} 1-n, & -p, & p+1 \\ & 2\nu+1, & 1 \end{matrix}\right). \tag{27}$$

These relations are "responsible" for the transformation between two different hypergeometric forms of the relativistic Coulomb integrals [35, 37]. The second named author was able to give only the proof of the last relation from the advanced theory of generalized hypergeometric functions.

With the `zb.m` package, one can not only prove but also find such relations, in the literature called also contiguous relations, automatically. We illustrate this by a computer derivation of (24).

```
In[1]:= <<zb.m
```

In[2] := $(a\_)_{k\_}$:= Pochhammer[a,k]; F0[k_] := $\dfrac{(1-n)_k\ (-p)_k\ (p+1)_k}{(2\ \nu+1)_k\ (1)_k\ k!}$;

F1[k_] := $\dfrac{(1-n)_k\ (-p-1)_k\ (p+2)_k}{(2\ \nu+2)_k\ (1)_k\ k!}$; F2[k_] := $\dfrac{(-n)_k\ (-p-1)_k\ (p+2)_k}{(2\ \nu)_k\ (1)_k\ k!}$;

Our goal is to compute rational function coefficients c_0, c_1, c_2, free of the summation variable k, such that

$$\sum_{k=0}^{\infty} (c_0 F0[k] + c_1 F1[k] + c_2 F2[k]) = 0.$$

With a parameterized version of Gosper's algorithm, similar to Zeilberger's extension of Gosper's algorithm, we compute such c_i together with a hypergeometric expression g such that for non-negative integer N:

$$\sum_{k=0}^{N}\left(F0[k]\left(c_0+c_1\frac{F1[k]}{F0[k]}+c_2\frac{F2[k]}{F0[k]}\right)\right)=g[N].$$

This is accomplished using the option `Parameterized`; finally we send N to infinity:

```
In[3]:= FullSimplify[{1, F1[k]/F0[k], F2[k]/F0[k]}]
```

$$\texttt{Out[3]=}\ \left\{1,\ -\frac{(1+k+p)\ (1+2\ \nu)}{(-1+k-p)\ (1+k+2\ \nu)},\ \frac{n\ (1+k+p)\ (k+2\ \nu)}{2\ (k-n)\ (-1+k-p)\ \nu}\right\}$$

$$\texttt{In[4]:= Gosper[F0[k], \{k, 0, N\},}\quad \texttt{Parameterized -> }\left\{1,\ -\frac{(1+k+p)\ (1+2\ \nu)}{(-1+k-p)\ (1+k+2\ \nu)},\ \frac{n\ (1+k+p)\ (k+2\ \nu)}{2\ (k-p)\ (-1+k-p)\ \nu}\right\}]$$

```
If 'N' is a natural number, then:
```

$$\texttt{Out[4]=}\ \Big\{\sum_{k=0}^{N} 4\ (1+p)\ \nu\ (n+\nu)\ (1+2\ \nu)\ F_0[k]-(1+2\ n+p)\ (n+2\ \nu)\ (1+p+2\ \nu)$$
$$(2+p+2\ \nu)\ F_1[k]+\ 2\ n\ \nu\ (1+2\ \nu)\ (1+2\ n+p+4\ \nu)\ F_2[k]\ ==$$
$$-((1+N+p)(1+2\ \nu)(2\ n+4\ n^2+n\ N+2\ n^2\ N+3\ n\ p+2\ n^2\ p+n\ N\ p+$$
$$n\ p^2+4\ \nu+8\ n\ \nu+8\ n^2\ \nu+2\ N\ \nu+6\ p\ \nu+8\ n\ p\ \nu+2\ N\ p\ \nu+$$
$$2\ p^2\ \nu+8\ \nu^2+8\ n\ \nu^2+8\ p\ \nu^2)$$
$$\texttt{Pochhammer[1 - n, N] Pochhammer[-p, N] Pochhammer[1 + p, N])/}$$
$$((1+N+2\ \nu)\ N!\ \texttt{Pochhammer[1, N] Pochhammer[}1+2\ \nu,\ N])\ \Big\}$$

For $N\to\infty$ this gives the desired relation because the right hand side is 0 when $N>p$.

The computerized proofs of (25)–(27) are similar and the corresponding *Mathematica* notebooks are available on the article's website.[3]

6 Virial Recurrence Relations

A general procedure of verification of the linear relations between the relativistic integrals can be formulated as follows. Start from the hypergeometric series representations for the integrals involved into the identity/relation in question, and find all linear dependencies between the corresponding hypergeometric series using the package `zb.m`. Substitute the integrals into the desired identity, eliminate the linear dependent sums/vectors from this equation, and then simplify the coefficients in front of the rest of the series to zero with the help of the standard identities among the quantum numbers of the relativistic Coulomb problem.

[3]See RelativisticCoulombIntegralsISupplementaryII.nb

One can easily see that the linear relation (11) is equivalent to (27), and that (12) follows from (11) and (13). To illustrate our strategy, we derive (13) directly from the hypergeometric representations for the relativistic Coulomb integrals (8)–(10). To this end, we input the hypergeometric summands involved in the relations (8)–(10) and (13):

```
In[1]:= <<zb.m
```

In[2]:= $(a\_)_{k\_}$:= Pochhammer[a,k]; F0[k_] := $\dfrac{(1-n)_k\ (-p)_k\ (p+1)_k}{(2\ \nu+1)_k\ (2)_k\ k!}$;

F1[k_] := $\dfrac{(1-n)_k\ (-p)_k\ (p+1)_k}{(2\ \nu+1)_k\ (1)_k\ k!}$; F2[k_] := $\dfrac{(-n)_k\ (-p)_k\ (p+1)_k}{(2\ \nu+1)_k\ (1)_k\ k!}$;

F3[k_] := $\dfrac{(1-n)_k\ (-p-1)_k\ (p+2)_k}{(2\ \nu+1)_k\ (1)_k\ k!}$; F4[k_] := $\dfrac{(-n)_k\ (-p-1)_k\ (p+2)_k}{(2\ \nu+1)_k\ (1)_k\ k!}$;

```
In[3]:= FullSimplify[{1, F1[k]/F0[k], F2[k]/F0[k], F3[k]/F0[k], F4[k]/F0[k]}]
```

Out[3]= $\left\{1,\ 1+k,\ -\dfrac{(1+k)\ n}{k-n},\ 1+k+\dfrac{2\ k\ (1+k)}{1-k+p},\ \dfrac{(1+k)\ n\ (1+k+p)}{(k-n)\ (-1+k-p)}\right\}$

In[4]:= Gosper[F0[k], {k, 0, N},

Parameterized -> $\left\{1,\ 1+k,\ -\dfrac{(1+k)\ n}{k-n},\ 1+k+\dfrac{2\ k\ (1+k)}{1-k+p},\ \dfrac{(1+k)\ n\ (1+k+p)}{(k-n)\ (-1+k-p)}\right\}$]

If 'N' is a natural number, then:

Out[4]= $\Big\{\sum_{k=0}^{N} -n\ F_1[k]+(1+n+p)\ F_2[k]-n\ F_3[k]+(-1+n-p)\ F_4[k]==0,$

$\sum_{k=0}^{N} 2\ n\ p\ F_0[k]+(1+2\ n+p+2\ \nu)\ F_2[k]+(-1-p-2\ \nu)\ F_4[k]==$

$\dfrac{2\ n\ (1+N+p)\ \text{Pochhammer}[1-n,\ N]\ \text{Pochhammer}[-p,\ N]\ \text{Pochhammer}[1+p,\ N]}{N!\ \text{Pochhammer}[2,\ N]\ \text{Pochhammer}[1+2\ \nu,\ N]},$

$\sum_{k=0}^{N} -2\ n\ (n+2\ \nu)\ F_1[k]+(1+2\ n+2\ n^2+2\ p+2\ n\ p+p^2+2\ \nu+$

$4\ n\ \nu+2\ p\ \nu)\ F_2[k]-(1+p)\ (1+p+2\ \nu)\ F_4[k]==$

$\dfrac{2\ n\ (1+N)(1+N+p)\ \text{Pochhammer}[1-n,\ N]\ \text{Pochhammer}[-p,\ N]\ \text{Pochhammer}[1+p,\ N]}{N!\ \text{Pochhammer}[2,\ N]\ \text{Pochhammer}[1+2\ \nu,\ N]}\Big\}$

Notice that for $N \to \infty$ all the right hand sides vanish because they are 0 when $N > p$.

Summarizing, the package has found the following three linear relations:

$$n\,(X+U)-(1+n+p)\,Y+(1-n+p)\,V=0, \tag{28}$$

$$2np\ Z+(1+2n+p+2\nu)\,Y-(1+p+2\nu)\,V=0, \tag{29}$$

$$\begin{aligned}2n\,(n+2\nu)\,X+(1+p)(1+p+2\nu)V&\\ =\left[(n+1)^2+2p+(n+p)^2+2(2n+p+1)\nu\right]Y&\end{aligned} \tag{30}$$

for the following five linear dependent vectors

$$X := {}_3F_2\left(\begin{matrix}1-n,\ -p,\ p+1\\ 2\nu+1,\ 1\end{matrix}\right), \qquad Y := {}_3F_2\left(\begin{matrix}-n,\ -p,\ p+1\\ 2\nu+1,\ 1\end{matrix}\right),$$

$$Z := {}_3F_2\left(\begin{matrix}1-n,\ -p,\ p+1\\ 2\nu+1,\ 2\end{matrix}\right), \qquad U := {}_3F_2\left(\begin{matrix}1-n,\ -p-1,\ p+2\\ 2\nu+1,\ 1\end{matrix}\right),$$

$$V := {}_3F_2\left(\begin{matrix}-n,\ -p-1,\ p+2\\ 2\nu+1,\ 1\end{matrix}\right).$$

We choose to present everything in terms of Y and V:

```
In[5]:= Lin1 = n*(X + U) - (1 + n + p)*Y + (1 - n + p)*V ;
        Lin2 = (2 n p)*Z + (1 + 2 n + p + 2 ν)*Y - (1 + p + 2 ν)*V ;
        Lin3 = -2 n (n + 2 ν)*X +
           (1 + 2 n + 2 n^2 + 2 p + 2 n p + p^2 + 2 ν + 4 n ν + 2 p ν)*Y -
           (1 + p) (1 + p + 2 ν)*V ;
        Solve[Lin1==0 && Lin2==0 && Lin3==0, {X, U, Z}];
        FullSimplify[%]
```

```
Out[5]= {{X → (-(1+p) V(1+p+2 ν)+Y (2 n^2+2 n (1+p+2 ν)+(1+p) (1+p+2 ν)))/(2 n (n+2 ν)),
         U → (V (2 n^2-2 n (1+p-2 ν)+(1+p) (1+p-2 ν))-(1+p) Y (1+p-2 ν))/(2 n (n+2 ν)),
         Z → (V (1+p+2 ν)-Y (1+2 n+p+2 ν))/(2 n p) }}
```

Introducing the Coulomb integrals,

```
In[6]:= Ap[X, Y, Z] :=

        (2 μ (2 a β)^(p) (Gamma[2 ν + 1])/(Gamma[2 ν + p + 1]))^(-1)*
           ((μ + a κ)*X + (μ - a κ)*Y + 2 p ε a n*Z) ;

        Bp[X, Y, Z] :=

         (2 μ (2 a β)^(p) (Gamma[2 ν + 1])/(Gamma[2 ν + p + 1]))^(-1)*
           (ε (μ + a κ)*X + ε (μ - a κ)*Y + 2 p a n*Z) ;

        Cplus1[U, V] := (4 μ (2 a β)^(p + 1) (Gamma[2 ν + 1])/
           (Gamma[2 ν + p + 2]))^(-1)*(a (μ + a κ)*U + a (μ - a κ)*V) ;
```

we express the desired relation in terms of $X, \ldots, V$:

```
In[7]:= (2 κ)*Bp[X, Y, Z] - (p + 1)*Ap[X, Y, Z] - (4 β)*Cplus1[U, V];
        % /. Gamma[2 + p + 2 ν] -> (1 + p + 2 ν)*Gamma[1 + p + 2 ν];
        FullSimplify[%]
```

```
Out[7]= - 1/(μ Gamma[1 + 2 ν]) 2^(-1-p) (a β)^(-p)
   (μ ((X + Y) (1 + p - 2 ε κ) + U (1 + p + 2 ν) - V (1 + p + 2 ν)) +
      a (2 n p Z (ε + p ε - 2 κ) + κ ((X - Y) (1 + p - 2 ε κ) +
      U (1 + p + 2 ν) + V (1 + p + 2 ν)))) Gamma[1 + p + 2 ν]
```

Next we rewrite the relevant part into a linear combination of $X, \ldots, V$:

```
In[8]:= ZERO = (μ ((X+Y) (1+ p - 2ε κ) + U (1 + p + 2ν) - V (1 + p + 2ν)) +
          a (2 n p Z (ε + p ε - 2 κ) +
          κ ((X-Y) (1 + p - 2εκ) + U (1 + p + 2ν) + V (1 + p + 2 ν)))) ;
        Collect[ZERO, {X, Y, Z, U, V}] ;

        FullSimplify[%]

Out[8]= 2 a n p Z (ε + p ε - 2 κ) + Y (1 + p - 2 ε κ) (-a κ + μ) +

   X(1 + p - 2ε κ)(aκ + μ) + V(aκ - μ)(1 + p + 2ν) + U(aκ + μ)(1 + p + 2ν)
```

Eliminating X, U and Z,

```
In[9]:= % /. {X -> (-(1+p)V(1+p+2ν)+Y(2n^2+2 n(1+p+2ν)+(1+p)(1+p+2 ν)))/(2 n (n+2 ν)),

        U -> (V (2 n^2 - 2 n(1+p-2 ν)+(1+p)(1+p-2 ν)) - (1+p) Y (1+p-2 ν))/(2 n (n+2 ν)),

        Z -> (V (1+p+2 ν) - Y (1+2 n+p+2 ν))/(2 n p) } ;

        FullSimplify[%] ;

        Collect[%, {Y, V}] ;

        FullSimplify[%]

Out[9]= ((1+p) V(1+p+2 ν) (-μ (n-ε κ+ν) + a (n^2 ε-n κ+ε κ^2+2 n ε ν-κ ν)))/(n (n+2 ν)) +
   Y ((1+p-2 ε κ) (-a κ+μ) - ((1+p) (a κ+μ) (1+p-2 ν) (1+p+2 ν))/(2 n (n+2 ν)) -
      a (ε+p ε-2 κ)(1+2 n+p+2 ν)+
      ((1+p-2 ε κ) (a κ+μ)(2 n^2+2 n (1+p+2 ν)+(1+p) (1+p+2 ν)))/(2 n (n+2 ν)))
```

Finally, we simplify the coefficients of V and Y:

```
In[10]:= ZeroV = - μ (n - ε κ + ν) + a (n^2 ε - n κ + ε κ^2 + 2 n ε ν - κ ν) ;
         ZeroY = 2 n (1 + p - 2ε κ) (-aκ + μ) (n + 2ν) -
           (1 + p) (aκ + μ) (1 + p - 2ν) (1 + p + 2ν) -
           2 a n (ε + pε -2κ) (n + 2ν) (1 + 2 n + p + 2ν) +
           (1+p-2ε κ) (aκ+μ) (2n^2 + 2n(1 + p + 2ν) + (1 + p)(1 + p + 2ν)) ;
         {ZeroV, ZeroY} ;
         FullSimplify[%] ;
         % /. n -> (ε μ - a ν)/a ;
         FullSimplify[%] ;
```

```
        % /. ε^2 - > 1 - a^2 ;
        FullSimplify[%] ;
        % /. κ^2 - > ν^2 + μ^2
Out[10]= {0, 0}
```

which is the name of the game.
Computerized proofs of (14) and some of its extensions work the same; they are available on the article's website.[4]
In a similar fashion, seeking for a more general linear combination of the corresponding integrals, with the zb.m package one can derive the following two-parameter relation:

$$\begin{aligned}&[D(p+1) - C\,(2\kappa + \varepsilon(p+1))]\,A_p \\ &\quad - [2D\,\kappa - C\,(2\varepsilon\kappa + p + 1)]\,B_p + 4\mu C\,C_p + 4\beta D\,C_{p+1} = 0, \qquad (31)\end{aligned}$$

where C and D are two arbitrary constants. The virial relations (11)–(13) are its special cases.
We would like to point out the following relation:

$$(p+1)\left(2\kappa C_p - \mu A_p\right) = \beta(p+2)\left(\varepsilon A_{p+1} - B_{p+1}\right), \qquad (32)$$

as another simple example.

Note. This relation is a linear combination of (12)–(14); see [1].

7 Conclusion

The relativistic Coulomb integrals (4)–(6) were recently evaluated in a hypergeometric form [35]. The corresponding system of the first order difference equations (15)–(16) has been solved in [37] in terms of linear combinations of the dual Hahn polynomials thus providing an independent proof. Here, with the help of the Fast Zeilberger package zb.m we give a direct derivation of these results.

One of the goals of this article is to demonstrate the power of symbolic computation for the study of relativistic Coulomb integrals. Namely, computer algebra methods related to Zeilberger's holonomic systems approach allow not only to verify some already known complicated relations, but also to derive new ones without making enormously time-consuming calculations by hands or with ad hoc usage of computer algebra procedures.

[4] See RelativisticCoulombIntegralsISupplementaryIII.nb

In a sequel to this article[5] we are planning to investigate the computer-assisted derivation of recurrences, e.g. the "birthday recurrences" from Sect. 4, by taking as a starting point the original definition of the Coulomb integrals (4)–(6). To this end, we will use Koutschan's package `HolonomicFunctions` [16]. This package, written in *Mathematica*, implements further ideas related to Zeilberger's holonomic systems paradigm [40]; for instance, it includes implementations of (variations of) Z's "slow algorithm", and algorithms by F. Chyzak (and B. Salvy), and N. Takayama. In this context we will have to exploit closure properties of classes of special (resp. holonomic) sequences and functions; an introduction to computer algebra methods for the univariate case can be found in [15].

Moreover, the `zb.m` package strongly suggests that there are, in fact, four linearly independent virial recurrence relations, see more details on the article's website, but only three of them (e.g., (12)–(14)) are available in the literature. Another next challenge is to study the off-diagonal matrix elements that are important in applications [22–24, 28, 29], and [32].

Acknowledgements We thank Doron Zeilberger for valuable discussions and encouragement. We are grateful to Christian Krattenthaler for his LaTeX style file for presenting *Mathematica* notebooks in this paper. This work has been partially supported by the Austrian Science Foundation FWF, P20162.

References

1. Adkins, G.S.: Dirac–Coulomb energy levels and expectation values. Am. J. Phys. **76**(6), 579–584 (2008)
2. Andrae, D.: Recursive evaluation of expectation values $\langle r^k \rangle$ for arbitrary states of the relativistic one-electron atom. J. Phys. B: At. Mol. Phys. **30**(20), 4435–4451
3. Beiersdorfer, P.: Testing QED and atomic-nuclear interactions with high-Z ions. J. Phys. B: At. Mol. Opt. Phys. **43**, 074032 (10pp) (2010). Special Issue on High Precision Atomic Physics
4. Bethe, H.A., Salpeter, E.E.: Quantum Mechanics of One- and Two-Electron Atoms. Springer, Berlin (1957)
5. Darwin, C.G.: The wave equations of the electron. Proc. R Soc. Lond. Ser. A **118**, 654–680 (1928)
6. Davis, L.: A note on the wave functions of the relativistic hydrogen atom. Phys. Rev. **56**, 186–187 (1939)
7. Epstein, J.H., Epstein, S.T.: Some applications of hypervirial theorems to the calculation of average values. Am. J. Phys. **30**(4), 266–268 (1962)
8. Gordon, W.: Die Energieniveaus des Wasserstoffatoms nach der Diracschen Quantentheorie des Elektrons. Zeitschrift für Physik **13**, 11–14 (1928)
9. Gosper, R.W., Jr.: Decision procedure for indefinite hypergeometric summation. Proc. Natl. Acad. Sci. U.S.A. **75**, 40–42 (1978)

[5] See recent preprint entitled "Relativistic Coulomb Integrals and Zeilberger's Holonomic Systems Approach. II" by C. Koutschan, P. Paule, and S.K. Suslov; arXiv:1306.1362v1 [quant-ph] 6 Jun 2013.

10. Gumberidze, A., et al.: Quantum electrodynamics in strong electric fields: the ground state Lamb shift in hydrogenlike uranium. Phys. Rev. Lett. **94**, 223001 (4pp) (2005).
11. Gumberidze A., et al.: Precision tests of QED in strong fields: experiments on hydrogen- and helium-like uranium. J. Phys. Conf. Ser. **58**, 87–92 (2007)
12. Karr, M.: Summation in finite terms. J. ACM **28**, 305–350 (1981)
13. Karshenboim, S.G., Smirnov, V.B. (eds.): Precision Physics of Simple Atomic Systems. Lecture Notes in Physics, vol. 627. Springer, Berlin/Heidelberg (2003)
14. Karshenboim, S.G., et al. (eds.): The Hydrogen Atom: Precision Physics of Simple Atomic Systems. Lecture Notes in Physics, vol. 570. Springer, Berlin/Heidelberg (2001)
15. Kauers, M., Paule, P.: The Concrete Tetrahedron. Symbolic Sums, Recurrence Equations, Generation Functions, Asymptotic Estimates. Springer, Wien/New York (2011)
16. Koutschan, C.: Advanced Applications of the Holonomic Systems Approach. Ph.D. thesis, RISC, Johannes Kepler University Linz (2009)
17. Mohr, P.J., Plunien, G., Soff, G.: QED corrections in heavy atoms. Phys. Rep. **293**, 227–369 (1998)
18. Nikiforov, A.F., Uvarov, V.B.: Special Functions of Mathematical Physics. Birkhäuser, Basel/Boston (1988)
19. Nikiforov, A.F., Suslov, S.K., Uvarov, V.B.: Classical Orthogonal Polynomials of a Discrete Variable. Springer, Berlin/New York (1991)
20. Olver, F.W.J., et al.: NIST Handbook of Mathematical Functions. Cambridge University Press, Cambridge (2010). Free electroninc access: NIST Digital Library of Mathematical Functions. http://dlmf.nist.gov/.
21. Petkovšek, M., Wilf, H.S., Zeilberger, D.: A=B. A. K. Peters, Wellesley (1997)
22. Puchkov, A.M.: The method of matrix elements' calculations for the Dirac equation in the Coulomb field. J. Phys. B: At. Mol. Opt. Phys. **44**, 045002 (6pp) (2010)
23. Puchkov, A.M., Labzovskiĭ, L.N.: Probabilities of forbidden magnetic-dipole transitions in the hydrogen atom and hydrogen-like ions. Opt. Spectrosc. **106**(2), 181–186 (2009)
24. Puchkov, A.M., Labzovskiĭ, L.N.: Parity violation effects in hydrogen atom in forbidden magnetic-dipole transitions. Opt. Spectrosc. **108**(5), 713–718 (2010)
25. Schiff, L.I.: Quantum Mechanics, 3rd edn. McGraw-Hill, New York (1968)
26. Schneider, C.: Symbolic summation assists combinatorics. Semin. Lothar. Comb. **56**, 1–36 (2007). (Article B56b)
27. Schneider, C.: Parameterized telescoping proves algebraic independence of sums. Ann. Comb. **14**, 533–552 (2010)
28. Shabaev, V.M.: Recurrence formulas and some exact relations for radial integrals with Dirac and Schrödinger wave functions. Vestnik Leningradskogo Universiteta, Seria 4: Fizika, Khimiya (1984) number 1, 15–19 (in Russian)
29. Shabaev, V.M.: Generalizations of the virial relations for the Dirac equation in a central field and their applications to the Coulomb field. J. Phys. B: At. Mol. Opt. Phys. **24**, 4479–4488 (1991)
30. Shabaev, V.M.: Relativistic recoil corrections to the atomic energy levels. In: The Hydrogen Atom: Precision Physics of Simple Atomic Systems. Lecture Notes in Physics, vol. 570, pp. 714–726. Springer, Berlin/Heidelberg (2001)
31. Shabaev, V.M.: Two-time Green's function method in quantum electrodynamics of high-Z few-electron atoms. Phys. Rep. **356**, 119–228 (2002)
32. Shabaev, V.M.: Virial relations for the Dirac equation and their applications to calculations of hydrogen-like atoms. In: Precision Physics of Simple Atomic Systems. Lecture Notes in Physics, vol. 627, pp. 97–113. Springer, Berlin/Heidelberg (2003)
33. Shabaev, V.M.: Quantum electrodynamics of heavy ions and atoms: current status and prospects. Phys. Usp. **178**(11), 1220–1225 (2008) (in Russian); Phys. Usp. **51**(11), 1175–1180 (2008)

34. Solovyev, D., Sharipov, V., Labzowsky, L., Plunien, G.: Influence of expernal electric fields on multy-photon transitions between the 2s, 2p and 1s levels for hydrogen and antihydrogen atoms and hydrogen-like ions. J. Phys. B: At. Mol. Opt. Phys. **43**, 074005 (16pp) (2010). Special Issue on High Precision Atomic Physics
35. Suslov, S.K.: Expectation values in relativistic Coulomb problems. J. Phys. B: At. Mol. Opt. Phys. **42**, 185003 (8pp) (2009)
36. Suslov, S.K.: Relativistic Kramers–Pasternack recurrence relations. J. Phys. B: At. Mol. Opt. Phys. **43**, 074006 (7pp) (2010). Special Issue on High Precision Atomic Physics
37. Suslov, S.K.: Mathematical structure of relativistic Coulomb integrals. Phys. Rev. A **81**, 032110 (8pp) (2010)
38. Suslov, S.K., Trey, B.: The Hahn polynomials in the nonrelativistic and relativistic Coulomb problems. J. Math. Phys. **49**(1), 012104 (51pp) (2008)
39. Vrscay, E.R., Hamidian, H.: Rayleigh–Schrö dinger perturbation theory at large order for radial relativistic Hamiltonians using hypervirial Hellmann–Feynman theorems. Phys. Lett. A **130**(3), 141–146 (1988)
40. Zeilberger, D.: A holonomic systems approach to special functions identities. J. Comput. Appl. Math. **32**, 321–368 (1990)
41. Zeilberger, D.: A fast algorithm for proving terminating hypergeometric identities. Discret. Math. **80**, 207–211 (1990)
42. Zeilberger, D.: The method of creative telescoping. J. Symb. Comput. **11**, 195–204 (1991)

Hypergeometric Functions in *Mathematica*®

Oleksandr Pavlyk

Abstract This paper is a short introduction to the generalized hypergeometric functions, with some theory, examples and notes on the implementation in the computer algebra system *Mathematica*®. (*Mathematica* is a registered trademark of Wolfram Research, Inc.)

1 Hypergeometric Series

A power series $\sum_{k=0}^{n} c_n$ is called hypergeometric if the ratio of successive c_n is rational in n, represented in its factored form:

$$\frac{c_{n+1}}{c_n} = \mathscr{R}(n) = \frac{(n+a_1)(n+a_2)\cdots(n+a_p)}{(n+1)(n+b_1)\cdots(n+b_q)}z. \tag{1}$$

Examples of such series include geometric series with $c_n = z^n$, exponential series with $c_n = \frac{z^n}{n!}$ and logarithmic series with $c_n = \frac{(-z)^n}{(n+1)}$. The general solution to the recurrence equation (1) is

$$c_n = c_0 \prod_{m=0}^{n-1} \frac{(m+a_1)(m+a_2)\cdots(m+a_p)}{(m+1)(m+b_1)\cdots(m+b_q)}z = c_0 \frac{z^n}{n!}\frac{(a_1)_n(a_2)_n\cdots(a_p)_n}{(b_1)_n\cdots(b_q)_n}, \tag{2}$$

where $(a)_n = a(a+1)\cdots(a+n-1)$ is the Pochhammer symbol, also known as raising factorial. The hypergeometric series need not be infinite. When the ratio (1)

O. Pavlyk (✉)
Wolfram Research, Inc. 100 Trade Centre Dr., Champaign, IL 61820, USA
e-mail: pavlyk@wolfram.com

C. Schneider and J. Blümlein (eds.), *Computer Algebra in Quantum Field Theory*, Texts & Monographs in Symbolic Computation, DOI 10.1007/978-3-7091-1616-6_10,

vanishes for some value of n, the hypergeometric series degenerates to a polynomial. Newton's generalized binomial theorem serves as an example:

$$(1-z)^{-\alpha} = \sum_{n=0}^{\infty} \frac{(\alpha)_n}{n!} z^n.$$

Convergence of the hypergeometric series can be easily established by the ratio test:

$$\frac{c_{n+1}}{c_n} = \frac{z}{n^{q+1-p}} \left(1 - \frac{1}{n}\left(1 + \sum_{j=1}^{q} b_j - \sum_{i=1}^{p} a_i\right) + o\left(\frac{1}{n}\right)\right). \tag{3}$$

When $q+1 > p$ the hypergeometric series converges for all $z \in \mathbb{C}$ and defines an entire function ${}_pF_q(z)$. When $p = q+1$ the series converges for $|z| < 1$ and for $|z| = 1$ if $\Re\left(\sum_{j=1}^{q} b_j - \sum_{i=1}^{p} a_i\right) > 0$. The hypergeometric series converges conditionally for $|z| = 1, z \neq 1$ if $\Re\left(\sum_{j=1}^{q} b_j - \sum_{i=1}^{p} a_i\right) = 0$. For $p > q+1$ the hypergeometric series diverges, unless it is a polynomial. An example of such a hypergeometric polynomial is provided by the Charlier polynomials [7]:

$$C_n(\mu; z) = {}_2F_0\left(\begin{matrix} -n, -z \\ - \end{matrix} \middle| -\frac{1}{\mu}\right)$$

Analytic continuation, whenever necessary, of the hypergeometric series defines a generalized hypergeometric function , denoted as

$${}_pF_q\left(\begin{matrix} a_1, a_2, \ldots, a_p \\ b_1, b_2, \ldots, b_q \end{matrix} \middle| z\right). \tag{4}$$

It is represented in *Mathematica* as

- `HypergeometricPFQ[{`$a_1, \ldots, a_p$`}, {`$b_1, \ldots, b_q$`},`z`]`

with special cases of p and q implemented as

- `HypergeometricOF1[`b`, `z`]`
- `Hypergeometric1F1[`a`, `b`, `z`]`
- `Hypergeometric2F1[`a_1`, `a_2`, `b`, `z`]`

for convenience. The family of the generalized hypergeometric functions encompasses many elementary and special functions and sequences. A compendium of such special cases can be found in [3, 5].

The hypergeometric function ${}_pF_q(z)$ is an analytic function of its parameters with poles at $b_j = -m$, $m \in \mathbb{Z}_{\geqslant 0}$. A *regularized* hypergeometric function, ${}_p\tilde{F}_q(z)$, is a multiple of hypergeometric functions with the property that it is an entire function of its parameters:

$$ {}_p\tilde{F}_q(z) = \prod_{j=1}^{q} \Gamma(b_j) \cdot {}_pF_q(z) = \sum_{m=0}^{\infty} \frac{(a_1)_m \cdots (a_p)_m}{\Gamma(b_1+m)\cdots\Gamma(b_q+m)} \frac{z^m}{m!}. \tag{5} $$

The regularization effect is the consequence of the reciprocal of the Γ-function vanishing at non-positive integers.

2 Parameter Shift Operators and Hypergeometric Differential Equation

Making use of the recurrence relation

$$ \alpha \cdot (\alpha+1)_n = (\alpha)_{n+1} = (\alpha)_n \cdot (\alpha+n) \tag{6} $$

for the Pochhammer symbol for $\alpha = a_1$, it is easily established that the differential operator $\hat{A}_{a_1} = z\frac{\mathrm{d}}{\mathrm{d}z} + a_1$ increases the parameter a_1 of the hypergeometric function by 1:

$$ \begin{aligned} \hat{A}_{a_1} \circ {}_pF_q\left(\left. \begin{matrix} a_1, \ldots, a_p \\ b_1 \ldots, b_q \end{matrix} \right| z \right) &= \sum_{n=0}^{\infty} \frac{\hat{A}_{a_1} \circ z^n}{n!} \frac{(a_1)_n \cdots (a_p)_n}{(b_1)_n \cdots (b_q)_n} \\ &= \sum_{n=0}^{\infty} (n+a_1) \frac{z^n}{n!} \frac{(a_1)_n \cdots (a_p)_n}{(b_1)_n \cdots (b_q)_n} \\ &= \sum_{n=0}^{\infty} a_1 \frac{z^n}{n!} \frac{(a_1+1)_n \cdots (a_p)_n}{(b_1)_n \cdots (b_q)_n} \\ &= a_1 \cdot {}_pF_q\left(\left. \begin{matrix} a_1+1, \ldots, a_p \\ b_1, \quad \ldots, b_q \end{matrix} \right| z \right). \end{aligned} \tag{7} $$

Clearly the choice of the parameter a_1 was arbitrary, θ_{a_p} will likewise raise the parameter a_p by 1. Using the recurrence equation (6) with $\alpha = b_1 - 1$, and introducing $\hat{B}_{b_1} = z\frac{\mathrm{d}}{\mathrm{d}z} + b_1 - 1$:

$$ \begin{aligned} \hat{B}_{b_1} \circ {}_pF_q\left(\left. \begin{matrix} a_1, \ldots, a_p \\ b_1 \ldots, b_q \end{matrix} \right| z \right) &= \sum_{n=0}^{\infty} \frac{\hat{B}_{b_1} \circ z^n}{n!} \frac{(a_1)_n \cdots (a_p)_n}{(b_1)_n \cdots (b_q)_n} \\ &= \sum_{n=0}^{\infty} (n+b_1-1) \frac{z^n}{n!} \frac{(a_1)_n \cdots (a_p)_n}{(b_1)_n \cdots (b_q)_n} \end{aligned} \tag{8} $$

$$= \sum_{n=0}^{\infty} (b_1 - 1) \frac{z^n}{n!} \frac{(a_1)_n \cdots (a_p)_n}{(b_1 - 1)_n \cdots (b_q)_n}$$

$$= (b_1 - 1) \cdot {}_pF_q \left(\begin{matrix} a_1, & \ldots, a_p \\ b_1 - 1, & \ldots, b_q \end{matrix} \middle| z \right).$$

The derivative of the hypergeometric function is also easy to find:

$$\frac{\mathrm{d}}{\mathrm{d}z} \circ {}_pF_q \left(\begin{matrix} a_1, \ldots, a_p \\ b_1 \ldots, b_q \end{matrix} \middle| z \right) = \sum_{n=0}^{\infty} \frac{n z^{n-1}}{n!} \frac{(a_1)_n \cdots (a_p)_n}{(b_1)_n \cdots (b_q)_n} \tag{9}$$

$$= \sum_{n=1}^{\infty} \frac{z^n}{(n-1)!} \frac{(a_1)_n \cdots (a_p)_n}{(b_1)_n \cdots (b_q)_n}$$

$$= \sum_{n=0}^{\infty} \frac{z^n}{n!} \frac{(a_1)_{n+1} \cdots (a_p)_{n+1}}{(b_1)_{n+1} \cdots (b_q)_{n+1}}$$

$$= \frac{a_1 \cdots a_p}{b_1 \cdots b_q} \cdot {}_pF_q \left(\begin{matrix} a_1 + 1, \ldots, a_p + 1 \\ b_1 + 1, \ldots, b_q + 1 \end{matrix} \middle| z \right),$$

where the first equality of (6) was used for each parameter of the hypergeometric function.

We are now in the position to derive the ordinary differential equation (ODE), satisfied by the hypergeometric function. To this end, note that

$$\hat{A}_{a_1} \circ \cdots \circ \hat{A}_{a_p} \circ {}_pF_q \left(\begin{matrix} \mathbf{a} \\ \mathbf{b} \end{matrix} \middle| z \right) = a_1 \cdots a_p \cdot {}_pF_q \left(\begin{matrix} \mathbf{a} + 1 \\ \mathbf{b} \end{matrix} \middle| z \right) \tag{10}$$

$$\hat{B}_{b_1} \circ \cdots \circ \hat{B}_{b_q} \circ {}_pF_q \left(\begin{matrix} \mathbf{a} \\ \mathbf{b} \end{matrix} \middle| z \right) = (b_1 - 1) \cdots (b_q - 1) \cdot {}_pF_q \left(\begin{matrix} \mathbf{a} \\ \mathbf{b} - 1 \end{matrix} \middle| z \right), \tag{11}$$

where $\mathbf{a} + 1$ denotes the vector with each element incremented by one. Differentiating (11), using (9) and comparing to (10) we arrive at the hypergeometric differential equation:

$$\left(z \cdot \hat{A}_{a_1} \circ \cdots \circ \hat{A}_{a_p} - \hat{B}_1 \circ \hat{B}_{b_1} \circ \cdots \circ \hat{B}_{b_q} \right) \circ {}_pF_q \left(\begin{matrix} \mathbf{a} \\ \mathbf{b} \end{matrix} \middle| z \right) = 0. \tag{12}$$

A few remarks are in order.

The hypergeometric differentiation equation is a linear ordinary differential equation with polynomial coefficients, and therefore the generalized hypergeometric function is a holonomic function [8].

Any solution of the hypergeometric differential equation is referred to as a hypergeometric function. Therefore, for example, the Tricomi's solution $U(a; b; z)$

of the ${}_1F_1$-differential equation (a.k.a. Kummer's ODE) $zy''(z) + (b - z)y'(z) - ay(z) = 0$ is also said to be a hypergeometric function.

In the differential equation parameters p and q can be arbitrary non-negative integers. The order of the differential equation equals $r = \max(p, q + 1)$. When $p \neq q + 1$, the leading term of the ODE is $z^r y^{(r)}$, meaning that $z = 0$ and $z = \infty$ are the only singular points of the hypergeometric differential equation. In the case of $p = q + 1$, the leading term of the ODE is $(1 - z)z^{r-1}y^{(r)}(z)$, so that the ODE has three singular points $z = 0$, $z = 1$ and $z = \infty$. One can show that $z = 1$ is always a regular singular point.

Let $f(z)$ be a solution of the hypergeometric ODE (12), and define $g(z) = z^\alpha f(\kappa z^m)$. Then $g(z)$ satisfies:

$$\left(\kappa z^m \prod_{i=1}^{p} \left(\frac{z}{m}\frac{\mathrm{d}}{\mathrm{d}z} + a_i - \frac{\alpha}{m}\right) - \prod_{j=0}^{q} \left(\frac{z}{m}\frac{\mathrm{d}}{\mathrm{d}z} + b_j - 1 - \frac{\alpha}{m}\right)\right) \circ g(z) = 0. \quad (13)$$

The equation obtained is similar in the structure to Eq. (12). Solutions of (13) are also said to be hypergeometric functions. Therefore the familiar Bessel's function of the first kind

$$J_\nu(z) = \frac{(z/2)^\nu}{\Gamma(\nu)} \cdot {}_0F_1\left(\begin{matrix} - \\ \nu + 1 \end{matrix} \middle| \frac{z^2}{4}\right)$$

can now be welcomed to the hypergeometric family, as well as the Bessel's function of the second kind $Y_\nu(z)$ which is another solution of the Bessel's differential equation, so that $J\nu(z)$ and $Y_\nu(z)$ form a non-degenerate basis on the solutions of the Bessel's differential equation for all regular points of the ODE.

3 Mellin Transform

The (direct) Mellin transform of a function $f(x)$ is defined, for $a < \Re(s) < b$, as the following integral transform

$$\mathscr{M}[f](s) = \int_0^\infty x^{s-1} f(x)\mathrm{d}x \quad (14)$$

The image of the Mellin transform is an analytic function of s in its definition domain. Therefore the region $\{s \in \mathbb{C} : a < \Re(s) < b\}$ is called the strip of analyticity. The inverse Mellin transform is defined as follows:

$$\mathscr{M}^{-1}[\varphi](x) = \frac{1}{2\pi i} \int_{\gamma - i\infty}^{\gamma + i\infty} \varphi(s)x^{-s}\mathrm{d}s \quad (15)$$

for arbitrary $a < \gamma < b$, and a and b define the strip of analyticity of $\varphi(s)$. The Mellin inversion theorem specifies when the inverse Mellin transform is defined, and when it recovers the original function.

The importance of the Mellin transform for the study of the hypergeometric differential equation lies in the following property:

$$\begin{aligned}\mathcal{M}\left[x\frac{\mathrm{d}}{\mathrm{d}x}f(x)\right](s) &= \int_0^\infty x^{s-1}\cdot x\frac{\mathrm{d}}{\mathrm{d}x}f(x)\mathrm{d}x \\ &= x^s f(x)\Big|_{x\downarrow 0}^{x\uparrow+\infty} - s\int_0^\infty x^{s-1}f(x)\mathrm{d}x \\ &= -s\mathcal{M}[f](s).\end{aligned}$$

Moreover,

$$\mathcal{M}[x^m f(x)](s) = \int_0^\infty x^{s+m-1}f(x)\mathrm{d}x = \mathcal{M}[f](s+m).$$

We now consider a suitable solution of the hypergeometric ODE (13), and apply these two facts about the Mellin transform to infer the constraints the ODE imposes on the Mellin image of its solution. We readily find

$$\kappa\prod_{i=1}^{p}\left(a_i - 1 - \frac{s+\alpha}{m}\right)\mathcal{M}[g](s+m) = \prod_{j=0}^{q}\left(b_j - 1 - \frac{s+\alpha}{m}\right)\mathcal{M}[g](s), \quad (16)$$

where we take $b_0 = 1$. The relation (16) is the rank 1 recurrence equation in $s_m = \frac{s}{m}$. Therefore the Mellin transform image of a solution of the hypergeometric differential equation (13) is an analytic solution of the rank 1 recurrence equation (16).

In order to simplify further analysis, we temporarily set $m = 1$.

Solutions of the recurrence relation (16) are easily constructed with the help of Euler's Γ-functions. Let Θ_s be the operator of taking the ratio of the unit shift of a function of s to the function itself:

$$\Theta_s h(s) = \frac{h(s+1)}{h(s)}.$$

Every factor of the rational function $\Theta_s\mathcal{M}[g](s)$ in (16) can be reproduced in 1 of 2 ways:

$$\Theta_s\Gamma(a-\alpha-s) = \frac{1}{a-1-\alpha-s} = \Theta_s\frac{(-1)^s}{\Gamma(1-a+\alpha+s)} \quad (17)$$

and

$$\Theta_s\frac{1}{\Gamma(b-\alpha-s)} = b-1-\alpha-s = \Theta_s(-1)^s\Gamma(1-b+\alpha+s). \quad (18)$$

A solution of (16) therefore involves combinations of these possibilities. Partition the set of parameters $\{a_i\}_{i=1}^{p}$ into two non-overlapping subsets, depending on whether the direct or the reciprocal Γ-function in (17) is used in the solution of (16). Denote these subsets A_d and A_r. Partition the set of parameters $\{b_j\}_{j=0}^{q}$ similarly into B_r and B_d. Then the solution of (16) reads

$$\mathcal{M}[g](s) = \frac{\prod_{a\in A_d} \Gamma\,(a-\alpha-s)}{\prod_{a'\in A_r} \Gamma\,(1-a'+\alpha+s)} \cdot \frac{\prod_{b'\in B_d} \Gamma\,(1-b'+\alpha+s)}{\prod_{b\in B_r} \Gamma\,(b-\alpha-s)} \cdot (-1)^{s(|A_r|+|B_d|)} \kappa^{-s}. \quad (19)$$

As an example, consider the solution for which $A_r = \emptyset$ and $B_d = \{b_0\}$, i.e.

$$g(z) = \frac{1}{2\pi i} \int_{\gamma-i\infty}^{\gamma+i\infty} \frac{\prod_{i=1}^{p} \Gamma(a_i-\alpha-s)}{\prod_{j=1}^{q} \Gamma(b_j-\alpha-s)} \Gamma(\alpha+s)(-z)^{-s} \mathrm{d}s,$$

where γ and the parameters a_i are such that $0 < \gamma + \alpha < \min_{1\leqslant i\leqslant p} \Re(a_i)$. It can be shown, that the integration contour can be closed to the left, encompassing all the poles of $\Gamma(s+\alpha)$. As these poles are all simple, finding residues at them is particularly simple:

$$\begin{aligned} \mathrm{Res}_{s=-\alpha-m} \frac{\prod_{i=1}^{p} \Gamma(a_i-\alpha-s)}{\prod_{j=1}^{q} \Gamma(b_j-\alpha-s)} \Gamma(\alpha+s)(-z)^{-s} &= \\ (-z)^{m+\alpha} \frac{\prod_{i=1}^{p} \Gamma(a_i+m)}{\prod_{j=1}^{q} \Gamma(b_j+m)} \mathrm{Res}_{s=-m}\, \Gamma(s) &= \\ (-z)^{m+\alpha} \frac{\prod_{i=1}^{p} \Gamma(a_i+m)}{\prod_{j=1}^{q} \Gamma(b_j+m)} \frac{(-1)^m}{\Gamma(m+1)}&, \end{aligned} \quad (20)$$

and hence

$$\begin{aligned} g(z) &= (-z)^{\alpha} \sum_{m=0}^{\infty} \frac{\prod_{i=1}^{p} \Gamma(a_i+m)}{\prod_{j=1}^{q} \Gamma(b_j+m)} \frac{z^m}{m!} \\ &= (-z)^{\alpha} \frac{\prod_{i=1}^{p} \Gamma(a_i)}{\prod_{j=1}^{q} \Gamma(b_j)} \cdot {}_pF_q\left(\begin{matrix} a_1 \dots a_p \\ b_1 \dots b_q \end{matrix} \middle| z \right) \quad (21) \\ &= (-z)^{\alpha} \prod_{i=1}^{p} \Gamma(a_i) \cdot {}_p\tilde{F}_q\left(\begin{matrix} a_1 \dots a_p \\ b_1 \dots b_q \end{matrix} \middle| z \right) \quad (22) \end{aligned}$$

which in the case of $\alpha = 0$ precisely corresponds to the generalized hypergeometric function we started with.

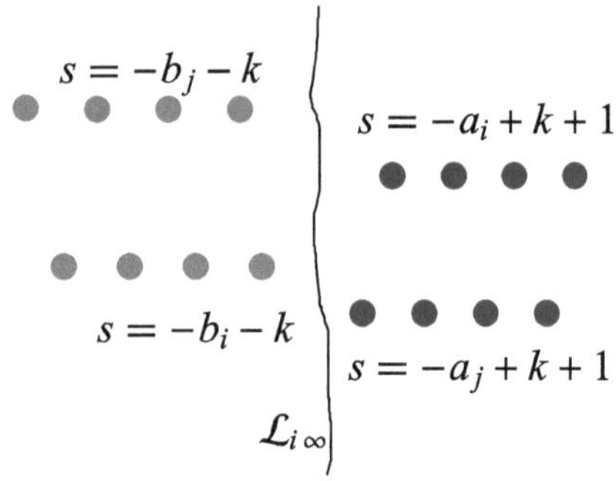

Fig. 1 The integration contour $\mathscr{L}_{i\infty}$ in the definition of the Meijer's G-function. It separates "left" poles (labeled as $s = -b - k$) of $\prod_{j=1}^{m} \Gamma(b_j + s)$ from the "right" poles (labeled as $s = -a + k + 1$) of $\prod_{i=1}^{n} \Gamma(1 - a_i - s)$

4 Meijer's G-Function

The Dutch mathematician Cornelis Meijer introduced [4] his famous G-function to be defined as follows:

$$G_{pq}^{mn}\left(\begin{matrix} a_1, \ldots, a_n, a_{n+1}, \ldots, a_p \\ b_1, \ldots, b_m, b_{m+1}, \ldots, b_q \end{matrix} \middle| z \right) =$$

$$\frac{1}{2\pi i} \int_{\mathscr{L}} \frac{\prod_{i=1}^{n} \Gamma(1 - a_i - s)}{\prod_{j=m+1}^{q} \Gamma(1 - b_j - s)} \cdot \frac{\prod_{j=1}^{m} \Gamma(b_j + s)}{\prod_{j=n+1}^{p} \Gamma(a_j + s)} \cdot z^{-s} \mathrm{d}s, \quad (23)$$

where the integration contour $\mathscr{L}$ starts with $|s| \to \infty$, separates "left" poles of $\prod_{j=1}^{m} \Gamma(b_j + s)$ from the "right" poles of $\prod_{i=1}^{n} \Gamma(1 - a_i - s)$, and goes back to $|s| \to \infty$. The Mellin-Barnes integration contour $\mathscr{L}_{i\infty}$ (see Fig. 1) starts from $-i\infty$ and continues to $+i\infty$, for example.

As discussed in the previous section, the G-function is the *general* hypergeometric function. It is implemented in *Mathematica* as

MeijerG[$\{\{ a_1, \ldots, a_n\}, \{a_{n+1}, \ldots, a_p\}\}, \{\{b_1, \ldots, b_m\}, \{b_{m+1}, \ldots, b_q\}\}, z]$.

The G-function satisfies an ordinary differential equation of hypergeometric type:

$$(-1)^{n+m-p} z \prod_{i=1}^{p} \left(z \frac{\mathrm{d}}{\mathrm{d}z} + 1 - a_i \right) \circ f(z) = \prod_{j=1}^{q} \left(z \frac{\mathrm{d}}{\mathrm{d}z} - b_j \right) \circ f(z). \quad (24)$$

By virtue of (23) several useful properties of the G-function follow easily:

$$z^{\alpha} \cdot G_{pq}^{mn}\left(\begin{matrix} a_1, \ldots, a_p \\ b_1, \ldots, b_q \end{matrix} \middle| z \right) \overset{s \to \tau + \alpha}{=} G_{pq}^{mn}\left(\begin{matrix} a_1 + \alpha, \ldots, a_p + \alpha \\ b_1 + \alpha, \ldots, b_q + \alpha \end{matrix} \middle| z \right) \quad (25)$$

$$G_{pq}^{mn}\left(\begin{matrix} a_1,\ldots,a_p \\ b_1,\ldots,b_q \end{matrix}\middle| z\right) \overset{s\to -\tau}{=} G_{qp}^{nm}\left(\begin{matrix} 1-b_1,\ldots,1-b_q \\ 1-a_1,\ldots,1-a_p \end{matrix}\middle| \frac{1}{z}\right). \tag{26}$$

Next we use Stirling's formula $\log\Gamma(s+a) = \left(a+s-\frac{1}{2}\right)\log(s) - s + \frac{1}{2}\log(2\pi) + \mathcal{O}(|s|^{-1})$ to study the convergence properties of the integral (23). Using $|\Gamma(a+s)| = \exp(\Re\log\Gamma(s+a))$ and taking care to carefully combine terms we get the asymptotic behavior of the integrand in (23) for large $|s|$:

$$\begin{aligned} \log\left(\left|z^{-s}\frac{\prod_{i=1}^{n}\Gamma(1-a_i-s)}{\prod_{i=1}^{q-m}\Gamma(1-d_i-s)}\cdot\frac{\prod_{i=1}^{m}\Gamma(s+c_i)}{\prod_{i=1}^{p-n}\Gamma(s+b_i)}\right|\right) &= \\ \Im(s)\,(\arg(s)\,(p-m-n) - \arg(-s)\,(q-m-n) + \arg(z)) &- \\ -\Re(s)\,((p-q)\,(\log|s|-1) + \log|z|) - \Re(\psi)\log|s| + \mathcal{O}(1) & \end{aligned} \tag{27}$$

where

$$\psi = \sum_{i=1}^{n}\left(a_i - \frac{1}{2}\right) + \sum_{i=1}^{p-n}\left(b_i - \frac{1}{2}\right) - \sum_{i=1}^{m}\left(c_i - \frac{1}{2}\right) - \sum_{i=1}^{p-m}\left(d_i - \frac{1}{2}\right).$$

The above result allows to establish the convergence of the integral (23). For integration along $\mathcal{L}_{i\infty}$, $\Re(s)$ is bounded and $\arg(s)$ approaches $\pm\frac{\pi}{2}$. Convergence for $s \to +i\infty$ and for $s \to -i\infty$ takes place for

$$|\arg(z)| < \theta_0 \text{ or } (|\arg(z)| = \theta_0 \text{ and } ((p-q)\Re(s) + \Re(\psi)) > 1), \tag{28}$$

where $\theta_0 = \pi\left(m + n - \frac{p+q}{2}\right)$. G-functions defined by the integration over the contour $\mathcal{L}_{i\infty}$ of Mellin-Barnes type admit a Mellin transform.

Since the integrand of (23) is meromorphic, the integration contour can be bent at our convenience so long as no poles are crossed in the process. We thus consider contours where $s \to |s|e^{\pm i\theta}$ for some $0 < \theta < \pi$.

We denote contours with $0 < \theta < \frac{\pi}{2}$ as $\mathcal{L}_{+\infty}$ as $\Re(s)$ is increasing to $+\infty$ along the contour and contours with $\frac{\pi}{2} < \theta < \pi$ as $\mathcal{L}_{-\infty}$.

The integral (23) with the contour $\mathcal{L}_{i\infty}$ replaced with $\mathcal{L}_{+\infty}$ converges when

$$(p > q) \text{ or } (p = q \text{ and } |z| > 1) \text{ or } (p = q \text{ and } |z| = 1 \text{ and } \Re(\psi) > 1). \tag{29}$$

The integral (23) with the contour $\mathcal{L}_{i\infty}$ replaced with $\mathcal{L}_{-\infty}$ converges when

$$(p < q) \text{ or } (p = q \text{ and } |z| < 1) \text{ or } (p = q \text{ and } |z| = 1 \text{ and } \Re(\psi) > 1). \tag{30}$$

Example 1. Consider an example of $f(z) = G_{01}^{10}\left(\begin{matrix} 0 \\ - \end{matrix}\middle| z\right)$ with $q = m = 1$ and $p = n = 0$. $f(z)$ satisfies $zf'(z) + zf(z) = 0$. The integral (23) over the contour

$\mathscr{L}_{i\infty}$ is convergent for $-\frac{\pi}{2} < \arg(z) < \frac{\pi}{2}$. Since $p < q$, the contour can be bent to the left. The integral (23) over the contour $\mathscr{L}_{-\infty}$ is convergent for all $z \in \mathbb{C}$. Of course, when $\Re(z) > 0$, these integrals define the same function. By the principle of analytic continuation the integral over $\mathscr{L}_{-\infty}$ gives the analytic continuation of the function, defined by the integral over $\mathscr{L}_{i\infty}$. It can be evaluated by the Cauchy integral theorem, as the sum of residues at the infinite sequence of poles of the $\Gamma(s)$:

$$\begin{aligned} G_{01}^{10}\left(\begin{matrix} 0 \\ - \end{matrix}\middle| z\right) &= \frac{1}{2\pi i}\int_{\mathscr{L}_{-\infty}} \Gamma(s) z^{-s} \mathrm{d}s = \sum_{n=0}^{\infty} \mathrm{Res}_{s=-n}\, \Gamma(s) z^{-s} \\ &= \sum_{n=0}^{\infty} \frac{(-1)^n}{n!} z^n = \mathrm{e}^{-z}. \end{aligned} \tag{31}$$

Note that the condition of integrability along $\mathscr{L}_{i\infty}$ coincides with the condition of absolute convergence of the Mellin transform in t for $f(tz)$.

Example 2. Consider $f(z) = G_{22}^{11}\left(\begin{matrix} 0\ 3/2 \\ 0\ 1/2 \end{matrix}\middle| z\right)$, which satisfies

$$2z(1-z)f''(z) + (1-3z)f'(z) + f(z) = 0.$$

The integral over the contour $\mathscr{L}_{i\infty}$ is *divergent*. Since $p = q = 2$, the integral over $\mathscr{L}_{-\infty}$ converges for $|z| < 1$, and the integral over $\mathscr{L}_{+\infty}$ converges for $|z| > 1$. It is therefore understood that function $f(z)$ is piecewise analytic, with the discontinuity at $|z| = 1$. For $|z| < 1$, $f(z)$ is defined as the sum of residues at left poles:

$$\begin{aligned} f_{<}(z) &= \sum_{n=0}^{\infty} \mathrm{Res}_{s=-n} \frac{\Gamma(1-s)\Gamma(s)}{\Gamma\left(\frac{1}{2}-s\right)\Gamma\left(\frac{3}{2}+s\right)} z^{-s} \\ &= \sum_{n=0}^{\infty} \mathrm{Res}_{s=-n} \frac{2z^{-s}\cot(\pi s)}{2s+1} = -\frac{2}{\pi}\sum_{n=0}^{\infty} \frac{z^n}{2n-1} \\ &= \frac{2}{\pi}\left(1 - \sqrt{z}\,\mathrm{arctanh}\left(\sqrt{z}\right)\right). \end{aligned} \tag{32}$$

For $|z| > 1$, $f(z)$ is defined as the negative of the sum of residues at right poles:

$$\begin{aligned} f_{>}(z) &= -\sum_{n=1}^{\infty} \mathrm{Res}_{s=n} \frac{\Gamma(1-s)\Gamma(s)}{\Gamma\left(\frac{1}{2}-s\right)\Gamma\left(\frac{3}{2}+s\right)} z^{-s} \\ &= -\sum_{n=1}^{\infty} \mathrm{Res}_{s=n} \frac{2z^{-s}\cot(\pi s)}{2s+1} = -\frac{2}{\pi}\sum_{n=1}^{\infty} \frac{z^n}{2n+1} \\ &= \frac{2}{\pi}\left(1 - \sqrt{z}\,\mathrm{arctanh}\left(\frac{1}{\sqrt{z}}\right)\right). \end{aligned} \tag{33}$$

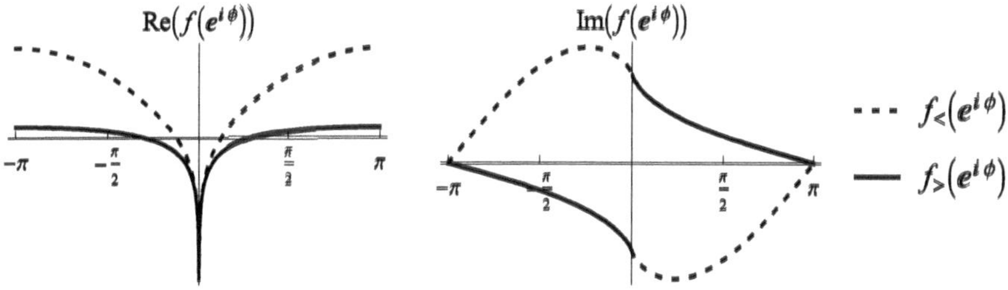

Fig. 2 Plots of $f_<\left(e^{i\phi}\right) = \lim_{t\uparrow 1} f_<\left(te^{i\phi}\right)$ and $f_>\left(e^{i\phi}\right) = \lim_{t\downarrow 1} f_>\left(te^{i\phi}\right)$

The discontinuity of $f(z)$ at the unit circle can be seen on the plot (see Fig. 2) of the real and imaginary parts of $f_<(z)$ and $f_>(z)$ as z approaches the unit circle from inside and from the outside respectively.

The G-functions such that the integral (23) over $\mathscr{L}_{i\infty}$ is divergent generically are not continuous at the unit circle. An example of exception to such a rule is

$$G_{22}^{11}\left(\begin{matrix} 0 \ 1/2 \\ 0 \ 1/2 \end{matrix} \middle| z\right) = \frac{1}{\pi}\frac{1}{1-z}. \tag{34}$$

Example 3. Consider $f(z) = G_{p,q+1}^{1,p}\left(\begin{matrix} 1-a_1,\ldots,1-a_p \\ 0, 1-b_1,\ldots,1-b_q \end{matrix} \middle| z\right)$. The integral (23) along $\mathscr{L}_{i\infty}$ converges for $|\arg(z)| < \frac{\pi}{2}(1+p-q)$. It converges along $\mathscr{L}_{-\infty}$ for $p < q+1$ or $p = q+1$ and $|z| < 1$, and under these conditions can be evaluated as a sum of residues at poles of $\Gamma(s)$ (compare to (22) for $\alpha = 0$):

$$\begin{aligned} f(z) &= \sum_{n=0}^{\infty} \mathrm{Ref}_{s=-n} \frac{\prod_{i=1}^{p} \Gamma(a_i - s)}{\prod_{j=1}^{q} \Gamma(b_j - s)} \Gamma(s) z^{-s} \\ &= \sum_{n=0}^{\infty} \frac{\prod_{i=1}^{p} \Gamma(a_i + n)}{\prod_{j=1}^{q} \Gamma(b_j + n)} \frac{(-z)^n}{n!} \\ &= \prod_{i=1}^{p} \Gamma(a_i) \sum_{n=0}^{\infty} \frac{\prod_{i=1}^{p} (a_i)_n}{\prod_{j=1}^{q} \Gamma(b_j + n)} \frac{(-z)^n}{n!} \\ &= \prod_{i=1}^{p} \Gamma(a_i) \cdot {}_p\tilde{F}_q\left(\begin{matrix} a_1,\ldots,a_p \\ b_1,\ldots,b_q \end{matrix} \middle| -z\right) \end{aligned} \tag{35}$$

where ${}_p\tilde{F}_q$ denotes the regularized hypergeometric function.
The integral (23) converges along the $\mathscr{L}_{+\infty}$ contour for $p > q+1$ or $p = q+1$ and $|z| > 1$. Under these conditions $f(z)$ is evaluated as the negative of the sum of residues at "right" poles of $\Gamma(\alpha_i - s)$. In the cases when these sequence poles $s = \alpha_i + n$ are all distinct, viz. none of coefficients a are integer apart from one another, we get

$$
\begin{aligned}
f(z) &= -\sum_{i=1}^{p} z^{-a_i} \sum_{n=0}^{\infty} \frac{\prod_{j=1,j\neq i}^{p} \Gamma(a_j - a_i - n)}{\prod_{k=1}^{q} \Gamma(b_k - a_i - n)} \Gamma(a_i + n) \frac{(-1/z)^n}{n!} \\
&= -\sum_{i=1}^{p} z^{-a_i} \frac{\pi^{p-1} \Gamma(a_i)}{\prod_{j=1,j\neq i}^{p} \sin\left(\pi\left(a_j - a_i\right)\right) \cdot \prod_{k=1}^{q} \Gamma(b_k - a_i)} \cdot \\
&\qquad {}_p\tilde{F}_q \left(\begin{matrix} a_i, \{1 + a_i - b_k\}_{k=1}^{q} \\ \{1 + a_i - a_j\}_{j\neq i} \end{matrix} \middle| \frac{(-1)^{p-q}}{z} \right).
\end{aligned} \tag{36}
$$

For $p = q + 1$, integration along $\mathscr{L}_{i\infty}$ is convergent for $|\arg(z)| < \pi$, and under this condition the formula (36) provides an analytic continuation of the hypergeometric function outside the region of convergence of the hypergeometric series. This motivates our next and last example, of a G-function with coincident poles.

Example 4. Consider $f(z) = G_{2,2}^{1,1}\left(\,|\,z\right)$. This is an instance of a G-function considered in the previous example, with $a_1 = a_2 = 1$ and $b_1 = 2$. Using (35) we have, for $|z| < 1$:

$$
\begin{aligned}
f(z) &= {}_2\tilde{F}_1(1,1;2;-z) = {}_2F_1(1,1;2;-z) = \sum_{n=0}^{\infty} \frac{(1)_n^2}{(2)_n} \frac{(-z)^n}{n!} = \\
&= \sum_{n=0}^{\infty} \frac{n!^2}{(n+1)!} \frac{(-z)^n}{n!} = \sum_{n=0}^{\infty} \frac{(-z)^n}{(n+1)} = \frac{\log(1+z)}{z}.
\end{aligned} \tag{37}
$$

For $|z| > 1$ the function is defined as a negative of the sum of residues over "right" poles, situated at positive integers:

$$
f(z) = -\sum_{n=1}^{\infty} \operatorname{Res}_{s=n} \frac{\Gamma(1-s)^2}{\Gamma(2-s)} \Gamma(s) z^{-s}. \tag{38}
$$

Only the pole at $s = 1$ is of order 2:

$$
\begin{aligned}
\operatorname{Res}_{s=1} \frac{\pi z^{-s}}{(s-1)\sin(\pi s)} &= \frac{1}{z} \operatorname{Res}_{s=1} \frac{1}{(s-1)^2} \frac{-\pi(s-1)}{\sin(\pi(s-1))} z^{-(s-1)} = \\
&= \frac{1}{z} \lim_{s\to 1} \frac{\mathrm{d}}{\mathrm{d}s} \left(\frac{-\pi(s-1)}{\sin(\pi(s-1))} z^{-(s-1)} \right) = \frac{\log(z)}{z}.
\end{aligned}
$$

The poles at $s = n, n \geqslant 2$ are all simple poles, due to cancelation by zeros of $\Gamma(2-s)$ in the denominator. Using $\Gamma(2-s) = (1-s)\Gamma(1-s)$ we get:

$$
\begin{aligned}
f(z) &= -\sum_{n=1}^{\infty} \operatorname{Res}_{s=n} \frac{\Gamma(1-s)}{(1-s)} \Gamma(s) z^{-s} \\
&= -\sum_{n=1}^{\infty} \operatorname{Res}_{s=n} \frac{\pi z^{-s}}{(1-s)\sin(\pi s)} = \\
&= \operatorname{Res}_{s=1} \frac{\pi z^{-s}}{(s-1)\sin(\pi s)} + \sum_{n=2}^{\infty} \frac{\pi z^{-s}}{(s-1)\sin(\pi s)} = \\
&= \frac{\log(z)}{z} + \sum_{n=2}^{\infty} \frac{(-1)^n z^{-n}}{n-1} = \frac{\log(z)}{z} + \frac{1}{z} \log\left(1 + \frac{1}{z}\right).
\end{aligned}
\tag{39}
$$

We thus established an equality of (37) and (39) for $-\pi < \arg(z) < \pi$.

5 Asymptotic Expansion of Meijer G-Functions at Infinity

As per (29), $z = \infty$ is a regular singular point of the differential equation (24) for $p > q$ and is an irregular singular point for $p < q$. Similarly, per (30), $z = 0$ is a regular singular point of (24) for $p < q$, and an irregular singular point for $p > q$. By virtue of (26) it suffices to consider the case of $p < q$. The sum of residues at left poles has infinite radius of convergence, and the sum of residues at right poles (if any) diverges. We now apply the technique from [1].

We start with (23) over $\mathscr{L}_{-\infty}$ and displace it to the right by an integer r, making sure to subtract the residue at every "right" pole that we crossed in the process:

$$
G_{pq}^{mn}(\bullet \mid z) = \frac{1}{2\pi i} \int_{\mathscr{L}_{-\infty}+r} \varphi(s) z^{-s} \mathrm{d}s - \sum_{k=0}^{r-1} \sum_{j=1}^{n} \operatorname{Res}_{s=1-a_j+k} \varphi(s) z^{-s}. \tag{40}
$$

The idea is to choose q large enough to warrant use of Stirling's formula for Γ-functions. We write:

$$
\begin{aligned}
\varphi(s) &= \frac{\prod_{i=1}^{n} \Gamma(1-a_i-s)}{\prod_{j=m+1}^{q} \Gamma(1-b_j-s)} \frac{\prod_{j=1}^{m} \Gamma(b_j+s)}{\prod_{i=n+1}^{p} \Gamma(a_i+s)} \\
&= \frac{\prod_{i=1}^{p} \Gamma(1-a_i-s)}{\prod_{j=1}^{q} \Gamma(1-b_j-s)} \cdot \frac{\prod_{i=n+1}^{p} \sin(\pi(a_i+s))}{\prod_{j=1}^{m} \sin(\pi(b_j+s))} \pi^{m+n-p} \\
&= \varphi_0(s) \cdot \varphi_1(s)
\end{aligned}
\tag{41}
$$

where $\varphi_0(s)$ denotes the ratio of products of Γ-functions and $\varphi_1(s)$ denotes the ratio of products of sines. By using Stirling's formula one obtains the following simple asymptotic expansion [1, 6]:

$$\varphi_0(s) = \frac{\kappa^{-\kappa s}}{\Gamma(1-\psi-\kappa s)} \frac{\kappa^{-\psi}}{(2\pi)^{\kappa-1}} \times$$
$$\times \left(1 + \frac{\mathscr{A}_1}{1-\psi-\kappa s} + \cdots + \frac{\mathscr{A}_\ell}{(1-\psi-\kappa s)_\ell} + \cdots\right), \quad (42)$$

where $\kappa = q - p > 0$ and $\psi = \sum_{i=1}^{p}\left(a_i - \frac{1}{2}\right) - \sum_{j=1}^{q}\left(b_j - \frac{1}{2}\right)$, and the coefficients $\mathscr{A}_\ell$ are polynomials in $\{a_i\}$ and $\{b_j\}$ and can be shown to satisfy a rank $q+1$ holonomic recurrence equation. See [6] for examples of computing $\mathscr{A}_\ell$ for small values of p and q. For the sake of convenience we set $\mathscr{A}_0 = 1$.

Substituting (42) into (40) one obtains

$$\begin{aligned} G_{pq}^{mn}(\bullet \mid z) = &-\sum_{k=0}^{r-1}\sum_{j=1}^{n} \mathrm{Res}_{s=1-a_j+k}\, \varphi(s) z^{-s} \\ &+ \sum_{j=1}^{m}\sum_{u=0}^{\infty} \mathrm{Res}_{s=-b_j+r-u}\, \tilde{\varphi}_0(s)\varphi_1(s) z^{-s}, \end{aligned} \quad (43)$$

where $\tilde{\varphi}_0(s)$ denotes the asymptotic expansion (42) of $\varphi_0(s)$. Assuming that all left poles are simple it follows that

$$\begin{aligned} \mathrm{Res}_{s=-b_1+r-u} \frac{\kappa^{-\kappa s-\psi}}{(2\pi)^{\kappa-1}\Gamma(1+\ell-\psi-\kappa s)} \varphi_1(s) z^{-s} = \\ \mathscr{C} \cdot \frac{\left((-1)^{m+n-p}\right)^{u-r} (\kappa^\kappa z)^{b_1+u-r}}{\Gamma\left(\kappa u + 1 + \ell - \psi + \kappa(b_1 - r)\right)}, \end{aligned} \quad (44)$$

where

$$\mathscr{C} = \frac{\kappa^{-\psi}}{(2\pi)^{\kappa-1}} \frac{\prod_{i=n+1}^{p} \sin(\pi(a_i - b_1))}{\prod_{j=2}^{m} \sin(\pi(b_j - b_1))} \pi^{m+n-p-1}. \quad (45)$$

Summation over u gives rise to the Mittag-Leffler function [2]:

$$\begin{aligned} &\sum_{u=0}^{\infty} \mathrm{Res}_{s=-b_1+r-u}\, \tilde{\varphi}_0(s)\varphi_1(s) z^{-s} = \\ &\qquad \mathscr{C}(-1)^{r(m+n-p)} (\kappa^\kappa z)^{b_1-r} \sum_{\ell=0}^{\ell_{\max}} \mathscr{A}_\ell E_{\kappa,\beta+\ell}\left((-1)^{m+n-p}\kappa^\kappa z\right) \end{aligned} \quad (46)$$

where $\beta = 1 - \psi + \kappa(b_1 - r)$. The Mittag-Leffler function $E_{\kappa,\beta}(z)$ is implemented in *Mathematica* since version 9 as

```
MittagLefflerE[κ,β,z].
```

Example 5. Consider $f(z) = G_{23}^{12}\left(\begin{matrix}1/2, 1/2\\ 0, 1, 1\end{matrix}\middle| -z\right) = \pi \cdot {}_2F_2\left(\frac{1}{2}, \frac{1}{2}; 1, 1; z\right)$. The corresponding values for φ_0 and $\varphi_1(s)$ are

$$\varphi_0(s) = \frac{\Gamma\left(\frac{1}{2} - s\right)^2}{\Gamma(1-s)^3} \qquad \varphi_1(s) = \frac{\pi}{\sin(\pi s)}. \tag{47}$$

Applying Stirling's formula we get

$$\tilde{\varphi}_0(s) = \sum_{\ell=0}^{\ell_{\max}} \frac{\mathscr{A}_\ell}{\Gamma(2+\ell-s)} + o\left(\frac{1}{\Gamma(2+\ell_{\max}-s)}\right) \tag{48}$$

where $\mathscr{A}_0 = 1$, $\mathscr{A}_1 = \frac{3}{4}$ and the coefficients $\mathscr{A}_\ell$ satisfy the following order 2 linear recurrence equation:

$$4(\ell+1)^3\mathscr{A}_\ell - (8\ell^2 + 24\ell + 19)\mathscr{A}_{\ell+1} + 4(\ell+2)\mathscr{A}_{\ell+2} = 0. \tag{49}$$

The right poles of $\varphi(s)$ are located at $s_k = \frac{1}{2} + k$ for $k \in \mathbb{Z}_{\geq 0}$ with the residue equal to

$$\begin{aligned} h_k(-z) = \mathrm{Res}_{s=s_k}\, \varphi(s) z^{-s} &= -\frac{\Gamma\left(k+\frac{1}{2}\right)}{z^{k+1/2}}\left(\frac{\Gamma\left(k+\frac{1}{2}\right)}{\pi k!}\right)^2 \times \\ &\left(\log(z) - 3\psi\left(k+\frac{1}{2}\right) + 2\psi(k+1)\right), \end{aligned} \tag{50}$$

where $\psi(z)$ denotes the digamma function. Thus

$$f(z) = -\sum_{k=0}^{r-1} h_k(z) + \sum_{\ell=0}^{\ell_{\max}} \mathscr{A}_\ell E_{1,2+\ell-r}(z). \tag{51}$$

A verification with *Mathematica* is obtained as follows. We first define the sequence of expansion coefficients $\mathscr{A}_\ell$:

```
A[ell_] := DifferenceRoot[ Function[{y, n},
  {4 (n + 1)^3 y[n] + (-19 - 24 n - 8 n^2) y[n + 1] +
    (8 + 4 n) y[2 + n]==0, y[0]==1, y[1]==3/4}]][ell]
```

We then define the function encoding the asymptotic expansion:

```
Asymp2F2[r_Integer, ellmax_Integer, z_] :=
  Sum[((-z)^(-1/2-n) Gamma[1/2 + n]^3 (Log[-z] -
      3 PolyGamma[0, 1/2 + n] + 2 PolyGamma[0, 1 + n]))/
    ((n!)^2 Pi^2), {n, 0, r - 1}] +
  Sum[A[ell] MittagLefflerE[1, 2 + ell - r, z],
        {ell, 0, ellmax}]/z^r
```

Numerical evaluation at some large argument gives a very good agreement:

```
In[534]:=
 With[{z = 2456` Exp[I Pi 1/3]}, {
    Pi HypergeometricPFQ[{1/2, 1/2}, {1, 1}, z],
    Asymp2F2[4, 3, z] }]

Out[534]= {
-4.89536461331929*10^529 + 6.80663165332857*10^529 I,
-4.89536461331808*10^529 + 6.80663165333123*10^529 I}
```

References

1. Braaksma, B.L.J.: Asymptotic expansions and analytic continuations for a class of Barnes-integrals. Compos. Math. **15**, 239–341 (1962–1964). http://www.numdam.org/item?id=CM_1962-1964__15__239_0
2. Haubold, H.J., Mathai, A.M., Saxena, R.K.: Mittag-Leffler Functions and Their Applications. arXiv:0909.0230 (math.CA)
3. Marichev, O., Trott, M.: The Wolfram Functions Site. http://functions.wolfram.com (2013)
4. Meijer, C.S.: Über Whittakersche bzw. Besselsche Funktionen und deren Produkte. Nieuw Arch. Wiskunde **18**(4), 10–39 (1936)
5. Olver F.W.J., et al.: NIST Handbook of Mathematical Functions. Cambridge University Press (2010). http://dlmf.nist.gov
6. Paris, R.B., Kaminski, D.: Asymptotics and Mellin-Barnes Integrals. Encyclopedia of Mathematics and Its Applications, vol. 85. Cambridge University Press, Cambridge/New York (2001)
7. Szegö, G.: Orthogonal Polynomials, 4th edn. American Mathematical Society, Providence (1975)
8. Zeilberger, D.: A holonomic systems approach to special functions identities. J. Comput. Appl. Math. **32**(3), 321–368 (1990)

Solving Linear Recurrence Equations with Polynomial Coefficients

Marko Petkovšek and Helena Zakrajšek

Abstract Summation is closely related to solving linear recurrence equations, since an indefinite sum satisfies a first-order linear recurrence with constant coefficients, and a definite proper-hypergeometric sum satisfies a linear recurrence with polynomial coefficients. Conversely, d'Alembertian solutions of linear recurrences can be expressed as nested indefinite sums with hypergeometric summands. We sketch the simplest algorithms for finding polynomial, rational, hypergeometric, d'Alembertian, and Liouvillian solutions of linear recurrences with polynomial coefficients, and refer to the relevant literature for state-of-the-art algorithms for these tasks. We outline an algorithm for finding the minimal annihilator of a given P-recursive sequence, prove the salient closure properties of d'Alembertian sequences, and present an alternative proof of a recent result of Reutenauer's that Liouvillian sequences are precisely the interlacings of d'Alembertian ones.

1 Introduction

Summation is related to solving linear recurrence equations in several ways. An indefinite sum

$$s(n) = \sum_{k=0}^{n-1} t(k)$$

M. Petkovšek (✉)
Faculty of Mathematics and Physics, University of Ljubljana, Jadranska 19, SI-1000 Ljubljana, Slovenia
e-mail: marko.petkovsek@fmf.uni-lj.si

H. Zakrajšek
Faculty of Mechanical Engineering, University of Ljubljana, Aškerčeva 6, SI-1000 Ljubljana, Slovenia
e-mail: helena.zakrajsek@fs.uni-lj.si

C. Schneider and J. Blümlein (eds.), *Computer Algebra in Quantum Field Theory*, Texts & Monographs in Symbolic Computation, DOI 10.1007/978-3-7091-1616-6_11,

satisfies the nonhomogeneous first-order recurrence equation

$$s(n+1) - s(n) = t(n); \quad s(0) = 0,$$

and also the homogeneous second-order recurrence equation

$$t(n)s(n+2) - (t(n) + t(n+1))s(n+1) + t(n+1)s(n) = 0; \quad s(0) = 0,\ s(1) = t(0).$$

A definite sum

$$s(n) = \sum_{k=0}^{n} F(n,k)$$

where the summand $F(n,k)$ is a proper hypergeometric term:

$$F(n,k) = P(n,k)\frac{\prod_{j=1}^{A}(\alpha_j)_{a_j n+\tilde{a}_j k}}{\prod_{j=1}^{B}(\beta_j)_{b_j n+\tilde{b}_j k}} z^k$$

with $P(n,k)$ a polynomial in n and k, $(z)_k$ the Pochhammer symbol, $a_j, b_j \in \mathbb{N}$, $\tilde{a}_j, \tilde{b}_j \in \mathbb{Z}$, and α_j, β_j, z constants such that $(\beta_j)_{b_j n+\tilde{b}_j k} \neq 0$ for all $k \in \{0,\ldots,n\}$, satisfies a linear recurrence equation with polynomial coefficients in n which can be computed with Zeilberger's algorithm (cf. [11, 29, 44, 45]). So the sum of interest may sometimes be found by solving a suitable recurrence equation.

The unknown object in a recurrence equation is a *sequence*, by which we mean a function mapping the nonnegative integers $\mathbb{N}$ to some algebraically closed field K of characteristic zero. Sequences can be represented in several different ways, among the most common of which are the following:

- *explicit* where a sequence $a : \mathbb{N} \to K$ is represented by an expression $e(x)$ such that $a(n) = e(n)$ for all $n \geq 0$,
- *recursive* where a sequence $a : \mathbb{N} \to K$ is represented by a function F and by some initial values $a(0), a(1), \ldots, a(d-1)$ such that

$$a(n) = F(n, a(n-1), a(n-2), \ldots, a(0)) \tag{1}$$

  for all $n \geq d$,
- by *Generating function* where a sequence $a : \mathbb{N} \to K$ is represented by the (formal) power series

$$G_a(z) = \sum_{n=0}^{\infty} a(n)z^n.$$

Each of these representations has several variants and special cases. In particular, if $a(n) = F(n, a(n-1), a(n-2), \ldots, a(n-d))$ for all $n \geq d$, the recursive representation (1) is said to be *of order at most d*.

Example 1 (Fibonacci numbers).

- Explicit representation: $a(n) = \frac{1}{\sqrt{5}}\left(\left(\frac{1+\sqrt{5}}{2}\right)^{n+1} - \left(\frac{1-\sqrt{5}}{2}\right)^{n+1}\right)$
- Recursive representation: $a(n) = a(n-1)+a(n-2)$ $(n \geq 2)$, $a(0) = a(1) = 1$
- Generating function: $G_a(z) = \dfrac{1}{1-z-z^2}$

From the viewpoint of representation of sequences, solving recurrence equations can be seen as the process of converting one (namely recursive) representation to another (explicit) representation.

In this paper we survey the properties of several important classes of sequences which satisfy linear recurrence equations with polynomial coefficients, and sketch algorithms for finding such solutions when they exist. In Sects. 2 and 3 we review the main results about C-recursive and P-recursive sequences, then we describe algorithms for finding polynomial, rational and hypergeometric solutions in Sects. 4 and 5. Difference rings and the Ore algebra of linear difference operators with rational coefficients, together with the outline of a factorization algorithm, are introduced in Sects. 6 and 7. In Sect. 8 we define d'Alembertian sequences and prove their closure properties. Finally, in Sect. 9, we give an alternative proof of the recent result of Reutenauer [30] that Liouvillian sequences are precisely the interlacings of d'Alembertian sequences by showing that the latter enjoy all the closure properties of the former.

2 C-Recursive Sequences

C-recursive sequences satisfy homogeneous linear recurrences with constant coefficients. Typical examples are geometric sequences of the form $a(n) = c\,q^n$ with $c, q \in K^*$, polynomial sequences, their products, and their linear combinations (such as the Fibonacci numbers of Example 1).

Definition 1. A sequence $a \in K^{\mathbb{N}}$ is *C-recursive* or *C-finite*[1] if there are $d \in \mathbb{N}$ and constants $c_1, c_2, \ldots, c_d \in K$, $c_d \neq 0$, such that

[1]C-recursive sequences are also called *linear recurrent (or: recurrence) sequences*. This neglects sequences satisfying linear recurrences with non-constant coefficients, and may lead to confusion.

$$a(n) = c_1\,a(n-1) + c_2\,a(n-2) + \cdots + c_d\,a(n-d)$$

for all $n \geq d$.

The following theorem describes the explicit and generating-function representations of C-recursive sequences. For a proof, see, e.g., [38].

Theorem 1. *Let $a \in K^{\mathbb{N}}$ and $G_a(z) = \sum_{n=0}^{\infty} a(n)z^n$. The following are equivalent:*

1. *a is C-recursive,*
2. $a(n) = \sum_{i=1}^{r} P_i(n)\,\alpha_i^n$ *for all $n \in \mathbb{N}$ where $P_i \in K[x]$ and $\alpha_i \in K$,*
3. $G_a(z) = \dfrac{P(z)}{Q(z)}$ *where $P, Q \in K[x]$, $\deg P < \deg Q$ and $Q(0) \neq 0$.*

The next two theorems are easy corollaries of Theorem 1.

Theorem 2. *The set of C-recursive sequences is closed under the following binary operations $(a, b) \mapsto c$:*

1. *Addition:* $c(n) = a(n) + b(n)$
2. *(Hadamard or termwise) multiplication:* $c(n) = a(n)b(n)$
3. *Convolution (Cauchy multiplication):* $c(n) = \sum_{i=0}^{n} a(i)b(n-i)$
4. *Interlacing:* $\langle c(0), c(1), c(2), c(3), \ldots\rangle = \langle a(0), b(0), a(1), b(1), \ldots\rangle$

Remark 1. These operations extend naturally to any nonzero number of operands.

Theorem 3. *The set of C-recursive sequences is closed under the following unary operations $a \mapsto c$:*

1. *Scalar multiplication:* $c(n) = \lambda a(n)$ $(\lambda \in K)$
2. *(Left) shift:* $c(n) = a(n+1)$
3. *Indefinite summation:* $c(n) = \sum_{k=0}^{n} a(k)$
4. *Multisection:* $c(n) = a(mn + r)$ $(m, r \in \mathbb{N},\ 0 \leq r < m)$

That (nonzero) C-recursive sequences are not closed under taking reciprocals is demonstrated, e.g., by $a(n) = n + 1$ which is C-recursive while its reciprocal $b(n) = 1/(n+1)$ is not, since its generating function $G_b(x) = -\ln(1-x)/x$ is not a rational function. Of course, there are C-recursive sequences whose reciprocals are C-recursive as well, such as all the geometric sequences.

Question 1. When are a and $1/a$ both C-recursive?

Theorem 4. *The sequences a and $1/a$ are both C-recursive iff a is the interlacing of one or more geometric sequences.*

For a proof, see [26].

3 P-Recursive Sequences

P-recursive sequences satisfy homogeneous linear recurrences with polynomial coefficients. While most of them lack a simple explicit representation, there exist several important subclasses of P-recursive sequences such as *polynomial, rational, hypergeometric* (Sect. 4), *d'Alembertian* (Sect. 8), and *Liouvillian* (Sect. 9) sequences which do have nice explicit representations. Figure 1 shows a hierarchy of these subclasses together with some examples. In the rest of the paper, we investigate their properties and sketch algorithms for finding such special solutions of linear recurrence equations with polynomial coefficients, whenever they exist.

Definition 2. A sequence $a \in K^{\mathbb{N}}$ is *P-recursive* if there are $d \in \mathbb{N}$ and polynomials $p_0, p_1, \ldots, p_d \in K[x]$, $p_0 p_d \neq 0$, such that

$$p_d(n)a(n+d) + p_{d-1}(n)a(n+d-1) + \cdots + p_0(n)a(n) = 0 \qquad (2)$$

for all $n \geq 0$.

Definition 3. A formal power series $f(z) = \sum_{n=0}^{\infty} a(n)z^n \in K[[z]]$ is *D-finite* if there exist $d \in \mathbb{N}$ and polynomials $q_0, q_1, \ldots, q_d \in K[x]$, $q_d \neq 0$, such that

$$q_d(z)f^{(d)}(z) + q_{d-1}(z)f^{(d-1)}(z) + \cdots + q_0(z)f(z) = 0.$$

Theorem 5. *Let $a \in K^{\mathbb{N}}$ and $G_a(z) = \sum_{n=0}^{\infty} a(n)z^n$. The following are equivalent:*

1. *a is P-recursive,*
2. *$G_a(z)$ is D-finite.*

For a proof, see [37] or [39].

Theorem 6. *P-recursive sequences are closed under the following operations:*

1. *addition,*
2. *multiplication,*
3. *convolution,*
4. *interlacing,*
5. *scalar multiplication,*
6. *shift,*
7. *indefinite summation,*
8. *multisection.*

For a proof, see [39].

Question 2. When are a and $1/a$ are both P-recursive?
The answer is given in Theorem 7.

Example 2. The sequences $a(n) = n!$ and $b(n) = 1/n!$ are both P-recursive since $a(n+1) - (n+1)a(n) = 0$ and $(n+1)b(n+1) - b(n) = 0$.

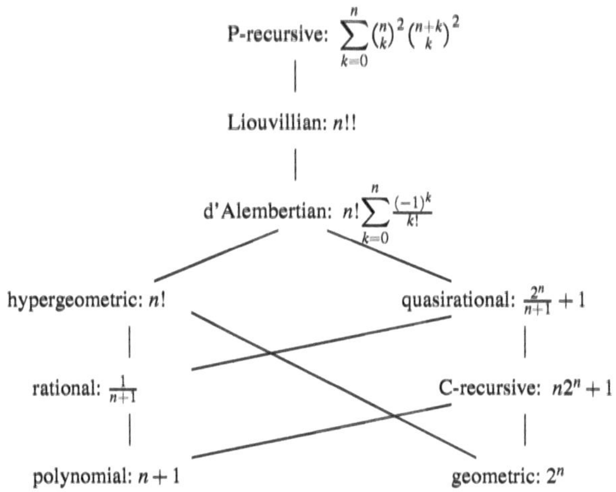

Fig. 1 A hierarchy of P-recursive sequences (with examples)

Example 3. The sequence $a(n) = 2^n + 1$ is P-recursive (even C-recursive) while its reciprocal $b(n) = 1/(2^n + 1)$ is not P-recursive.

Proof. We use the fact that a D-finite function can have at most finitely many singularities in the complex plane (see, e.g., [39]). The generating function

$$G_b(z) = \sum_{n=0}^{\infty} b(n) z^n = \sum_{n=0}^{\infty} \frac{z^n}{2^n + 1}$$

obviously has radius of convergence equal to two. We can rewrite

$$\begin{aligned} G_b(2z) &= \sum_{n=0}^{\infty} \frac{2^n}{2^n + 1} z^n = \sum_{n=0}^{\infty} \left(1 - \frac{1}{2^n + 1}\right) z^n \\ &= \frac{1}{1-z} - G_b(z). \end{aligned} \tag{3}$$

At $z = 1$ the function $1/(1-z)$ is singular, G_b is regular, so G_b is singular at $z = 2$. At $z = 2$ the function $1/(1-z)$ is regular, G_b is singular, so G_b is singular at $z = 4$. At $z = 4$ the function $1/(1-z)$ is regular, G_b is singular, so G_b is singular at $z = 8$, and so on. By induction on k it follows that $G_b(z)$ is singular at $z = 2^k$ for all $k \in \mathbb{N}$, $k \geq 1$, hence G_b is not D-finite, and b is not P-recursive. □

4 Hypergeometric Sequences

Hypergeometric sequences are P-recursive sequences which satisfy homogeneous linear recurrence equations with polynomial coefficients of order one. They can be represented explicitly as products of rational functions, Pochhammer symbols, and geometric sequences. The algorithm for finding hypergeometric solutions of linear recurrence equations with polynomial coefficients plays an important role in other, more involved computational tasks such as finding d'Alembertian or Liouvillian solutions, and factoring linear recurrence operators.

Definition 4. A sequence $a \in K^{\mathbb{N}}$ is *hypergeometric*[2] if there is an $N \in \mathbb{N}$ such that $a(n) \neq 0$ for all $n \geq N$, and there are polynomials $p, q \in K[n] \setminus \{0\}$ such that

$$p(n)\, a(n+1) \;=\; q(n)\, a(n) \tag{4}$$

for all $n \geq 0$. We denote by $\mathscr{H}(K)$ the set of all hypergeometric sequences in $K^{\mathbb{N}}$.

Clearly, each hypergeometric sequence is P-recursive.

Proposition 1. *The set $\mathscr{H}(K)$ is closed under the following operations:*

1. *multiplication,*
2. *reciprocation,*
3. *nonzero scalar multiplication,*
4. *shift,*
5. *multisection.*

Proof. For 1–4, see [29]. For multisection, let $a \in \mathscr{H}(K)$ satisfy (4) and let $b(n) = a(mn + r)$ where $m \in \mathbb{N}$, $m \geq 2$, and $0 \leq r < m$. For $i = 0, 1, \ldots, m-1$, substituting $mn + r + i$ for n in (4) yields

$$p(mn+r+i)\, a(mn+r+i+1) \;=\; q(mn+r+i)\, a(mn+r+i). \tag{5}$$

Multiply (5) by $\prod_{j=0}^{i-1} p(mn+r+j) \prod_{j=i+1}^{m-1} q(mn+r+j)$ on both sides to obtain $lhs_i = rhs_i$ for $i = 0, 1, \ldots, m-1$, where

$$lhs_i = \prod_{j=0}^{i} p(mn+r+j) \prod_{j=i+1}^{m-1} q(mn+r+j)\, a(mn+r+i+1),$$

$$rhs_i = \prod_{j=0}^{i-1} p(mn+r+j) \prod_{j=i}^{m-1} q(mn+r+j)\, a(mn+r+i).$$

[2] A hypergeometric sequence is also called a *hypergeometric term*, because the nth term of a hypergeometric series, considered as a function of n, is a hypergeometric sequence in our sense.

Note that $lhs_i = rhs_{i+1}$ for $i = 0, 1, \ldots, m-2$, hence, by induction on i,

$$rhs_0 = lhs_i \quad \text{for } i = 0, 1, \ldots, m-1.$$

In particular, $lhs_{m-1} = rhs_0$, so

$$\prod_{j=0}^{m-1} p(mn + r + j) b(n + 1) = \prod_{j=0}^{m-1} q(mn + r + j) b(n),$$

hence $b \in \mathscr{H}(K)$. □

Theorem 7. *The sequences a and $1/a$ are both P-recursive iff a is the interlacing of one or more hypergeometric sequences.*

For a proof, see [40].

5 Closed-Form Solutions

In this section, we sketch algorithms for finding polynomial, rational, and hypergeometric solutions of linear recurrence equations with polynomial coefficients.

5.1 Recurrence Operators

Let $E : K^{\mathbb{N}} \to K^{\mathbb{N}}$ be the (left) shift operator acting on sequences by $(Ea)(n) = a(n+1)$, so that $(E^k a)(n) = a(n+k)$ for $k \in \mathbb{N}$. For a given $d \in \mathbb{N}$ and polynomials $p_0, p_1, \ldots, p_d \in K[n]$ such that $p_d \neq 0$, the operator $L : K^{\mathbb{N}} \to K^{\mathbb{N}}$ defined by

$$L = \sum_{k=0}^{d} p_k(n)\, E^k$$

is a *linear recurrence operator of order d with polynomial coefficients*, acting on a sequence a by $(La)(n) = \sum_{k=0}^{d} p_k(n)\, a(n+k)$. We denote by $K[n]\langle E\rangle$ the algebra of linear recurrence operators with polynomial coefficients. The commutation rule $E \cdot p(n) = p(n+1)\, E$ induces the rule for composition of operators:

$$\sum_{k=0}^{d} p_k(n) E^k \cdot \sum_{j=0}^{e} q_j(n) E^j = \sum_{k=0}^{d} \sum_{j=0}^{e} p_k(n) q_j(n+k) E^{j+k}.$$

5.2 *Polynomial Solutions*

Given: $L \in K[n]\langle E\rangle,\ L \neq 0$
Find: a basis of the space $\{y \in K[n];\ Ly = 0\}$

Outline of algorithm

1. Find an upper bound for $\deg y$.
2. Use the method of undetermined coefficients.

For more details, see [3, 9, 29].

5.3 *Rational Solutions*

Given: $L \in K[n]\langle E\rangle,\ L \neq 0$
Find: a basis of the space $\{y \in K(n);\ Ly = 0\}$

Outline of algorithm

1. Find a universal denominator for y.
2. Find a basis of the space of polynomial solutions of the equation satisfied by the numerator of y.

For more details, see [4, 6, 7, 41].

5.4 *Hypergeometric Solutions*

Given: $L = \sum_{k=0}^{d} p_k E^k \in K[n]\langle E\rangle,\quad L \neq 0$
Find: a generating set for the linear hull of $\{y \in \mathcal{H}(K);\ Ly = 0\}$

Outline of algorithm

1. Construct the "Riccati equation" for $r = \frac{Ey}{y} \in K(n)$:

$$\sum_{k=0}^{d} p_k \prod_{j=0}^{k-1} E^j r \;=\; 0 \tag{6}$$

2. Use the ansatz

$$r \;=\; z\frac{a}{b}\frac{Ec}{c}$$

with $z \in K^*, a, b, c \in K[n]$ monic, a, c coprime, b, Ec coprime, $a, E^k b$ coprime for all $k \in \mathbb{N}$ to obtain

$$\sum_{k=0}^{d} z^k p_k \left(\prod_{j=0}^{k-1} E^j a\right) \left(\prod_{j=k}^{d-1} E^j b\right) E^k c = 0. \tag{7}$$

3. Construct a finite set of candidates for (a, b, z) using the following consequences of (7):
   - $a \mid p_0$,
   - $b \mid E^{1-d} p_d$,
   - $\sum_{\substack{0 \le k \le d \\ \deg P_k = m}} \mathrm{lc}(P_k) z^k = 0$

   where $P_k = p_k \left(\prod_{j=0}^{k-1} E^j a\right) \left(\prod_{j=k}^{d-1} E^j b\right)$, $\quad m = \max_{0 \le k \le d} \deg P_k$.
4. For each candidate triple (a, b, z), find all polynomial solutions c of the equation

$$\sum_{k=0}^{d} z^k P_k E^k c = 0.$$

For more details, see [28] or [29]. A much more efficient algorithm (although still exponential in $\deg p_0 + \deg p_d$ in the worst case) is given in [42] and [18].

Example 4 (Amer. Math. Monthly problem no. 10375). Solve

$$y(n+2) - 2(2n+3)^2 y(n+1) + 4(n+1)^2(2n+1)(2n+3)y(n) = 0. \tag{8}$$

Denote $p_2(n) = 1$, $p_1(n) = -2(2n+3)^2$, and $p_0(n) = 4(n+1)^2(2n+1)(2n+3)$. In search of hypergeometric solutions we follow the four steps described above:

1. Riccati equation:

$$p_2(n)\, r(n+1) r(n) + p_1(n)\, r(n) + p_0(n) = 0$$

2. Plug in the ansatz:

$$\begin{aligned} &z^2\ p_2(n)\ a(n+1)\, a(n)\ c(n+2) \\ +\ &z\ \ p_1(n)\ a(n)\, b(n+1)\ c(n+1) \\ +\ &\ \ \ \ p_0(n)\ b(n+1)\, b(n)\ c(n) \quad = 0 \end{aligned}$$

3. Candidates for (a, b, z):
   - $a(n) \mid 4(n+1)^2(2n+1)(2n+3)$
   - $b(n) \mid 1$

   Take, e.g., $a(n) = (n+1)(n+\frac{1}{2})$, $b(n) = 1$, Then $z^2 - 8z + 16 = (z-4)^2 = 0$, so $z = 4$.

4. Equation for c:

$$(n+2)c(n+2) - (2n+3)c(n+1) + (n+1)c(n) = 0$$

Polynomial solution: $c(n) = 1$

We have found

$$\frac{y(n+1)}{y(n)} = r(n) = z\frac{a(n)}{b(n)}\frac{c(n+1)}{c(n)} = (2n+1)(2n+2),$$

therefore $y(n) = (2n)!$ is a hypergeometric solution of equation (8).

6 Difference Rings

Definition 5. A *difference ring* is a pair (K, σ) where K is a commutative ring with multiplicative identity and $\sigma : K \to K$ is a ring automorphism. If, in addition, K is a field, then (K, σ) is a *difference field*.

Example 5. • $(K[x], \sigma)$ with $\sigma x = x + 1, \sigma|_K = \mathrm{id}_K$ is a difference ring.
- $(K(x), \sigma)$ with $\sigma x = x + 1, \sigma|_K = \mathrm{id}_K$ is a difference field.
- $(K^{\mathbb{N}}, E)$ is *not* a difference ring since the shift operator E is not injective on $K^{\mathbb{N}}$.

For $a, b \in K^{\mathbb{N}}$ define $a \sim b$ if there is an $N \in \mathbb{N}$ such that $a(n) = b(n)$ for all $n \geq N$. The ring $\mathscr{S}(K) = K^{\mathbb{N}}/\sim$ of equivalence classes is the ring of *germs of sequences*. Let $\varphi : K^{\mathbb{N}} \to \mathscr{S}(K)$ be the canonical projection, and $\sigma : \mathscr{S}(K) \to \mathscr{S}(K)$ the unique automorphism of $\mathscr{S}(K)$ s.t. $\sigma \circ \varphi = \varphi \circ E$. Then $(\mathscr{S}(K), \sigma)$ is a difference ring.

Henceforth we work in $(\mathscr{S}(K), \sigma)$ rather than $(K^{\mathbb{N}}, E)$ (but still call the elements of $\mathscr{S}(K)$ just "sequences" for short). One important advantage of this setting is that the dimension of the solution space of Eq. (2) (with $a(n+i)$ replaced by $\sigma^i a(n)$) is precisely equal to its order, d (this follows from [19, p. 272, Theorem XII]). Another advantage is that sequences with, possibly, finitely many undefined terms (such as the values of a rational function at nonnegative integers) also have their germs in $\mathscr{S}(K)$ (e.g., if $a(n)$ is defined for all $n \geq N$, its germ can be represented by the sequence $\langle 0, 0, \ldots, 0, a(N), a(N+1), \ldots \rangle$). Thus $K[n]$, $K(n)$, and $\mathscr{H}(K)$ can all be naturally embedded into $\mathscr{S}(K)$.

7 An Ore Algebra of Operators

Instead of linear recurrence operators with polynomial coefficients from $K[n]\langle E \rangle$, we will henceforth use *linear difference operators with rational coefficients* from the algebra $K(n)\langle \sigma \rangle$. The rule for composition of these operators follows from the commutation rule $\sigma \cdot r(n) = r(n+1)\sigma$ for all $r \in K(n)$.

The identity

$$r(n)\,\sigma^k \;=\; \left(\frac{r(n)}{s(n+k-j)}\sigma^{k-j}\right)\cdot s(n)\,\sigma^j$$

describes how to perform right division of $r(n)\,\sigma^k$ by $s(n)\,\sigma^j$. Hence there is an *algorithm for right division* in $K(n)\langle\sigma\rangle$:

Theorem 8. *For $L_1, L_2 \in K(n)\langle\sigma\rangle$, $L_2 \neq 0$, there are $Q, R \in K(n)\langle\sigma\rangle$ such that*

- $L_1 \;=\; QL_2 + R$,
- $\operatorname{ord} R \;<\; \operatorname{ord} L_2$.

As a consequence, the right extended Euclidean algorithm (REEA) can be used to compute a greatest common right divisor (gcrd) and a least common left multiple (lclm) of operators in $K(n)\langle\sigma\rangle$, which is therefore a left Ore algebra. In particular, given $L_1, L_2 \in K(n)\langle\sigma\rangle$, REEA yields $S, T, U, V \in K(n)\langle\sigma\rangle$ such that

- $SL_1 + TL_2 \;=\; \operatorname{gcrd}(L_1, L_2)$,
- $UL_1 \;=\; VL_2 \;=\; \operatorname{lclm}(L_1, L_2)$.

Definition 6. Let a be P-recursive. The unique monic operator $M_a \in K(n)\langle\sigma\rangle\setminus\{0\}$ of least order such that $M_a a = 0$ is the *minimal operator* of a.

Example 6. Let $h \in \mathscr{H}(K)$ where $\sigma h/h = r \in K(n)^*$. Then $M_h = \sigma - r$.

Question 3. How to compute M_a for a given P-recursive a?
The outline of an algorithm for solving this problem is given on page 271.

Proposition 2. *Let a be P-recursive, and $L \in K(n)\langle\sigma\rangle$ such that $La = 0$. Then L is right-divisible by M_a.*

Proof. Divide L by M_a. Then:

$$L = QM_a + R \quad\Longrightarrow\quad La = QM_a a + Ra \quad\Longrightarrow\quad 0 = Ra \quad\Longrightarrow\quad R = 0$$

□

Corollary 1. *Let $L \in K(n)\langle\sigma\rangle$ and $h \in \mathscr{H}(K)$ be such that $Lh = 0$. Then there is $Q \in K(n)\langle\sigma\rangle$ such that $L = Q(\sigma - r)$ where $r = \sigma h/h \in K(n)^*$.*

Hence there is a one-to-one correspondence between hypergeometric solutions of $Ly = 0$ and first-order right factors of L having the form $\sigma - r$ with $r \neq 0$.

Example 7 (Amer. Math. Monthly problem no. 10375 – continued from Example 4).

$$L \;=\; \sigma^2 - 2(2n+3)^2\sigma + 4(n+1)^2(2n+1)(2n+3)$$

We saw in Example 4 that $Ly = 0$ is satisfied by $y(n) = (2n)!$, hence $L = QL_1$ where

$$L_1 = \sigma - (2n+1)(2n+2),$$
$$Q = \sigma - (2n+2)(2n+3).$$

The operator factorization problem

Given: $L \in K(n)\langle\sigma\rangle$ and $r \in \mathbb{N}$
Find: all $L_1 \in K(n)\langle\sigma\rangle$ s.t.

- ord $L_1 = r$,
- $L = QL_1$ for some $Q \in K(n)\langle\sigma\rangle$

Suppose such L_1 exists, and let $y^{(1)}, y^{(2)}, \ldots, y^{(r)}$ be linearly independent solutions of $L_1 y = 0$ in $\mathscr{S}(K)$. Define the *Casoratian* of $y^{(1)}, y^{(2)}, \ldots, y^{(r)}$ by

$$\mathrm{Cas}(y^{(1)}, y^{(2)}, \ldots, y^{(r)}) = \det \begin{bmatrix} y^{(1)} & y^{(2)} & \cdots & y^{(r)} \\ \sigma y^{(1)} & \sigma y^{(2)} & \cdots & \sigma y^{(r)} \\ \vdots & \vdots & & \vdots \\ \sigma^{r-1} y^{(1)} & \sigma^{r-1} y^{(2)} & \cdots & \sigma^{r-1} y^{(r)} \end{bmatrix}$$

as usual. Then (see [14] or [15]):

1. $\mathrm{Cas}(y^{(1)}, y^{(2)}, \ldots, y^{(r)}) \in \mathscr{H}(K)$.
2. From L and r one can construct a linear recurrence with polynomial coefficients satisfied by $\mathrm{Cas}(y^{(1)}, y^{(2)}, \ldots, y^{(r)})$.
3. From L and r one can construct linear recurrences with polynomial coefficients satisfied by the coefficients of L_1, multiplied by $\mathrm{Cas}(y^{(1)}, y^{(2)}, \ldots, y^{(r)})$.

Outline of an algorithm to solve the operator factorization problem:

1. Construct a recurrence satisfied by $\mathrm{Cas}(y^{(1)}, y^{(2)}, \ldots, y^{(r)})$.
2. Find all hypergeometric solutions of this recurrence.
3. Construct recurrences satisfied by the coefficients of L_1.
4. Find all rational solutions of these recurrences.
5. Select candidates for L_1 which right-divide L.

Outline of an algorithm to find the minimal operator of a P-recursive sequence:

Given: $L \in K(n)\langle\sigma\rangle$ and $a \in \mathscr{S}(K)$ s.t. $La = 0$
Find: minimal operator M_a of a

`for` $r = 1, 2, \ldots,$ ord L `do:`
  find all monic $L_1 \in K(n)\langle\sigma\rangle$ of order r s.t. $\exists Q \in K(n)\langle\sigma\rangle$: $L = QL_1$
  `for` every such L_1 `do:`
    `if` $(L_1\, a)(n) = 0$ for ord Q consecutive values of n `then` return L_1.

In the last line, the ord Q consecutive values of n must be greater than any integer root of the leading coefficient of Q.

8 D'Alembertian Solutions

Write $\Delta = \sigma - 1$ for the forward difference operator as usual. If $y = a$ satisfies $Ly = 0$, then substituting $y \leftarrow az$ where z is a new unknown sequence yields

$$L' \Delta z = 0$$

where $\operatorname{ord} L' = \operatorname{ord} L - 1$. This is known as *reduction of order* or *d'Alembert substitution* [5]. By finding hypergeometric solutions and using d'Alembert substitution repeatedly we obtain a set of solutions which can be written as nested indefinite sums with hypergeometric summands. These so-called *d'Alembertian sequences* include harmonic numbers and their generalizations, and play an important role in the theory of Padé approximations (cf. [20, 21]), in combinatorics (cf. [27, 33]) and in particle physics (cf. [1, 2, 12]).

8.1 Definition and Representation

Definition 7. A sequence $a \in \mathscr{S}(K)$ is *d'Alembertian* if there are first-order operators $L_1, L_2, \ldots, L_d \in K(n)\langle\sigma\rangle$ such that

$$L_d \cdots L_2 L_1 \, a = 0. \tag{9}$$

We denote by $\mathscr{A}(K)$ the set of all d'Alembertian elements of $\mathscr{S}(K)$, and write $\operatorname{nd}(a)$ for the least $d \in \mathbb{N}$ for which (9) holds (the *nesting depth* of a). For a set $S \subseteq \mathscr{A}(K)$ we write $\operatorname{nd}(S)$ for $\max_{a \in S} \operatorname{nd}(a)$.

Remark 2. Let $a \in \mathscr{A}(K)$. Then:

1. $\operatorname{nd}(a) = 0$ if and only if $a = 0$,
2. $\operatorname{nd}(a) = 1$ if and only if $a \in \mathscr{H}(K)$.

Example 8.

- Harmonic numbers $H(n) = \sum_{k=1}^{n} \frac{1}{k}$ are d'Alembertian because

$$\left(\sigma - \frac{n+1}{n+2}\right)(\sigma - 1)\, H(n) = \left(\sigma - \frac{n+1}{n+2}\right) \frac{1}{n+1} = 0.$$

- Derangement numbers $d(n) = n! \sum_{k=0}^{n} \frac{(-1)^k}{k!}$ are d'Alembertian because

$$(\sigma+1)(\sigma-(n+1))\, d(n) = (\sigma+1)(n+1)! \frac{(-1)^{n+1}}{(n+1)!} = (\sigma+1)(-1)^{n+1} = 0.$$

Notation: For $a \in \mathscr{S}(K)$ and $A, B \subseteq \mathscr{S}(K)$ we write

$$
\begin{aligned}
\Sigma a &= \{b \in \mathscr{S}(K);\ \Delta b = a\} = \sum_{k=0}^{n-1} a(k) + K, \\
\Sigma A &= \{b \in \mathscr{S}(K);\ \Delta b \in A\}, \\
a + B &= \{a + b;\ b \in B\}, \quad aB = \{ab;\ b \in B\}, \\
A + B &= \{a + b;\ a \in A, b \in B\}
\end{aligned}
$$

and identify $a \in \mathscr{S}(K)$ with $\{a\} \subseteq \mathscr{S}(K)$ if suitable.

Remark 3.

$$
\begin{array}{ll}
1.\ \Delta + 1 = \sigma, & 3.\ \sigma\Sigma = \Sigma + 1, \\
2.\ \Delta\Sigma = 1, & 4.\ \Sigma 0 = K.
\end{array}
$$

Proposition 3. *Let $r \in K(n)$, $\sigma h = rh$, and $f \in \mathscr{S}(K)$. Then*

$$\{y \in \mathscr{S}(K);\ (\sigma - r)y = f\} = h\,\Sigma \frac{f}{\sigma h}.$$

Proof. Assume that $(\sigma - r)y = f$ and write $y = h\,z$. Then

$$f = (\sigma - r)y = (\sigma - r)h\,z = \sigma h\,\sigma z - r\,h\,z = \sigma h\,\Delta z,$$

hence $\Delta z = \frac{f}{\sigma h}$, so $z \in \Sigma \frac{f}{\sigma h}$ and $y = h\,z \in h\,\Sigma \frac{f}{\sigma h}$. – Conversely,

$$(\sigma - r)h\,\Sigma \frac{f}{\sigma h} = \sigma h\,\sigma\Sigma \frac{f}{\sigma h} - r h\,\Sigma \frac{f}{\sigma h} = \sigma h\,\Delta\Sigma \frac{f}{\sigma h} = f. \qquad \square$$

Corollary 2.

$$\operatorname{Ker}\,(\sigma - r_d)\cdots(\sigma - r_2)(\sigma - r_1) = h_1\,\Sigma \frac{h_2}{\sigma h_1}\,\Sigma \frac{h_3}{\sigma h_2} \cdots \Sigma \frac{h_d}{\sigma h_{d-1}}\,\Sigma 0 \tag{10}$$

where Ker $L = \{y \in \mathscr{S}(K);\ L\,y = 0\}$ *and* $\sigma h_i = r_i h_i$ *for* $i = 1, 2, \ldots, d$.

It turns out that for any $L \in K(n)\langle\sigma\rangle$, the space of all d'Alembertian solutions of $Ly = 0$ is of the form

$$h_1\,\Sigma h_2\,\Sigma h_3 \cdots \Sigma h_d\,\Sigma 0 \tag{11}$$

for some $d \leq \operatorname{ord} L$ and $h_1, h_2, \ldots, h_d \in \mathscr{H}(K)$.

Example 9 (Amer. Math. Monthly problem no. 10375 – continued from Example 7).

$$\begin{aligned} L &= \sigma^2 - 2(2n+3)^2\sigma + 4(n+1)^2(2n+1)(2n+3), \\ L &= L_2 L_1, \\ L_1 &= \sigma - (2n+1)(2n+2), \\ L_2 &= \sigma - (2n+2)(2n+3). \end{aligned}$$

Since $L_1(2n)! = 0$ and $L_2(2n+1)! = 0$, it follows from (10) that

$$\begin{aligned} \operatorname{Ker} L &= (2n)!\,\Sigma \frac{(2n+1)!}{(2n+2)!} \Sigma\, 0 \;=\; (2n)!\,\Sigma \frac{K}{n+1} \\ &= (2n)! \left(K \sum_{k=0}^{n-1} \frac{1}{k+1} + K \right) \;=\; K\,(2n)!\,H(n) + K\,(2n)!. \end{aligned}$$

8.2 Closure Properties of $\mathscr{A}(K)$

Definition 8. For operators $L, R \in K(n)\langle\sigma\rangle$, denote by L/R the right quotient of $\operatorname{lclm}(L, R)$ by R.

Remark 4. Clearly, $(L/R)R \;=\; \operatorname{lclm}(L, R) \;=\; (R/L)L.$

Example 10. Let $L_1 = \sigma - r_1$ and $L_2 = \sigma - r_2$ be first-order operators. If $r_1 = r_2$ then $L_1/L_2 = L_2/L_1 = 1$. If $r_1 \neq r_2$ it is straightforward to check that

$$\left(\sigma - \frac{\sigma r_1 - \sigma r_2}{r_1 - r_2} r_1\right)(\sigma - r_2) \;=\; \left(\sigma - \frac{\sigma r_1 - \sigma r_2}{r_1 - r_2} r_2\right)(\sigma - r_1),$$

hence

$$L_1/L_2 \;=\; \sigma - \frac{\sigma r_1 - \sigma r_2}{r_1 - r_2} r_1, \qquad L_2/L_1 \;=\; \sigma - \frac{\sigma r_1 - \sigma r_2}{r_1 - r_2} r_2.$$

Lemma 1. *Let $L_1, L_2, \ldots, L_k, R \;\in\; K(n)\langle\sigma\rangle$ be monic first-order operators. Then there are monic operators $N_1, N_2, \ldots, N_k, M \in K(n)\langle\sigma\rangle \setminus \{0\}$ of order ≤ 1 such that*

$$M L_k L_{k-1} \cdots L_1 \;=\; N_k N_{k-1} \cdots N_1 R.$$

Proof. By induction on k.

$k = 1$: Take $N_1 = L_1/R$, $M = R/L_1$. Then $ML_1 = (R/L_1)L_1 = (L_1/R)R = N_1 R$.

$k > 1$: By the induction hypothesis, there are monic operators $N_1, N_2, \ldots, N_{k-1}$, $\tilde{M} \in K(n)\langle\sigma\rangle \setminus \{0\}$ of order ≤ 1 such that

$$\tilde{M} L_{k-1} L_{k-2} \cdots L_1 = N_{k-1} N_{k-2} \cdots N_1 R. \quad (12)$$

Take $N_k = L_k / \tilde{M}$, $M = \tilde{M} / L_k$. Then, using (12) in the last equality, we obtain

$$\begin{aligned} M L_k L_{k-1} \cdots L_1 &= (\tilde{M}/L_k) L_k L_{k-1} L_{k-2} \cdots L_1 = (L_k/\tilde{M}) \tilde{M} L_{k-1} L_{k-2} \cdots L_1 \\ &= N_k \tilde{M} L_{k-1} L_{k-2} \cdots L_1 = N_k N_{k-1} N_{k-2} \cdots N_1 R. \end{aligned}$$

□

Lemma 2. *Let $L_1, L_2, \ldots, L_k, R_1, R_2, \ldots, R_m \in K(n)\langle\sigma\rangle$ be monic first-order operators. Then there are monic operators $M_1, M_2, \ldots, M_m, N_1, N_2, \ldots, N_k \in K(n)\langle\sigma\rangle \setminus \{0\}$ of order ≤ 1 such that*

$$M_m M_{m-1} \cdots M_1 L_k L_{k-1} \cdots L_1 = N_k N_{k-1} \cdots N_1 R_m R_{m-1} \cdots R_1.$$

Proof. By induction on m.

$m = 1$: By Lemma 1 applied to $L_1, L_2, \ldots, L_k, R_1$, there are $N_1, N_2, \ldots, N_k$ and M_1 such that $M_1 L_k L_{k-1} \cdots L_1 = N_k N_{k-1} \cdots N_1 R_1$.

$m > 1$: By the induction hypothesis applied to $R_1, R_2, \ldots, R_{m-1}$, there are monic operators $M_1, M_2, \ldots, M_{m-1}, \tilde{N}_1, \tilde{N}_2, \ldots, \tilde{N}_k \in K(n)\langle\sigma\rangle \setminus \{0\}$ of order ≤ 1 such that

$$M_{m-1} M_{m-2} \cdots M_1 L_k L_{k-1} \cdots L_1 = \tilde{N}_k \tilde{N}_{k-1} \cdots \tilde{N}_1 R_{m-1} R_{m-2} \cdots R_1. \quad (13)$$

By Lemma 1 applied to $\tilde{N}_1, \tilde{N}_2, \ldots, \tilde{N}_k, R_m$, there are $N_1, N_2, \ldots, N_k$ and M_m such that

$$M_m \tilde{N}_k \tilde{N}_{k-1} \cdots \tilde{N}_1 = N_k N_{k-1} \cdots N_1 R_m,$$

hence, by multiplying (13) with M_m from the left, we obtain

$$\begin{aligned} M_m M_{m-1} \cdots M_1 L_k L_{k-1} \cdots L_1 &= M_m \tilde{N}_k \tilde{N}_{k-1} \cdots \tilde{N}_1 R_{m-1} R_{m-2} \cdots R_1 \\ &= N_k N_{k-1} \cdots N_1 R_m R_{m-1} \cdots R_1. \end{aligned}$$

□

Proposition 4. *$\mathscr{A}(K)$ is closed under addition.*

Proof. Let $a, b \in \mathscr{A}(K)$. Then there are monic first-order operators $L_1, L_2, \ldots, L_k$, $R_1, R_2, \ldots, R_m \in K(n)\langle\sigma\rangle$ such that

$$L_k L_{k-1} \cdots L_1 a = R_m R_{m-1} \cdots R_1 b = 0.$$

By Lemma 2, there are monic operators $M_1, \ldots, M_m, N_1, \ldots, N_k \in K(n)\langle\sigma\rangle \setminus \{0\}$ of order ≤ 1 such that

$$L := M_m M_{m-1} \cdots M_1 L_k L_{k-1} \cdots L_1 = N_k N_{k-1} \cdots N_1 R_m R_{m-1} \cdots R_1.$$

Then $La = Lb = 0$, so $L(a + b) = 0$ and $a + b \in \mathscr{A}(K)$. □

Proposition 5. *$\mathscr{A}(K)$ is closed under multiplication.*

Proof. Let $a, b \in \mathscr{A}(K)$. We show that $ab \in \mathscr{A}(K)$ by induction on the sum of their nesting depths $\mathrm{nd}(a) + \mathrm{nd}(b)$.

(a) $\mathrm{nd}(a) = 0$ or $\mathrm{nd}(b) = 0$: In this case one of a, b is 0, hence $ab = 0 \in \mathscr{A}(K)$.
(b) $\mathrm{nd}(a), \mathrm{nd}(b) \geq 1$: By (11) we can write $a \in h_1\, \Sigma h_2\, \Sigma h_3 \cdots \Sigma h_d\, \Sigma 0$ and $b \in g_1\, \Sigma g_2\, \Sigma g_3 \cdots \Sigma g_e\, \Sigma 0$ where $h_i, g_j \in \mathscr{H}(K)$, $d = \mathrm{nd}(a)$, and $e = \mathrm{nd}(b)$. Let $a_1 = h_2\, \Sigma h_3 \cdots \Sigma h_d\, \Sigma 0$ and $b_1 = g_2\, \Sigma g_3 \cdots \Sigma g_e\, \Sigma 0$, so that $a \in h_1\, \Sigma a_1$ and $b \in g_1\, \Sigma b_1$ with $a_1, b_1 \subseteq \mathscr{A}(K)$, $\mathrm{nd}(a_1) < \mathrm{nd}(a)$ and $\mathrm{nd}(b_1) < \mathrm{nd}(b)$. Clearly $ha \in \mathscr{A}(K)$ whenever $h \in \mathscr{H}(K)$ and $a \in \mathscr{A}(K)$, hence it suffices to show that $(\sum a_1) g_1 \sum b_1 \subseteq \mathscr{A}(K)$. Using the product rule of difference calculus

$$\Delta uv = u\Delta v + \Delta u\, \sigma v$$

and Remark 3 repeatedly, we obtain

$$\begin{aligned}
\Delta\left((\textstyle\sum a_1) g_1 \sum b_1\right) &= (\textstyle\sum a_1) g_1 \Delta \sum b_1 + \Delta((\sum a_1) g_1)\sigma \sum b_1 \\
&= (\textstyle\sum a_1) g_1 b_1 + ((\sum a_1)\Delta g_1 + a_1 \sigma g_1)(\sum b_1 + b_1) \\
&= \Delta g_1 (\textstyle\sum a_1) \sum b_1 + (g_1 + \Delta g_1) b_1 \sum a_1 + a_1 \sigma g_1 (\sum b_1 + b_1) \\
&= \Delta g_1 (\textstyle\sum a_1) \sum b_1 + \sigma g_1 (a_1 \sum b_1 + b_1 \sum a_1 + a_1 b_1).
\end{aligned}$$

Assume first that $g_1 = 1$. Then $\Delta\left((\sum a_1) \sum b_1\right) = a_1 \sum b_1 + b_1 \sum a_1 + a_1 b_1$. By the induction hypothesis and from Proposition 4 it follows that $a_1 \sum b_1 + b_1 \sum a_1 + a_1 b_1 \subseteq \mathscr{A}(K)$. Therefore there are first-order operators $L_1, L_2, \ldots, L_k \in K(n)\langle\sigma\rangle$ such that

$$L_k L_{k-1} \cdots L_1 \Delta\left((\textstyle\sum a_1) \sum b_1\right) = 0,$$

hence $(\sum a_1) \sum b_1 \subseteq \mathscr{A}(K)$. In the general case, $\Delta g_1, \sigma g_1 \in \mathscr{H}(K) \cup \{0\}$ now implies $\Delta\left((\sum a_1) g_1 \sum b_1\right) \subseteq \mathscr{A}(K)$. Again we conclude that $(\sum a_1) g_1 \sum b_1 \subseteq \mathscr{A}(K)$. □

Proposition 6. *$\mathscr{A}(K)$ is closed under σ and σ^{-1}.*

Proof. Let $a \in \mathscr{A}(K)$. Then there are monic first-order operators $L_1, L_2, \ldots, L_k \in K(n)\langle\sigma\rangle$ such that $L_k L_{k-1} \cdots L_1 a = 0$.

By Lemma 1 with $R = \sigma$, there are monic operators $N_1, N_2, \ldots, N_k, M \in K(n)\langle\sigma\rangle \setminus \{0\}$ of order ≤ 1 such that $M L_k L_{k-1} \cdots L_1 = N_k N_{k-1} \cdots N_1 \sigma$. Hence

$$N_k N_{k-1} \cdots N_1 \sigma a = M L_k L_{k-1} \cdots L_1 a = 0,$$

so $\sigma a \in \mathscr{A}(K)$.

From $L_k L_{k-1} \cdots L_1 a = 0$ it follows that $L_k L_{k-1} \cdots L_1 \sigma(\sigma^{-1} a) = 0$, hence $\sigma^{-1} a \in \mathscr{A}(K)$ as well. □

Theorem 9. *$\mathscr{A}(K)$ is a difference ring.*

Proof. This follows from Propositions 4, 5 and 6. □

Corollary 3. *$\mathscr{A}(K)$ is the least subring of $\mathscr{S}(K)$ which contains $\mathscr{H}(K)$ and is closed under σ, σ^{-1}, and Σ.*

Proof. Denote by $HS(K)$ the least subring of $\mathscr{S}(K)$ which contains $\mathscr{H}(K)$ and is closed under σ, σ^{-1}, and Σ.

By Corollary 2, every $a \in \mathscr{A}(K)$ is obtained from 0 by using Σ and multiplication with elements from $\mathscr{H}(K)$. Hence $\mathscr{A}(K) \subseteq HS(K)$.

Conversely, $\mathscr{A}(K)$ is closed under σ and σ^{-1} by Proposition 6, and under Σ by Corollary 2. Since $\mathscr{A}(K)$ is a subring of $\mathscr{S}(K)$ containing $\mathscr{H}(K)$, it follows that $HS(K) \subseteq \mathscr{A}(K)$. □

Proposition 7. *$\mathscr{A}(K)$ is closed under multisection.*

Proof. Let $a \in \mathscr{A}(K)$. We show that any multisection of a belongs to $\mathscr{A}(K)$ by induction on the nesting depth $\mathrm{nd}(a)$ of a.

(a) $\mathrm{nd}(a) = 0$: In this case $a = 0$, so the assertion holds.
(b) $\mathrm{nd}(a) \geq 1$: By (11) we can write $a \in h_1 \Sigma h_2 \Sigma h_3 \cdots \Sigma h_d \Sigma 0$ where $d = \mathrm{nd}(a)$ and $h_1, h_2, \ldots, h_d \in \mathscr{H}(K)$. Let $h = h_1$ and $b = h_2 \Sigma h_3 \cdots \Sigma h_d \Sigma 0$, so that $a \in h \Sigma b$ where $b \subseteq \mathscr{A}(K)$ and $\mathrm{nd}(b) < \mathrm{nd}(a)$.

Let $c \in \mathscr{S}(K)$, defined by $c(n) = a(mn + r)$ for all $n \in \mathbb{N}$, where $m, r \in \mathbb{N}$, $m \geq 2$, $0 \leq r < m$, be a multisection of a. Then for all $n \in \mathbb{N}$

$$\begin{aligned} c(n) = a(mn+r) &= h(mn+r) \sum_{k=0}^{mn+r-1} b(k) \\ &= h(mn+r) \left(\sum_{i=0}^{m-1} \sum_{j=0}^{n-1} b(mj+i) + \sum_{i=0}^{r-1} b(mn+i) \right) \\ &= h_{m,r}(n) \left(\sum_{i=0}^{m-1} \sum_{j=0}^{n-1} b_{m,i}(j) + \sum_{i=0}^{r-1} b_{m,i}(n) \right) \end{aligned}$$

where $h_{m,r}(n) = h(mn + r)$ and $b_{m,i}(n) = b(mn + i)$ for $0 \leq i < m$. Hence

$$c = h_{m,r} \left(\sum_{i=0}^{m-1} \Sigma b_{m,i} + \sum_{i=0}^{r-1} b_{m,i} \right)$$

where $h_{m,r} \in \mathscr{H}(K) \subseteq \mathscr{A}(K)$ by Proposition 1, and $b_{m,i} \in \mathscr{A}(K)$ by the induction hypothesis as a multisection of b. Since $\mathscr{A}(K)$ is closed under Σ, addition and multiplication, it follows that $c \in \mathscr{A}(K)$. □

8.3 Finding d'Alembertian Solutions

The following theorem provides a way to find all d'Alembertian solutions of $Ly = 0$.

Theorem 10. *$Ly = 0$ has a nonzero d'Alembertian solution if and only if $Ly = 0$ has a hypergeometric solution.*

For a proof, see [10].

Outline of an algorithm for finding the space of all d'Alembertian solutions:

1. Find a hypergeometric solution h_1 of $Ly = 0$.
   If none exists then return 0 and stop.
2. Let $L_1 = \sigma - \frac{\sigma h_1}{h_1}$. Right-divide L by L_1 to obtain $L = QL_1$.
3. Recursively use the algorithm on $Qy = 0$. Let the output be a.
4. Return $h_1 \Sigma \frac{a}{\sigma h_1}$ and stop.

A much more general algorithm which finds solutions in $\Pi\Sigma^*$-difference extension fields of $(K(n), \sigma)$ is presented in [31]. For the relevant theory, see [32, 34–36].

9 Liouvillian Solutions

Definition 9. $\mathscr{L}(K)$ is the least subring of $\mathscr{S}(K)$ containing $\mathscr{H}(K)$, closed under

- σ, σ^{-1},
- Σ,
- interlacing of an arbitrary number of sequences.

A sequence $a \in \mathscr{S}(K)$ is *Liouvillian* iff $a \in \mathscr{L}(K)$.

Example 11. The sequence

$$n!! = \begin{cases} 2^k k!, & n = 2k, \\ \frac{(2k+1)!}{2^k k!}, & n = 2k+1 \end{cases}$$

is Liouvillian (as an interlacing of two hypergeometric sequences).

The following theorem provides a way to find Liouvillian solutions of $Ly = 0$.

Theorem 11. *$Ly = 0$ has a nonzero Liouvillian solution if and only if $Ly = 0$ has a solution which is an interlacing of at most* ord L *hypergeometric sequences.*

For a proof, see [24]. For algorithms to find Liouvillian solutions, see [8, 13, 16, 17, 22, 23, 25, 43].

Theorem 12. *A sequence in $\mathscr{S}(K)$ is Liouvillian if and only if it is an interlacing of d'Alembertian sequences.*

This is proved in [30] as a corollary of the results of [24] obtained by means of Galois theory of difference equations. Here we give a self-contained proof based on closure properties of interlacings of d'Alembertian sequences.

Let $\Lambda(a_0, a_1, \ldots, a_{k-1})$, or $\Lambda_{j=0}^{k-1} a_j$, denote the interlacing of $a_0, a_1, \ldots, a_{k-1}$. By definition of interlacing we have

$$\left(\Lambda_{j=0}^{k-1} a_j\right)(n) = \Lambda(a_0, a_1, \ldots, a_{k-1})(n) = a_{n \bmod k}(n \text{ div } k)$$

for all $n \in \mathbb{N}$, where

$$n \text{ div } k = \left\lfloor \frac{n}{k} \right\rfloor, \quad n \bmod k = n - \left\lfloor \frac{n}{k} \right\rfloor k.$$

Denote temporarily the set of all interlacings of (one or more) d'Alembertian sequences by $AL(K)$. The goal is to prove that $AL(K) = \mathscr{L}(K)$.

Proposition 8. *$AL(K) \subseteq \mathscr{L}(K)$.*

Proof. Since $\mathscr{H}(K) \subseteq \mathscr{L}(K)$ and $\mathscr{L}(K)$ is a ring closed under Σ, we have $\mathscr{A}(K) \subseteq \mathscr{L}(K)$. Since $\mathscr{L}(K)$ is closed under interlacing, $AL(K) \subseteq \mathscr{L}(K)$. □

Lemma 3. *$AL(K)$ is closed under addition and multiplication.*

Proof. Let $\odot$ denote either addition or multiplication in K and $\mathscr{S}(K)$. We claim that, for $k, m \in \mathbb{N}$ and $a_0, a_1, \ldots, a_{k-1}, b_0, b_1, \ldots, b_{m-1} \in \mathscr{A}(K)$, we have

$$\left(\Lambda_{j=0}^{k-1} a_j\right) \odot \left(\Lambda_{j=0}^{m-1} b_j\right) = \Lambda_{\ell=0}^{km-1} (a_{\ell,k,m} \odot b_{\ell,k,m}) \tag{14}$$

where for all $n \in \mathbb{N}$,

$$a_{\ell,k,m}(n) = a_{\ell \bmod k}(mn + \ell \text{ div } k),$$
$$b_{\ell,k,m}(n) = b_{\ell \bmod m}(kn + \ell \text{ div } m).$$

Indeed,

$$\left(\Lambda_{\ell=0}^{km-1} (a_{\ell,k,m} \odot b_{\ell,k,m})\right)(n)$$
$$= a_{n \bmod km,k,m}(n \text{ div } km) \odot b_{n \bmod km,k,m}(n \text{ div } km) = u \odot v$$

where

$$u = a_{(n \bmod km) \bmod k}(m(n \text{ div } km) + (n \bmod km) \text{ div } k),$$
$$v = b_{(n \bmod km) \bmod m}(k(n \text{ div } km) + (n \bmod km) \text{ div } m).$$

From

$$(n \bmod km) \bmod k = \left(n - \left\lfloor \frac{n}{km} \right\rfloor km\right) \bmod k = n \bmod k,$$
$$(n \bmod km) \bmod m = \left(n - \left\lfloor \frac{n}{km} \right\rfloor km\right) \bmod m = n \bmod m,$$
$$m(n \text{ div } km) + (n \bmod km) \text{ div } k = m\left\lfloor \frac{n}{km} \right\rfloor + \left\lfloor \frac{n - \lfloor \frac{n}{km} \rfloor km}{k} \right\rfloor = \left\lfloor \frac{n}{k} \right\rfloor = n \text{ div } k,$$
$$k(n \text{ div } km) + (n \bmod km) \text{ div } m = k\left\lfloor \frac{n}{km} \right\rfloor + \left\lfloor \frac{n - \lfloor \frac{n}{km} \rfloor km}{m} \right\rfloor = \left\lfloor \frac{n}{m} \right\rfloor = n \text{ div } m$$

it follows that

$$\begin{aligned} u \odot v &= a_{n \bmod k}(n \text{ div } k) \odot b_{n \bmod m}(n \text{ div } m) \\ &= \left(\Lambda_{j=0}^{k-1} a_j\right)(n) \odot \left(\Lambda_{j=0}^{m-1} b_j\right)(n) = \left(\left(\Lambda_{j=0}^{k-1} a_j\right) \odot \left(\Lambda_{j=0}^{m-1} b_j\right)\right)(n), \end{aligned}$$

proving (14). By Proposition 7, the sequences $a_{\ell,k,m}$ and $b_{\ell,k,m}$ belong to $\mathscr{A}(K)$. Since $\mathscr{A}(K)$ is a ring, the right-hand side of (14) is an interlacing of d'Alembertian sequences, and hence so is the left-hand side. □

Lemma 4. *$AL(K)$ is closed under σ and σ^{-1}.*

Proof. Let $a_0, a_1, \ldots, a_{k-1}$ be d'Alembertian sequences. Then:

$$\begin{aligned} \left(\sigma\left(\Lambda_{j=0}^{k-1} a_j\right)\right)(n) &= \left(\Lambda_{j=0}^{k-1} a_j\right)(n+1) \\ &= a_{(n+1) \bmod k}((n+1) \text{ div } k) \\ &= \begin{cases} a_{n \bmod k+1}(n \text{ div } k), & n \bmod k \neq k-1, \\ a_0(n \text{ div } k + 1), & n \bmod k = k-1 \end{cases} \\ &= \begin{cases} a_{n \bmod k+1}(n \text{ div } k), & n \bmod k \neq k-1, \\ (\sigma a_0)(n \text{ div } k), & n \bmod k = k-1 \end{cases} \\ &= \left(\Lambda_{j=0}^{k-1} b_j\right)(n) \end{aligned}$$

where

$$b_j = \begin{cases} a_{j+1}, & j \neq k-1, \\ \sigma a_0, & j = k-1. \end{cases}$$

By Proposition 6, $b_0, b_1, \ldots, b_{k-1}$ are d'Alembertian. So $\sigma\left(\Lambda_{j=0}^{k-1} a_j\right) = \Lambda_{j=0}^{k-1} b_j$ is an interlacing of d'Alembertian sequences.

Similarly,

$$\begin{aligned}
\left(\sigma^{-1}\left(\Lambda_{j=0}^{k-1} a_j\right)\right)(n) &= \left(\Lambda_{j=0}^{k-1} a_j\right)(n-1) \\
&= a_{(n-1) \bmod k}((n-1) \operatorname{div} k) \\
&= \begin{cases} a_{n \bmod k-1}(n \operatorname{div} k), & n \bmod k \neq 0, \\ a_{k-1}(n \operatorname{div} k - 1), & n \bmod k = 0 \end{cases} \\
&= \begin{cases} a_{n \bmod k-1}(n \operatorname{div} k), & n \bmod k \neq 0, \\ (\sigma^{-1} a_{k-1})(n \operatorname{div} k), & n \bmod k = 0 \end{cases} \\
&= \left(\Lambda_{j=0}^{k-1} c_j\right)(n)
\end{aligned}$$

where

$$c_j = \begin{cases} a_{j-1}, & j \neq 0, \\ \sigma^{-1} a_{k-1}, & j = 0. \end{cases}$$

By Proposition 6, $c_0, c_1, \ldots, c_{k-1}$ are d'Alembertian. So $\sigma^{-1}\left(\Lambda_{j=0}^{k-1} a_j\right) = \Lambda_{j=0}^{k-1} c_j$ is an interlacing of d'Alembertian sequences. □

Lemma 5. *$AL(K)$ is closed under Σ.*

Proof. Let $a_0, a_1, \ldots, a_{k-1}$ be d'Alembertian sequences. We claim that

$$\Sigma\left(\Lambda_{j=0}^{k-1} a_j\right) = \Lambda_{j=0}^{k-1}\left(\sum_{i=0}^{j-1} \sigma \Sigma a_i + \sum_{i=j}^{k-1} \Sigma a_i\right). \tag{15}$$

Indeed, for all $n \in \mathbb{N}$,

$$\begin{aligned}
&\left(\Sigma\left(\Lambda_{j=0}^{k-1} a_j\right)\right)(n) \\
&= \sum_{\ell=0}^{n-1}\left(\Lambda_{j=0}^{k-1} a_j\right)(\ell) = \sum_{\ell=0}^{n-1} a_{\ell \bmod k}(\ell \operatorname{div} k) \\
&= \sum_{i=0}^{(n-1) \bmod k} \sum_{j=0}^{\left\lfloor \frac{n-1}{k} \right\rfloor} a_i(j) + \sum_{i=(n-1) \bmod k+1}^{k-1} \sum_{j=0}^{\left\lfloor \frac{n-1}{k} \right\rfloor - 1} a_i(j)
\end{aligned} \tag{16}$$

$$= \sum_{i=0}^{n \bmod k - 1} \sum_{j=0}^{\lfloor \frac{n}{k} \rfloor} a_i(j) \ + \sum_{i=n \bmod k}^{k-1} \ \sum_{j=0}^{\lfloor \frac{n}{k} \rfloor - 1} a_i(j) \tag{17}$$

$$= \sum_{i=0}^{n \bmod k - 1} (\sigma \Sigma a_i)\,(n \text{ div } k) \ + \sum_{i=n \bmod k}^{k-1} (\Sigma a_i)\,(n \text{ div } k)$$

$$= \left(\Lambda_{j=0}^{k-1} \left(\sum_{i=0}^{j-1} \sigma \Sigma a_i + \sum_{i=j}^{k-1} \Sigma a_i \right) \right) (n),$$

proving (15). Here equality in (16) follows by mapping each $\ell \in \{0, 1, \ldots, n-1\}$ to the pair $(i, j) = (\ell \bmod k,\ \ell \text{ div } k)$ and summing over all the resulting pairs, and equality in (17) follows by noting that when $n \bmod k \neq 0$, we have

$$(n-1) \bmod k = n \bmod k - 1,$$
$$(n-1) \text{ div } k = n \text{ div } k,$$

while for $n \bmod k = 0$, both (16) and (17) are equal to $\sum_{i=0}^{k-1} \sum_{j=0}^{\frac{n}{k}-1} a_i(j)$.

Since $\mathscr{A}(K)$ is closed under Σ, σ and addition, the right-hand side of (15) is an interlacing of d'Alembertian sequences, and hence so is the left-hand side. □

Lemma 6. *$AL(K)$ is closed under interlacing.*

Proof. An arbitrary interlacing can be obtained by using addition, shifts, and interlacing of zero sequences with a single non-zero sequence by the formula

$$\Lambda(a_0, a_1, \ldots, a_{k-1}) = \sum_{i=0}^{k-1} \sigma^i \Lambda(0, 0, \ldots, 0, a_{k-1-i}).$$

Hence, by Propositions 3 and 4, it suffices to show that $AL(K)$ is closed under interlacing of zero sequences with a single non-zero sequence from $AL(K)$. But this is immediate: Let $a_0, a_1, \ldots, a_{k-1}$ be d'Alembertian sequences. Then the interlacing of m zero sequences with $\Lambda(a_0, a_1, \ldots, a_{k-1})$

$$\Lambda(0, 0, \ldots, 0, \Lambda(a_0, a_1, \ldots, a_{k-1}))$$
$$= \Lambda(0, 0, \ldots, 0, a_0, 0, 0, \ldots, 0, a_1, \ldots, 0, 0, \ldots, 0, a_{k-1})$$

is an interlacing of $mk + k$ d'Alembertian sequences. □

Proof of Theorem 12. By Proposition 8, it suffices to show that $\mathscr{L}(K) \subseteq AL(K)$. This is true since by Lemmas 3–6, $AL(K)$ is a subring of $\mathscr{S}(K)$ containing $\mathscr{H}(K)$ and closed under σ, σ^{-1}, Σ, and interlacing, while $\mathscr{L}(K)$ is the least such subring. □

References

1. Ablinger, J., Blümlein, J., Schneider, C.: Harmonic sums and polylogarithms generated by cyclotomic polynomials. J. Math. Phys. **52**, 1–52 (2011)
2. Ablinger, J., Blümlein, J., Schneider, C.: Analytic and algorithmic aspects of generalized harmonic sums and polylogarithms. J. Math. Phys. (2013, to appear)
3. Abramov, S.A.: Problems in computer algebra that are connected with a search for polynomial solutions of linear differential and difference equations. Moscow Univ. Comput. Math. Cybern. **3**, 63–68 (1989). Transl. from Vestn. Moskov. univ. Ser. XV. Vychisl. mat. kibernet. **3**, 56–60 (1989)
4. Abramov, S.A.: Rational solutions of linear differential and difference equations with polynomial coefficients. U.S.S.R. Comput. Maths. Math. Phys. **29**, 7–12 (1989). Transl. from Zh. vychisl. mat. mat. fyz. **29**, 1611–1620 (1989)
5. Abramov, S.A.: On d'Alembert substitution. In: Proceedings of the ISSAC'93, Kiev, pp. 20–26 (1993)
6. Abramov, S.A.: Rational solutions of linear difference and q-difference equations with polynomial coefficients. Progr. Comput. Softw. **21**, 273–278 (1995). Transl. from Programmirovanie **21**, 3–11 (1995)
7. Abramov, S.A., Barkatou, M.A.: Rational solutions of first order linear difference systems. In: Proceedings of the ISSAC'98, Rostock, pp. 124–131 (1998)
8. Abramov, S.A., Barkatou, M.A., Khmelnov, D.E.: On m-interlacing solutions of linear difference equations. In: Proceedings of the CASC'09, Kobe. LNCS 5743, pp. 1–17. Springer, Berlin/Heidelberg/New York (2009)
9. Abramov, S.A., Bronstein, M., Petkovšek, M.: On polynomial solutions of linear operator equations. In: Proceedings of the ISSAC'95, Montreal, pp. 290–296 (1995)
10. Abramov, S.A., Petkovšek, M.: D'Alembertian solutions of linear operator equations. In: Proceedings of the ISSAC'94, Oxford, pp. 169–174 (1994)
11. Apagodu, M., Zeilberger, D.: Multi-variable Zeilberger and Almkvist-Zeilberger algorithms and the sharpening of Wilf-Zeilberger theory. Adv. Appl. Math. **37**, 139–152 (2006)
12. Blümlein, J., Klein, S., Schneider, C., Stan, F.: A symbolic summation approach to Feynman integral calculus. J. Symb. Comput. **47**, 1267–1289 (2012)
13. Bomboy, R.: Liouvillian solutions of ordinary linear difference equations. In: Proceedings of the CASC'02, Simeiz (Yalta), pp. 17–28 (2002)
14. Bronstein, M., Petkovšek, M.: Ore rings, linear operators and factorization. Progr. Comput. Softw. **20**, 14–26 (1994)
15. Bronstein, M., Petkovšek, M.: An introduction to pseudo-linear algebra. Theoret. Comput. Sci. **157**, 3–33 (1996)
16. Cha, Y., van Hoeij, M.: Liouvillian solutions of irreducible linear difference equations. In: Proceedings of the ISSAC'09, Seoul, pp. 87–94 (2009)
17. Cha, Y., van Hocij, M., Levy, G.: Solving recurrence relations using local invariants. In: Proceedings of the ISSAC'10, München, pp. 303–309 (2010)
18. Cluzeau, T., van Hoeij, M.: Computing hypergeometric solutions of linear recurrence equations. Appl. Algebra Eng. Comm. Comput. **17**, 83–115 (2006)
19. Cohn, R.M.: Difference Algebra. Interscience Publishers/Wiley, New York/London/Sydney (1965)
20. Driver, K., Prodinger, H., Schneider, C., Weideman, J.A.C.: Padé approximations to the logarithm. II. Identities, recurrences, and symbolic computation. Ramanujan J. **11**, 139–158 (2006)
21. Driver, K., Prodinger, H., Schneider, C., Weideman, J.A.C.: Padé approximations to the logarithm. III. Alternative methods and additional results, Ramanujan J. **12**, 299–314 (2006)
22. Feng, R., Singer, M.F., Wu, M.: Liouvillian solutions of linear difference-differential equations. J. Symb. Comput. **45**, 287–305 (2010)

23. Feng, R., Singer, M.F., Wu, M.: An algorithm to compute Liouvillian solutions of prime order linear difference-differential equations. J. Symb. Comput. **45**, 306–323 (2010)
24. Hendriks, P.A., Singer, M.F.: Solving difference equations in finite terms. J. Symb. Comput. **27**, 239–259 (1999)
25. Khmelnov, D.E.: Search for Liouvillian solutions of linear recurrence equations in MAPLE computer algebra system. Progr. Comput. Softw. **34**, 204–209 (2008)
26. Larson, R.G., Taft, E.J.: The algebraic structure of linearly recursive sequences under Hadamard product. Israel J. Math. **72**, 118–132 (1990)
27. Paule, P., Schneider, C.: Computer proofs of a new family of harmonic number identities. Adv. Appl. Math. **31**, 359–378 (2003)
28. Petkovšek, M.: Hypergeometric solutions of linear recurrences with polynomial coefficients. J. Symb. Comput. **14**, 243–264 (1992)
29. Petkovšek, M., Wilf, H.S., Zeilberger, D.: A=B. A K Peters, Wellesley (1996)
30. Reutenauer, C.: On a matrix representation for polynomially recursive sequences. Electron. J. Combin. **19**(3), P36 (2012)
31. Schneider, C.: The summation package Sigma: underlying principles and a rhombus tiling application. Discret. Math. Theor. Comput. Sci. **6**, 365–386 (2004). (electronic)
32. Schneider, C.: Solving parameterized linear difference equations in terms of indefinite nested sums and products. J. Differ. Equ. Appl. **11**, 799–821 (2005)
33. Schneider, C.: Symbolic summation assists combinatorics. Sém. Lothar. Combin. **56** (2006/07). Art. B56b (electronic), 36 pp
34. Schneider, C.: Simplifying sums in $\Pi\Sigma^*$-extensions. J. Algebra Appl. **6**, 415–441 (2007)
35. Schneider, C.: A refined difference field theory for symbolic summation. J. Symb. Comput. **43**, 611–644 (2008)
36. Schneider, C.: Structural theorems for symbolic summation. Appl. Algebra Eng. Comm. Comput. **21**, 1–32 (2010)
37. Stanley, R.P.: Differentiably finite power series. Eur. J. Combin. **1**, 175–188 (1980)
38. Stanley, R.P.: Enumerative Combinatorics, vol. 1. Cambridge University Press, Cambridge (1997) (corrected reprint of the 1986 original)
39. Stanley, R.P.: Enumerative Combinatorics, vol. 2. Cambridge University Press, Cambridge (1999)
40. van der Put, M., Singer, M.F.: Galois Theory of Difference Equations. Springer, Berlin (1997)
41. van Hoeij, M.: Rational solutions of linear difference equations. In: Proceedings of the ISSAC'98, Rostock, pp. 120–123 (1998)
42. van Hoeij, M.: Finite singularities and hypergeometric solutions of linear recurrence equations. J. Pure Appl. Algebra **139**, 109–131 (1999)
43. van Hoeij, M., Levy, G.: Liouvillian solutions of irreducible second order linear differential equations. In: Proceedings of the ISSAC'10, München, pp. 297–301 (2010)
44. Zeilberger, D.: A fast algorithm for proving terminating hypergeometric identities. Discret. Math. **80**, 207–211 (1990)
45. Zeilberger, D.: The method of creative telescoping. J. Symb. Comput. **11**, 195–204 (1991)

Generalization of Risch's Algorithm to Special Functions

Clemens G. Raab

Abstract Symbolic integration deals with the evaluation of integrals in closed form. We present an overview of Risch's algorithm including recent developments. The algorithms discussed are suited for both indefinite and definite integration. They can also be used to compute linear relations among integrals and to find identities for special functions given by parameter integrals. The aim of this presentation is twofold: to introduce the reader to some basic ideas of differential algebra in the context of integration and to raise awareness in the physics community of computer algebra algorithms for indefinite and definite integration.

1 Introduction

In earlier times large tables of integrals were compiled by hand [19, 20, 30, 36]. Nowadays, computer algebra tools play an important role in the evaluation of definite integrals and we will mention some approaches below. Tables of integrals are even used in modern software as well. Algorithms for symbolic integration in general proceed in three steps. First, in computer algebra the functions typically are modeled by algebraic structures. Then, the computations are done in the algebraic framework and, finally, the result needs to be interpreted in terms of functions again. Some considerations concerning the first step, i.e., algebraic representation of functions, will be part of Sect. 2. A brief overview of some approaches and corresponding algorithms will be given below. We will focus entirely on the approach using differential fields. Other introductory texts on this subject include [12, 32]. Manuel Bronstein's book on symbolic integration [13] is a standard reference. The interested reader is referred to one of these for more information.

C.G. Raab (✉)
Deutsches Elektronen-Synchrotron, DESY, Platanenallee 6, 15738 Zeuthen, Germany
e-mail: clemens.raab@desy.de

C. Schneider and J. Blümlein (eds.), *Computer Algebra in Quantum Field Theory*, Texts & Monographs in Symbolic Computation, DOI 10.1007/978-3-7091-1616-6_12,

A recent version of Risch's algorithm will be presented in Sect. 3. The subtle issues of the last step, i.e., translating the algebraic result to a valid statement in the world of functions, will not be dealt with here.

1.1 Parametric Integration

Integration of functions can be done in two variants: indefinite and definite integration, which are closely related via the fundamental theorem of calculus. On the one hand, an indefinite integrals still is a function in the variable of integration and is nothing else than the antiderivative of a given function $f(x)$. On the other hand, a definite integral is the value

$$\int_a^b f(x)\,dx$$

resulting from integrating the function $f(x)$ over the given interval (a,b). Another difference between the two is that in general it is easy to verify an indefinite integral just by differentiating it, whereas in general it is hard to verify the result of a definite integral without recomputing it.

For the evaluation of definite integrals many tools may be applied to transform them to simpler integrals which are known or can be evaluated easily: change of variable, series expansion of the integrand, integral transforms, etc. As mentioned above by the fundamental theorem of calculus it is obvious that we can use indefinite integrals for the evaluation of definite integrals. It is well known that for a function $g(x)$ with $g'(x) = f(x)$ we have

$$\int_a^b f(x)\,dx = g(b) - g(a).$$

This fact has also been exploited in order to evaluate definite integrals for which a corresponding indefinite integral is not available in nice form. We give an overview of this method, which will be the main focus for computing definite integrals in the present paper. If the integral depends on a parameter, we can differentiate the parameter integral with respect to this parameter and obtain an integral that might be evaluated more easily. Under suitable assumptions on the integrand we have

$$\frac{d}{dy}\int_a^b f(x,y)\,dx = \int_a^b \frac{df}{dy}(x,y)\,dx,$$

which is called *differentiating under the integral sign*. A related paradigm, known as *creative telescoping*, is used in symbolic summation to compute recurrences for parameter dependent sums, see [48] for instance. Based on these two principles Almkvist and Zeilberger [6] were the first to propose a completely systematic way for treating parameter integrals by differentiating under the integral sign by giving an algorithm to compute differential equations for parameter integrals with holonomic integrands. They gave a fast variant of it for hyperexponential integrands, which may also be used for computing recurrences for such parameter integrals. From a very general point of view the underlying principle might be understood as combination of the fundamental theorem of calculus and the linearity of the integral in the following way. If for integrable functions $f_0(x), \dots, f_m(x)$ and constants $c_0, \dots, c_m$ the function $g(x)$ is an antiderivative such that

$$c_0 f_0(x) + \cdots + c_m f_m(x) = g'(x),$$

then we can deduce the relation

$$c_0 \int_a^b f_0(x)\,dx + \cdots + c_m \int_a^b f_m(x)\,dx = g(b) - g(a)$$

among the definite integrals $\int_a^b f_i(x)\,dx$ provided they exist. Both the functions $f_i(x)$ and the constants c_i may depend on additional parameters, which are not shown here. In order that this works the important point is that the c_i do not depend on the variable of integration. In practice, the functions $f_i(x)$ are chosen to be derivatives or shifts in the parameter(s) of the integrand $f(x)$ if we are interested in differential equations or recurrences for the definite integral.

The main task for finding such relations of definite integrals of given functions $f_i(x)$ consists in finding suitable choices for the constants c_i which allow a closed form of the antiderivative $g(x)$ to be computed. We will call this *parametric integration* as it can be viewed as making suitable choices for the parameters c_i occurring in the combined integrand $c_0 f_0(x) + \cdots + c_m f_m(x)$.

The approach above also addresses the issue of verifiability. When given such a linear relation of integrals

$$c_0 \int_a^b f_0(x)\,dx + \cdots + c_m \int_a^b f_m(x)\,dx = r$$

the function $g(x)$ may act as a proof certificate of it as we just need to verify

$$c_0 f_0(x) + \cdots + c_m f_m(x) = g'(x) \qquad \text{and} \qquad r = g(b) - g(a),$$

where the left hand sides are directly read off from the integral relation we want to verify.

1.2 Symbolic Integration

Algorithms to compute indefinite integrals of rational integrands are known for a long time already and many other integrals were computed analytically by hand as mentioned above. Especially in the last century algorithms have been developed capable of dealing with more general classes of integrands in a completely systematic way. In the following we want to give an overview of three different approaches that were taken. We also mention some relevant cornerstones but we do not include a fully comprehensive survey of the corresponding literature, many other contributions were made. Note that all of those approaches extend to definite integration in one way or the other.

The differential algebra approach represents functions as elements of differential fields and differential rings. These are algebraic structures not only capturing the arithmetic properties of functions but also their differential properties by including derivation as an additional unary operation. In general terms, starting with a prescribed differential field one is interested in indefinite integrals in the same field or in extensions of that field constructed in a certain way. Based on a book by Joseph F. Ritt [42] using differential fields Robert H. Risch gave a decision procedure [10, 41] for computing elementary integrals of elementary functions by closely investigating the structure of the derivatives of such functions. Since then this result has been extended in various directions. A parametric version was discussed in [29]. Michael F. Singer et al. generalized this to a parametric algorithm computing elementary integrals over regular Liouvillian fields in the appendix of [45] and Manuel Bronstein gave partial results for more general differential fields constructed by monomials [9, 13]. The author's thesis [37] can be seen as a continuation of this line of research. In [33] Arthur C. Norman published a variant of Risch's algorithm avoiding its recursive structure, which therefore is sometimes also called the parallel Risch algorithm. The Risch-Norman algorithm can be used in even more general differential fields and has proven to be a rather powerful heuristic in practice, see [8, 13, 14] and references therein. Most results mentioned so far restrict to the case where the generators of the differential fields are algebraically independent. The presence of algebraic relations causes new situations and requires more involved algebraic tools, see [8,10,12,23] and references therein. Another type of generalization is to search also for certain types of non-elementary integrals over certain differential fields. Some results for this problem have been achieved in [45], see also [7] and the references to the work of Cherry and Knowles in [13].

Indefinite integrals of products of special functions that satisfy homogeneous second-order differential equations were considered by Jean C. Piquette. His ansatz for the integral in terms of linear combinations of such products led to a differential system, which after uncoupling he solved by heuristic methods, see [34, 35] and references therein. The holonomic systems approach was initiated by Doron Zeilberger in [47] and puts this on more general and more algorithmic grounds. Functions are represented by the differential and difference operators that annihilate them. The notion of D-finite functions is closely related and refers to functions satisfying homogeneous linear differential equations with rational functions as coefficients. Hence, the derivatives of a D-finite function generate a finite-dimensional vector space over the rational functions. Frédéric Chyzak [16] presented an efficient algorithm for computing indefinite integrals of such functions in the same vector space. The algorithm handles also parametric integration and summation and utilizes Ore algebras to represent the operators corresponding to functions. For extensions and improvements see [17, 24].

The rule-based approach operates on the syntactic presentation of the integral by a table of transformation rules. This comes close to what is done when integrating by hand based on integral tables such as [19,20,36]. Also most computer algebra systems make at least partial use of transformations and table look-up. These tables may contain rules for virtually any special function, which makes such algorithms easily extensible in principle. This approach is recently being investigated systematically by Albert D. Rich and David J. Jeffrey [40], who point out several subtle issues related to efficiency.

1.3 Risch's Algorithm

When computing elementary integrals the paradigm followed by Risch's algorithm and many of its generalizations is that the computation proceeds recursively, focusing one by one on a particular function, which is involved in the integrands, at a time. For each of these functions the computation is organized in several main steps, where each step computes a part of the integral and subtracts its derivative from the integrand to obtain the remaining integrand to proceed with. The part of the integral that is computed in each step is chosen in such a way that the remaining integrand is simpler than the previous one in some suitable sense.

Before we discuss the computation for rather general types of integrands in a bit more detail, it will be instructive to consider the simplest case first, namely rational functions. The main steps of the full algorithm will be used to work out a closed form of the following integral of a rational function.

$$\int \frac{x^4 + 2x^3 - x^2 + 3}{x^3 + 5x^2 + 8x + 4}\,dx$$

For rational functions three steps are relevant. First, we will apply Hermite reduction to reduce the task to an integrand that does not have poles of order greater than one. Then, we will compute the residues at the simple poles of the remaining integrand to obtain the logarithmic part of the integral. Finally, the remaining integrand will be a polynomial, which is easily integrated.

Let us start by outlining the main idea of Hermite reduction [21], which repeats as needed what can be summarized as a suitably chosen additive splitting of the integrand followed by integration by parts of one of the two summands. Each time the order of some poles of the integrand is reduced. We will see later how such splittings are determined, for now we just emphasize that no partial fraction decomposition is required. In our example the denominator factors as

$(x+1)(x+2)^2$. This means that we have to reduce the order of the pole at $x = -2$, which is achieved by the following splitting.

$$\int \frac{x^4+2x^3-x^2+3}{(x+1)(x+2)^2}\,dx = \int \frac{1}{(x+2)^2}\,dx + \int \frac{x^3-x+1}{(x+1)(x+2)}\,dx$$
$$= -\frac{1}{x+2} + \int \frac{x^3-x+1}{(x+1)(x+2)}\,dx.$$

The remaining integrand has only simple poles, so we proceed by computing the residues at its poles, from which we obtain the logarithmic part of the integral. For an integrand $\frac{a(x)}{b(x)}$, where $a(x)$ and $b(x)$ are polynomials and $b(x)$ is squarefree, the residue at a root x_0 of $b(x)$ is given by $z_0 := \frac{a(x_0)}{b'(x_0)}$ and we get a contribution $z_0 \log(x - x_0)$ to the integral. Instead of determining the residue in dependence of the location of the pole, there are algorithms which first compute the set of values occurring as residues and then determine the appropriate logands for each residue. One such algorithm relying on resultants has its roots in the work of Rothstein and Trager, see [25,31,43,46], and another algorithm using Gröbner bases was proposed by Czichowski, see [18]. The main idea to compute the residues directly, without computing the roots of $b(x)$, is to characterize them as those values z_0 such that the equations $a(x) - z_0 \cdot b'(x) = 0$ and $b(x) = 0$ are satisfied for some x at the same time. So the Rothstein-Trager resultant $r(z) = \operatorname{res}_x(a(x) - z \cdot b'(x), b(x))$ is a polynomial in the new variable z having the residues of the integrand as its roots. For each root z_0 of $r(z)$ we need to compute $\gcd(a(x) - z_0 \cdot b'(x), b(x))$, which is the corresponding logand. A modern variant, the Lazard-Rioboo-Trager algorithm, computes the subresultant polynomial remainder sequence of $a(x) - z \cdot b'(x)$ and $b(x)$, from which both $r(z)$ and the logands can be read off. Similarly, Czichowski's algorithm computes a Gröbner basis of $\{a(x) - z \cdot b'(x), b(x)\}$ w.r.t. $z <_{lex} x$, from which both the squarefree part of $r(z)$ and the logands can be read off. In our present example we determine the polynomials

$$r(z) = (z-1)(z-5) \quad \text{and} \quad s(z,x) = x + \frac{1}{4}z + \frac{1}{4},$$

where the roots of $r(z)$ are the residues and $s(z,x)$ gives the corresponding logands, which give rise to the following logarithmic part of the integral:

$$\sum_{r(z)=0} z \cdot \log(s(z,x)) = \log(x+1) + 5\log(x+2).$$

Therefore, subtracting its derivative from the integrand we obtain the polynomial

$$\frac{x^3-x+1}{(x+1)(x+2)} - \sum_{r(z)=0} z \frac{\frac{d}{dx}s(z,x)}{s(z,x)} = x - 3$$

as remaining integrand or, in other words,

$$\int \frac{x^3 - x + 1}{(x+1)(x+2)}\,dx = \log(x+1) + 5\log(x+2) + \int x - 3\,dx.$$

The integral of a polynomial is determined via an appropriate ansatz, based on the fact that the derivative of a non-constant polynomial is a polynomial with degree exactly one less. The coefficients in the integral are then determined by equating the coefficients in the derivative of the ansatz to those in the integrand. In our case the ansatz for the integral is $a_2x^2 + a_1x$ and comparing coefficients of the powers of x in

$$\frac{d}{dx}\left(a_2x^2 + a_1x\right) = x - 3$$

yields $2a_2 = 1$ and $a_1 = -3$. Plugging the solution of these equations into the ansatz we obtain the integral

$$\int x - 3\,dx = \frac{1}{2}x^2 - 3x.$$

Altogether, we obtained the following closed form of the integral.

$$\int \frac{x^4 + 2x^3 - x^2 + 3}{(x+1)(x+2)^2}\,dx = -\frac{1}{x+2} + \ln(x+1) + 5\ln(x+2) + \frac{1}{2}x^2 - 3x$$

2 Algebraic Representation of Functions

In differential algebra functions are represented as elements of differential fields and differential rings. These are algebraic structures not only capturing the arithmetic relations of functions but also their differential properties by including derivation as an additional operation. For more information on differential algebra, see [22] for example.

Definition 1. Let F be a field and let $D : F \to F$ be a unary operation on it, which is additive and satisfies the product rule, i.e.,

$$D(f+g) = Df + Dg \qquad \text{and} \qquad D(fg) = fDg + (Df)g.$$

Then D is called a *derivation* on F and (F, D) is called a *differential field*. The set of *constants* is denoted by $\mathrm{const}_D(F) := \{f \in F \mid Df = 0\}$.

It follows from the definition that the set $\mathrm{const}_D(F)$ is closed under the basic arithmetic operations and hence is a field. Note that while for $f, g \in F$ sums

$f + g$, products fg, and derivatives Df by definition are in F again, powers f^g, compositions $f \circ g$ and antiderivatives $\int f$ need not be in F in general. The same statements apply for differential rings where every occurrence of the word field is to be replaced by the word ring.

The basic example for a differential field is the field of rational functions $(F, D) = (C(x), \frac{d}{dx})$, where $Dx = 1$ and $\mathrm{const}_D(F) = C$. Note that C, the field of constants, may not only consist of numbers, but it may also contain elements depending on variables other than x. For example, $(F, D) = (\mathbb{R}(n, x, x^n, \ln(x)), \frac{d}{dx})$, where the notation is meant to imply $Dn = 0$, $Dx = 1$, $Dx^n = \frac{n}{x}x^n$, and $D\ln(x) = \frac{1}{x}$, is a differential field with $\mathrm{const}_D(F) = \mathbb{R}(n)$. In principle the definition of a differential field does not require the existence of an element $x \in F$ with $Dx = 1$. For example, $(\mathbb{Q}(e^x), \frac{d}{dx})$ is a differential field since the derivative of any rational expression in e^x is again a rational expression in e^x. In practice, however, such cases are not very important.

In general, we will consider finitely generated differential fields of the form $(F, D) = (C(t_1, \ldots, t_n), D)$, where $C = \mathrm{const}_D(F)$ and $t_1, \ldots, t_n$ represent some functions. Algebraically, an element $f \in F$ is a rational expression in $t_1, \ldots, t_n$ with coefficients in C and, resembling the chain rule, the derivative can be expressed as

$$Df = \frac{\partial f}{\partial t_1} \cdot Dt_1 + \cdots + \frac{\partial f}{\partial t_n} \cdot Dt_n.$$

When given some differential field (K, D), by adjoining additional elements $t_1, \ldots, t_n$ under the condition that we can extend D to a derivation on $K(t_1, \ldots, t_n)$ we obtain a differential field extension $(K(t_1, \ldots, t_n), D)$, i.e., a differential field containing (K, D) as a differential subfield. The following theorem makes the choice explicit which we have when extending the derivation from (K, D) to a differential field extension of the form $(K(t), D)$.

Theorem 1 ([13, Theorems 3.2.2, 3.2.3]). *Let (K, D) be a differential field and let $K(t)$ be the field generated by a new element t.*

1. *If t is algebraic over K, then D can be uniquely extended to a derivation on $K(t)$.*
2. *If t is transcendental over K, then, for any $w \in K(t)$, D can be uniquely extended to a derivation on $K(t)$ such that $Dt = w$.*

In our presentation we will focus on transcendental extensions, in which case the notion of a (differential) monomial, introduced by Bronstein [9], is very important for practical algorithms.

Definition 2. Let (F, D) be a differential field, (K, D) a differential subfield, and $t \in F$. Then t is called a *monomial* over (K, D) if

1. t is transcendental over K and
2. $Dt \in K[t]$.

If $\deg_t(Dt) \geq 2$, we call t *nonlinear*.

There are many similarities between rational functions $(C(x), \frac{d}{dx})$ and monomial extensions $(K(t), D)$, but there are, of course, some important differences as well. The derivative of polynomials $p \in C[x]$ is a polynomial again, likewise any polynomial $p \in K[t]$ has its derivative Dp in $K[t]$. However, unlike the degree of the derivative of $p \in C[x]$, the degree need not drop when applying D to a $p \in K[t]$, it may stay the same or even increase depending on the degree in t of Dt. An irreducible polynomial $p \in C[x]$ never divides its derivative, this need not be true for polynomials $p \in K[t]$. More generally, a squarefree polynomial $p \in K[t]$ need not be coprime with Dp, while it always is if $p \in C[x]$.

Definition 3. Let (K, D) be a differential field and let t be a monomial over (K, D). We call a polynomial $p \in K[t]$ *normal*, if p and Dp are coprime, or *special*, if p divides Dp.

A squarefree polynomial $p \in K[t]$ is normal if and only if it does not contain a factor of degree at least 1 which is special.

These properties of the derivation on $K[t]$ deserve to be exemplified, for which we consider $(K, D) = (C(x), \frac{d}{dx})$. The transcendental functions $\ln(x)$, $\exp(x)$, and $\tan(x)$ satisfy $\frac{d}{dx}\ln(x) = \frac{1}{x}$, $\frac{d}{dx}\exp(x) = \exp(x)$, and $\frac{d}{dx}\tan(x) = \tan(x)^2+1$, respectively. Hence they can be represented by monomials over (K, D). Representing the logarithm by a monomial t with $Dt = \frac{1}{x} \in K$ it can be proven that the degree of a polynomial $p \in K[t]$ is always at least as large as that of its derivative Dp. The degrees are unequal if and only if the leading coefficient of p is in C, which is true for $p = t^2 + xt$ with $Dp = \frac{x+2}{x}t + 1$, for example, but not for $p = xt^2 - \frac{2x^2-x}{x+1}t$ with $Dp = t^2 + \frac{3}{(x+1)^2}t - \frac{2x-1}{x+1}$. In addition, every squarefree polynomial is indeed coprime with its derivative. For a monomial t with $Dt = t$ we always have that $p \in K[t]$ and Dp have the same degree if p is not constant. There are polynomials which divide their derivative, and all of them are of the form $p = at^n$ where $a \in K$ and $n \in \mathbb{N}$. Finally, a monomial with $Dt = t^2 + 1$ has the property that the degree of Dp is strictly greater than that of $p \in K[t]$ as long as the degree of p is at least one, e.g., $p = t(t^2 + 1)$ has derivative $Dp = (3t^2 + 1)(t^2 + 1)$, and a squarefree polynomial is normal if and only if it is coprime with $t^2 + 1$.

Furthermore, in monomial extensions $(K(t), D)$ we will rely on the *canonical representation*

$$f = p + \frac{a_s}{b_s} + \frac{a_n}{b_n}$$

of elements $f \in K(t)$, where $p, a_s, a_n, b_s, b_n \in K[t]$ with $\deg_t(a_s) < \deg_t(b_s)$ and $\deg_t(a_n) < \deg_t(b_n)$ are such that b_s is special and every irreducible factor of b_n is normal, cf. [13, p. 103]

2.1 Relevant Classes of Functions

Apart from rational functions $(C(x), \frac{d}{dx})$ and algebraic functions $(\overline{C(x)}, \frac{d}{dx})$, elementary functions are a very basic class of functions as well and were among the first to be considered algorithmically. The *elementary functions* are those which can be constructed from rational functions by the following operations in addition to the basic arithmetic operations: taking the logarithm, applying the exponential function, and solving algebraic equations with elementary functions as coefficients. Elementary functions include rational and algebraic functions, logarithms, c^x and x^c, trigonometric functions and their inverses, as well as hyperbolic functions and their inverses. Recall that trigonometric and hyperbolic functions can be expressed in terms of exponentials and their inverses can be expressed in terms of logarithms of algebraic functions. Note that compositions $f(g(x))$ and powers $f(x)^{g(x)}$ of elementary functions are elementary functions again. When representing elementary functions in differential fields we make use of the following relations:

$$\frac{d}{dx}\ln(a(x)) = \frac{a'(x)}{a(x)} \tag{1}$$

$$\frac{d}{dx}\exp(a(x)) = a'(x)\exp(a(x)). \tag{2}$$

Definition 4. Let (K, D) be a differential field and let t be a monomial over (K, D). Then we call t an *elementary monomial* over (K, D) if it is either

1. a *logarithm* over (K, D), i.e., there exists $a \in K$ such that $Dt = \frac{Da}{a}$, or
2. an *exponential* over (K, D), i.e., there exists $a \in K$ such that $\frac{Dt}{t} = Da$.

Let $(F, D) = (K(t_1, \dots, t_n), D)$ be a differential field extension of (K, D). Then (F, D) is called *elementary extension* of (K, D), if each t_i is either algebraic or an elementary monomial over $(K(t_1, \dots, t_{i-1}), D)$.

An elementary function is a function representable as an element of some elementary extension of $(C(x), \frac{d}{dx})$. Note that an elementary extension of some differential field (K, D) does not only contain elementary functions unless K does.

The notion of elementary functions is generalized naturally to give *Liouvillian functions* by considering differential equations of the form

$$\frac{d}{dx}y(x) = a(x) \tag{3}$$

$$\frac{d}{dx}y(x) = a(x)y(x) \tag{4}$$

instead of their special cases for logarithms and exponentials above. In other words, Liouvillian functions are the functions obtained from rational functions by the basic arithmetic operations, by taking primitive functions $\int a(x)\,dx$, by taking hyperexponential functions $e^{\int a(x)\,dx}$, and by solving algebraic equations with Liouvillian functions as coefficients. Again, the composition of Liouvillian functions as well as powers $f(x)^{g(x)}$ of Liouvillian functions are Liouvillian. Several special functions can be found in the class of Liouvillian functions, e.g., logarithmic and exponential integrals, error functions, Fresnel integrals, incomplete Beta and Γ functions, polylogarithms, harmonic polylogarithms [39], and hyperlogarithms [15].

Definition 5. Let (K, D) be a differential field and let t be a monomial over (K, D). Then we call t a *Liouvillian monomial* over (K, D) if it is either

1. *primitive* over (K, D), i.e., there exists $a \in K$ such that $Dt = a$, or
2. *hyperexponential* over (K, D), i.e., there exists $a \in K$ such that $\frac{Dt}{t} = a$.

Let $(F, D) = (K(t_1, \ldots, t_n), D)$ be a differential field extension of (K, D). Then (F, D) is called *Liouvillian extension* of (K, D), if each t_i is either algebraic or a Liouvillian monomial over $(K(t_1, \ldots, t_{i-1}), D)$.

A Liouvillian function is a function representable as an element of some Liouvillian extension of $(C(x), \frac{d}{dx})$. Note that there are a few equivalent definitions of the class of Liouvillian functions. For instance, we need not start the construction from the rational functions but it suffices to start from the set of constants because the rational functions are obtained by the basic arithmetic operations from constants and the identity function, which in turn is a primitive function of the constant 1. Similarly, we may also choose to keep the operation of applying the exponential function instead of replacing it by taking hyperexponential functions as the latter operation can obviously be decomposed into applying the exponential function to a primitive function. Alternatively, we may also summarize taking primitive and hyperexponential functions into taking solutions of linear first-order differential equations. More precisely, the class of Liouvillian functions may also be constructed from the set of constants by the basic arithmetic operations and taking particular solutions of

$$y'(x) = a(x)y(x) + b(x) \tag{5}$$

and of algebraic equations with Liouvillian coefficients each. Note that the solutions of (5) may be expressed in terms of primitives and (hyper)exponentials by $y(x) = e^{\int a(x)\,dx} \int \frac{b(x)}{e^{\int a(x)\,dx}}\,dx$.

For the sake of completeness we also give the definition of hyperexponential and d'Alembertian functions [4, 5], although they are not so relevant in our considerations. They are continuous analogues of hypergeometric and d'Alembertian sequences, respectively. An algorithm for integration of hyperexponential functions is given in [6].

Definition 6. Let (F, D) be a differential field, (K, D) a differential subfield, and $t \in F$. Then t is called

1. *hyperexponential* over (K, D) if $\frac{Dt}{t} \in K$, or
2. *d'Alembertian* over (K, D) if there exist $n \in \mathbb{N}$ and $r_1, \ldots, r_n \in K$ such that t is a solution of the homogeneous linear differential equation obtained from the composition of differential operators $D - r_i$, i.e., $(D - r_n)\ldots(D - r_1)t = 0$.

The *hyperexponential functions* are functions $h(x)$ being hyperexponential over $(C(x), \frac{d}{dx})$, i.e., their logarithmic derivative $\frac{h'(x)}{h(x)}$ is a rational function. Typical examples of hyperexponential functions are $c^{f(x)}$ and $f(x)^c$, where $f(x)$ is a rational function. Note that the product and the quotient of hyperexponential functions are hyperexponential again, but the sum of hyperexponential functions is not hyperexponential in general. So, in contrast to the classes of elementary and Liouvillian functions, the class of hyperexponential functions is not closed under the basic arithmetic operations.

Similarly, *d'Alembertian functions* are the functions that are d'Alembertian over $(C(x), \frac{d}{dx})$. The class of d'Alembertian functions is not closed under the basic arithmetic operations either, as the sum and the product of d'Alembertian functions are d'Alembertian again, but the quotient of d'Alembertian functions is not d'Alembertian in general. Most of the special functions listed above as being Liouvillian functions are in fact even d'Alembertian functions: exponential integrals, error functions, Fresnel integrals, incomplete Beta and Γ functions, polylogarithms, harmonic polylogarithms, and hyperlogarithms. Note that hyperexponential functions are d'Alembertian as well, and d'Alembertian functions are Liouvillian. An equivalent characterization of d'Alembertian functions is that they can be written as iterated integrals over hyperexponential functions

$$h_1(x) \int h_2(x) \int \ldots \int h_n(x)\, dx \ldots dx.$$

The relation to the previous definition is that the product $h_1(x)\ldots h_i(x)$ is a solution of $y'(x) - r_i(x)y(x) = 0$.

2.2 Liouville's Theorem

Liouville [26–28] was the first to prove an observation on the structure of elementary integrals. In the language of differential fields it can be stated as follows.

Theorem 2 (Liouville's Theorem [13, Theorem 5.5.3]). *Let (F, D) be a differential field and $C := \text{const}(F)$. If $f \in F$ has an elementary integral over (F, D), then there are $v \in F$, $c_1, \ldots, c_n \in \overline{C}$, and $u_1, \ldots, u_n \in F(c_1, \ldots, c_n)^*$ such that*

$$f = Dv + \sum_{i=1}^{n} c_i \frac{Du_i}{u_i}. \tag{6}$$

In view of this theorem we always can express an elementary integral $\int f$ as the sum of two parts: a $v \in F$, which then is called the *rational part*, and a sum of logarithms $\sum c_i \log(u_i)$, which is called the *logarithmic part* of the integral. This theorem and its refinements [13, 37] which consider a special structure of the integrand are the main theoretical foundation for algorithms computing elementary integrals. There are even generalizations of Liouville's theorem dealing also with non-elementary integrals, e.g. [7, 45].

3 Risch's Algorithm in Monomial Extensions

As already explained earlier, we are interested in parametric integration. In terms of differential fields this problem can be formulated as follows.

Problem 1 (parametric elementary integration). Given: a differential field (F, D) and $f_0, \ldots, f_m \in F$.

Find: a C-vector space basis $\mathbf{c}_1, \ldots, \mathbf{c}_n \in C^{m+1}$, where $C := \text{const}(F)$, of all coefficient vectors $(c_0, \ldots, c_m) \in C^{m+1}$ such that $c_0 f_0 + \cdots + c_m f_m \in F$ has an elementary integral over (F, D) and compute corresponding integrals $g_1, \ldots, g_n$ from some elementary extensions of (F, D).

We consider this problem over towers of monomial extensions, i.e., $(F, D) = (C(t_1, \ldots, t_n), D)$ where each t_i is a monomial over $(C(t_1, \ldots, t_{i-1}), D)$ subject to some technical conditions. For details see [13, 37]. A big part of the common special functions can be represented in such differential fields. In addition to Liouvillian functions, most importantly functions satisfying (possibly inhomogeneous) linear second-order differential equations can be fit into this framework. Concrete examples include orthogonal polynomials, associated Legendre functions, Bessel functions, Airy functions, complete elliptic integrals, Whittaker functions, Mathieu functions, hypergeometric functions, Heun functions, Struve functions, Anger functions, Weber functions, Lommel functions, Scorer functions, etc. How this can be done is explained in [37].

As mentioned above Risch's algorithm proceeds recursively, thereby exploiting the structure of the underlying differential field that is used to model the functions occurring. The focus of the computation always is on the topmost generator of the differential field and everything else is regarded as part of the coefficients. In essence, the steps dealing with expressions from $C(x)$ outlined above are generalized to work with expressions from $K(t)$ where some monomial t, cf. Definition 2, takes the role of x and coefficients appearing in rational or polynomial expressions in t do not necessarily have zero derivative. Moreover, we do not consider the poles of the integrand by interpreting it as a function of x, we will work on a syntactic level instead by considering the factors of the denominator in the representation of the integrand in terms of t. The algorithms outlined above carry over as long as they are applied to the normal part of the denominator only.

If present, the special part of the denominator needs to be treated differently, which is done similarly to integrating the polynomial part.

Along with the main ideas of the algorithm in monomial extensions we present a specific example to illustrate how the integrand is processed. For the explicit computation we consider the integral

$$\int \frac{x^2e^{5x} - 2xe^{4x} + (2x^3 + 5x + 1)e^{3x} - (6x^3 + x + 1)e^x + 4x^3}{x^2e^{2x}(e^x - 1)^2}\,dx.$$

The integrand can be represented in the differential field $(C(x,t), D)$ with $Dx = 1$ and $Dt = t$ as

$$\frac{x^2t^5 - 2xt^4 + (2x^3 + 5x + 1)t^3 - (6x^3 + x + 1)t + 4x^3}{x^2t^2(t - 1)^2}.$$

In the general setting Hermite reduction requires some preprocessing, since it only deals with terms for which all irreducible factors of the denominator are normal. To this end, we compute the canonical representation mentioned earlier. We ignore any terms with special polynomials in the denominator for the moment.

In our example we have that the polynomial t is special and the polynomial $t - 1$ is normal. So the canonical representation is given by

$$t + \frac{2x-2}{x} + \frac{\frac{2x^3-x-1}{x^2}t + 4x}{t^2} + \frac{\frac{3x^2+2x+2}{x^2}t - \frac{2x^2+2}{x^2}}{(t-1)^2},$$

where the last fraction is the one we will focus on now.

Hermite reduction repeatedly splits the integrand and applies integration by parts to one of the two summands each time. More precisely, if the integrand is of the form $\frac{a}{uv^m}$, where $a, u, v \in K[t]$ are pairwise relatively prime polynomials with v being normal and $m \in \{2, 3, \dots\}$, then there are unique polynomials $r, s \in K[t]$ such that $\deg_t(r) < \deg_t(v)$ and

$$a = (1 - m)ruDv + sv.$$

Such polynomials can be readily computed by the extended euclidean algorithm, for instance. With this splitting of the numerator we have

$$\int \frac{(1-m)ruDv + sv}{uv^m} = \frac{r}{v^{m-1}} + \int \frac{s - uDr}{uv^{m-1}}, \tag{7}$$

where the power of v in the denominator of the remaining integrand has dropped by (at least) one. Note that the polynomial v is merely required to be normal, so all normal irreducible factors in the denominator of the integrand occurring with power m can be treated at once.

Hermite reduction repeats the above step until an integrand with a normal denominator is obtained. Starting from an integrand $\frac{a}{b}$ with $a, b \in K[t]$ and every irreducible factor of b being normal, we first compute a squarefree factorization of the denominator $b = b_1 b_2^2 \dots b_n^n$ and then after at most $n - 1$ reduction steps going from $m = n$ down to $m = 2$, reducing the highest-order poles in each step, we arrive at an integrand with a normal denominator.

There is also a variant of the Hermite reduction where at each reduction step the order of all poles of order greater than one is reduced, instead of the highest-order poles only. This has the additional advantage that no squarefree factorization needs to be computed at the beginning.

In our example the denominator $(t - 1)^2$ is already given in factored form. This means that we have $m = 2$, $u = 1$, and $v = t - 1$. With these values we need to find the polynomials $r, s \in C(x)[t]$ satisfying

$$\frac{3x^2 + 2x + 2}{x^2} t - \frac{2x^2 + 2}{x^2} = r \cdot (-t) + s \cdot (t - 1)$$

and $\deg_t(r) < 1$. We compute $r(x) = -\frac{x+2}{x}$ and $s = \frac{2x^2+2}{x^2}$, so by (7) we obtain

$$\int \frac{\frac{3x^2+2x+2}{x^2} t - \frac{2x^2+2}{x^2}}{(t - 1)^2} = -\frac{x + 2}{x(t - 1)} + \int \frac{2}{t - 1}.$$

The remaining integrand has a normal denominator and we still focus on the part of the integrand which has normal irreducible factors in its denominator only, which just occur with multiplicity one now. For such integrands the notion of a residue can be defined appropriately in monomial extensions, which we do not detail here. We proceed by computing the logarithmic part of the integral, which will be of the form

$$\sum_i \sum_{r_i(z)=0} z \cdot \log(s_i(z, t))$$

with $r_i \in C[z]$ squarefree and $s_i \in K[z, t]$. This means that the residues are the roots of the polynomials r_i and the polynomials s_i give the corresponding logands. In general it may happen that the residue is not a constant, i.e., potentially we have $r_i \in K[z]$ only. If this happens, it can be shown that the integral is not elementary over $(K(t), D)$. This gives a necessary condition on the coefficients of the linear combination of several integrands in the parametric integration problem. An algorithm to ensure that we will consider only linear combinations which actually have $r_i \in C[z]$ can be found in [37], a different algorithm was already used in [45]. Once this is done we compute the corresponding polynomials r_i and s_i

via generalizations of the algorithms mentioned earlier that originally were designed for rational functions, see [13, 38]. Note that subtracting the derivative

$$\sum_{l}\sum_{r_l(z)=0} z\cdot\frac{D(s_l(z,t))}{s_l(z,t)}$$

of the logarithmic part of the integral from the integrand may also change the polynomial part of the integrand in the general case, in particular this happens if t is nonlinear.

In our case we simply have one polynomial $r_1 = z - 2$ and $s_1 = t - 1$ each, which give rise to the logarithmic part

$$2\log(t-1)$$

Subtracting its derivative $D(2\log(t-1)) = 2 + \frac{2}{t-1}$ from the integrand we obtain

$$t + \frac{2x-2}{x} + \frac{\frac{2x^3-x-1}{x^2}t + 4x}{t^2} + \frac{2}{t-1} - \left(2 + \frac{2}{t-1}\right) = t - \frac{2}{x} + \frac{\frac{2x^3-x-1}{x^2}t + 4x}{t^2}.$$

At this point the remaining integrands are such that their denominator is special. Depending on the specific properties of t this condition admits only a very restricted form of the denominator and in many cases even implies that the denominator is in K. The aim is to reduce the integrands to lie in K. In short, the idea how to proceed is to make an appropriate ansatz for part of the integral based on the partial fraction decomposition of the integrands. Comparing coefficients then leads to differential equations with coefficients in K, for which solutions have to be found in K. While setting up the ansatz and solving for the coefficients was the easiest part in the integration of rational functions, it is the most difficult part in the general setting and algorithms exist only for certain types of monomials t and underlying differential fields (K, D). Under certain technical assumptions on t the following ansatz for the part of the integrands having special denominators can be justified. The integrand on the left hand side has only irreducible polynomials $p_j \in K[t]$ in its denominator which are special and it is given by its partial fraction decomposition, i.e.,

$$\sum_{j=1}^{n}\sum_{k=1}^{l_j}\frac{f_{j,k}}{p_j^k} = D\left(\sum_{j=1}^{n}\sum_{k=1}^{l_j}\frac{g_{j,k}}{p_j^k}\right).$$

After rewriting the right hand side in its partial fraction decomposition we can compare coefficients in order to obtain differential equations for $g_{j,k} \in K[t]$. Note that the derivative

$$D\frac{g_{j,k}}{p_j^k} = \frac{Dg_{j,k} - k\frac{Dp_j}{p_j}g_{j,k}}{p_j^k}$$

again has the same power p_j^k in the denominator since $\frac{Dp_j}{p_j} \in K[t]$ for special polynomials. Roughly speaking, upon comparing coefficients of p_j^{-k} we obtain differential equations relating each $g_{j,k}$ to $f_{j,k}$. This leads to the problem of finding solutions of certain type to differential equations, which may or may not exist. If no solution of the correct type exists, then it can be shown that the integral is not elementary over $(K(t), D)$. This again restricts the possible linear combinations in the parametric integration problem. There is a lot more to this, but we do not go into detail here. Instead we refer to [13] where relevant results are given. Not all cases can be dealt with algorithmically so far, this depends on the structure of (K, D) as well as on t. The main difficulty lies in the algorithmic solution of the differential equations arising, for which we also refer to [3, 11, 44] for example. This can be skipped if t is such that $K[t]$ does not contain any special irreducible polynomial. In practice this is often the case, the most notable exception are hyperexponential monomials t.

The above ansatz deals with the remaining denominators in the integrands. Similarly, for the remaining polynomial parts we can set up an ansatz of the form

$$\sum_{j=1}^{n} f_j t^j = D\left(\sum_{j=1}^{n+1-d} g_j t^j\right)$$

where $d := \deg_t(Dt)$ and $g_j \in K$. After expanding the right hand side in powers of t, we compare coefficients of $t^{\max(d,1)}, \dots, t^{n+\max(1-d,0)}$. The degree d of Dt determines the main features of the action of the derivation on polynomials from $K[t]$. If t is nonlinear, i.e., $d \geq 2$, then we can directly solve for g_j one by one. Otherwise, this leads to differential equations for g_j, which again impose restrictions on the possible linear combinations of integrands. As above, depending on the structure of (K, D) as well as on t the algorithms for computing solutions to these differential equations given in [3, 11, 13, 44] apply. There are large classes relevant in practice, which can be solved completely algorithmically. Remaining integrands are polynomials in $K[t]$ of degree less than $\max(d, 1)$, which can be reduced further to integrands in K under certain assumptions on t.

Our running example is such that complete algorithms exist. The fractional part has partial fraction decomposition

$$\frac{\frac{2x^3-x-1}{x^2}t + 4x}{t^2} = \frac{2x^3 - x - 1}{x^2 t} + \frac{4x}{t^2}$$

with respect to t. The ansatz $\frac{g_1}{t} + \frac{g_2}{t^2}$ has the derivative $\frac{Dg_1 - g_1}{t} + \frac{Dg_2 - 2g_2}{t^2}$ and hence leads to the differential equations

$$Dg_1 - g_1 = \frac{2x^3 - x - 1}{x^2}$$

$$Dg_2 - 2g_2 = 4x$$

with solutions $g_1 = -\frac{2x^2+2x-1}{x} \in C(x)$ and $g_2 = -2x-1 \in C(x)$. The polynomial part is just t, for which the ansatz $g_1 t$ for the integral trivially leads to $g_1 = 1$. Altogether, we have the remaining integrand

$$t - \frac{2}{x} + \frac{\frac{2x^3-x-1}{x^2}t + 4x}{t^2} - D\left(t - \frac{2x^2+2x-1}{xt} - \frac{2x+1}{t^2}\right) = -\frac{2}{x} \in C(x).$$

Now we reduced to integrands in K, still we want to find integrals which are elementary over $(K(t), D)$. If t is elementary over (K, D), then this obviously is equivalent to finding integrals elementary over (K, D). In order to apply our algorithm recursively we have to reduce this to a problem of finding elementary integrals over (K, D) also in the case where t is non-elementary over (K, D). Various refinements of Liouville's theorem are needed to solve this issue. For details we refer to [37], we just mention that this may lead to an increase in the number of integrands we have to consider in the recursive application of the algorithm.

In case of our example t is elementary over $(K, D) = (C(x), \frac{d}{dx})$, so we just need to apply the algorithm recursively to the remaining integrand $-\frac{2}{x}$. This yields $-2\log(x)$ as elementary integral over $(C(x), \frac{d}{dx})$. Now, collecting all the parts of the integral we computed, we obtain the following closed form

$$-2\log(x) + t - \frac{2x^2+2x-1}{xt} - \frac{2x+1}{t^2} + 2\log(t-1) - \frac{x+2}{x(t-1)}.$$

In other words we computed

$$\int \frac{x^2e^{5x} - 2xe^{4x} + (2x^3+5x+1)e^{3x} - (6x^3+x+1)e^x + 4x^3}{x^2e^{2x}(e^x-1)^2}\,dx =$$
$$2\ln\left(\frac{e^x-1}{x}\right) + \frac{xe^{4x} - xe^{3x} - (2x^2+3x+1)e^{2x} + (x-1)e^x + 2x^2 + x}{xe^{2x}(e^x-1)}.$$

3.1 Non-monomial Extensions

To a certain extent the algorithm can also be applied even in situations where the differential field does not meet all the requirements. Depending on which properties are violated the computation still may make sense, for instance if some algebraic relations among the generators of the differential field exist. Then it is just not guaranteed to find all possible solutions. Recently this heuristic has proven to be quite effective in the computation of massive Feynman diagrams at three-loops [1] where new iterated integrals involving square-root terms emerged [2].

Acknowledgements The author was supported by the Austrian Science Fund (FWF), grant no. W1214-N15 project DK6, by the strategic program "Innovatives OÖ 2010 plus" of the Upper Austrian Government, by DFG Sonderforschungsbereich Transregio 9 "Computergestützte Theoretische Teilchenphysik", and by the Research Executive Agency (REA) of the European Union under the Grant Agreement number PITN-GA-2010-264564 (LHCPhenoNet).

References

1. Ablinger, J., Blümlein, J., Raab, C.G., Schneider, C., Wißbrock, F.: DESY 13-063 (preprint)
2. Ablinger, J., Blümlein, J., Raab, C.G., Schneider, C.: Iterated Binomial Sums (in preparation)
3. Abramov, S.A.: Rational solutions of linear differential and difference equations with polynomial coefficients. USSR Comput. Math. Math. Phys. **29**(6), 7–12 (1989). (English translation of Zh. vychisl. Mat. mat. Fiz. 29, pp. 1611–1620, 1989)
4. Abramov, S.A.: On d'Alembert substitution. In: Proceedings of ISSAC'93, Kiev, pp. 20–26 (1993)
5. Abramov, S.A., Petkovšek, M.: D'Alembertian solutions of linear differential and difference equations. In: Proceedings of ISSAC'94, Oxford, pp. 169–174 (1994)
6. Almkvist, G., Zeilberger, D.: The method of differentiating under the integral sign. J. Symb. Comput. **10**, 571–591 (1990)
7. Baddoura, J.: Integration in finite terms with elementary functions and dilogarithms. J. Symb. Comput. **41**, 909–942 (2006)
8. Boettner, S.T.: Mixed transcendental and algebraic extensions for the Risch-Norman algorithm. PhD thesis, Tulane University, New Orleans (2010)
9. Bronstein, M.: A unification of Liouvillian extensions. Appl. Algebra Eng. Commun. Comput. **1**, 5–24 (1990)
10. Bronstein, M.: Integration of elementary functions. J. Symb. Comput. **9**, 117–173 (1990)
11. Bronstein, M.: On solutions of linear ordinary differential equations in their coefficient field. J. Symb. Comput. **13**, 413–439 (1992)
12. Bronstein, M.: Symbolic Integration Tutorial. Course Notes of an ISSAC'98 Tutorial, Rostock. Available at http://www-sop.inria.fr/cafe/Manuel.Bronstein/publications/issac98.pdf (1998)
13. Bronstein, M.: Symbolic Integration I – Transcendental Functions, 2nd edn. Springer, Berlin/New York (2005)
14. Bronstein, M.: Structure theorems for parallel integration. J. Symb. Comput. **42**, 757–769 (2007)
15. Brown, F.C.S.: The massless higher-loop two-point function. Commun. Math. Phys. **287**, 925–958 (2009). arXiv:0804.1660
16. Chyzak, F.: An extension of Zeilberger's fast algorithm to general holonomic functions. Discret. Math. **217**, 115–134 (2000)
17. Chyzak, F., Kauers, M., Salvy, B.: A non-holonomic systems approach to special function identities. In: Proceedings of ISSAC'09, Seoul, pp. 111–118 (2009)
18. Czichowski, G.: A note on Gröbner bases and integration of rational functions. J. Symb. Comput. **20**, 163–167 (1995)
19. Gradshteyn, I.S., Ryzhik, I.M.: In: Jeffrey, A., Zwillinger, D. (eds.) Table of Integrals, Series, and Products, 7th edn. Academic, New York (2007)
20. Gröbner, W., Hofreiter, N.: Integraltafel – Zweiter Teil: Bestimmte Integrale, 4th edn. Springer, Wien (1966)
21. Hermite, C.: Sur l'intégration des fractions rationelles. Nouvelles annales de mathématiques (2e série) **11**, 145–148 (1872)
22. Kaplansky, I.: An Introduction to Differential Algebra. Hermann, Paris (1957)
23. Kauers, M.: Integration of algebraic functions: a simple heuristic for finding the logarithmic part. In: Proceedings of ISSAC'08, Hagenberg, pp. 133–140 (2008)

24. Koutschan, C.: Advanced applications of the holonomic systems approach. PhD thesis, Johannes Kepler Universität Linz (2009)
25. Lazard, D., Rioboo, R.: Integration of rational functions: rational computation of the logarithmic part. J. Symb. Comput. **9**, 113–115 (1990)
26. Liouville, J.: Premier mémoire sur la detérmination des intégrales dont la valeur est algébrique. J. l'Ecole Polytechnique **14**, 124–148 (1833)
27. Liouville, J.: Second mémoire sur la detérmination des intégrales dont la valeur est algébrique. J. l'Ecole Polytechnique **14**, 149–193 (1833)
28. Liouville, J.: Mémoire sur l'intégration d'une classe de fonctions transcendantes. J. reine und angewandte Mathematik **13**, 93–118 (1835)
29. Mack, C.: Integration of affine forms over elementary functions. Computational Physics Group Report UCP-39, University of Utah (1976)
30. Magnus, W., Oberhettinger, F., Soni, R.P.: Formulas and Theorems for the Special Functions of Mathematical Physics. Springer, Berlin (1966)
31. Mulders, T.: A note on subresultants and the Lazard/Rioboo/Trager formula in rational function integration. J. Symb. Comput. **24**, 45–50 (1997)
32. Norman, A.C.: Integration in Finite Terms. In: Buchberger, B., Collins, G.E., Loos, R., Albrecht R. (eds.) Computer Algebra: Symbolic and Algebraic Computation, pp. 57–69. Springer, Wien (1983)
33. Norman, A.C., Moore, P.M.A.: Implementing the new Risch Integration algorithm. In: Proceedings of the 4th International Colloquium on Advanced Computing Methods in Theoretical Physics, Saint-Maximin, pp. 99–110 (1977)
34. Piquette, J.C.: A method for symbolic evaluation of indefinite integrals containing special functions or their products. J. Symb. Comput. **11**, 231–249 (1991)
35. Piquette, J.C., Van Buren, A.L.: Technique for evaluating indefinite integrals involving products of certain special functions. SIAM J. Math. Anal. **15**, 845–855 (1984)
36. Prudnikov, A.P., Brychkov, Y.A., Marichev, O.I.: Integrals and Series, vols. 1–3. Gordon & Breach, New York (1986–1990)
37. Raab, C.G.: Definite integration in differential fields. PhD thesis, JKU Linz (2012)
38. Raab, C.G.: Using Gröbner bases for finding the logarithmic part of the integral of transcendental functions. J. Symb. Comput. **47**, 1290–1296 (2012)
39. Remiddi, E., Vermaseren, J.A.M.: Harmonic polylogarithms. Int. J. Mod. Phys. **A15**, 725–754 (2000). arXiv:hep-ph/9905237
40. Rich, A.D., Jeffrey, D.J.: A knowledge repository for indefinite integration based on transformation rules. In: Proceedings of Calculemus/MKM 2009, Grand Bend, pp. 480–485 (2009)
41. Risch, R.H.: The problem of integration in finite terms. Trans. Am. Math. Soc. **139**, 167–189 (1969)
42. Ritt, J.F.: Integration in Finite Terms – Liouville's Theory of Elementary Methods. Columbia University Press, New York (1948)
43. Rothstein, M.: Aspects of symbolic integration and simplification of exponential and primitive functions. PhD thesis, Univ. of Wisconsin-Madison (1976)
44. Singer, M.F.: Liouvillian solutions of linear differential equations with Liouvillian coefficients. J. Symb. Comput. **11**, 251–273 (1991)
45. Singer, M.F., Saunders, B.D., Caviness, B.F.: An extension of Liouville's theorem on integration in finite terms. SIAM J. Comput. **14**, 966–990 (1985)
46. Trager, B.M.: Algebraic factoring and rational function integration. In: Proceedings of SYMSAC'76, Yorktown Heights, pp. 219–226 (1976)
47. Zeilberger, D.: A holonomic systems approach to special functions identities. J. Comp. Appl. Math. **32**, 321–368 (1990)
48. Zeilberger, D.: A fast algorithm for proving terminating hypergeometric identities. Discret. Math. **80**, 207–211 (1990)

Multiple Hypergeometric Series: Appell Series and Beyond

Michael J. Schlosser

Abstract This survey article provides a small collection of basic material on multiple hypergeometric series of Appell-type and of more general series of related type.

1 Introduction

Hypergeometric series and its various generalizations, in particular such involving *multiple* series, appear in various branches of mathematics and its applications. This survey article features a small collection of selected material on multiple hypergeometric series of *Appell*-type and of more general series of closely related type.

These types of series appear very naturally in quantum field theory, in particular in the computation of analytic expressions for Feynman integrals (for which we kindly refer to other relevant chapters in this volume). Such integrals can be obtained and computed in different ways – which may lead to identities for Appell series (see e.g. M.A. Shpot [30]). On the other hand, the application of known relations for Appell series may lead to simplifications, help to solve problems or lead to more insight in quantum field theory. Therefore it is of importance that people working in this area have a basic understanding of the existing theory for such series.[1] This survey is meant to provide a very digestible, easy introduction to

[1]Researchers working with Feynman integrals who are in demand of effective manipulation of Appell-type series including differential reductions and ϵ-expansions may find HYPERDIRE (located at https://sites.google.com/site/loopcalculations/) useful, which is a set of Wolfram

M.J. Schlosser (✉)
Fakultät für Mathematik, Universität Wien, Nordbergstrasse 15, A-1090 Vienna, Austria
e-mail: michael.schlosser@univie.ac.at

C. Schneider and J. Blümlein (eds.), *Computer Algebra in Quantum Field Theory*, Texts & Monographs in Symbolic Computation, DOI 10.1007/978-3-7091-1616-6_13,

Appell-type series. Besides of recalling some known results, some of the standard mathematical techniques which are used to prove and derive these identities are illustrated. We highlight some of the most fundamental properties and relations for Appell hypergeometric series and further give hints of similar relations for the series which are (slightly) beyond the hierarchy of Appell series. All the series we consider admit very explicit series and integral representations.

To warn the reader: There exist various different types of multivariate hypergeometric series which are not covered in this survey. In particular, here we do not treat multiple hypergeometric series associated with *root systems* (cf. [16, 20, 25, 29]), hypergeometric series of *matrix argument* [15], and other types of multivariate hypergeometric series such as those which appear in the study of *orthogonal polynomials* of several variables (often also associated with root systems) [11, 22].

A very important extension of Appell-type series which is just beyond the scope of this basic survey article are the multivariate hypergeometric functions considered by I.M. Gelfand, M.M. Kapranov, and A.V. Zelevinsky [14], developed in the late 1980s. These A-hypergeometric functions (or GKZ-hypergeometric functions) are fundamental objects in the theory of integrable systems as they are the holonomic solutions of a (certain) A-hypergeometric system of partial differential equations. Natural questions regarding algebraic solutions and monodromy for A-hypergeometric functions have been recently addressed by F. Beukers [4, 5].

For *basic* (or *q-series*) analogues of Appell functions, see [13, Chap. 10].

2 Appell Series

Appell series are a natural two-variable extension of hypergeometric series. They are treated with detail in Érdelyi et al. [12], the classical reference for special functions.

In the following, we follow to great extent the expositions from the classical texts of W.N. Bailey [3], and L.J. Slater [31] (both contain a great amount of material on hypergeometric series).

For convenience, we use the Pochhammer symbol notation for the shifted factorial,

$$(a)_n := \begin{cases} a(a+1)\dots(a+n-1) & \text{if } n = 1,2,\dots, \\ 1 & \text{if } n = 0. \end{cases} \tag{1a}$$

Accordingly, we have

$$(a)_n = \frac{\Gamma(a+n)}{\Gamma(a)} \tag{1b}$$

Mathematica based programs for differential reduction of Horn-type hypergeometric functions, see V. Bytev et al. [9].

which is used as a definition for the shifted factorial in case n is not necessarily a nonnegative integer.

The goal is to generalize the Gauß hypergeometric function

$$ {}_2F_1\left(\begin{matrix} a,\ b \\ c \end{matrix}; x\right) = \sum_{n\geq 0} \frac{(a)_n\,(b)_n}{n!\,(c)_n} x^n $$

to a double series depending on two variables.

The easiest is to consider the simple product

$$ {}_2F_1\left(\begin{matrix} a,\ b \\ c \end{matrix}; x\right) {}_2F_1\left(\begin{matrix} a',\ b' \\ c' \end{matrix}; y\right) = \sum_{m\geq 0}\sum_{n\geq 0} \frac{(a)_m\,(a')_n\,(b)_m\,(b')_n}{m!\,n!\,(c)_m\,(c')_n} x^m y^n, $$

where on the right-hand side the indices m, n appear uncoupled.

To consider a genuine double series instead (which does not factor into a simple product of two series), we now deliberately choose to replace one or more of the three products $(a)_m\,(a')_n$, $(b)_m\,(b')_n$, $(c)_m\,(c')_n$ by products of coupled type $(a)_{m+n}$ (other choices such as $(a)_{m-n}$ or $(a)_{2m-n}$, etc., instead, may be sensible as well; they lead to Horn-type series, see Sect. 3.1).

There are five different possibilities, one of which by application of the binomial theorem gives the series

$$ \sum_{m\geq 0}\sum_{n\geq 0} \frac{(a)_{m+n}\,(b)_{m+m}}{m!\,n!\,(c)_{m+n}} x^m y^n = {}_2F_1\left(\begin{matrix} a,\ b \\ c \end{matrix}; x+y\right), $$

i.e., an ordinary hypergeometric series.

The other four remaining possibilities are classified as F_1-, F_2-, F_3-, and F_4-series (cf. P. Appell [1] and P. Appell and M.-J. Kampé de Fériet [2]):

$$ F_1(a;b,b';c;x,y) := \sum_{m\geq 0}\sum_{n\geq 0} \frac{(a)_{m+n}\,(b)_m\,(b')_n}{m!\,n!\,(c)_{m+n}} x^m y^n, \qquad |x|,|y|<1. \tag{2a} $$

$$ F_2(a;b,b';c,c';x,y) := \sum_{m\geq 0}\sum_{n\geq 0} \frac{(a)_{m+n}\,(b)_m\,(b')_n}{m!\,n!\,(c)_m\,(c')_n} x^m y^n, \qquad |x|+|y|<1. \tag{2b} $$

$$ F_3(a,a';b,b';c;x,y) := \sum_{m\geq 0}\sum_{n\geq 0} \frac{(a)_m\,(a')_n\,(b)_m\,(b')_n}{m!\,n!\,(c)_{m+n}} x^m y^n, \qquad |x|,|y|<1. \tag{2c} $$

$$ F_4(a;b;c,c';x,y) := \sum_{m\geq 0}\sum_{n\geq 0} \frac{(a)_{m+n}\,(b)_{m+n}}{m!\,n!\,(c)_m\,(c')_n} x^m y^n, \qquad |x|^{\frac{1}{2}}+|y|^{\frac{1}{2}}<1. \tag{2d} $$

One immediately observes the following simple identities:

$$F_1(a;b,b';c;x,y) = \sum_{m\geq 0} \frac{(a)_m\,(b)_m}{m!\,(c)_m} x^m\, {}_2F_1\left(\begin{matrix} a+m,\, b' \\ c+m \end{matrix}; y\right). \tag{3}$$

$$F_1(a;b,b';c;x,0) = F_2(a;b,b';c,c';x,0) = F_3(a,a';b,b';c;x,0) \tag{4a}$$

$$= F_4(a;b;c,c';x,0) = {}_2F_1\left(\begin{matrix} a,\, b \\ c \end{matrix}; x\right). \tag{4b}$$

$$F_1(a;b,0;c;x,y) = F_2(a;b,0;c,c';x,y) = F_3(a,a';b,0;c;x,y) \tag{5a}$$

$$= {}_2F_1\left(\begin{matrix} a,\, b \\ c \end{matrix}; x\right). \tag{5b}$$

Using ideas of N.Ja. Vilenkin [32], W. Miller, Jr. [23] has given a Lie theoretic interpretation of the Appell functions F_1. In particular, he showed that $sl(5,\mathbb{C})$ is the dynamical symmetry algebra for the F_1.

2.1 Contiguous Relations and Recursions

All contiguous relations for the F_1 function can be derived from these four relations:

$$(a-b-b')\,F_1(a;b,b';c;x,y) - a\,F_1(a+1;b,b';c;x,y)$$
$$+b\,F_1(a;b+1,b';c;x,y) + b'\,F_1(a;b,b'+1;c;x,y) = 0, \tag{6a}$$

$$c\,F_1(a;b,b';c;x,y) - (c-a)\,F_1(a;b,b';c+1;x,y)$$
$$-a\,F_1(a+1;b,b';c+1;x,y) = 0, \tag{6b}$$

$$c\,F_1(a;b,b';c;x,y) + c(x-1)\,F_1(a;b+1,b';c;x,y)$$
$$-(c-a)x\,F_1(a;b+1,b';c+1;x,y) = 0, \tag{6c}$$

$$c\,F_1(a;b,b';c;x,y) + c(y-1)\,F_1(a;b,b'+1;c;x,y)$$
$$-(c-a)y\,F_1(a;b,b'+1;c+1;x,y) = 0. \tag{6d}$$

Similar sets of relations exist for the other Appell functions, see R.G. Buschman [8].

Recently, X. Wang [35] has used contiguous relations and induction to derive various recursion formulae for all the Appell functions F_1, F_2, F_3, F_4. (Some of the recursions for F_2 were previously given by S.B. Opps, N. Saad and

H.M. Srivastava [26].) For $n = 1$ these recursions reduce to equivalent forms of the known contiguous relations.

In particular, for F_1 we have

$$F_1(a+n;b,b';c;x,y) = F_1(a;b,b';c;x,y) + \frac{bx}{c}\sum_{k=1}^{n} F_1(a+k;b+1,b';c+1;x,y) + \frac{b'y}{c}\sum_{k=1}^{n} F_1(a+k;b,b'+1;c+1;x,y), \tag{7a}$$

$$F_1(a-n;b,b';c;x,y) = F_1(a;b,b';c;x,y) - \frac{bx}{c}\sum_{k=1}^{n-1} F_1(a-k;b+1,b';c+1;x,y) - \frac{b'y}{c}\sum_{k=1}^{n-1} F_1(a-k;b,b'+1;c+1;x,y), \tag{7b}$$

$$F_1(a;b+n,b';c;x,y) = F_1(a;b,b';c;x,y) + \frac{ax}{c}\sum_{k=1}^{n} F_1(a+1;b+k,b';c+1;x,y), \tag{7c}$$

$$F_1(a;b-n,b';c;x,y) = F_1(a;b,b';c;x,y) - \frac{ax}{c}\sum_{k=1}^{n-1} F_1(a+1;b-k,b';c+1;x,y), \tag{7d}$$

$$F_1(a;b,b';c-n;x,y) = F_1(a;b,b';c;x,y) + abx\sum_{k=1}^{n} \frac{F_1(a+1;b+1,b';c-k+2;x,y)}{(c-k)(c-k+1)} + ab'y\sum_{k=1}^{n} \frac{F_1(a+1;b,b'+1;c-k+2;x,y)}{(c-k)(c-k+1)}. \tag{7e}$$

For F_2 we have

$$F_2(a+n;b,b';c,c';x,y) = F_2(a;b,b';c,c';x,y) + \frac{bx}{c}\sum_{k=1}^{n} F_2(a+k;b+1,b';c+1,c';x,y) + \frac{b'y}{c'}\sum_{k=1}^{n} F_2(a+k;b,b'+1;c,c'+1;x,y), \tag{8a}$$

$$F_2\bigl(a-n;b,b';c,c';x,y\bigr) = F_2\bigl(a;b,b';c,c';x,y\bigr) - \frac{bx}{c}\sum_{k=1}^{n-1} F_2\bigl(a-k;b+1,b';c+1,c';x,y\bigr) - \frac{b'y}{c'}\sum_{k=1}^{n-1} F_2\bigl(a+k;b,b'+1;c,c'+1;x,y\bigr), \tag{8b}$$

$$F_2\bigl(a;b+n,b';c,c';x,y\bigr) = F_2\bigl(a;b,b';c,c';x,y\bigr) + \frac{ax}{c}\sum_{k=1}^{n} F_2\bigl(a+1;b+k,b';c+1,c';x,y\bigr), \tag{8c}$$

$$F_2\bigl(a;b-n,b';c,c';x,y\bigr) = F_2\bigl(a;b,b';c,c';x,y\bigr) - \frac{ax}{c}\sum_{k=1}^{n-1} F_2\bigl(a+1;b-k,b';c+1,c';x,y\bigr), \tag{8d}$$

$$F_2\bigl(a;b,b';c-n,c';x,y\bigr) = F_2\bigl(a;b,b';c,c';x,y\bigr) + abx\sum_{k=1}^{n} \frac{F_2\bigl(a+1;b+1,b';c-k+2,c';x,y\bigr)}{(c-k)(c-k+1)}. \tag{8e}$$

For F_3 we have

$$F_3\bigl(a+n,a';b,b';c;x,y\bigr) = F_3\bigl(a,a';b,b';c;x,y\bigr) + \frac{bx}{c}\sum_{k=1}^{n} F_3\bigl(a+k,a';b+1,b';c+1;x,y\bigr), \tag{9a}$$

$$F_3\bigl(a-n,a';b,b';c;x,y\bigr) = F_3\bigl(a,a';b,b';c;x,y\bigr) - \frac{bx}{c}\sum_{k=1}^{n-1} F_3\bigl(a-k,a';b+1,b';c+1;x,y\bigr), \tag{9b}$$

$$F_3\bigl(a,a';b,b';c-n;x,y\bigr) = F_3\bigl(a,a';b,b';c;x,y\bigr) + abx\sum_{k=1}^{n} \frac{F_3\bigl(a+1,a';b+1,b';c-k+2;x,y\bigr)}{(c-k)(c-k+1)} + a'b'y\sum_{k=1}^{n} \frac{F_3\bigl(a,a'+1;b,b'+1;c-k+2;x,y\bigr)}{(c-k)(c-k+1)}. \tag{9c}$$

Finally, for F_4 we have

$$F_4\big(a+n;b;c,c';x,y\big) = F_4\big(a;b;c,c';x,y\big) + \frac{bx}{c}\sum_{k=1}^{n} F_4\big(a+k;b+1;c+1,c';x,y\big) + \frac{by}{c'}\sum_{k=1}^{n} F_4\big(a+k;b+1;c,c'+1;x,y\big), \tag{10a}$$

$$F_4\big(a-n;b;c,c';x,y\big) = F_4\big(a;b;c,c';x,y\big) - \frac{bx}{c}\sum_{k=1}^{n-1} F_4\big(a-k;b+1;c+1,c';x,y\big) - \frac{by}{c'}\sum_{k=1}^{n-1} F_4\big(a-k;b+1;c,c'+1;x,y\big), \tag{10b}$$

$$F_4\big(a;b;c-n,c';x,y\big) = F_4\big(a;b;c,c';x,y\big) + abx\sum_{k=1}^{n-1} \frac{F_4\big(a+1;b+1;c-k+1,c';x,y\big)}{(c-k)(c-k-1)}. \tag{10c}$$

Most of these recursions can be extended to elegant recursions involving more terms. For instance,

$$F_1\big(a+n;b,b';c;x,y\big) = \sum_{i=0}^{n}\sum_{k=0}^{n-i}\binom{n}{i}\binom{n-i}{k}\frac{(b)_i(b')_k}{(c)_{k+i}} \times x^i y^j F_1\big(a+i+k;b+i,b'+k;c+i+k;x,y\big), \tag{11a}$$

$$F_1\big(a-n;b,b';c;x,y\big) = \sum_{i=0}^{n}\sum_{k=0}^{n-i}\binom{n}{i}\binom{n-i}{k}\frac{(b)_i(b')_k}{(c)_{k+i}} \times (-x)^i(-y)^j F_1\big(a;b+i,b'+k;c+i+k;x,y\big), \tag{11b}$$

$$F_1\big(a;b+n,b';c;x,y\big) = \sum_{k=0}^{n}\binom{n}{k}\frac{(a)_k}{(c)_k}x^k F_1\big(a+k;b+k,b';c+k;x,y\big), \tag{11c}$$

$$F_1\big(a;b-n,b';c;x,y\big)=\sum_{k=0}^{n}\binom{n}{k}\frac{(a)_k}{(c)_k}(-x)^k F_1\big(a+k;b,b';c+k;x,y\big), \tag{11d}$$

or

$$F_4\big(a;b;c-n,c';x,y\big)=\sum_{k=0}^{n}\binom{n}{k}\frac{(a)_k(b)_k}{(c)_k(c-n)_k}x^k F_4\big(a+k;b+k;c+k,c';x,y\big). \tag{12}$$

2.2 Partial Differential Equations

Let

$$z=F_1\big(a;b,b';c;x,y\big)=\sum_{m\geq 0}\sum_{n\geq 0}A_{m,n}x^m y^n.$$

Then

$$A_{m+1,n}=\frac{(a+m+n)(b+m)}{(1+m)(c+m+n)}A_{m,n},$$

and

$$A_{m,n+1}=\frac{(a+m+n)(b'+n)}{(1+n)(c+m+n)}A_{m,n}.$$

Denoting the partial differential operators by

$$\theta=x\frac{\partial}{\partial x}\qquad\text{and}\qquad\phi=y\frac{\partial}{\partial y},$$

we readily see that $z=F_1$ satisfies the partial differential equations

$$\Big[(\theta+\phi+a)(\theta+b)-\frac{1}{x}\theta(\theta+\phi+c-1)\Big]z=0, \tag{13a}$$

$$\Big[(\theta+\phi+a)(\phi+b')-\frac{1}{x}\phi(\theta+\phi+c-1)\Big]z=0. \tag{13b}$$

Now let

$$p=\frac{\partial z}{\partial x},\quad q=\frac{\partial z}{\partial y},\quad r=\frac{\partial z}{\partial x}\frac{\partial z}{\partial y},\quad s=\frac{\partial z^2}{\partial x^2},\quad t=\frac{\partial z^2}{\partial y^2}.$$

Then $z = F_1$ satisfies the partial differential equations

$$x(1-x)r + y(1-x)s + [c-(a+b+1)x]p - byq - abz = 0, \tag{14a}$$

$$y(1-y)t + x(1-y)s + [c-(a+b'+1)y]q - b'xp - ab'z = 0. \tag{14b}$$

Similarly, $z = F_2$ satisfies the partial differential equations

$$x(1-x)r - xys + [c-(a+b+1)x]p - byq - abz = 0, \tag{15a}$$

$$y(1-y)t - xys + [c'-(a+b'+1)y]q - b'xp - ab'z = 0. \tag{15b}$$

Similarly, $z = F_3$ satisfies the partial differential equations

$$x(1-x)r + ys + [c-(a+b+1)x]p - abz = 0, \tag{16a}$$

$$y(1-y)t + xs + [c-(a'+b'+1)y]q - a'b'z = 0. \tag{16b}$$

Finally, $z = F_4$ satisfies the partial differential equations

$$x(1-x)r - y^2t - 2xys + cp - (a+b+1)(xp+yq) - abz = 0, \tag{17a}$$

$$y(1-y)t - x^2r - 2xys + c'q - (a+b+1)(xp+yq) - abz = 0. \tag{17b}$$

2.3 *Integral Representations*

Integral representations for Appell series are very useful. Substitution of variables in theses integrals lead to equivalent integrals. This provides an effective and easy method to derive transformation formulae for Appell series, see Sect. 2.4.

Consider the integral

$$I = \iint u^{b-1} v^{b'-1} (1-u-v)^{c-b-b'-1} (1-ux-vy)^{-a}\, du\, dv,$$

taken over the triangular region $u \geq 0,\ v \geq 0,\ u+v \leq 1$. (We implicitly assume suitable conditions of the parameters a, b, b', c such that the integral is well-defined and converges.)

Now, provided $|vy/(1-ux)| < 1$, we have, by binomial expansion,

$$(1-ux-vy)^{-a} = (1-ux)^{-a} \sum_{m \geq 0} \frac{(a)_m}{m!} \left(\frac{vy}{1-ux}\right)^m$$

$$= \sum_{m\geq 0} \frac{(a)_m}{m!} v^m y^m (1-ux)^{-a-m}$$

$$= \sum_{m\geq 0} \frac{(a)_m}{m!} v^m y^m \sum_{n\geq 0} \frac{(a+m)_n}{n!} u^n x^n.$$

Thus,

$$I = \sum_{m\geq 0}\sum_{n\geq 0} \frac{(a)_{m+n}}{m!n!} x^n y^m \iint u^{b-1+n} v^{b'-1+m} (1-u-v)^{c-b-b'-1}\, du\, dv$$

$$= \sum_{m\geq 0}\sum_{n\geq 0} \frac{(a)_{m+n}}{m!n!} x^n y^m\, \Gamma\begin{bmatrix} b+n, b'+m, c-b-b' \\ c+m+n \end{bmatrix},$$

which yields

$$I = \Gamma\begin{bmatrix} b, b', c-b-b' \\ c \end{bmatrix} F_1\big(a; b, b'; c; x, y\big). \tag{18}$$

While I is a double integral, a *single integral* for F_1 even exists, see (22).

Similarly,

$$\int_0^1\int_0^1 u^{b-1} v^{b'-1} (1-u)^{c-b'-1} (1-v)^{c'-b'-1} (1-ux-vy)^{-a}\, du\, dv$$

$$= \Gamma\begin{bmatrix} b, b', c-b, c'-b' \\ c,\ c' \end{bmatrix} F_2\big(a; b, b'; c, c'; x, y\big), \tag{19}$$

and

$$\iint u^{b-1} v^{b'-1} (1-u-v)^{c-b-b'-1} (1-ux)^{-a} (1-vy)^{-a'}\, du\, dv$$

$$= \Gamma\begin{bmatrix} b, b', c-b-b' \\ c' \end{bmatrix} F_3\big(a, a'; b, b'; c'; x, y\big), \tag{20}$$

the last integral taken over the triangular region $u \geq 0$, $v \geq 0$, $u+v \leq 1$.

The double integral for F_4 is more complicated:

$$\int_0^1\int_0^1 u^{a-1} v^{b-1} (1-u)^{c-a-1} (1-v)^{c'-b-1} (1-ux)^{-b} (1-vy)^{-a}$$

$$\times \left(1 - \frac{uvxy}{(1-ux)(1-vy)}\right)^{c+c'-a-b-1} du\, dv$$

$$= \Gamma\begin{bmatrix} a, b, c-a, c'-b \\ c,\ c' \end{bmatrix} F_4\big(a; b; c, c'; x(1-y), y(1-x)\big). \tag{21}$$

In 1881, É. Picard [27] discovered a single integral for F_1. Let

$$I' = \int_0^1 u^{a-1}(1-u)^{c-a-1}(1-ux)^{-b}(1-uy)^{-b'}\,du,$$

where $\Re c > \Re a > 0$. Then

$$\begin{aligned} I' &= \sum_{m\geq 0}\sum_{n\geq 0}\int_0^1 u^{a-1}(1-u)^{c-a-1}\frac{(b)_m}{m!}u^m x^m \frac{(b')_n}{n!}u^n y^n\,du \\ &= \sum_{m\geq 0}\sum_{n\geq 0}\frac{(b)_m(b')_n}{m!n!}x^m y^n \int_0^1 u^{a+m+n-1}(1-u)^{c-a-1}\,du \\ &= \sum_{m\geq 0}\sum_{n\geq 0}\frac{(b)_m(b')_n}{m!n!}x^m y^n\, \Gamma\begin{bmatrix} a+m+n, c-a \\ c+m+n \end{bmatrix}, \end{aligned}$$

hence

$$I' = \Gamma\begin{bmatrix} a, c-a \\ c \end{bmatrix} F_1\big(a;b,b';c;x,y\big). \tag{22}$$

2.3.1 Incomplete Elliptic Integrals

As immediate consequences of (22), it follows that the incomplete elliptic integrals F and E and the complete elliptic integral Π can all be expressed in terms of special cases of the Appell F_1 function:

$$\begin{aligned} F(\phi,k) :&= \int_0^\phi \frac{d\theta}{\sqrt{1-k^2\sin^2\theta}} \\ &= \sin\phi\; F_1\!\left(\frac{1}{2};\frac{1}{2},\frac{1}{2};\frac{3}{2};\sin^2\phi, k^2\sin^2\phi\right), \qquad |\Re\phi| < \frac{\pi}{2}, \end{aligned} \tag{23a}$$

$$\begin{aligned} E(\phi,k) :&= \int_0^\phi \sqrt{1-k^2\sin^2\theta}\,d\theta \\ &= \sin\phi\; F_1\!\left(\frac{1}{2};\frac{1}{2},-\frac{1}{2};\frac{3}{2};\sin^2\phi, k^2\sin^2\phi\right), \qquad |\Re\phi| < \frac{\pi}{2}, \end{aligned} \tag{23b}$$

$$\Pi(n,k) := \int_0^{\pi/2} \frac{d\theta}{(1-n\sin^2\theta)\sqrt{1-k^2\sin^2\theta}} = \frac{\pi}{2}\, F_1\!\left(\frac{1}{2};1,\frac{1}{2};1;n,k^2\right). \tag{23c}$$

2.4 Transformations

In the single integral for the F_1 series,

$$F_1(a;b,b';c;x,y) = \Gamma\begin{bmatrix} c \\ a, c-a \end{bmatrix} \int_0^1 u^{a-1}(1-u)^{c-a-1}(1-ux)^{-b}(1-uy)^{-b'}\,du,$$

one may use the substitution of variables $u = 1 - v$ to prove

$$F_1(a;b,b';c;x,y) = (1-x)^{-b}(1-y)^{-b'} F_1\left(c-a;b,b';c;\frac{x}{x-1},\frac{y}{y-1}\right). \tag{24}$$

For $b' = 0$ this reduces to the well-known *Pfaff–Kummer transformation* for the ${}_2F_1$:

$${}_2F_1\left(\begin{matrix} a, b \\ c \end{matrix};x\right) = (1-x)^{-b}\,{}_2F_1\left(\begin{matrix} c-a, b \\ c \end{matrix};\frac{x}{x-1}\right).$$

Similarly, the substitution of variables $u = \frac{v}{1-x+vx}$ can be used to prove

$$F_1(a;b,b';c;x,y) = (1-x)^{-a} F_1\left(a;-b-b'+c,b';c;\frac{x}{x-1},\frac{y-x}{1-x}\right). \tag{25}$$

For $b' = 0$ this reduces again to the Pfaff–Kummer transformation for the ${}_2F_1$ series.

On the other hand, if $c = b + b'$, then

$$F_1(a;b,b';b+b';x,y) = (1-x)^{-a}\,{}_2F_1\left(\begin{matrix} a, b' \\ b+b' \end{matrix};\frac{y-x}{1-x}\right) \tag{26a}$$

$$= (1-y)^{-a}\,{}_2F_1\left(\begin{matrix} a, b \\ b+b' \end{matrix};\frac{x-y}{1-y}\right). \tag{26b}$$

Similarly,

$$F_1(a;b,b';c;x,y) = (1-y)^{-a} F_1\left(a;b,c-b-b';c;\frac{x-y}{1-y},\frac{y}{y-1}\right), \tag{27}$$

$$F_1(a;b,b';c;x,y) = (1-x)^{c-a-b}(1-y)^{-b'} F_1\left(c-a;c-b-b',b';c;x,\frac{x-y}{1-y}\right), \tag{28}$$

$$F_1(a;b,b';c;x,y) = (1-x)^{-b}(1-y)^{c-a-b'} F_1\left(c-a;b,c-b-b';c;\frac{y-x}{1-x},y\right). \tag{29}$$

Further,

$$F_2\big(a;b,b';c,c';x,y\big) = (1-x)^{-a} F_2\Big(a;c-b,b';c,c';\frac{x}{x-1},\frac{y}{1-x}\Big), \tag{30}$$

$$F_2\big(a;b,b';c,c';x,y\big) = (1-y)^{-a} F_2\Big(a;b,c'-b';c,c';\frac{x}{1-y},\frac{y}{y-1}\Big), \tag{31}$$

$$F_2\big(a;b,b';c,c';x,y\big) = (1-x-y)^{-a} F_2\Big(a;c-a,c'-b';c,c';\tfrac{x}{x+y-1},\tfrac{y}{x+y-1}\Big). \tag{32}$$

Also *quadratic transformations* are known for Appell functions, see B.C. Carlson [10].

2.5 Reduction Formulae

The transformations of Sect. 2.4 readily imply the following reduction formulae (typically a double series being reduced to a single series):

- $y = x$ in F_1:

$$F_1\big(a;b,b';c;x,x\big) = (1-x)^{c-a-b-b'}\,{}_2F_1\left(\begin{matrix}c-a,\ c-b-b'\\ c\end{matrix};x\right). \tag{33a}$$

  By Euler's transformation this is

$$F_1\big(a;b,b';c;x,x\big) = {}_2F_1\left(\begin{matrix}a,\ b+b'\\ c\end{matrix};x\right). \tag{33b}$$

- $c = b + b'$ in F_1:

$$F_1\big(a;b,b';b+b';x,y\big) = (1-y)^{-a}\,{}_2F_1\left(\begin{matrix}a,\ b\\ b+b'\end{matrix};\frac{x-y}{1-y}\right). \tag{34}$$

- $c = b$ in F_2:

$$F_2\big(a;b,b';b,c';x,y\big) = (1-x)^{-a}\,{}_2F_1\left(\begin{matrix}a,\ b'\\ c'\end{matrix};\frac{y}{1-x}\right). \tag{35}$$

- $y = 1$ in F_1:
  Since

$$F_1\big(a;b,b';c;x,y\big) = \sum_{m\ge 0} \frac{(a)_m\,(b)_m}{m!\,(c)_m} x^m\,{}_2F_1\left(\begin{matrix}a+m,\ b'\\ c+m\end{matrix};y\right)$$

and

$$ {}_2F_1\left(\begin{matrix} a, b \\ c \end{matrix}; 1\right) = \Gamma\left[\begin{matrix} c, c-a-b \\ c-a, c-b \end{matrix}\right], \qquad \Re(c-a-b) > 0, $$

we have

$$ F_1\big(a; b, b'; c; x, 1\big) = \Gamma\left[\begin{matrix} c, c-a-b' \\ c-a, c-b' \end{matrix}\right] {}_2F_1\left(\begin{matrix} a, b \\ c-b' \end{matrix}; x\right), \tag{36} $$

for $\Re(c-a-b') > 0$.

- An $F_1 \leftrightarrow F_3$ transformation:

  Since

$$ F_1\big(a; b, b'; c; x, y\big) = \sum_{m \geq 0} \frac{(a)_m (b)_m}{m!\,(c)_m} x^m \, {}_2F_1\left(\begin{matrix} a+m, b' \\ c+m \end{matrix}; y\right) $$

and

$$ {}_2F_1\left(\begin{matrix} a, b \\ c \end{matrix}; y\right) = (1-y)^{-b} \, {}_2F_1\left(\begin{matrix} c-a, b \\ c \end{matrix}; \frac{y}{y-1}\right), $$

we have

$$ \begin{aligned} F_1\big(a; b, b'; c; x, y\big) &= (1-y)^{-b'} \sum_{m \geq 0} \frac{(a)_m (b)_m}{m!\,(c)_m} x^m \, {}_2F_1\left(\begin{matrix} c-a, b' \\ c+m \end{matrix}; \frac{y}{y-1}\right) \\ &= (1-y)^{-b'} F_3\left(a, c-a; b, b'; c; x, \frac{y}{y-1}\right). \end{aligned} \tag{37} $$

Hence, any F_1 function can be expressed in terms of an F_3 function. The converse is only true when $c = a + a'$.

- $a' = c - a$ and $b' = c - b$ in F_3:

  Since by Eq. (34) the F_1 function reduces to an ordinary ${}_2F_1$ function when $c = b + b'$, we have

$$ F_3\left(a, c-a; b, c-b; c; x, \frac{y}{y-1}\right) = (1-x)^{-a}(1-y)^{c-b} \, {}_2F_1\left(\begin{matrix} a, c-b \\ c \end{matrix}; \frac{y-x}{1-x}\right). \tag{38} $$

- $c' = a$ in F_2:

$$ F_2\big(a; b, b'; c, a; x, y\big) = (1-y)^{-b'} F_1\left(b; a-b', b'; c; x, \frac{x}{1-y}\right). \tag{39} $$

Conversely, any F_1 function can be expressed in terms of an F_2 function where $c' = a$.

If further $c = a$, then

$$F_2\big(a; b, b'; a, a; x, y\big) = (1-x)^{-b}(1-y)^{-b'}\,{}_2F_1\!\left(\begin{matrix} b, b' \\ a \end{matrix}; \frac{xy}{(1-x)(1-y)}\right). \tag{40}$$

2.6 An Expansion of an F_4 Series

In 1940 and 1941, J.L. Burchnall and T.W. Chaundy [6, 7] gave the following expansion of an F_4 series in terms of products of two hypergeometric ${}_2F_1$ series:

$$\begin{aligned} &F_4\big(a; b; c, c'; x(1-y), y(1-x)\big) \\ &\qquad = \sum_{m\geq 0} \frac{(a)_m\,(b)_m\,(1+a+b-c-c')_m}{m!\,(c)_m\,(c')_m} x^m y^m \\ &\qquad\quad \times {}_2F_1\!\left(\begin{matrix} a+m, b+m \\ c+m \end{matrix}; x\right) {}_2F_1\!\left(\begin{matrix} a+m, b+m \\ c'+m \end{matrix}; y\right). \end{aligned} \tag{41}$$

This expansion has applications to classical orthogonal polynomials. It can also be used to deduce the double integral representation for F_4. Various special cases are interesting enough to state separately:

- $c' = 1 + a + b - c$ in F_4:
  We have the product formula

$$F_4\big(a; b; c, 1+a+b-c; x(1-y), y(1-x)\big) = {}_2F_1\!\left(\begin{matrix} a, b \\ c \end{matrix}; x\right) {}_2F_1\!\left(\begin{matrix} a, b \\ c' \end{matrix}; y\right). \tag{42}$$

- $c' = b$ in F_4:
  Here we have the reduction formula

$$\begin{aligned} &F_4\big(a; b; c, b; x(1-y), y(1-x)\big) \\ &= (1-x)^{-a}(1-y)^{-a}\,F_1\!\left(a; 1+a-c, c-b; c; \frac{xy}{(1-x)(1-y)}, \frac{x}{x-1}\right). \end{aligned} \tag{43}$$

- $c' = b$ and $c = a$ in F_4:
  Further specialization of (43) gives the quite attractive summation formula

$$F_4\big(a; b; a, b; x(1-y), y(1-x)\big) = (1-x)^{1-b}(1-y)^{1-a}(1-x-y)^{-1}. \tag{44a}$$

Written out in explicit terms, this is

$$\sum_{m\geq 0}\sum_{n\geq 0}\frac{(a)_{m+n}\,(b)_{m+n}}{m!\,n!\,(a)_m\,(b)_n}x^m(1-y)^m y^n(1-x)^n = \frac{(1-x)^{1-b}(1-y)^{1-a}}{(1-x-y)}. \tag{44b}$$

For $y=0$ this reduces to I. Newton's binomial expansion formula

$${}_1F_0\left(\begin{matrix}b\\ -\end{matrix};x\right)=(1-x)^{-b}.$$

3 Related Series and Extensions of Appell Series

3.1 Horn Functions

In 1931, Jacob Horn [17] studied convergent bivariate hypergeometric functions $\sum_{m,n} f_{m,n}x^m y^n$ with certain (degree and other) restrictions on the two ratios of consecutive terms

$$\frac{f_{m+1,n}}{f_{m,n}}, \qquad \frac{f_{m,n+1}}{f_{m,n}}.$$

He arrived at a complete set of 34 different functions among which are the Appell functions F_1, F_2, F_3, F_4.

They include series such as

$$G_1(a,b,b';x,y):=\sum_{m\geq 0}\sum_{n\geq 0}\frac{(a)_{m+n}(b)_{n-m}(b')_{m-n}}{m!n!}x^m y^n, \tag{45}$$

$$H_3(a,b,c;x,y):=\sum_{m\geq 0}\sum_{n\geq 0}\frac{(a)_{2m+n}(b)_n}{(c)_{m+n}\,m!n!}x^m y^n, \tag{46}$$

and

$$H_7(a,b,b',c;x,y):=\sum_{m\geq 0}\sum_{n\geq 0}\frac{(a)_{2m-n}(b)_n(b')_n}{(c)_m\,m!n!}x^m y^n. \tag{47}$$

3.2 Kampé de Fériet Series

In 1937, J. Kampé de Fériet [18] introduced the following bivariate extension of the generalized hypergeometric series:

$$F^{p:q}_{r:s}\left(\begin{matrix}a_1,\ldots,a_p : b_1,b'_1;\ldots;b_q,b'_q;\\ c_1,\ldots,c_r : d_1,d'_1;\ldots;d_s,d'_s;\end{matrix}\, x,y\right)$$

$$=\sum_{m\ge 0}\sum_{n\ge 0}\frac{(a_1)_{m+n}\ldots(a_p)_{m+n}}{(c_1)_{m+n}\ldots(c_r)_{m+n}}\frac{(b_1)_m(b'_1)_n\ldots(b_q)_m(b'_q)_n}{(d_1)_m(d'_1)_n\ldots(d_s)_m(d'_s)_n}\frac{x^m y^n}{m!n!}. \tag{48}$$

Numerous identities exist for special instances of such series. For illustration, we list three summation formulae.

- P.W. Karlsson [19], 1994:

$$F^{0:3}_{1:1}\left(\begin{matrix}- : a,d-a;b,d-b;c,-c;\\ d : e,d+e-a-b-c;\end{matrix}\,1,1\right)=\Gamma\left[\begin{matrix}e,\ e+d-a-b-c\\ e-c,\ e+d-a-b\end{matrix}\right], \tag{49}$$

where $\Re(e)>0$ and $\Re(d+e-a-b-c)>0$.

- S.N. Pitre and J. Van der Jeugt [28], 1996:

$$F^{0:3}_{1:1}\left(\begin{matrix}- : a,d-a;b,d-b;c,d-c;\\ d : e,d+e-a-b-c;\end{matrix}\,1,1\right)=\Gamma\left[\begin{matrix}e,e+d-a-b-c,e-d\\ e-a,\ e-b,\ e-c\end{matrix}\right], \tag{50}$$

where $\Re(e-d)>0$ and $\Re(d+e-a-b-c)>0$. Further

$$F^{0:3}_{1:1}\left(\begin{matrix}- : a,d-a;b,d-b;c,e-c-1;\\ d : e,d+e-a-b-c;\end{matrix}\,1,1\right)$$

$$=\Gamma\left[\begin{matrix}1-a,\ 1-b,\ e,\ e-d,\ d+e-a-b-c\\ 1-d,\ e-a,\ e-b,\ e-c,\ 1+d-a-b\end{matrix}\right], \tag{51}$$

where $\Re(d+e-a-b-c)>0$, and $d-a$ or $d-b$ is a negative integer.

3.3 Lauricella Series

In 1893, G. Lauricella [21] investigated properties of the following four series $F_A^{(n)}$, $F_B^{(n)}$, $F_C^{(n)}$, $F_D^{(n)}$, of n variables:

$$F_A^{(n)}\big(a;b_1,\dots,b_n;c_1,\dots,c_n;x_1,\dots,x_n\big)$$
$$=\sum_{m_1\geq 0}\cdots\sum_{m_n\geq 0}\frac{(a)_{m_1+\cdots+m_n}(b_1)_{m_1}\dots(b_n)_{m_n}}{(c_1)_{m_1}\dots(c_n)_{m_n}\,m_1!\dots m_n!}x_1^{m_1}\dots x_n^{m_n}, \tag{52}$$

where $|x_1|+\cdots+|x_n|<1$.

$$F_B^{(n)}\big(a_1,\dots,a_n;b_1,\dots,b_n;c;x_1,\dots,x_n\big)$$
$$=\sum_{m_1\geq 0}\cdots\sum_{m_n\geq 0}\frac{(a_1)_{m_1}\dots(a_n)_{m_n}(b_1)_{m_1}\dots(b_n)_{m_n}}{(c)_{m_1+\cdots+m_n}\,m_1!\dots m_n!}x_1^{m_1}\dots x_n^{m_n}, \tag{53}$$

where $|x_1|,\dots,|x_n|<1$.

$$F_C^{(n)}\big(a;b;c_1,\dots,c_n;x_1,\dots,x_n\big)$$
$$=\sum_{m_1\geq 0}\cdots\sum_{m_n\geq 0}\frac{(a)_{m_1+\cdots+m_n}(b)_{m_1+\cdots+m_n}}{(c_1)_{m_1}\dots(c_n)_{m_n}\,m_1!\dots m_n!}x_1^{m_1}\dots x_n^{m_n}, \tag{54}$$

where $|x_1|^{\frac{1}{2}}+\cdots+|x_n|^{\frac{1}{2}}<1$.

$$F_D^{(n)}\big(a;b_1,\dots,b_n;c;x_1,\dots,x_n\big)$$
$$=\sum_{m_1\geq 0}\cdots\sum_{m_n\geq 0}\frac{(a)_{m_1+\cdots+m_n}(b_1)_{m_1}\dots(b_n)_{m_n}}{(c)_{m_1+\cdots+m_n}\,m_1!\dots m_n!}x_1^{m_1}\dots x_n^{m_n}, \tag{55}$$

where $|x_1|,\dots,|x_n|<1$.

Certainly, we have

$$F_A^{(2)}=F_2,\qquad F_B^{(2)}=F_3,\qquad F_C^{(2)}=F_4,\qquad F_D^{(2)}=F_1.$$

Many properties for Lauricella functions, such as integral representations and partial differential equations, are given by Appell and Kampé de Fériet [2]. From the vast amount of material, we single out the following integral representation of the Lauricella $F_D^{(n)}$ series as a specific example.

3.3.1 Integral Representation of $F_D^{(n)}$

The formula

$$F_D^{(n)}\big(a;b_1,\dots,b_n;c;x_1,\dots,x_n\big)$$
$$=\Gamma\begin{bmatrix}c\\a,c-a\end{bmatrix}\int_0^1 u^{a-1}(1-u)^{c-a-1}(1-ux_1)^{-b_1}\dots(1-ux_n)^{-b_n}\,\mathrm{d}u, \tag{56}$$

where $\Re c > \Re a > 0$, is very useful for deriving relations for F_D series. It can be easily verified by Taylor expansion of the integrand, followed by termwise integration.

3.3.2 Group Theoretic Interpretations

A group theoretic interpretation of the Lauricella $F_A^{(n)}$ functions corresponding to the most degenerate principal series representations of $\mathrm{SL}(n,\mathbb{R})$ was given by N.Ja. Vilenkin [33] (see also [34, Sect. 16.3.4]). Similarly, W. Miller, Jr. [24] has shown that the Lauricella $F_D^{(n)}$ functions transform as basis vectors corresponding to irreducible representations of the Lie algebra $sl(n+3,\mathbb{C})$ (by which he generalized his previous observation in [23] for the $n = 2$ case, corresponding to the Appell functions F_1).

Acknowledgements I would like to thank Tom Koornwinder for pointing out references on group theoretic interpretations of Lauricella series. This work was partially supported by FWF Austrian Science Fund grants S9607 & F50-08.

References

1. Appell, P.: Sur les séries hypergéométriques de deux variables et sur des équations différentielles linéaires aux dérivées partielles. Comptes rendus hebdomadaires des séances de l'Académie des sciences **90**, 296–298 & 731–735 (1880)
2. Appell, P., Kampé de Fériet, J.: Fonctions hypergéométriques et hypersphériques; Polynômes d'Hermite. Gauthier–Villars, Paris (1926)
3. Bailey, W.N.: Generalized Hypergeometric Series. Cambridge Mathematical Tract, vol. 32, 2nd edn. Cambridge University Press, Cambridge (1964)
4. Beukers, F.: Algebraic A-hypergeometric functions. Invent. Math. **180**, 589–610 (2010)
5. Beukers, F.: Monodromy of A-hypergeometric functions (2011, preprint). arXiv:1101.0493
6. Burchnall, J.L., Chaundy, T.W.: Expansions of Appell's double hypergeometric functions, I. Q. J. Math. (Oxford) **11**, 249–270 (1940)
7. Burchnall, J.L., Chaundy, T.W.: Expansions of Appell's double hypergeometric functions, II. Quart. J. Math. (Oxford) **12**, 112–128 (1941)
8. Buschman, R.G.: Contiguous relations for Appell functions. J. Indian Math. Soc. **29**, 165–171 (1987)
9. Bytev, V.V., Kalmykov M.Yu., Kniehl B.A.: HYPERDIRE – HYPERgeometric functions DIfferential REduction MATHEMATICA based packages for differential reduction of generalized hypergeometric functions: now with ${}_pF_{p-1}, F_1, F_2, F_3, F_4$ (2011, preprint). arXiv:1105.3565
10. Carlson, B.C.: Quadratic transformations of Appell functions. SIAM J. Math. Anal. **7**, 291–304 (1976)
11. Dunkl, C.F., Xu, Y.: Orthogonal Polynomials of Several Variables. Encyclopedia of Mathematics and Its Applications, vol. 81. Cambridge University Press, Cambridge (2001)
12. Erdélyi, A., Magnus, W., Oberhettinger, F., Tricomi, F.G.: Higher Transcendental Functions. Vol. I. Based on Notes Left by Harry Bateman. With a Preface by Mina Rees. With a Foreword by E.C. Watson. Reprint of the 1953 original. Robert E. Krieger Publishing, Melbourne (1981)

13. Gasper, G., Rahman, M.: Basic Hypergeometric Series, 2nd edn. Encyclopedia of Mathematics and Its Applications, vol. 96. Cambridge University Press, Cambridge (2004)
14. Gelfand, I.M., Kapranov, M.M., Zelevinsky, A.V.: Hypergeometric functions and toric manifolds. Funct. Anal. Appl. **23**, 94–106 (1989)
15. Gross, K.I., Richards, D.St.P.: Total positivity, spherical series, and hypergeometric functions of matrix argument. J. Approx. Theory **59**(2), 224–246 (1989)
16. Heckman, G.J., Opdam, E.M.: Root systems and hypergeometric functions I. Compos. Math. **64**, 329–352 (1987)
17. Horn, J.: Hypergeometrische Funktionen zweier Veränderlichen. Math. Ann. **105**, 381–407 (1931)
18. Kampé de Fériet, J.: La fonction hypergéométrique. Gauthier–Villars, Paris (1937)
19. Karlsson, P.W.: Two hypergeometric summation formulae related to 9-j coefficients. J. Phys. A **27** (20), 6943–6945 (1994)
20. Koornwinder, T.H.K.: Special functions associated with root systems: a first introduction for non-specialists. In: Srinivasa Rao, K., Jagannathan, R., Vanden Berghe, G., Van der Jeugt, J. (eds.) Special Functions and Differential Equations, pp. 10–24. Allied Publishers, New Delhi (1998)
21. Lauricella, G.: Sulle funzioni ipergeometriche a più variabili. Rediconti del Circolo Matematico di Palermo **7**(S1), 111–158 (1893)
22. Macdonald I.G.: Symmetric Functions and Orthogonal Polynomials. University Lecture Series, vol. 12. American Mathematical Society, Providence (1998)
23. Miller, W. Jr.: Lie theory and the Appell functions F_1. J. Math. Phys. **13**(9), 1393–1399 (1972)
24. Miller, W. Jr.: Lie theory and the Lauricella functions F_D. SIAM J. Math. Anal. **4**(4), 638–655 (1973)
25. Milne, S.C.: Transformations of $U(n+1)$ multiple basic hypergeometric series. In: Kirillov, A.N., Tsuchiya, A., Umemura, H. (eds.) Physics and Combinatorics: Proceedings of the Nagoya 1999 International Workshop, Nagoya University, Nagoya, 23–27 Aug 1999, pp. 201–243. World Scientific, Singapore (2001)
26. Opps, S.B., Saad, N., Srivastava, H.M.: Recursion formulas for Appell's hypergeometric function F_2 with some applications to radiation field problem. Appl. Math. Comput. **207**, 545–558 (2009)
27. Picard, É.: Sur une extension aux fonctions de deux variables du problème de Riemann relatif aux fonctions hypergéométriques. Annales scientifiques de l'É.N.S. 2e série **10**, 305–322 (1881)
28. Pitre, S.N., Van der Jeugt, J.: Transformation and summation formulas for Kampé de Fériet series $F_{1:1}^{0:3}(1,1)$. J. Math. Anal. Appl. **202**(1), 121–132 (1996)
29. Rosengren, H.: Elliptic hypergeometric series on root systems. Adv. Math. **181**, 417–447 (2004)
30. Shpot, M.A.: A massive Feynman integral and some reduction relations for Appell functions. J. Math. Phys. **48**(12), 123512, 13 (2007)
31. Slater, L.J.: Generalized Hypergeometric Functions. Cambridge University Press, Cambridge (1966)
32. Vilenkin, N.Ja.: Special Functions and the Theory of Group Representations. American Mathematical Society Translations of Mathematical Monographs, vol. 22. American Mathematical Society, Providence (1968)
33. Vilenkin, N.Ja.: Hypergeometric functions of several variables, and degenerate representations of the group SL(n, R). Izv. Vyssh. Uchebn. Zaved. Mat. **4**(95), 50–55 (1970)
34. Vilenkin, N.Ja., Klimyk A.U.: Representation of Lie Groups and Special Functions, Vol. 3. Classical and Quantum Groups and Special Functions (Translated from the Russian by V.A. Groza and A.A. Groza). Mathematics and Its Applications (Soviet Series), vol. 75. Kluwer, Dordrecht (1992)
35. Wang, X.: Recursion formulas for Appell functions. Integral Transforms Spec. Funct. **23**(6), 421–433 (2012)

Simplifying Multiple Sums in Difference Fields

Carsten Schneider

Abstract In this survey article we present difference field algorithms for symbolic summation. Special emphasize is put on new aspects in how the summation problems are rephrased in terms of difference fields, how the problems are solved there, and how the derived results in the given difference field can be reinterpreted as solutions of the input problem. The algorithms are illustrated with the Mathematica package `Sigma` by discovering and proving new harmonic number identities extending those from Paule and Schneider, 2003. In addition, the newly developed package `EvaluateMultiSums` is introduced that combines the presented tools. In this way, large scale summation problems for the evaluation of Feynman diagrams in QCD (Quantum ChromoDynamics) can be solved completely automatically.

1 Introduction

We will elaborate a symbolic summation toolbox based on up–to-date algorithms in the setting of difference fields. It contains hypergeometric and q–hypergeometric summation, see, e.g., [28, 34, 37, 38, 52, 55, 57, 59, 82] and [16, 43, 53] respectively, and it can deal with multiple sums covering big parts of (q–)hypergeometric multi-summation [15, 60, 79, 80] and (q–)holonomic sequences [30, 44, 68, 81].

This difference field approach started with Karr's theory of $\Pi\Sigma^*$-fields and his indefinite summation algorithm [39, 40]; for the continuous analogue of indefinite integration see [61]. In principle, the algorithm solves the telescoping problem in a given field generated by indefinite nested sums and products. In this article we restrict this input class to *nested hypergeometric sum expressions* (see Definition 4), i.e., expressions where the arising products represent hypergeometric sequences,

C. Schneider (✉)
Research Institute for Symbolic Computation (RISC), Johannes Kepler University Linz, Linz, Austria
e-mail: Carsten.Schneider@risc.jku.at

C. Schneider and J. Blümlein (eds.), *Computer Algebra in Quantum Field Theory*, Texts & Monographs in Symbolic Computation, DOI 10.1007/978-3-7091-1616-6_14,

and the sums and products occur only as polynomial expressions in the numerators; evaluating such expressions produces d'Alembertian sequences [12] and [58, this book], a subclass of Liouvillian sequences [36]. We point out that exactly this restriction covers all the summation problems that have been relevant in practical problem solving so far. There we solve the following fundamental problem: given a nested hypergeometric sum expression, calculate an alternative expression such that the occurring sums are algebraically independent [76]; for related work see [9,18,35, 42,69]. In addition, the found representation should be given in terms of sums and products that are as simple as possible; for a general framework we refer to [49]. This is possible by representing the sums and products in $\Pi\Sigma^*$-fields reflecting certain optimality properties: We will exploit simplifications taking into account, e.g., the minimal nesting depth [65,67,74,75,77] or minimal degrees [13,69,72].

Besides indefinite summation, we aim at the transformation of a definite multiple sum to nested hypergeometric sums. As for the special case of hypergeometric summation [59,82] one looks for a recurrence of such a sum [62]. If one succeeds, one computes all solutions of the found recurrence that are expressible in terms of nested hypergeometric sum expressions; for solvers of recurrences in terms of polynomials and $\Pi\Sigma^*$-fields see [12, 36, 57] and [14, 26, 62, 70], respectively. Finally, one tries to combine the solutions to an expression that equals the input sum.

All these algorithms (also for the q-hypergeometric and mixed case) are available in the summation package Sigma [73] and have been used to discover and prove demanding identities from combinatorics or related fields, like, e.g., in [32,51,56, 71]. A typical example is the sum

$$A_\alpha(a) = \sum_{k=0}^{a} (1 + \alpha(n - 2k)S_1(k)) \binom{n}{k}^{\alpha} \quad \text{with } S_1(k) = \sum_{i=1}^{k} \frac{1}{i}$$

which is connected to supercongruences of the Apéry numbers. For the treatment of the cases $\alpha \in \{1, 2, \ldots, 5\}$ and $\alpha > 5$ we refer to [54] and [45], respectively. As running example we will discover and prove the following identities[1]

$$A_{-1}(n) = (n + 1)S_1(n) + 1, \tag{1}$$

$$A_{-2}(n) = \frac{(n + 1)^2}{(n + 2)^2} + \frac{\left(n + 2\left(n^2 + 3n + 2\right) S_1(n) + 3\right)(n + 1)}{(n + 2)^2}, \tag{2}$$

$$A_{-3}(n) = (-1)^n (5S_{-3}(n)(n + 1)^3 - 6S_{-2,1}(n)(n + 1)^3) + 6S_1(n)(n + 1) + 1, \tag{3}$$

[1]For identities (1) and (2) we point also to [29]; for their indefinite versions see (10) and (11) below.

$$A_{-4}(n) = \frac{(-1)^n \binom{2n}{n}^{-1}(n+1)^5}{(4n(n+2)+3)} \left(\frac{7}{2} \sum_{i=1}^{n} \frac{(-1)^i \binom{2i}{i}}{i^3} - 5 \sum_{i=1}^{n} \frac{(-1)^i \binom{2i}{i} S_1(i)}{i^2} \right) + \\ + \frac{(10(n+1)S_1(n)+3)(n+1)}{2n+3} \tag{4}$$

where the harmonic sums [20, 78] are defined by

$$S_{m_1,\dots,m_k}(n) = \sum_{i_1=1}^{n} \frac{\operatorname{sign}(m_1)^{i_1}}{i_1^{|m_1|}} \cdots \sum_{i_k=1}^{i_{k-1}} \frac{\operatorname{sign}(m_k)^{i_k}}{i_k^{|m_k|}}, \qquad m_i \in \mathbb{Z} \setminus \{0\}. \tag{5}$$

We emphasize that exactly this type of nested hypergeometric sums is related to summation problems coming from QCD like., e.g., in [21, 48] or in [4, 7, 17, 25]. More precisely, two- and three-loop Feynman integrals with at most one mass and with operator insertion can be transformed to multiple sums [24] depending on a discrete Mellin parameter n. Then these sums must be simplified in terms of special functions [2, this book], such as harmonic sums (5), their infinite versions of multiple zeta values [22] and generalizations like S-sums [9,47] and cyclotomic harmonic sums [5]. In recent calculations [6] also binomial sums as in (4) arose. For certain sum classes we point to efficient tools like [47, 78]. For harder sums such as [7]

$$\sum_{j=0}^{n-3}\sum_{k=0}^{j}\sum_{l=0}^{k}\sum_{q=0}^{-j+n-3}\sum_{s=1}^{-l+n-q-3} \\ \times \sum_{r=0}^{-l+n-q-s-3} \frac{\binom{j+1}{k+1}\binom{k}{l}\binom{n-1}{j+2}\binom{-j+n-3}{q}\binom{-l+n-q-3}{s}\binom{-l+n-q-s-3}{r} r!(-l+n-q-r-s-3)!(s-1)!}{(-l+n-q-2)!(-j+n-1)(n-q-r-s-2)(q+s+1)} \\ (-1)^{-j+k-l+n-q-3}\Big[4S_1(-j+n-1) - 4S_1(-j+n-2) - 2S_1(k) - (S_1(-l+n-q-2) \\ + S_1(-l+n-q-r-s-3) - 2S_1(r+s)) + 2S_1(s-1) - 2S_1(r+s)\Big] \tag{6}$$

the summation techniques under consideration work successfully and are applied automatically in the newly developed package `EvaluateMultiSums` [23]. In this way, millions of multiple sums [4, 25] could be treated. In addition, the presented packages and algorithms are used by an enhanced version [6, 7] of the method of hyperlogarithms [27] and by new algorithms for the calculation of ε-expansions [8, 24] utilizing multi-summation and integration methods [15, 68, 79].

The outline of this article is as follows. In Sect. 2 we present the basic mechanism how expressions in terms of indefinite nested sums and products can be rephrased in a difference field. For interested readers details are given in Sect. 3. Readers, that are primarily interested in the summation tools and how they can be applied with the summation package `Sigma`, can jump directly to Sect. 4. Finally, in Sect. 5 the new package `EvaluateMuliSums` is introduced that combines all the presented summation methods. It enables one to simplify definite nested sums to indefinite nested hypergeometric sums completely automatically.

2 Indefinite Summation: The Basic Mechanism

We will work out the basic principles how indefinite summation can be carried out in the setting of difference fields. This will be illustrated by the task to simplify

$$A_{-1}(a)=\sum_{k=0}^{a}F(k)=\sum_{k=0}^{a}(1-(n-2k)S_1(k))\binom{n}{k}^{-1}, \tag{7}$$

to be more precise, by the task to solve the following problem.

Problem T: Telescoping. *Given* a summand $F(k)$. *Find* an expression $G(k)$ such that

$$G(k+1)=G(k)+F(k+1) \tag{8}$$

and such that $G(k)$ is not "more complicated" than $F(k)$.

For our given $F(k)$ in (7) we will compute for $k\geq 0$ the solution

$$G(k)=((k+1)S_1(k)+1)\tbinom{n}{k}^{-1}+c,\qquad c\in\mathbb{Q}(n). \tag{9}$$

Since $A_{-1}(a)$ and $G(a)$ satisfy both the recurrence $A(a+1)=A(a)+F(a+1)$, they are equal for $a\geq 0$ if they agree at $a=0$; this is the case with $c=0$. Hence we get for $A_{-1}(a)$, and with the same technique for $A_{-2}(a)$, the simplifications

$$A_{-1}(a)=((a+1)S_1(a)+1)\binom{n}{a}^{-1}, \tag{10}$$

$$A_{-2}(a)=\frac{(n+1)^2}{(n+2)^2}+\frac{(a+1)(-a+2n+2(a+1)(n+2)S_1(a)+3)}{(n+2)^2}\binom{n}{a}^{-2}. \tag{11}$$

Note that for the special case $a=n$ this simplifies to (1) and (2), respectively.

Subsequently, we give more details how this solution $G(k)$ for (8) can be derived automatically. First observe that the occurring sums can be written in terms of indefinite sums and products: for all $k \in \mathbb{N}$,

$$S_1(k) = \sum_{i=1}^{k} \frac{1}{i} \quad \text{and} \quad \binom{n}{k} = \prod_{i=1}^{k} \frac{n-i+1}{k};$$

here n is considered as a variable. Now let $\mathcal{S}_k$ be the shift operator w.r.t. k. Then using the shift behavior of the summand objects, namely

$$\mathcal{S}_k n = n, \quad \mathcal{S}_k k = k+1, \quad \mathcal{S}_k \binom{n}{k} = \tfrac{n-k}{k+1}\binom{n}{k}, \quad \mathcal{S}_k S_1(k) = S_1(k) + \tfrac{1}{k+1}, \tag{12}$$

we can write, e.g., $F(k+1)$ again in terms of $n, k, \binom{n}{k}$ and $S_1(k)$:

$$F(k+1) = \mathcal{S}_k F(k) = \big(1-(n-2(k+1))\big(S_1(k)+\tfrac{1}{k+1}\big)\big)\tfrac{n-k}{k+1}\binom{n}{k}^{-1}. \tag{13}$$

We will utilize this property, but instead of working with the summand objects k, $\binom{n}{k}$ and $S_1(k)$ we will represent the objects by the variables x, b, h, respectively; n is also considered as a variable. Here we start with the rational numbers and construct the rational function field[2] $\mathbb{F} := \mathbb{Q}(n)(x)(b)(h)$, i.e., the field of quotients of polynomials in the variables n, x, b, h. In this way, (13) is represented by

$$f = \big(1-(n-2(x+1))(h+\frac{1}{x+1})\big)\frac{n-x}{x+1}b^{-1} \in \mathbb{F}. \tag{14}$$

Finally, we model the shift operator $\mathcal{S}_k$ by a field automorphism $\sigma : \mathbb{F} \to \mathbb{F}$.

Definition 1. Let $\mathbb{F}$ be a field (resp. ring). A bijective map $\sigma : \mathbb{F} \to \mathbb{F}$ is called *field* (resp. *ring*) *automorphism* if $\sigma(a \circ b) = \sigma(a) \circ \sigma(b)$ for all $a, b \in \mathbb{F}$ and $\circ \in \{+, \cdot\}$.

Remark. If $\mathbb{F}$ is a ring, it follows that $\sigma(0) = 0$, $\sigma(1) = 1$ and $\sigma(-a) = -\sigma(a)$ for all $a \in \mathbb{F}$. In addition, if $\mathbb{F}$ is a field, this implies that $\sigma(1/a) = 1/\sigma(a)$ for all $a \in \mathbb{F}^*$.

Namely, looking at the shift behavior of the summand objects (12) the automorphism is constructed as follows. We start with the rational function field $\mathbb{Q}(n)$ and define $\sigma : \mathbb{Q}(n) \to \mathbb{Q}(n)$ with $\sigma(c) = c$ for all $c \in \mathbb{Q}(n)$. Next, we extend σ to $\mathbb{Q}(n)(x)$ such that $\sigma(x) = x + 1$. We note that this construction is unique:

Lemma 1. *Let $\mathbb{F}(t)$ be a rational function field, $\sigma : \mathbb{F} \to \mathbb{F}$ be a field automorphism, and $a, b \in \mathbb{F}$ with $a \neq 0$. Then there is exactly one way how the field automorphism is extended to $\mathbb{F}(t)$ subject to the relation $\sigma(t) = a\,t + b$. Namely,*

[2] $\mathbb{Z}$ are the integers, $\mathbb{N} = \{0, 1, 2, \dots\}$ are the non-negative integers, and all fields (resp. rings) contain the rational numbers $\mathbb{Q}$ as a subfield (resp. subring). For a set A we define $A^* := A \setminus \{0\}$.

for $f = \sum_{i=0}^{n} f_i t^i \in \mathbb{F}[t]$, $\sigma(f) = \sum_{i=0}^{n} \sigma(f_i)(a\,t+b)^i$. *And for* $p, q \in \mathbb{F}[t]$ *with* $q \neq 0$, $\sigma(\frac{p}{q}) = \frac{\sigma(p)}{\sigma(q)}$.

As a consequence, by iterative application we extend σ uniquely from $\mathbb{Q}(n)$ to $\mathbb{Q}(n)(x)(b)(h)$ subject to the shift relations (compare (12))

$$\sigma(x) = x + 1, \qquad \sigma(b) = \frac{n-x}{x+1} b, \qquad \sigma(h) = h + \frac{1}{x+1}. \tag{15}$$

In summary, we represent the summand $F(k+1)$ given in (13) by (14) in the rational function field $\mathbb{F} := \mathbb{Q}(n)(x)(b)(h)$ together with its field automorphism $\sigma : \mathbb{F} \to \mathbb{F}$ subject to the shift relations (15). Exactly this construction is called difference field; for a general theory see [31, 46].

Definition 2. A *difference field* (resp. *difference ring*) $(\mathbb{F}, \sigma)$ is a field (resp. ring) $\mathbb{F}$ together with a field automorphism (resp. ring automorphism) $\sigma : \mathbb{F} \to \mathbb{F}$. Here we define the set of constants by $\text{const}_\sigma\mathbb{F} := \{c \in \mathbb{F} | \sigma(c) = c\}$.

Remark. For a difference field $(\mathbb{F}, \sigma)$ the set $\text{const}_\sigma\mathbb{F}$ forms a subfield $\mathbb{F}$ which is also called *constant field* of $(\mathbb{F}, \sigma)$. Since $\mathbb{Q}$ is always kept invariant under σ (this is a consequence of $\sigma(1) = 1$), $\mathbb{Q}$ is always contained in $\text{const}_\sigma\mathbb{F}$ as a subfield.

We continue with our concrete example. Given the difference field $(\mathbb{F}, \sigma)$ in which $F(k+1)$ is represented by (14), we search for a rational function $g \in \mathbb{F}$ such that

$$\sigma(g) = g + f. \tag{16}$$

Namely, we activate the algorithm from Sect. 3.2 below and calculate the solution

$$g = ((x+1)h + 1)b^{-1} + c, \qquad c \in \mathbb{Q}(n) \tag{17}$$

which rephrased in terms of the summation objects gives the solution (9) for (8).

In a nutshell, the proposed simplification tactic consists of the following steps.

1. Construct a difference field in which the summand objects can be rephrased.
2. Find a solution g of (16) in this difference field (or a suitable extension).
3. Reformulate g to a solution $G(k)$ of (8) in terms of product-sum expressions.

The algorithms in the next section deliver tools to attack this problem for the class of indefinite nested product-sum expression; for a more formal framework see [75].

Definition 3. Let $\mathbb{K}$ be a field and the variable k be algebraically independent over $\mathbb{K}$. An expression is called *indefinite nested product-sum expression w.r.t.* k iff it can be built by k, a finite number of constants from $\mathbb{K}$, the four operations $(+, -, \cdot, /)$, and sums and products of the type $\sum_{i=l}^{k} f(i)$ or $\prod_{i=l}^{k} f(i)$ where $l \in \mathbb{N}$ and where $f(i)$ is an indefinite nested product-sum expression w.r.t. i which is free of k. In particular, we require that there is a $\lambda \in \mathbb{N}$ such that for any integer $n \geq \lambda$

the expression evaluates for $k = n$ (to an element form $\mathbb{K}$) without entering in any pole.

In this recursive definition the sums and products can be arbitrarily composed and can arise also as polynomial expressions in the denominators. Here we restrict ourselves to those expressions that occurred in practical problem solving so far.

Definition 4. A sequence $\langle h_u \rangle_{u\geq 0} \in \mathbb{K}^{\mathbb{N}}$ is called *hypergeometric* if there are $\alpha(x) \in \mathbb{K}(x)$ and $l \in \mathbb{N}$ such that $h_{u+1}/h_u = \alpha(u)$ for all $u \geq l$. I.e., $h_u = c \prod_{i=l+1}^{u} \alpha(i-1)$ for all $u \geq l$ for some $c \in \mathbb{K}^*$. Such a symbolic product (u replaced by a variable k) is called *hypergeometric product* w.r.t. k. An expression is called *nested hypergeometric sum expression* w.r.t. k if it is an indefinite nested product-sum expression w.r.t. k (see Definition 3) such that the arising products are hypergeometric and the arising sums and products occur only as polynomial expressions in the numerators. The arising sums (with upper bound k) are called *nested hypergeometric sums (w.r.t. k)*.

E.g., the harmonic sums (5) and their generalizations [5, 47] fall into this class. In particular, the right hand sides of (1)–(4) are covered. These expressions evaluate exactly to the d'Alembertian sequences [12] and [58, this book].

Remark 1. In the difference field approach also the q-hypergeometric and mixed case [16] can be handled. All what will follow generalizes to this extended setting.

Subsequently, we will derive a full algorithm that treats the three steps from above automatically for the class of nested hypergeometric sum expressions.

3 Details of the Difference Field Machinery

We will work out how the Steps 1–3 from above (covering also the more general paradigms of creative telescoping and recurrence solving) can be carried out automatically. As an important consequence we will obtain tools to compactify nested hypergeometric sum expressions, i.e., the occurring sums in the derived expression are algebraically independent (see also Sect. 4.1).

3.1 *Step 1: From Indefinite Nested Sums and Products to $\Pi\Sigma^*$-Fields*

In the previous section the construction of a difference field for a given summand in terms of indefinite nested product-sums was as follows. We start with a constant difference field $(\mathbb{K}, \sigma)$, i.e., $\sigma(c) = c$ for all $c \in \mathbb{K}$ or equivalently $\text{const}_\sigma \mathbb{K} = \mathbb{K}$. Then we adjoin step by step new variables, say $t_1, \ldots, t_e$ to $\mathbb{K}$ which gives the rational function field $\mathbb{F} := \mathbb{K}(t_1)(t_2)\ldots(t_e)$ and extend the field automorphism from $\mathbb{K}$ to $\mathbb{F}$ subject to the shift relations $\sigma(t_i) = a_i t_i$ or $\sigma(t_i) = t_i + a_i$ for some $a_i \in \mathbb{F}(t_1)\dots(t_{i-1})^*$. Subsequently, we restrict this construction to $\Pi\Sigma^*$-fields; for a slightly more general but rather technical definition of Karr's $\Pi\Sigma$-fields see [39, 40].

Definition 5. $(\mathbb{F}, \sigma)$ as given above is called *$\Pi\Sigma^*$-field over* $\mathbb{K}$ if $\text{const}_\sigma\mathbb{F} = \mathbb{K}$. The adjoined elements $(t_1, \dots, t_e)$ are also called generators of the $\Pi\Sigma^*$-field.

E.g., our difference field $(\mathbb{Q}(n)(x)(b)(h), \sigma)$ with (15) is a $\Pi\Sigma^*$-field over $\mathbb{Q}(n)$. To see that the constants are just $\mathbb{Q}(n)$, the following result is crucial [39, 62].

Theorem 1. *[Karr's theorem] Let $(\mathbb{F}, \sigma)$ be a difference field, take a rational function field $\mathbb{F}(t)$, and extend the automorphism σ from $\mathbb{F}$ to $\mathbb{F}(t)$ subject to the relation $\sigma(t) = a\,t + f$ for some $a \in \mathbb{F}^*$ and $f \in \mathbb{F}$. Then the following holds.*

1. *Case $a = 1$: $\text{const}_\sigma\mathbb{F}(t) = \text{const}_\sigma\mathbb{F}$ iff there is no $g \in \mathbb{F}$ with $\sigma(g) = g + f$.*
2. *Case $f = 0$: $\text{const}_\sigma\mathbb{F}(t) = \text{const}_\sigma\mathbb{F}$ iff there is no $g \in \mathbb{F}^*$, $r > 0$ with $\sigma(g) = a^r g$.*

Example 1. Using Theorem 1 we represent the sum (7) in a $\Pi\Sigma^*$-field parsing the occurring objects in the following order: $\xrightarrow{1.} n \xrightarrow{2.} k \xrightarrow{3.} \binom{n}{k} \xrightarrow{4.} S_1(k) \xrightarrow{5.} A_{-1}(a)$.

1. We start with $(\mathbb{Q}(n), \sigma)$ setting $\sigma(c) = c$ for all $c \in \mathbb{Q}(n)$.
2. Then we construct the difference field $(\mathbb{Q}(n)(x), \sigma)$ subject to the shift relation $\sigma(x) = x + 1$. Since there is no $g \in \mathbb{Q}(n)$ such that $\sigma(g) = g + 1$, it follows by Karr's Theorem that $\text{const}_\sigma\mathbb{Q}(n)(x) = \mathbb{Q}(n)$, i.e., $(\mathbb{Q}(n)(x), \sigma)$ is a $\Pi\Sigma^*$-field over $\mathbb{Q}(n)$.
3. One can check by an algorithm of Karr [39] that there is no $r > 0$ and $g \in \mathbb{Q}(n)(x)^*$ such that $\sigma(g) = \left(\frac{n-x}{x+1}\right)^r g$. Thus for our difference field $(\mathbb{Q}(n)(x)(b), \sigma)$ with $\sigma(b) = \frac{n-x}{x+1} b$ we have that $\text{const}_\sigma\mathbb{Q}(n)(x)(b) = \text{const}_\sigma\mathbb{Q}(n)(x) = \mathbb{Q}(n)$ by Theorem 1, i.e., $(\mathbb{Q}(n)(x)(b), \sigma)$ is a $\Pi\Sigma^*$-field over $\mathbb{Q}(n)$.
4. Next, we extend the $\Pi\Sigma^*$-field $(\mathbb{Q}(n)(x)(b), \sigma)$ to $(\mathbb{Q}(n)(x)(b)(h), \sigma)$ subject to the shift relation $\sigma(h) = h + \frac{1}{x+1}$. There is no $g \in \mathbb{Q}(n)(x)(b)$ with $\sigma(g) = g + \frac{1}{x+1}$; this can be checked by the algorithm given in Sect. 3.2 below. Thus the constants remain unchanged by Theorem 1, and $(\mathbb{Q}(n)(x)(b)(h), \sigma)$ is a $\Pi\Sigma^*$-field over $\mathbb{Q}(n)$.
5. Given f in (14), that represents $F(k+1)$ in (7), we find (17) such that $\sigma(g) = g + f$. In other words, g reflects the shift behavior of $A_{-1}(k) = \sum_{i=1}^{k} F(i)$ with $\mathcal{S}_k A_{-1}(k) = A_{-1}(k) + F(k+1)$. Reformulating g in terms of sums and products yields (9) and choosing $c = 0$ delivers the identity (10). In other words, g (for $c = 0$) can be identified with the sum $A_{-1}(k)$. This construction will be done more precise in Sect. 3.4; in particular, we refer to Remark 4.

In general, Theorem 1 yields the following telescoping tactic to represent a given indefinite nested product-sum expression (see Definition 3) in terms of a $\Pi\Sigma^*$-field. One starts with the constant field $(\mathbb{K}, \sigma)$ with $\sigma(c) = c$ for all $c \in \mathbb{K}$. Then one parses all the summation objects. Suppose one treats in the next step a sum of the form $\sum_{i=1}^{k} F(i)$ where one can express $F(k)$ in the so far constructed $\Pi\Sigma^*$-field $(\mathbb{F},\sigma)$, say $F(k+1)$ can be rephrased by $f \in \mathbb{F}$. Then there are two cases: one finds a $g \in \mathbb{F}$ such $\sigma(g) = g + f$ and one can model the sum $\sum_{i=1}^{k} F(i)$ with its shift behavior

$$\mathscr{S}_k \sum_{i=1}^{k} F(i) = \sum_{i=1}^{k} F(i) + F(k+1) \tag{18}$$

by $g + c$ (for some properly chosen $c \in \mathbb{K}$). If this fails, one can adjoin a new variable, say t, to $\mathbb{F}$ and extends the automorphism to $\sigma : \mathbb{F}(t) \to \mathbb{F}(t)$ subject to the shift relation $\sigma(t) = t + f$. By Theorem 1 the constants remain unchanged, i.e., $(\mathbb{F}(t),\sigma)$ is a $\Pi\Sigma^*$-field over $\mathbb{K}$, and $t \in \mathbb{F}(t)$ models accordingly the shift behavior (18) of our sum. The product case can be treated similarly; see also Problem RP on page 336.

3.2 Step 2: Solving the Telescoping Problem in a Given $\Pi\Sigma^*$-Field

Karr's algorithm [39] solves the telescoping problem within a fixed $\Pi\Sigma^*$-field exploiting its recursive nature: it tries to solve the problem for the top most generator and reduces the problem to the subfield (i.e., without the top generator). This reduction is possible by solving the following more general problem.

Problem FPLDE: First-order Parameterized Linear Difference Equations.
Given a $\Pi\Sigma^*$-field $(\mathbb{F},\sigma)$ over $\mathbb{K}$, $\alpha_0,\alpha_1 \in \mathbb{F}^*$ and $f_0,\dots,f_d \in \mathbb{F}$.
Find all[a] $c_0,\dots,c_d \in \mathbb{K}$ and $g \in \mathbb{F}$ such that $\alpha_1\sigma(g) + \alpha_0 g = c_0 f_0 + \dots + c_d f_d$.

[a] The solution set $V = \{(c_0,\dots,c_d,g) \in \mathbb{K}^{d+1} \times \mathbb{F} | \alpha_1\sigma(g) + \alpha_0 g = \sum_{i=0}^{d} c_i f_i\}$ forms a $\mathbb{K}$-vector space of dimension $\leq d+2$ and the task is to get an explicit basis of V.

Remark. Problem FPLDE contains not only the summation paradigm of telescoping, but also of creative (resp. parameterized) telescoping (40) for a fixed $\Pi\Sigma^*$-field.

Subsequently, we sketch a simplified version of Karr's algorithm applied to our concrete problem: Given the $\Pi\Sigma^*$-field $(\mathbb{F},\sigma)$ with $\mathbb{F} = \mathbb{Q}(n)(x)(b)(h)$ and the shift relations (15) and given the summand (14), calculate (if possible) $g \in \mathbb{F}$ such that $\sigma(g)-g = f$ holds. The algorithm is recursive: it treats the top most variable h and needs to solve FPLDEs in the smaller $\Pi\Sigma^*$-field $(\mathbb{H}, \sigma)$ with $\mathbb{H} = \mathbb{Q}(n)(x)(b)$.

Denominator bounding: Calculate a polynomial $q \in \mathbb{H}[h]^*$ such that for any $g \in \mathbb{H}(h)$ with (16) we have that $g\,q \in \mathbb{H}[h]$, i.e., q contains the denominators of all the solutions as a factor. For a general $\Pi\Sigma^*$-field and f such a universal denominator q can be calculated; see [26, 39, 63]. Then given such a q, it suffices to search for a polynomial $p \in \mathbb{H}[h]$ such that the first order difference equation

$$\frac{1}{\sigma(q)}\sigma(p) - \frac{1}{q}p = f \tag{19}$$

holds (which is covered by Problem FPLDE). In our concrete example the algorithm outputs that we can choose $q = 1$, i.e., we have to search for a $p \in \mathbb{H}[h]$ such that $\sigma(p) - p = f$ holds.

Degree bounding: Calculate b such that for any $p \in \mathbb{H}[h]$ with (19) we have that $\deg(p) \leq b$. For a general $\Pi\Sigma^*$-field and f such a b can be calculated; see [39,66]. In our concrete example we get $b = 2$. Hence, any solution $p \in \mathbb{H}[h]$ of $\sigma(p)-p = f$ is of the form $p = p_2h^2 + p_1h + p_0$ and it remains to determine $p_2, p_1, p_0 \in \mathbb{H}$.

Degree reduction: By coefficient comparison of h^2 in

$$\sigma(p_2h^2 + p_1h^1 + p_0) - (p_2h^2 + p_1h^1 + p_0) = f \tag{20}$$

we obtain the constraint $\sigma(p_2) - p_2 = 0$ on p_2. Since $(\mathbb{H}, \sigma)$ is a $\Pi\Sigma^*$-field, $p_2 \in \mathbb{Q}(n)$. Hence we can choose $p_2 = d$ where $d \in \mathbb{Q}(n)$ is (at this point) free to choose. Now we move $p_2h^2 = d\,h^2$ in (20) to the other side and get the equation

$$\sigma(p_1h^1 + p_0) - (p_1h^1 + p_0) = f - d\,\tfrac{2h(x+1)+1}{(x+1)^2}. \tag{21}$$

Note that we accomplished a simplification: the degree of h in the difference equation is reduced (with the price to introduce the constant d). Now we repeat this degree reduction process. By coefficient comparison of h^1 in (21) we get the constraint $\sigma(p_1) - p_1 = \frac{(x+1)(2x-n+2)}{b(n-x)} + d\frac{-2}{x+1}$ on p_1. Again we succeeded in a reduction: we have to solve Problem FPLDE in $\mathbb{H}$. Applying the sketched method recursively, gives the generic solution $d = 0$ and $p_1 = \frac{x+1}{b} + e$ with $e \in \mathbb{Q}(n)$. Plugging this solution into (21) and bringing $\sigma(p_1\,h) - p_1\,h$ to the right hand side reduce the problem to $\sigma(p_0) - p_0 = \frac{-2x+n-1}{b(x-n)} + e\frac{-1}{x+1}$; note that we decreased the degree of h from 1 to 0, i.e., we have to solve again Problem FPLDE in $\mathbb{H}$. Recursive application of the algorithm calculates the generic solution $e = 0$ and $p_0 = b^{-1} + c$ with $c \in \mathbb{Q}(n)$. Putting everything together gives the solution (17).

The technical details of the sketched algorithm for Problem FPLDE can be found in [26, 65]. More generally, this algorithm can be extended to a method from the first-order case to the mth-order case ($m \in \mathbb{N}$) as described in [70]. Furthermore, taking results from [14] we obtain a full algorithm that solves the following

Key problem PLDE: Parameterized Linear Difference Equations.
Given a $\Pi\Sigma^*$-field $(\mathbb{F},\sigma)$ over $\mathbb{K}$, $\alpha_0,\dots,\alpha_m \in \mathbb{F}$ (not all zero) and $f_0,\dots,f_d \in \mathbb{F}$.
Find all[a] $c_0,\dots,c_d \in \mathbb{K}$ and $g \in \mathbb{F}$ such that

$$\alpha_m\sigma^m(g) + \dots + \alpha_0 g = c_0 f_0 + \dots + c_d f_d. \quad (22)$$

[a]The solution set $V = \{(c_0,\dots,c_d,g) \in \mathbb{K}^{d+1} \times \mathbb{F} \mid (22) \text{ holds}\}$ forms a $\mathbb{K}$-vector space of dimension $\leq m+d+1$ and the task is to get an explicit basis of V.

Remark. Problem PLDE covers telescoping (see (16)), creative telescoping (see (40)) and recurrence solving (see (46)) for a given $\Pi\Sigma^*$-field. In particular, it is a crucial building block for the enhanced summation paradigms given below. Furthermore, it allows to deal with holonomic sequences in the setting of difference fields [68, 73].

3.3 Restriction to Polynomial $\Pi\Sigma^*$-Fields

We described how the summation objects can be rephrased in a $\Pi\Sigma^*$-field (Step 1) and how the telescoping problem, and more generally Problems FPLDE and PLDE can be solved there (Step 2). Subsequently, we restrict to polynomial $\Pi\Sigma^*$-fields. This will allow us to reformulate the found result completely automatically in terms of the given summation objects (Step 3) in Sect. 3.4.

Definition 6. A $\Pi\Sigma^*$-field $(\mathbb{F},\sigma)$ over $\mathbb{K}$ with $\mathbb{F} = \mathbb{K}(x)(p_1)\dots(p_r)(s_1)\dots(s_e)$ is called *polynomial* if $\sigma(x) = x+1$,

- $\sigma(p_i) = a_i\, p_i$ with $a_i \in \mathbb{K}(x)^*$ for all $1 \leq i \leq r$, and
- $\sigma(s_i) = s_i + f_i$, with[3] $f_i \in \mathbb{K}(x)[p_1, p_1^{-1},\dots,p_r,p_r^{-1}][s_1,\dots,s_{i-1}]$ for all $1 \leq i \leq e$.

Related to Remark 1 we note that in [76] a more general definition is used that covers also the q-hypergeometric and mixed case [16]. All what will follow generalizes to this general setting. Let $(\mathbb{F},\sigma)$ be a polynomial $\Pi\Sigma^*$-field over $\mathbb{K}$ as in Definition 6 and define the ring

$$R = \mathbb{K}(x)[p_1, p_1^{-1},\dots,p_r,p_r^{-1}][s_1,\dots,s_e]. \quad (23)$$

[3]$\mathbb{K}(x)[p_1, p_1^{-1},\dots,p_r,p_r^{-1}]$ stands for the polynomial Laurent ring in the variables $p_1,\dots,p_r$, i.e., an element is of the form $\sum_{(i_1,\dots,i_r)\in S} f_{(i_1,\dots,i_r)} p_1^{i_1}\dots p_r^{i_r}$ where $f_{(i_1,\dots,i_r)} \in \mathbb{K}(x)$ and $S \subseteq \mathbb{Z}^r$ is finite.

Note that for all $g \in R$ and $k \in \mathbb{Z}$ we have that $\sigma^k(g) \in R$. Thus restricting σ to R gives a ring automorphism (see Definition 1). Therefore (R, σ) is a difference ring and the set of constants is the field $\mathbb{K}$.

Example 2 (See Example 1). $(\mathbb{Q}(n)(x)(b)(h), \sigma)$ is a polynomial $\Pi\Sigma^*$-field over $\mathbb{Q}(n)$. In particular, we get the difference ring (R, σ) with constant field $\mathbb{Q}(n)$ for the polynomial (Laurent) ring

$$R = \mathbb{Q}(n)(x)[b, b^{-1}][h]. \tag{24}$$

We highlight that polynomial $\Pi\Sigma^*$-fields cover (up to the alternating sign) all *nested hypergeometric sums* (see Definition 4). This can be seen as follows.

- *Hypergeometric sequences.* Consider, e.g., the hypergeometric products

$$H_1(k) = \prod_{i=l_1}^{k} \alpha_1(i-1), \; \dots \;, H_v(k) = \prod_{i=l_v}^{k} \alpha_v(i-1) \qquad \text{with } \alpha_j(x) \in \mathbb{K}(x), \tag{25}$$

$(H_j(k) \neq 0$ for all $k \geq 0)$ where $\mathbb{K} = \mathbb{Q}$ or $\mathbb{K} = \mathbb{Q}(n_1, \dots, n_u)$ is a rational function field. Then there is an algorithm [69, Theorem 6.10] based on Theorem 1 that solves

Problem RP: Represent Products. *Given* a $\Pi\Sigma^*$-field $(\mathbb{K}(x), \sigma)$ over $\mathbb{K}$ with $\sigma(x) = x + 1$; $\alpha_1, \dots, \alpha_v \in \mathbb{K}(x)^*$.
Find a $\Pi\Sigma^*$-field $(\mathbb{F}, \sigma)$ over $\mathbb{K}$ with $\mathbb{F} = \mathbb{K}(x)(p_1)\dots(p_r)$ and $\sigma(p_i)/p_i \in \mathbb{K}(x)$ for $1 \leq i \leq r$ together with $g_j \in \mathbb{K}(x)[p_1, p_1^{-1}, \dots, p_r, p_r^{-1}]^*$ and $b_j \in \{-1, 1\}$ for $1 \leq j \leq v$ such that $\sigma(g_j) = b_j \alpha_j g_j$.

Namely, given $\alpha_1, \dots, \alpha_v \in \mathbb{K}(x)$, let $(\mathbb{F}, \sigma)$ with $R := \mathbb{K}(x)[p_1, p_1^{-1}, \dots, p_r, p_r^{-1}]$ together with $g_j \in R$ and $b_j \in \{-1, 1\}$ for $1 \leq j \leq v$ be the output of Problem RP.
If $b_j = 1$ for all $1 \leq j \leq v$, the products $H_j(k)$ in (25) can be expressed with $c_j\, g_j$ for appropriate $c_j \in \mathbb{K}^*$ in the polynomial $\Pi\Sigma^*$-field $(\mathbb{F}, \sigma)$.
Otherwise, construct the difference ring $(\mathbb{F}[m], \sigma)$ with $\sigma(m) = -m$, $m^2 = 1$ and $\text{const}_\sigma \mathbb{F}[m] = \mathbb{K}$; see [62]. Here m models $(-1)^k$ with $\mathscr{S}_k(-1)^k = -(-1)^k$. Then the $H_j(k)$ in (25) are rephrased with $c_j\, m^{(1-b_j)/2}\, g_j$ for appropriate $c_j \in \mathbb{K}^*$.
- *Indefinite nested sums.* Take an expression in terms of nested hypergeometric sums, i.e., the sums do not occur in a denominator. Moreover, suppose that all the arising hypergeometric products can be expressed in a polynomial $\Pi\Sigma^*$-field. Thus it remains to deal only with summation signs and to extend the given

polynomial $\Pi\Sigma^*$-field using Theorem 1.1. Suppose that during this construction it was so far possible to obtain a polynomial $\Pi\Sigma^*$-field, say it is of the form $(\mathbb{F},\sigma)$ as given in Definition 6 with the difference ring (R,σ) with R as in (23), and let $f \in R$ be the summand of the next sum under consideration. Then there are two cases. If we fail to find a $g \in \mathbb{F}$ such that $\sigma(g) = g + f$ then we can construct the $\Pi\Sigma^*$-field $(\mathbb{F}(t),\sigma)$ with $\sigma(t) = t+f$ by Theorem 1. In particular, this $\Pi\Sigma^*$-field is again polynomial. Otherwise, if we find such a $g \in \mathbb{F}$ with $\sigma(g) - g = f \in R$, we can apply the following result; the proof is a slight extension of the one given in [76, Theorem 2.7].

Theorem 2. *Let $(\mathbb{F},\sigma)$ be a polynomial $\Pi\Sigma^*$-field over $\mathbb{K}$ and consider the difference ring (R,σ) as above. Let $g \in \mathbb{F}$. If $\sigma(g) - g \in R$ then $g \in R$.*

Example 3 (Cont. Example 2). For $f \in R$ with (14) it follows that any solution $g \in \mathbb{F}$ with $\sigma(g) = g + f$ is in R. Indeed, we calculated (17).

Thus we always have $g \in R$. As a consequence we can express the sum over f with $g + c$ for some properly chosen $c \in \mathbb{K}$ (see Remark 4). Hence by iterative application of the above construction we never enter in the case that sums occur in the denominators. Consequently, a nested hypergeometric sum expression can be rephrased in a polynomial $\Pi\Sigma^*$-field up to the following technical aspect.

Remark 2. If the hypergeometric products (25) cannot be expressed in a $\Pi\Sigma^*$-field solving Problem RP, one needs in addition the alternating sign $(-1)^k$ in the setting of difference rings; note that here we cannot work anymore with fields, since zero divisors pop up: $(1 + (-1)^k)(1 - (-1)^k) = 0$. For simplicity, we restrict ourselves to polynomial $\Pi\Sigma^*$-fields; the described techniques and algorithms in this article can be extended for the more technical case allowing also $(-1)^k$; see [33, 62].

3.4 Step 3: Evaluating Elements from a $\Pi\Sigma^*$-Field to Sequences

Let $(\mathbb{F},\sigma)$ be a polynomial $\Pi\Sigma^*$-field over $\mathbb{K}$ as in Definition 6 and define R by (23). In this section we make the step precise how elements from R can be reformulated as a nested hypergeometric sum expression. I.e., how such an element $f \in R$ can be mapped via an evaluation function $\text{ev} : R \times \mathbb{N} \to \mathbb{K}$ to a sequence $\langle \text{ev}(f,i)\rangle_{i\geq 0}$ by using an explicitly given nested hypergeometric sum expression w.r.t. a variable k.

Before we start with a concrete example, we emphasize that this map ev should respect the ring structure R and the ring automorphism σ as follows.

Definition 7. A map $\text{ev} : R \times \mathbb{N} \to \mathbb{K}$ for a difference ring (R,σ) with constant field $\mathbb{K}$ is called *evaluation function* if the following evaluation properties hold: For all $c \in \mathbb{K}$ and all $i \geq 0$ we have that $\text{ev}(c,i) = c$, for all $f, g \in R$ there is a $\delta \geq 0$ with

$$\forall i \geq \delta : \mathrm{ev}(f\, g, i) = \mathrm{ev}(f, i)\, \mathrm{ev}(g, i), \tag{26}$$

$$\forall i \geq \delta : \mathrm{ev}(f + g, i) = \mathrm{ev}(f, i) + \mathrm{ev}(g, i); \tag{27}$$

and for all $f \in R$ and $j \in \mathbb{Z}$ there is a $\delta \geq 0$ with

$$\forall i \geq \delta : \mathrm{ev}(\sigma^j(f), i) = \mathrm{ev}(f, i + j). \tag{28}$$

Example 4 (See Example 1). Take the polynomial $\Pi\Sigma^*$-field $(\mathbb{Q}(n)(x)(b)(h), \sigma)$ with (15) and consider (R, σ) with (24) and constant field $\mathbb{Q}(n)$. We construct an evaluation map $\mathrm{ev} : R \times \mathbb{N} \to \mathbb{Q}(n)$ as follows. For $f \in R$ and $k \in \mathbb{N}$, $\mathrm{ev}(f, k)$ is calculated by evaluating an explicitly given nested hypergeometric sum expression. The construction is performed iteratively following the tower of extensions in R.

1. We define $\mathrm{ev} : \mathbb{Q}(n)(x) \times \mathbb{N} \to \mathbb{Q}(n)$ as follows. For $\frac{p}{q} \in \mathbb{Q}(n)(x)$ with $p, q \in \mathbb{Q}(n)[x]$ and $\gcd(p, q) = 1$,

$$\mathrm{ev}(\tfrac{p}{q}, k) = \begin{cases} \frac{p(k)}{q(k)} & \text{if } q(k) \neq 0 \\ 0 & \text{if } q(k) = 0 \quad \text{(pole case)}; \end{cases} \tag{29}$$

   here $p(k), q(k)$ with $k \in \mathbb{N}$ denotes the evaluation of the polynomials at $x = k$. Note that the properties in Definition 7 are satisfied for δ chosen sufficiently large: that is the case when one does not enter in the pole case in (29) for elements $f, g \in R$ as given in Definition 7; see also Example 5.
2. Next, we extend ev from $\mathbb{Q}(n)(x)$ to $\mathbb{Q}(n)(x)[b, b^{-1}]$. We set

$$\mathrm{ev}(b, k) = c_1 \prod_{i=l_1}^{k} \frac{n + 1 - i}{i}, \qquad l_1 \in \mathbb{N}, c_1 \in \mathbb{Q}(n)^* \tag{30}$$

   and prolong the ring structure as follows: for $f = \sum_{j=u}^{d} f_j b^j \in \mathbb{Q}(n)(x)[b, b^{-1}]$ with $f_j \in \mathbb{Q}(n)(x)$ and $u, d \in \mathbb{Z}$ we define $\mathrm{ev}(f, k) = \sum_{j=u}^{d} \mathrm{ev}(f_j, k)\mathrm{ev}(b, k)^j$; this implies that (26) and (27) hold for some $\delta \in \mathbb{N}$ sufficiently large; see Example 5. Note that for any choice of $l_1 \in \mathbb{N}$ and $c \in \mathbb{Q}(n)^*$ also (28) is valid. Since we want to model $\binom{n}{k} = \prod_{i=1}^{k} \frac{n+1-i}{i}$, a natural choice is $l_1 = 1, c_1 = 1$.
3. Finally, we set

$$\mathrm{ev}(h, k) = \sum_{i=\lambda_1}^{k} \frac{1}{i} + d_1, \qquad \lambda_1 \in \mathbb{N}, d_1 \in \mathbb{Q}(n). \tag{31}$$

   Again properties (26) and (27) hold for some $\delta \in \mathbb{N}$ (see Example 5) if we extend ev as follows: for $f = \sum_{j=0}^{d} f_j h^j \in R$ with $f_j \in \mathbb{Q}(n)(x)[b, b^{-1}]$ we

set $\text{ev}(f,k) = \sum_{j=0}^{d} \text{ev}(f_j,k)\text{ev}(h,k)^j$. In addition, property (28) holds for any choice of $\lambda_1 \in \mathbb{N}$ and $d_1 \in \mathbb{Q}(n)$. Since we want to model $S_1(k)$, we take, e.g., $\lambda_1 = 1$ and $d_1 = 0$.
For instance, for $f \in R$ as in (14) the evaluation is given by the nested hypergeometric sum expression (cf. (13))

$$\text{ev}(f,k) = \Big(1 - (n - 2(k+1))\Big(S_1(k) + \frac{1}{k+1}\Big)\Big)\frac{n-k}{k+1}\binom{n}{k}^{-1}; \qquad (32)$$

the usage of $S_1(k)$, $\binom{n}{k}$ is just pretty printing and stands for (30) and (31), respectively.

Besides the function ev we aim at the calculation of the bounds δ in Definition 7.

Example 5. For the evaluation function $\text{ev} : R \times \mathbb{N} \to \mathbb{Q}(n)$ from Example 4 the bounds can be extracted by the computable map $\beta : R \to \mathbb{N}$ defined as follows. For $f \in R$ let $d \in \mathbb{N}$ be minimal such that for all $k \geq d$ the calculation of $\text{ev}(f,k)$ does not enter in the pole case in (29). More precisely, write f in the form $f = \sum_{i\in\mathbb{Z},j\in\mathbb{N}} f_{i,j} b^i h^j$ with finitely many $f_{i,j} \in \mathbb{Q}(n)(x)$ being non-zero, and choose $d \in \mathbb{N}$ such that for all i, j and all $k \geq d$ the denominator of $f_{i,j}$ has no pole at $x = k$. This d can be calculated explicitly for any $f \in R$ and defines the function β with $\beta(f) := d$. Now let $f, g \in R$. Then for $\delta := \max(\beta(f), \beta(g))$ we have that (26) and (27). In addition, for all $j \in \mathbb{Z}$ choose $\delta := \beta(f) + \max(0, -j)$ and we get (28).

This example motivates the following definition [76].

Definition 8. Let (R, σ) be a difference ring with constant field $\mathbb{K}$ and consider an evaluation function $\text{ev} : R \times \mathbb{N} \to \mathbb{K}$. $\beta : R \to \mathbb{N}$ is called *bounding function* of ev if for all $f, g \in R$ we can take $\delta := \max(\beta(f), \beta(g))$ such that (26) and (27) holds, and for all $f \in R$ and $j \in \mathbb{Z}$ we can take $\delta := \beta(f) + \max(0, -j)$ such that (28) holds.

The concrete construction above carries over to the general case. Let $(\mathbb{F}, \sigma)$ be a polynomial $\Pi\Sigma^*$-field over $\mathbb{K}$ as given in Definition 6 and define R by (23). Then one obtains an evaluation map $\text{ev} : R \times \mathbb{N} \to \mathbb{K}$ in terms of explicitly given nested hypergeometric sum expressions as follows. We start with $\text{ev} : \mathbb{K}(x) \times \mathbb{N} \to \mathbb{K}$ defined by (29) for $\frac{p}{q} \in \mathbb{K}(x)$ with $p, q \in \mathbb{K}[x]$ and $\gcd(p,q) = 1$. Next we define how the map acts on the p_i:

$$\text{ev}(p_i, k) = c_i \prod_{j=l_i}^{k} a_i(j-1), \quad (1 \leq i \leq r); \qquad (33)$$

here we are free to choose $c_i \in \mathbb{K}^*$, and l_i is chosen such that the numerator and denominator of a_i evaluated at j is non-zero for all $j \geq l_i$. Then this map is extended to $\mathrm{ev} : \bar{R}_0 \times \mathbb{N} \to \mathbb{K}$ with $\bar{R}_0 := \mathbb{K}(x)[p_1, p_1^{-1}, \dots, p_r, p_r^{-1}]$ as follows. For $f = \sum_{(i_1,\dots,i_r)\in\mathbb{Z}^r} f_{(i_1,\dots,i_r)} p_1^{i_1} \dots p_r^{i_r} \in \bar{R}_0$ with $f_{(i_1,\dots,i_r)} \in \mathbb{K}(x)$ we set

$$\mathrm{ev}(f, k) = \sum_{(i_1,\dots,i_r)\in\mathbb{Z}^r} \mathrm{ev}(f_{(i_1,\dots,i_r)}, k)\mathrm{ev}(p_1, k)^{i_1} \dots \mathrm{ev}(p_r, k)^{i_r}.$$

Finally, we extend iteratively this map from $\bar{R}_0$ to $R := \bar{R}_e$. Suppose that we are given already the map for $\bar{R}_i = \bar{R}[s_i, \dots, s_{i-1}]$ with $1 \le i < e$. Then we define

$$\mathrm{ev}(s_i, k) = \sum_{j=\lambda_i}^{k} \mathrm{ev}(f_i, j-1) + d_i \tag{34}$$

where $d_i \in \mathbb{K}$ can be arbitrarily chosen, and $\lambda_i \in \mathbb{N}$ is sufficiently large in the following sense: it is larger than the lower bounds of the arising sums and products of the explicitly given nested hypergeometric sum expression for $\mathrm{ev}(f_i, j-1)$ and such that during the evaluation one never enters in the pole case in (29). In a nutshell, the underlying expression for $\mathrm{ev}(s_i, k)$ (for symbolic k) can be written as an indefinite nested sum without entering poles that are captured via (29). Finally, we extend this construction to $\bar{R}_{i-1}[s_i]$: for $f = \sum_{j=0}^{v} f_j s_i^j \in \bar{R}_{i-1}[s_i]$ with $f_j \in \bar{R}_{i-1}$ we define $\mathrm{ev}(f, k) = \sum_{j=0}^{v} \mathrm{ev}(f_j, k)\mathrm{ev}(s_i, k)^j$. To this end, by iteration on i $(1 \le i \le e)$ we obtain $\mathrm{ev} : R \times \mathbb{N} \to \mathbb{K}$ with $R := \bar{R}_e$ that satisfies the properties in Definition 7 and which is explicitly given in terms of nested hypergeometric sum expressions.

Remark. Note that the products in (33) and sums in (34) are just nested hypergeometric sum expressions w.r.t. k; see Definition 4.

Moreover, we can define explicitly a bounding function $\beta : R \to \mathbb{N}$ that produces the required bounds δ in Definition 7 following the construction in Example 5. In short, consider $f \in R$ as a polynomial in the variables p_i, s_i and take all its coefficients from $\mathbb{K}(x)$. Then $\beta(f)$ is the minimal value $d \in \mathbb{N}$ such that the evaluation of the coefficients does not enter in the pole case of (29); note that the positive integer roots of the denominators can be detected if $\mathbb{K}$ is computable (in particular, if one can factorize polynomials over $\mathbb{K}$). We can summarize this construction as follows; cf. [76].

Lemma 2. *Let* $(\mathbb{F}, \sigma)$ *be a polynomial* $\Pi\Sigma^*$*-field over* $\mathbb{K}$ *with* R *as in* (23). *Then* $\mathrm{ev} : R \times \mathbb{N} \to \mathbb{K}$ *defined above in terms of nested hypergeometric sum expressions is an evaluation function, and* $\beta : R \to \mathbb{N}$ *given above is a corresponding bounding function of* ev. *If* $\mathbb{K}$ *is computable, such functions can be calculated explicitly.*

Remark 3. Solving Problem PLDE. Take a polynomial $\Pi\Sigma^*$-field $(\mathbb{F}, \sigma)$ as in Definition 6 and define R by (23). Let $\mathrm{ev} : R \times \mathbb{N} \to \mathbb{K}$ be an evaluation function with a bounding function $\beta : R \to \mathbb{N}$ as constructed above by means of hypergeometric sum expressions. Let $\alpha_i, f_i \in R$ and take $g \in R$ such that Eq. (22) holds.

Now calculate $\delta := \max(\beta(\alpha_0), \ldots, \beta(\alpha_m), \beta(f_0), \ldots, \beta(f_d), \beta(g))$. Then for any $k \geq \delta$,

$$\mathrm{ev}(\alpha_m, k)\mathrm{ev}(g, k+m) + \cdots + \mathrm{ev}(\alpha_0, k)\mathrm{ev}(g, k) = c_0\mathrm{ev}(f_0, k) + \cdots + c_d\mathrm{ev}(f_d, k). \tag{35}$$

Hence a solution $g \in R$ of (22) produces a solution of (35) in terms of nested hypergeometric sum expressions. In particular, a lower bound δ for its validity can be computed (if $\mathbb{K}$ is computable). We emphasize that this property is crucial for the automatic execution of the summation paradigms given in Sect. 4 below.

Remark 4. Simultaneous construction of a $\Pi\Sigma^*$-field and its evaluation function. In order to model the summation problem accordingly (see, e.g., Problem EAR on page 344) the construction of the $\Pi\Sigma^*$-field (Step 1) and the evaluation function with its bounding function should be performed simultaneously. Here the choice of the lower bounds and constants in (33) and (34) are adjusted such that the evaluation of the introduced products and sums agrees with the objects of the input expression; for a typical execution see Example 4. In particular the evaluation function is crucial if a sum can be represented in the already given $\Pi\Sigma^*$-field by telescoping. This is, e.g., the case in (5) of Example 1. We succeeded in representing the summand $F(k)$ in (13) by the element $f \in R$ as in (14). Namely, using the evaluation function from Example 4 we have (32) for all $k \geq 0$. Then we calculate the solution (17) of $\sigma(g) = g + f$. In particular, we obtain $\mathrm{ev}(g, k) = ((k+1)S_1(k)+1)\binom{n}{k}^{-1} + c$ with $c \in \mathbb{Q}(n)$ such that for all $k \geq 0$ we have that $\mathrm{ev}(g, k+1) = \mathrm{ev}(g, k) + \mathrm{ev}(f, k+1)$; see Remark 3. Now we follow the same arguments as in the beginning of Sect. 2: $A_{-1}(a)$ and $\mathrm{ev}(g, a)$ satisfy the same recurrence $A(a+1) = A(a) + \mathrm{ev}(f, a+1)$ for all $a \geq 0$ and thus $\mathrm{ev}(g, a) = A_{-1}(a)$ when choosing $c = 0$. In this way, we represent $A_{-1}(a)$ precisely with g ($c = 0$) in the given $\Pi\Sigma^*$-field and its evaluation function.

3.5 Crucial Property: Algebraic Independence of Sequences

Take the elements from a polynomial $\Pi\Sigma^*$-field, rephrase them as nested hypergeometric sum expressions, and evaluate the derived objects to sequences. The main result of this subsection is that the sequences of the generators of the $\Pi\Sigma^*$-field are algebraically independent over each other. To see this, let $(\mathbb{F}, \sigma)$ be a polynomial $\Pi\Sigma^*$-field over $\mathbb{K}$ with $\mathbb{F} = \mathbb{K}(x)(p_1)\ldots(p_r)(s_1)\ldots(s_e)$ and define R by (23). Moreover, take an evaluation function $\mathrm{ev} : R \times \mathbb{N} \to \mathbb{K}$ following the construction of the previous subsection. Then we can define the map $\tau : R \to \mathbb{K}^{\mathbb{N}}$ with

$$\tau(f) = \langle \mathrm{ev}(f, k) \rangle_{k \geq 0} = \langle \mathrm{ev}(f, 0), \mathrm{ev}(f, 1), \mathrm{ev}(f, 2), \ldots \rangle. \tag{36}$$

Now we explore the connection between the difference ring (R, σ) and the set of sequences $\tau(R) = \{\tau(f) | f \in R\}$. First, we introduce the following notions.

Definition 9. Let (R_1, σ_1) and (R_2, σ_2) be difference rings.

- If R_1 is a subring of R_2 and $\sigma_1(f) = \sigma_2(f)$ for all $f \in R_1$ then (R_1, σ) is called *sub-difference ring* of (R_2, σ).
- A map $\tau : R_1 \to R_2$ is called ring homomorphism if $\tau(f\,g) = \tau(f)\tau(g)$ and $\tau(f + g) = \tau(f) + \tau(g)$ for all $f, g \in R_1$. If τ is in addition injective (resp. bijective), τ is called *ring embedding* (resp. *ring isomorphism*). Note: if τ is an isomorphism, the rings R_1 and R_2 are the same up to renaming of the elements with τ.
- A map $\tau : R_1 \to R_2$ is called *difference ring homomorphism* (resp. *embedding/isomorphism*) if it is a ring homomorphism (resp. embedding/isomorphism) and for all $n \in \mathbb{Z}$, $f \in R_1$ we have $\tau(\sigma_1^n(f)) = \sigma_2^n(\tau(f))$. Note: if τ is an isomorphism, (R_1, σ) and (R_2, σ) are the same up to renaming of the elements by τ.

With component-wise addition and multiplication of the elements from $\mathbb{K}^{\mathbb{N}}$ we obtain a commutative ring where the multiplicative unit is $\mathbf{1} = \langle 1, 1, 1, \dots \rangle$; the field $\mathbb{K}$ can be naturally embedded by mapping $k \in \mathbb{K}$ to $\mathbf{k} = \langle k, k, k, \dots \rangle$.

Example 6. Let $\mathbb{K}(x)$ be a rational function field, take the evaluation function $\text{ev} : \mathbb{K}(x) \times \mathbb{N} \to \mathbb{K}(x)$ defined by (29), and define $\tau : \mathbb{K}(x) \to \mathbb{K}^{\mathbb{N}}$ as in (36). Now define the set

$$F := \tau(\mathbb{K}(x)) = \{\langle \text{ev}(f, k) \rangle_{k \geq 0} | f \in \mathbb{K}(x)\}.$$

Observe that F is a subring of $\mathbb{K}^{\mathbb{N}}$. However, it is not a field. E.g., if we multiply $\text{ev}(x, k))_{k \geq 0}$ with $\text{ev}(1/x, k))_{k \geq 0}$, we obtain $\langle 0, 1, 1, 1, \dots \rangle$ which is not the unit $\mathbf{1}$. But, we can turn it to a field by identifying two sequences if they agree from a certain point on. Then the inverse of $\text{ev}(x, k))_{k \geq 0}$ is $\text{ev}(1/x, k))_{k \geq 0}$. More generally, for $f \in \mathbb{K}(x)^*$ we get $\text{ev}(f, k))_{k \geq 0}\, \text{ev}(1/f, k))_{k \geq 0} = \mathbf{1}$.

To be more precise, we follow the construction from [59, Sect. 8.2]: We define an equivalence relation $\sim$ on $\mathbb{K}^{\mathbb{N}}$ by $\langle a_n \rangle_{n \geq 0} \sim \langle b_n \rangle_{n \geq 0}$ if there exists a $\delta \geq 0$ such that $a_n = b_n$ for all $n \geq \delta$. The equivalence classes form a ring which is denoted by $S(\mathbb{K})$; the elements of $S(\mathbb{K})$ (also called germs) will be denoted, as above, by sequence notation. Finally, define the shift operator $\mathscr{S} : S(\mathbb{K}) \to S(\mathbb{K})$ with

$$\mathscr{S}(\langle a_0, a_1, a_2, \dots \rangle) = \langle a_1, a_2, a_3, \dots \rangle.$$

In this ring the shift is invertible with $\mathscr{S}^{-1}(\langle a_1, a_2, \dots \rangle) = \langle 0, a_1, a_2, a_3, \dots \rangle = \langle a_0, a_1, a_2, \dots \rangle$. It is immediate that $\mathscr{S}$ is a ring automorphism and thus $(S(\mathbb{K}), \mathscr{S})$ is a difference ring. In short, we call this difference ring also *ring of sequences*.

Example 7 (Cont. Example 6). Consider our subring F of $S(\mathbb{K})$. Restricting $\mathscr{S}$ to F gives a bijective map and thus it is again a ring automorphism. Even more, since F is a field, it is a field automorphism, and $(F, \mathscr{S})$ is a difference field. In particular, $(F, \mathscr{S})$ is a sub-difference ring of $(S(\mathbb{K}), \mathscr{S})$.

More generally, consider the map $\tau : R \to S(\mathbb{K})$ as in (36). Since $\mathrm{ev} : R \times \mathbb{N} \to \mathbb{K}$ has the properties as in Definition 7, it follows that for all $f, g \in R$ we have $\tau(f\,g) = \tau(f)\,\tau(g)$ and $\tau(f+g) = \tau(f) + \tau(g)$. Hence τ is a ring homomorphism. Moreover, for all $f \in R$ and all $n \in \mathbb{Z}$,

$$\mathcal{S}^n(\langle \mathrm{ev}(f,k)\rangle_{k\geq 0}) = \langle \mathrm{ev}(f,k+n)\rangle_{k\geq 0} = \langle \mathrm{ev}(\sigma^n(f),k)\rangle_{k\geq 0}.$$

Thus τ is a difference ring homomorphism between (R,σ) and $(S(\mathbb{K}),\mathcal{S})$. Since $\tau(R)$ is a subring of $S(\mathbb{K})$ and $\mathcal{S}$ restricted to $\tau(R)$ is a ring automorphism, $(\tau(R),\mathcal{S})$ is a difference ring, and it is a sub-difference ring of $(S(\mathbb{K}),\mathcal{S})$.

Example 8. Take the polynomial $\Pi\Sigma^*$-field $(\mathbb{Q}(n)(x)(b)(h),\sigma)$ over $\mathbb{Q}(n)$ with (15) and (24), and let $\mathrm{ev} : R \times \mathbb{N} \to \mathbb{Q}(n)$ be the evaluation function from Example 4; define $\tau : R \to S(\mathbb{Q}(n))$ with (36). Then τ is a difference ring homomorphism. In particular, $(\tau(R),\mathcal{S})$ is a difference ring and a sub-difference ring of $(S(\mathbb{Q}(n)),\mathcal{S})$.

Now we can state the crucial property proven in [76]: our map (36) is injective.

Theorem 3. *Let* $(\mathbb{F},\sigma)$ *be a polynomial* $\Pi\Sigma^*$*-field over* $\mathbb{K}$*, define* R *by* (23)*, and take an evaluation function* $\mathrm{ev} : R \times \mathbb{N} \to \mathbb{K}$ *as given in Lemma 2. Then* $\tau : R \to S(\mathbb{K})$ *with* (36) *is a difference ring embedding.*

Example 9. $(\mathbb{Q}(n)(x),\sigma)$ and $(\tau(\mathbb{Q}(n)(x)),\sigma)$ are isomorphic. In addition, the rings (R,σ) and $(\tau(R),\sigma)$ with $R := \mathbb{Q}(n)(x)[b,b^{-1}][h]$ are isomorphic. Thus $\tau(R) = \tau(\mathbb{Q}(n)(x))[\tau(b),\tau(b^{-1})][\tau(h)]$ is a polynomial ring and there are no algebraic relations among the sequences $\tau(b)$, $\tau(b^{-1})$, $\tau(h)$ with coefficients from $\tau(\mathbb{Q}(n)(x))$.

In general, the difference rings (R,σ) and $(\tau(R),\mathcal{S})$ are isomorphic: they are the same up to renaming of the elements by τ. In particular, we get the polynomial ring

$$\tau(R) = \tau(\mathbb{K}(x))[\tau(p_1),\tau(p_1^{-1}),\dots,\tau(p_r),\tau(p_r^{-1})][\tau(s_1),\dots,\tau(s_e)] \tag{37}$$

with coefficients form the field $\tau(\mathbb{K}(x))$. I.e., there are no algebraic relations among the sequences $\tau(p_i)$, $\tau(p_i^{-1})$ and $\tau(s_i)$ with coefficients from $\tau(\mathbb{K}(x))$.

4 The Symbolic Summation Toolbox of `Sigma`

In the following we will give an overview of the symbolic summation toolbox that is available in the Mathematica package `Sigma` [73]. Here we focus on *nested hypergeometric sum expressions (w.r.t. k)* as given in Definition 4: the products are hypergeometric expressions (for more general classes see Remark 1) and the sums and products do not arise in the denominators.

Concerning *indefinite summation* it is shown how a nested hypergeometric sum expression can be compactified such that the arising sums are algebraically independent and such that the sums are simplified concerning certain optimality criteria. Concerning *definite summation* the package Sigma provides the following toolkit. In Sect. 4.2 it is worked out how a recurrence can be computed with creative telescoping for a definite sum over a nested hypergeometric sum expression. Moreover, in Sect. 4.3 it is elaborated how such a recurrence can be solved in terms of nested hypergeometric sum expressions which evaluate to d'Alembertian sequences. Usually the derived solutions are highly nested, and thus indefinite summation is heavily needed. Finally, given sufficiently many solutions their combination gives an alternative representation of the definite input sum. Summarizing, the following "summation spiral" is applied [64]:

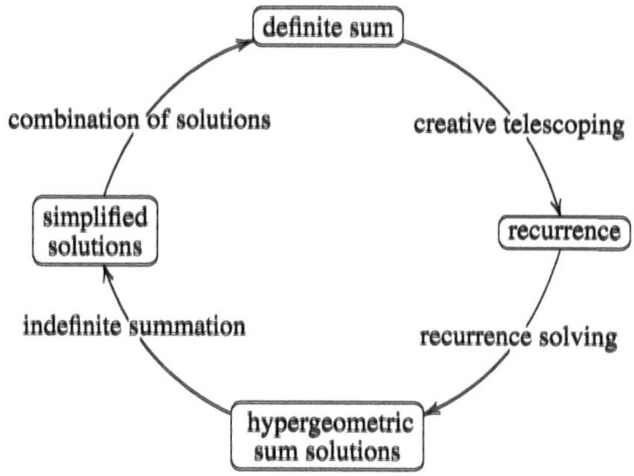

Remark. We give details how these summation paradigms are solved in the setting of polynomial $\Pi\Sigma^*$-fields introduced in Sect. 3. These technical parts marked with * can be ignored if one is mostly interested in applying the summation tools.

4.1 Simplification of Nested Hypergeometric Sum Expressions

All of the simplification strategies of Sigma solve the following basic problem.

Problem EAR: Elimination of algebraic relations.
Given a nested hypergeometric sum expression $F(k)$.
Find a nested hypergeometric sum expression $\bar{F}(k)$ and $\lambda \in \mathbb{N}$ such that $F(k) = \bar{F}(k)$ for all $k \geq \lambda$ and such that the occurring sums are algebraically independent.

The following solution relies on Sect. 3 utilizing ideas from [39, 70, 76].

*Solution**. Compute a polynomial $\Pi\Sigma^*$-field[4] $(\mathbb{F}, \sigma)$ over $\mathbb{K}$ as in Definition 23 with R defined as in (6) together with an evaluation function $\mathrm{ev} : R \times \mathbb{N} \to \mathbb{K}$ in which one obtains an explicit $f \in R$ with $\lambda \in \mathbb{N}$ such that $\mathrm{ev}(f, k) = F(k)$ for all $k \geq \lambda$. Output the nested hypergeometric sum expression $\bar{F}(k)$ that encodes the evaluation $\mathrm{ev}(f, k)$. Concerning the algebraic independence note that the sub-difference ring (37) of the ring of sequences $(S(\mathbb{K}), \mathscr{S})$ forms a polynomial ring; here the difference ring embedding τ is defined by (36). The sequences given by the objects occurring in $\bar{F}(k)$ are just the the generators of the polynomial ring (37).

Remark. $\mathbb{K}$ is the smallest field that contains the values of $F(r)$ for all $r \in \mathbb{N}$ with $r \geq \lambda$. Here extra parameters are treated as variables. However, in most examples these parameters are assumed to be integer valued within a certain range. In such cases it might be necessary to adjust the summation bounds accordingly.

A typical instance of Problem EAR is the simplification of the sum (7): Using the above technologies, see Examples 1 and 4 for further details, we can reduce the sum $A_{-1}(a)$ in terms of $\binom{n}{a}$ and $S_1(a)$ and obtain the simplification (10). After loading in

```
In[1]:= << Sigma.m
        Sigma − A summation package by Carsten Schneider© RISC
```

this task can be accomplished with the function call

$$\text{In[2]:= SigmaReduce}[\sum_{k=0}^{a}(1-(n-2k)S[1,k])\binom{n}{k}^{-1}, a]$$

$$\text{Out[2]=}\ ((a+1)S[1,a]+1)\binom{n}{a}^{-1}$$

Note that $S[m_1, \ldots, m_k, n]$ stands for the harmonic sums (5). More generally, one gets reduced representations for nested hypergeometric sum expressions such as

$$\text{In[3]:= SigmaReduce}[\sum_{k=1}^{a}k^4\binom{2k}{k}^2+\frac{249}{20}\sum_{k=1}^{a}k^3\binom{2k}{k}^2+\frac{259}{20}\sum_{k=1}^{a}k^2\binom{2k}{k}^2+\sum_{k=1}^{a}\binom{2k}{k}^2+2\sum_{k=1}^{a}k\binom{2k}{k}^2, a]$$

$$\text{Out[3]=}\ \sum_{i_1=1}^{a}\binom{2i_1}{i_1}^2-\sum_{i_1=1}^{a}\binom{2i_1}{i_1}^2 i_1+\frac{1}{15}a(2a+1)^2(4a+45)\binom{2a}{a}^2$$

4.1.1 Simplification with Improved Difference Field Theory

The solution of Problem EAR is obtained by calculating a set of algebraic independent sums (the generators of the $\Pi\Sigma^*$-field) in which the occurring sums of the

[4] As observed in Remark 2 one might need in addition the alternating sign to represent all hypergeometric products. The underlying solution works analogously by adapted algorithms.

input expression can be rephrased. In order to guarantee that the output expression consists of sums and products that are simpler (or at least not more complicated) than the input expression, the generators of the $\Pi\Sigma^*$-field must be constructed such that certain optimality criteria are fulfilled. In short, we refine Problem EAR using improved $\Pi\Sigma^*$-difference field theory and enhanced algorithms for Problem T. The most useful features of `SigmaReduce` can be summarized as follows.

- *Atomic representation.* By default all sums are split into atomic parts (using partial fraction decomposition) and an algebraic independent representation of the arising sums and products is calculated. In addition, `Sigma` outputs sums such that the denominators have minimal degrees w.r.t. the summation index (i.e., if possible, the denominator w.r.t. the summation index is linear). A typical example is

$$\text{In[4]:= SigmaReduce}[\sum_{k=1}^{a}\Big(\frac{-2+k}{10(1+k^2)}+\frac{(1-4k-2k^2)S[1,k]}{10(1+k^2)(2+2k+k^2)}+\frac{(1-4k-2k^2)S[3,k]}{5(1+k^2)(2+2k+k^2)}\Big),a]$$

$$\text{Out[4]= } \frac{a^2+4a+5}{10(a^2+2a+2)}S[1,a]-\frac{(a-1)(a+1)}{5(a^2+2a+2)}S[3,a]_3(a)-\tfrac{2}{5}\sum_{k=1}^{a}\frac{1}{k^2}$$

  This feature relies on algorithms refining those given in [72]; for the special case of rational sums see, e.g., [11, 52]. By default this refinement is activated; it can be switched off by using the option `SimpleSumRepresentation-> False`.

- By default the following fundamental problem is solved:

Problem DOS: Depth Optimal Summation. *Given* a nested hypergeometric sum expression. *Find* an alternative representation of a nested hypergeometric sum expression whose nesting depth is minimal. Moreover, each derived sum cannot be expressed by a nested hypergeometric sum expression with lower depth.

  The solution to this problem is possible by the enhanced difference field theory of depth-optimal $\Pi\Sigma^*$-fields and the underlying telescoping algorithms; see [74, 75]. E.g., we can flatten the harmonic sum $S_{3,2,1}(a)$ of depth 3 to sums of depth$\leq$ 2:

$$\text{In[5]:= SigmaReduce}[\sum_{i=1}^{a}\frac{1}{i^3}\sum_{j=1}^{i}\frac{1}{j^2}\sum_{k=1}^{j}\frac{1}{k},a]$$

$$\text{Out[5]= } \sum_{i_1=1}^{a}\frac{1}{i_1^5}\sum_{i_2=1}^{i_1}\frac{1}{i_2}+\Big(\sum_{i_1=1}^{a}\frac{1}{i_1^3}\Big)\Big(\sum_{i_1=1}^{a}\frac{1}{i_1^2}\sum_{i_2=1}^{i_1}\frac{1}{i_2}\Big)-\sum_{i_1=1}^{a}\frac{1}{i_1^2}\Big(\sum_{i_2=1}^{i_1}\frac{1}{i_2^3}\Big)\Big(\sum_{i_2=1}^{i_1}\frac{1}{i_2}\Big)$$

  This depth-optimal $\Pi\Sigma^*$-field theory yields various structural theorems [77], i.e., gives a priori certain properties how the telescoping solution looks like. In particular, this leads to very efficient algorithms (for telescoping but also for creative telescoping and recurrence solving given below) where we could work with more than 500 sums in a depth-optimal $\Pi\Sigma^*$-field. The naive (and usually less efficient) $\Pi\Sigma^*$-field approach is used with the option `SimplifyByExt→None`.

Example 10. For the 2,186 harmonic sums (5) with weight $\sum_{i=1}^{k} |m_i| \leq 7$ all algebraic relations are determined [10]. More precisely, using their quasi-shuffle algebra the sums could be reduced by the `HarmonicSums` package [1] to 507 basis sums. Then using the algorithms above we showed that they are algebraic independent.

- *Reducing the number of objects and the degrees in the summand.* The depth-optimal representation can be refined further as follows.

  Given a nested hypergeometric sum expression, *find* an alternative sum representation such that for the outermost summands the number of occurring objects is as small as possible (more precisely, concerning a given tower of a $\Pi\Sigma^*$-field the smallest subfield is searched in which the summand can be represented); see [65].

  E.g., in the following example we can eliminate $S_1(k)$ from the summand:

In[6]:= $\textbf{SigmaReduce}[\sum_{k=0}^{a}(-1)^k S_1(k)^2\binom{n}{k}, a, \textbf{SimplifyByExt} \to \textbf{DepthNumber}]$

Out[6]= $-(a-n)(n^2 S_1(a)^2+2nS_1(a)+2)\frac{(-1)^a\binom{n}{a}}{n^3}-\frac{2}{n^2}-\frac{1}{n}\sum_{i_1=1}^{a}\frac{(-1)^{i_1}}{i_1}\binom{n}{i_1}$

Furthermore, one can calculate representations such that the degrees (w.r.t. the top extension of a $\Pi\Sigma^*$-field) in the numerators and denominators of the summands are minimal [72]. For algorithms dealing with the product case we point to [13,69].

4.2 *Finding Recurrence Relations for Definite Sums*

Given a sum, say[5] $A(n) = \sum_{k=0}^{L(m_1,\dots,m_u,n)} F_n(k)$ where $F_n(k)$ is a nested hypergeometric sum expression depending on a discrete parameter n, *find* polynomials $c_0(n), \dots, c_d(n)$ (not all zero) and an expression $h(n)$ in terms of sums that are simpler (see below) than the sum $A(n)$ such that the following linear recurrence holds:

$$c_0(n)A(n) + \dots + c_d(n)A(n+d) = h(n). \tag{38}$$

We treat this problem by the following variation of creative telescoping [82].

[5] $L(m_1,\dots,m_u,n)$, $u \geq 0$, stands for a linear combination of the m_i and n with integer coefficients.

Problem CT: Creative telescoping (general paradigm). *Given* $d \in \mathbb{N}$ and $F_n(k)$ such that $F_{n+i}(k)$ with $i \in \mathbb{N}$ $(0 \leq i \leq d)$ can be written as nested hypergeometric sum expression. *Find* $\lambda \in \mathbb{N}$, $c_0, \ldots, c_d \in \mathbb{K}(n)$, not all zero, and $G(k)$ such that for all $k \geq \lambda$ we have

$$c_0 F_n(k) + c_1 F_{n+1}(k) + \cdots + c_d F_{n+d}(k) = \mathcal{S}_k G(k) - G(k) \qquad (39)$$

and such that the summands of the occurring sums in $G(k)$ are simpler (depending on the chosen strategy, see below) than $F_{n+i}(k)$; if this is not possible, return $\bot$.

The following solution relies on [39, 70, 76].

*Solution**. We consider the parameter n as variable. Compute an "appropriate" polynomial $\Pi\Sigma^*$-field $(\mathbb{F}, \sigma)$ over the constant field $\mathbb{K}(n)$ as in Definition 23 with R defined as in (6) together with an evaluation function $\text{ev} : R \times \mathbb{N} \to \mathbb{K}$ in which one obtains explicitly $f_0, \ldots, f_d \in R$ with $\lambda' \in \mathbb{N}$ such that $\text{ev}(f_i, k) = F_{n+i}(k)$ for all $k \geq \lambda'$; again we point to Footnote 4. Compute, if possible, a solution $c_0, \ldots, c_d \in \mathbb{K}(n)$ (not all zero) and $g \in R$ (or an extension of (R, σ) with an extended evaluation function ev) such that

$$c_0 f_0 + \cdots + c_d f_d = \sigma(g) - g \qquad (40)$$

holds; in addition we require that in g the summands of the occurring sum extensions are simpler (depending on the chosen strategy, see below) than each of the given f_i. If there is not such a solution, return $\bot$. Otherwise extract a nested hypergeometric sum expression $G(k)$ such that $G(k) = \text{ev}(g, k)$ and compute $\lambda \in \mathbb{N}$ such that (39) holds for all $k \geq \lambda$; see Remark 3 with $m = 1, \alpha_1 = 1, \alpha_0 = -1$. Then return $(c_0, \ldots, c_d, G(k))$ and λ.

Application. Usually, one loops over $d = 0, 1, \ldots$ until a solution for (39) is found; for termination issues see Remark 1 on page 355. Then summing (39) over a valid range, e.g., from λ to a, gives

$$c_0(n) \sum_{k=\lambda}^{a} F_n(k) + \cdots + c_d(n) \sum_{k=\lambda}^{a} F_{n+d}(k) = G(a+1) - G(\lambda) \qquad (41)$$

where by construction the summands of the arising sums in $\bar{h}(a) := G(a+1) - G(\lambda)$ are simpler than $F_{n+i}(k)$. This implies that also the arising sums $\bar{h}(a)$ are simpler than $\sum_{k=\lambda}^{a} F_{n+i}(k)$. Note that so far n is considered as an indeterminate. We remark that in many applications n itself is an integer valued parameter and extra caution is necessary to avoid poles when summing up (39). Finally, when setting $a = L(m_1, \ldots, m_u, n)$ in (41) (if $a = \infty$, a limit has to be performed) and taking care of missing summands yields (38) for $A(n)$; see Example 11 for details.

Proof certificate. The correctness of (38) for a given sum $A(n)$ is usually hard to prove. However, given the proof certificate $(c_0, \ldots, c_d, G(k))$ it can be easily verified that (39) holds within the required summation range. Then summing this equation over this range yields the verified result (38).

With `Sigma` one can calculate for $A_{-3}(n) =$ `SUM[n]` a recurrence[6] as follows:

In[7]:= $\text{mySum} = \sum_{k=0}^{n} (1 - 3(n - 2k)S_1(k)) \binom{n}{k}^{-3}$;

In[8]:= rec = GenerateRecurrence[mySum, n][[1]]

Out[8]= $(n+2)^4(n+3)^2 \text{SUM}[n] + (n+1)^3(n+3)^2(2n+5)\text{SUM}[n+1] + (n+1)^3(n+2)^3 \text{SUM}[n+2]$
$== (20n^3 + 138n^2 + 311n + 229)(n+1)^2 + 6(n+2)^2(n+3)(2n+5)(n+1)^3 S_1(n)$

The essential calculation steps are given in the following example.

Example 11. * Take $A_{-3}(n) = \sum_{k=0}^{n} F_n(k)$ with $F_n(k) = (1 - 3(n - 2k)S_1(k))\binom{n}{k}^{-3}$. We calculate a recurrence for $A_{-3}(n)$ in n by the techniques described above. First, we search for a solution of (39) with $d = 0$ (which amounts to telescoping). I.e., we construct the polynomial $\Pi\Sigma^*$-field $(\mathbb{Q}(n)(x)(b)(h), \sigma)$ and evaluation function $ev : R \times \mathbb{N} \to \mathbb{Q}(n)$ with $R := \mathbb{Q}(n)(x)[b, b^{-1}][h]$ as in Example 4. There we take $f_0 = (1 - 3(2n - x)h)b^{-3} \in R$ such that $\text{ev}(f_0) = F_n(k)$ for all $k \geq 0$. Unfortunately, our telescoping algorithm fails to find a $g \in R$ such that $\sigma(g) - g = f_0$ holds. So we try to find a solution of (39) with $d = 1$. Since $\binom{n+1}{k} = \frac{n+1}{n-k+1}\binom{n}{k}$, we can rephrase $F_{n+1}(k)$ by $f_1 = (1 - 3(2n - x)h)\frac{(n-x+1)^3}{(n+1)^3}b^{-3}$, i.e., $\text{ev}(f_1, k) = F_{n+1}(k)$ for all $k \geq 0$. Then we activate the algorithm for Problem FPLDE and search for $c_0, c_1 \in \mathbb{Q}(n)$ (not both zero) and $g \in R$ such that (40) holds with $d = 1$. Again there is no solution. We continue our search and take $f_2 = (1 - 3(2n - x)h)\frac{(n-x+1)^3(n-x+2)^3}{(n+1)^3(n+2)^3}b^{-3}$ with $\text{ev}(f_2, k) = F_{n+2}(k)$ and look for $c_0, c_1, c_2 \in \mathbb{Q}(n)$ (not all zero) and $g \in R$ such that (40) holds with $d = 2$. This time our algorithm for Problem FPLDE outputs $c_0 = (n+2)^4(n+3)^2$, $c_1 = (n+1)^3(n+3)^2(2n+5)$, $c_2 = (n+1)^3(n+2)^3$ and $g = (p_1(n,x) + p_2(n,x)h)b^{-3}$ for polynomials $p_1(n,x), p_2(n,x) \in \mathbb{Q}[n,x]$. Hence we get (39) with $G(k) = \text{ev}(g,k) = (p(n,k) + p_2(n,k)S_1(k))\binom{n}{k}^{-3}$. We emphasize that the correctness of (39) for the given solution for all k with $0 \leq k \leq n$ can be verified easily. Finally, summing (39) over k from 0 to n one gets

$$
\begin{aligned}
&c_0(n)A_3(n) + c_1(n)(A_3(n+1) - F_{n+1}(n+1)) \\
&\quad + c_2(n)(A_3(n+2) - F_{n+2}(n+1) - F_{n+2}(n+2)) = G(n+1) - G(n);
\end{aligned}
$$

moving the $F_{n+i}(n+j)$ terms to the right hand side gives the recurrence Out[8].

[6]For a rigorous verification the proof certificate $(c_0, \ldots, c_d, G(k))$ of (41) with $d = 2$ is returned with the function call `CreativeTelescoping[mySum,n]`.

Note that creative telescoping is only a slight extension of telescoping, in particular, all the enhanced telescoping algorithms from Sect. 4.1 carry over to creative telescoping. In all variations, a polynomial $\Pi\Sigma^*$-field $(\mathbb{F}, \sigma)$ (more precisely a depth-optimal $\Pi\Sigma^*$-field [67] for efficiency reasons) is constructed in which the summands $F_{n+i}(k)$ $(0 \leq i \leq d)$ can be expressed. Starting from there, the following tactics are most useful to search for a solution of (40). They are activated by using the option `SimplifyByExt->Mode` where `Mode` is chosen as follows.

- *None:* The solution $G(k)$ is searched in $(\mathbb{F}, \sigma)$, i.e., only objects occurring in $F_{n+i}(k)$ are used. Here a special instance of FPLDE is solved; see Example 11.
- *MinDepth:* The solution $G(k)$ is searched in terms of sum extensions which are not more nested than the objects in $F_{n+i}(k)$ and which have minimal depth among all the possible choices [74]. This is the default option.
- *DepthNumber:* The solution is given in terms of sum extensions which are not more nested than $\sum_{k=0}^{n} F_{n+i}(k)$, however, if the nesting depth is the same, the number of the objects in the summands must be smaller than in $F_{n+i}(k)$. If such a recurrence exists, the machinery from [62] computes it. Using this refined version for our example, one finds a recurrence of order 1 (instead of 2)

$$\begin{aligned}&(n+1)^3 A_{-3}(n+1) + (n+2)^3 A_{-3}(n)\\ &= 6(n+2)(n+1)^3 S_1(n) + (7n+13)(n+1)^2 + 3(n+2)^2 \sum_{i=0}^{n} (n-2i) \binom{n}{i}^{-3}\end{aligned} \tag{42}$$

where the sum $E(n) = \sum_{i=0}^{n} (n-2i)\binom{n}{i}^{-3}$ does not contain $S_1(i)$; it turns out that $E(n) = 0$ (using again our tools) and the recurrence simplifies further to

$$(n+1)^3 A_{-3}(n+1) + (n+2)^3 A_{-3}(n) = 6(n+2)(n+1)^3 S_1(n) + (7n+13)(n+1)^2. \tag{43}$$

4.3 Solving Recurrence Relations

Next, we turn to recurrence solving in terms of nested hypergeometric sum expressions, i.e., expressions that evaluate to d'Alembertian sequence solutions.

Example 12. Given the recurrence `rec` in Out[8] of $A_{-3}(n) =$ `SUM[n]`, all nested hypergeometric sum solutions are calculated with the following `Sigma` command:

```
In[9]:= recSol = SolveRecurrence[rec, SUM[n]]
Out[9]= {{0, −(−1)^n (n + 1)^3}, {0, (−1)^n (− S1(n)(n + 1)^3 − (n + 1)^2)},
        {1, 6(n + 1)S1(n) + (−1)^n (5(n + 1)^3 S−3(n) − 6(n + 1)^3 S−2,1(n)) + 1}}
```

The output means that we calculated two linearly independent solutions $H_1(n) = -(-1)^n(n+1)^3$ and $H_2(n) = (-1)^n\big(-S_1(n)(n+1)^3 - (n+1)^2\big)$ (for $n \geq 0$)

of the homogeneous version of the recurrence and a particular solution $P(n) = 6(n+1)S_1(n) + (-1)^n\big(5(n+1)^3 S_{-3}(n) - 6(n+1)^3 S_{-2,1}(n)\big) + 1$ (for $n \geq 0$) of the recurrence itself; since the solutions are indefinite nested, the verification of the correctness can be verified easily by rational function arithmetic. Note that $\{c_1 H_1(n) + c_2 H_2(n) + P(n) | c_1, c_2 \in \mathbb{Q}\}$ produces all sequence solutions whose entries are from $\mathbb{Q}$. Since also $A_{-3}(n)$ is a solution of the recurrence, there is an element in L that evaluates to $A_{-3}(n)$ for all $n \geq 0$. Using, e.g., the first two initial values $A_{-3}(0) = 1$ and $A_{-3}(1) = 5$ the c_1, c_2 are uniquely determined: $c_1 = c_2 = 0$. Thus we arrive at $A_{-3}(n) = P(n)$, i.e., we discovered and proved the identity (3) for $n \geq 0$ (recall that we verified that both sides satisfy the same recurrence and that both sides agree with the first two initial values). This last step is executed by taking `recSol` and `mySum` $= A_{-3}(n)$ (to get two initial values) as follows.

In[10]:= **FindLinearCombination[recSol, mySum, n, 2]**

Out[10]= $6(n+1)S_1(n) + (-1)^n\big(5(n+1)^3 S_{-3}(n) - 6(n+1)^3 S_{-2,1}(n)\big) + 1$

In general, `Sigma` can solve the following problem [12, 50, 62].

Problem RS: Recurrence solving. *Given* polynomials $a_0(n), \dots, a_m(n) \in \mathbb{K}(n)$ and a nested hypergeometric sum expression $f(n)$. *Find* the full solution set of the mth-order linear recurrence

$$a_0(n)G(n) + \cdots + a_m(n)G(n+m) = f(n) \tag{44}$$

in terms of nested hypergeometric sum expressions. I.e., return $\perp$ if there is no particular solution. Otherwise, find $\lambda \in \mathbb{N}$ and nested hypergeometric sum expressions $((1, P(n)), (0, H_1(n)), \dots, (0, H_l(n)))$ where $P(n)$ is a particular solution and $H_1(n), \dots H_l(n)$ are solutions of the homogeneous version of (44) for $n \geq \lambda$; the sequences (in $S(\mathbb{K})$) produced by $H_1(n), \dots H_l(n)$ are linearly independent. In addition, all sequences $(G(n))_{n\geq 0} \in \mathbb{K}^{\mathbb{N}}$, that are solutions of (44) for all $n \geq \lambda$ and that can be given by nested hypergeometric sum expressions, can be also produced by

$$L = \{P(n) + c_1 H_1(n) + \dots c_l H_l(n) | c_i \in \mathbb{K}\} \tag{45}$$

starting from $n \geq \lambda$.

The following solution relies on [39, 57, 62, 70, 76].

*Solution**. Construct a polynomial $\Pi\Sigma^*$-field $(\mathbb{F}, \sigma)$ as in Definition 23 with R defined as in (6) together with an evaluation function $\mathrm{ev} : R \times \mathbb{N} \to \mathbb{K}$ in which one obtains explicitly a $\Phi \in R$ with $\lambda' \in \mathbb{N}$ such that $\mathrm{ev}(\Phi, n) = f(n)$ for all $n \geq \lambda'$; again Footnote 4 applies. In other words, with $\alpha_i := a_i(x) \in \mathbb{K}(x)$ we can reformulate (44) with

$$\alpha_0 g + \alpha_1 \sigma(g) + \cdots + \alpha_m \sigma^m(g) = \Phi. \tag{46}$$

Factorize the homogeneous recurrence (written as linear operator) as much as possible in linear right factors using Hyper [57]. Each linear factor describes a hypergeometric solution which is adjoined to our $\Pi\Sigma^*$-field (see Problem RP); for simplicity we exclude the possible case that $(-1)^n$ is needed for this task. Applying Algorithm [62, Algorithm 4.5.3] to this recurrence returns the output $((1,p),(0,h_1),\ldots,(0,h_l))$ in a polynomial $\Pi\Sigma^*$-field $(\mathbb{E},\sigma)$ that contains $(\mathbb{F},\sigma)$ with the following property: p is a particular solution of (46) and the h_i are l linearly independent solutions of the homogeneous version of (46). We omit details here and remark only that it is crucial to solve (22) as subproblem. Then extend the evaluation function from $\mathbb{F}$ to $\mathbb{E}$ (we are free to choose appropriate lower bounds and constants of the sums/products), and let $H_i(n)$ $(1 \leq i \leq l)$ and $P(n)$ be the nested hypergeometric sum expressions that define the evaluations $\text{ev}(h_i,n)$ and $\text{ev}(p,n)$, respectively. Compute λ such that the $H_i(n)$ are solutions of the homogeneous version and $P(n)$ is a particular solution of (44) for all $n \geq \lambda$; see Remark 3 with $d = 0$, $f_0 := \Phi$.

Remark. 1. If one computes $m = l$ linearly independent solutions plus a particular solution, the set (45) gives all solutions. If this is not the case, the completeness of the method, i.e., that no solution in terms of nested hypergeometric sum expressions is missed, needs further justification: it can be deduced from [62, Corollary 4.5.2] and Remark 3; for deep insight and alternative proofs see [36] and [58, this book].

2. The derived solutions are highly nested: For each additional solution one needs one extra indefinite sum on top. In most examples the simplification of these solutions (see Sect. 4.1) is the most challenging task; see, e.g., Example 14.
3. Since the solutions are indefinite nested, the shifted versions can be expressed by the non-shifted versions. Using this property and considering the sums and products as variables, the correctness can be verified by rational function arithmetic.
4. Also the $a_i(n)$ in (46) can be from a $\Pi\Sigma^*$-field and one can factorize the difference operator in linear factors; this is based on work by [14, 26, 70].

Example 13. * We construct the polynomial $\Pi\Sigma^*$-field $(\mathbb{Q}(x)(h),\sigma)$ with $\sigma(x) = x+1$ and $\sigma(h) = h + \frac{1}{x+1}$ over $\mathbb{Q}$ and interpret the elements with the evaluation function $\text{ev} : \mathbb{Q}(x)[h] \times \mathbb{N} \to \mathbb{Q}(x)[h]$ canonically defined by (29) and $\text{ev}(h,n) = S_1(n)$. In this way, we can reformulate the recurrence Out[8] with

$$(x+2)^4(x+3)^2 g + (x+1)^3(x+3)^2(2x+5)\sigma(g) + (x+1)^3(x+2)^3\sigma^2(g) =$$
$$\big(20x^3 + 138x^2 + 311x + 229\big)(x+1)^2 + 6(x+2)^2(x+3)(2x+5)(x+1)^3 h. \tag{47}$$

Then we execute the recurrence solver in this $\Pi\Sigma^*$-field and get as output the difference ring $(\mathbb{Q}(x)[m][h][s][H], \sigma)$ with $\sigma(m) = -m$ where $m^2 = 1$, $\sigma(s) = s + \frac{-m}{(x+1)^3}$ and $\sigma(H) = H + \frac{-m(h+\frac{1}{x+1})}{(x+1)^2}$ such that $\text{const}_\sigma \mathbb{Q}(x)[m][h][H] = \mathbb{Q}$. There it returns the linearly independent solutions $h_1 = m(x+1)^3$ and $h_2 = m\big(h(x+1)^3 + (x+1)^2\big)$ of the homogeneous version of (47) and the particular solution $p = 6(x+1)h + m\big(5(x+1)^3 s - 6(x+1)^3 H\big) + 1$ of (47) itself. Note that the solutions (coming from the factorization of the recurrence) have been simplified already using the technologies presented in Sect. 4.1. Finally, we extend the evaluation function from $\mathbb{Q}(x)[h]$ to $\mathbb{Q}(x)[m][h][H]$ by $\text{ev}(m,n) = (-1)^n$, $\text{ev}(s,n) = S_{-3}(n)$ and $\text{ev}(H,n) = S_{-2,1}(n)$. This choice yields $H_1(n) = \text{ev}(h_1, n)$, $H_2(n) = \text{ev}(h_2, n)$ and $P(n) = \text{ev}(p, n)$ as given in Out[9].

Example 14. In [21] (see also [41, this book]) recurrences are guessed with minimal order that contain as solutions the massless Wilson coefficients to three-loop order for individual color coefficients [48]. Afterwards the recurrences have been solved. The largest recurrence of order 35 could be factorized completely into linear factors in about 1 day. This yields 35 linearly independent solutions in terms of sums up to nesting depth 34. Then their simplifications in terms of harmonic sums took 5 days.

5 Simplification of Multiple Sums with `EvaluateMultiSums`

In Sect. 4 we transformed the definite sum $A_{-3}(n)$ to a nested hypergeometric sum expression given in (3) by calculating a recurrence and solving it. Applying this tactic iteratively leads to a successful method to transform certain classes of definite multiple sums to nested hypergeometric sum expressions. Consider, e.g.,

$$F(n) := \sum_{j=0}^{n-2} (-j+n-2)! \overbrace{\sum_{r=0}^{j+1} \frac{(-1)^r \binom{j+1}{r} r!}{(-j+n+r)!} \underbrace{\sum_{s=0}^{-j+n+r-2} \frac{(-1)^s \binom{-j+n+r-2}{s}}{(n-s)(s+1)}}_{=:F_0(n,j,r)}}^{=:F_1(n,j)} \tag{48}$$

which arose in QCD calculations needed in [4]; see also [3]. We zoom into the sum $F_0(n,j,r)$, a definite sum over a hypergeometric sequence. Calculating a recurrence, solving the recurrence, and combining the solutions leads to the simplification

$$F_0(n,j,r) = \frac{1}{(n+1)(-j+n+r-1)} + \frac{(-1)^n (j+1)!(-j+n-1)_r}{(n-1)n(n+1)(-j-1)_r (2-n)_j}.$$

This closed form could be also derived by hypergeometric summation [59]: the double sum $F_1(n,j)$ turns out to be a single sum. Next, finding a recurrence for this sum and solving the recurrence lead to a nested hypergeometric sum expression w.r.t. j:

$$F_1(n,j) = (-1)^j (j+1)!\Big[\frac{1}{n!}\Big(\frac{(-1)^n (j+2)}{(n+1)^2(-j+n-1)} + \frac{n^2+1}{(n-1)n(n+1)^2}\Big) + \frac{1}{n+1}\sum_{i=1}^{j}\frac{(-1)^i}{(n-i)!(i+1)!(n-i-1)}\Big].$$

In other words, $F(n)$ can be written as a definite sum where the summand is a nested hypergeometric sum expression. Therefore we are again in the position to apply our technologies from Sect. 4. Computing a recurrence and solving it yields

$$F(n) = \frac{-n^2-n-1}{n^2(n+1)^3} + \frac{(-1)^n(n^2+n+1)}{n^2(n+1)^3} + \frac{S_1(n)}{(n+1)^2} - \frac{S_2(n)}{n+1} - \frac{2S_{-2}(n)}{n+1}.$$

Summarizing, we transformed a definite nested sum from inside to outside to a nested hypergeometric sum expression. More generally, we deal with the following

Problem EMS: EvaluateMultiSum. *Given* a definite multiple sum

$$F(\mathbf{m},n) = \sum_{k=l}^{L_0(\mathbf{m},n)} \overbrace{\sum_{k_1=l_1}^{L_1(\mathbf{m},n,k)} \cdots \sum_{k_v=l_v}^{L_v(\mathbf{m},n,k,k_1,\dots,k_{v-1})} \overline{f}(\mathbf{m},n,k,k_1,\dots,k_v)}^{f(\mathbf{m},n,k)} \tag{49}$$

with a nested hypergeometric sum expression $\overline{f}$ w.r.t. k_v, integer parameters n and $\mathbf{m}=(m_1\dots,m_r)$, and $L_i(\dots)$ being integer linear (see Footnote 5) or ∞. *Find* $\lambda\in\mathbb{N}$ and a nested hypergeometric sum expression $\bar{F}(\mathbf{m},n)$ such that $F(\mathbf{m},n)=\bar{F}(\mathbf{m},n)$ for $n\geq\lambda$.

Method. Apply the techniques of Sect. 4 recursively as follows [23].

1. Transform the outermost summand $f(\mathbf{m},n,k)$ to a nested hypergeometric sum expression w.r.t. k by applying the proposed method recursively to all the arising definite sums (i.e., the parameter vector $\mathbf{m}$ is replaced by $(\mathbf{m},n)$ and the role of n is k). Note that the sums in f are simpler than $F(\mathbf{m},n)$ (one definite sum less). If the summand f is free of sums, nothing has to be done.
2. Solve Problem CT: Compute a recurrence (38) for the sum $A(n)=F(\mathbf{m},n)$; if this fails, ABORT. If successful (say it is of order o), the right hand side might

be again an expression in terms of definite sums, but their summands are simpler than f (see, e.g., the recurrence (42)). Apply the method recursively to these sums such that the right hand side is transformed to a nested hypergeometric sum expression w.r.t. n (see, e.g., recurrence (43)).

3. Solve Problem RS: Compute all nested hypergeometric sum solutions of the recurrence (38) and simplify the solutions using the techniques from Sect. 4.1.
4. Compute o initial values, i.e., specialize the parameter n to appropriate values from $\mathbb{N}$, say $n = l, l+1, \ldots, l+o-1$, and apply the method recursively to the arising sums where m_1 takes over the role of n and the remaining parameters are $(m_2, \ldots, m_r)$. If no parameter is left, the expression is a constant. It is usually from $\mathbb{Q}$ (if no sum is left) or it simplifies, e.g., to [22] or infinite versions of S–sums [9] and cyclotomic sums [5].
5. Try to combine the solutions to find a nested hypergeometric sum expression w.r.t. n of $F(\mathbf{m}, n)$. If this fails, ABORT. Otherwise return the solution.

Remark. 1. The *existence* of a recurrence in Step 2 is guaranteed in many cases (in particular for sums coming from Feynman integrals [24]) by using arguments, e.g., form [15, 59, 79, 80]. Here often computation issues are a bottleneck. Usually, we succeed in finding recurrences when f consists of up to 100 nested hypergeometric sums. If f is more complicated (or if it seems appropriate), the sum is split into several parts and the method is applied to each sum separately.

2. *Termination:* The method is applied recursively to sums which are always simpler than the original sum (less summation quantifiers, less parameters, or less objects in the summand). Hence eventually one arrives at the base case.
3. *Success:* If the method does not abort in one of the executions of step 2 or step 5, it terminates and outputs a nested hypergeometric sum expression w.r.t. n. Note that finding not sufficiently many solutions of a given recurrence in step 5 is the main reason why the method might fail. For general multiple sums this failure would happen all over. However, e.g., in the context of Feynman integrals, the recurrence is usually completely solvable (i.e., we find m linearly independent solutions of the homogeneous version of (38) and one particular solution of the recurrence itself).

We emphasize that three-loop Feynman integrals with at most 1 mass [24] can be transformed to multiple sums and that the simplification of these sums is covered exactly by Problem EMS. The described method is implemented in the following new package which uses the summation algorithms in `Sigma`:

In[11]:= << **EvaluateMultiSums.m**

`EvaluateMultiSums by Carsten Schneider – –© RISC`

In addition it uses (some of the many) functions from J. Ablinger's package `HarmonicSums` [1,5,9,19,20,78] to transform – if possible – the arising indefinite sums to harmonic sums, S-sums, cyclotomic sums or their infinite versions, to find algebraic relations among these sums, and to calculate asymptotic expansions of these sums for limit computations (this is needed if upper bounds in (49) are ∞).

In[12]:= **<< HarmonicSums.m**

HarmonicSums by Jakob Ablinger − −© RISC

Then inserting the summand with the summation ranges of (48) and the information that there is the extra integer parameter n with $2 \leq n \leq \infty$ we can activate the simplification of the sum (48) to a nested hypergeometric sum expression as follows.

In[13]:= $\mathbf{EvaluateMultiSum}[\frac{(-j+n-2)!(-1)^{r+s}\binom{j+1}{r}r!}{(-j+n+r)!}\frac{\binom{-j+n+r-2}{s}}{(n-s)(s+1)},$
$\{\{s,0,-j+n+r-2\},\{r,0,j+1\},\{j,0,n-2\}\},\{n\},\{2\},\{\infty\}]$

Out[13]= $$\frac{-n^2-n-1}{n^2(n+1)^3}+\frac{(-1)^n(n^2+n+1)}{n^2(n+1)^3}+\frac{S_1(n)}{(n+1)^2}-\frac{S_2(n)}{n+1}-\frac{2S_{-2}(n)}{n+1}$$

Similarly, we can calculate the simplification given in identity (4):

In[14]:= $\mathbf{EvaluateMultiSum}[(1-4(n-2k)S_1(k))\binom{n}{k}^{-4},\{\{k,0,n\}\},\{n\},\{0\},\{\infty\}]$

Out[14]= $$\frac{(10(n+1)S_1(n)+3)(n+1)}{2n+3}+\frac{(-1)^n\binom{2n}{n}^{-1}(n+1)^5}{(4n(n+2)+3)}\left(\frac{7}{2}\sum_{i=1}^{n}\frac{(-1)^i\binom{2i}{i}}{i^3}-5\sum_{i=1}^{n}\frac{(-1)^i\binom{2i}{i}S_1(i)}{i^2}\right).$$

As mentioned already above, the multiple sums coming from many two-loop and three-loop Feynman integrals fit into the input class of the package `EvaluateMultiSums`. Here two extremes occurred: In [4, 25] about a million multiple sums (mostly triple and quadruple sums) were simplified. Using the package `SumProduction` [23] we merged the sums to several 100 basis sums where each of the summands required up to 20 MB memory. The other extreme are sums whose summands are in compact size, but the number of summations is large; one of the most complicated input sums from [25] is, e.g., (6). In both setups the transformed summands during the `EvaluateMultiSum` method became rather large containing complicated nested hypergeometric sums. Only in the last step these nasty sums vanished and the expected nice result popped up; note that already for the transformation of the sum (48) this effect is visible. Summarizing, the summation algorithms based on enhanced difference field theory, presented in this article, were indispensable to master the challenging calculations as given, e.g., in [4, 7, 17, 25].

Acknowledgements Supported by the Austrian Science Fund (FWF) grants P20347-N18 and SFB F50 (F5009-N15) and by the EU Network LHCPhenoNet PITN-GA-2010-264564.

References

1. Ablinger, J.: Computer algebra algorithms for special functions in particle physics. Ph.D. thesis, J. Kepler University Linz (2012)
2. Ablinger, J., Blümlein, J.: Harmonic sums, polylogarithms, special numbers, and their generalizations. In: Blümlein, J., Schneider, C. (eds.) Computer Algebra in Quantum Field Theory: Integration, Summation and Special Functions. Springer, Vienna (2013)

3. Ablinger, J., Blümlein, J., Klein, S., Schneider, C.: Modern summation methods and the computation of 2- and 3-loop Feynman diagrams. Nucl. Phys. B (Proc. Suppl.) **205–206**, 110–115 (2010). ArXiv::1006.4797 (math-ph)
4. Ablinger, J., Blümlein, J., Klein, S., Schneider, C., Wissbrock, F.: The $O(\alpha_s^3)$ massive operator matrix elements of $O(n_f)$ for the structure function $F_2(x, Q^2)$ and transversity. Nucl. Phys. B **844**, 26–54 (2011). ArXiv:1008.3347 (hep-ph)
5. Ablinger, J., Blümlein, J., Schneider, C.: Harmonic sums and polylogarithms generated by cyclotomic polynomials. J. Math. Phys. **52**(10), 1–52 (2011, to appear). arXiv:1007.0375 (hep-ph)
6. Ablinger, J., Blümlein, J., Freitas, A.D., Hasselhuhn, A., Klein, S., Raab, C., Round, M., Schneider, C., Wissbrock, F.: Three-loop contributions to the gluonic massive operator matrix elements at general values of N. In: Proceedings of the Loops and Legs in Quantum Field Theory 2012, PoS(LL2012)033, Wernigerode, pp. 1–12 (2012)
7. Ablinger, J., Blümlein, J., Hasselhuhn, A., Klein, S., Schneider, C., Wissbrock, F.: Massive 3-loop ladder diagrams for quarkonic local operator matrix elements. Nucl. Phys. B **864**, 52–84 (2012). ArXiv:1206.2252v1 (hep-ph)
8. Ablinger, J., Blümlein, J., Round, M., Schneider, C.: Advanced computer algebra algorithms for the expansion of Feynman integrals. In: Loops and Legs in Quantum Field Theory 2012, PoS(2012), Wernigerode, pp. 1–14 (2012)
9. Ablinger, J., Blümlein, J., Schneider, C.: Analytic and algorithmic aspects of generalized harmonic sums and polylogarithms (2013). ArXiv:1302.0378 (math-ph)
10. Ablinger, J., Blümlein, J., Schneider, C.: Structural Relations of Harmonic Sums (2013, In preparation)
11. Abramov, S.A.: On the summation of rational functions. Zh. vychisl. mat. Fiz. **11**, 1071–1074 (1971)
12. Abramov, S., Petkovšek, M.: D'Alembertian solutions of linear differential and difference equations. In: von zur Gathen, J. (ed.) Proceedings of the ISSAC'94, Oxford, pp. 169–174. ACM (1994)
13. Abramov, S., Petkovšek, M.: Polynomial ring automorphisms, rational (w, σ)-canonical forms, and the assignment problem. J. Symb. Comput. **45**(6), 684–708 (2010)
14. Abramov, S., Bronstein, M., Petkovšek, M., Schneider, C.: (2012, In preparation)
15. Apagodu, M., Zeilberger, D.: Multi-variable Zeilberger and Almkvist-Zeilberger algorithms and the sharpening of Wilf-Zeilberger theory. Adv. Appl. Math. **37**, 139–152 (2006)
16. Bauer, A., Petkovšek, M.: Multibasic and mixed hypergeometric Gosper-type algorithms. J. Symb. Comput. **28**(4–5), 711–736 (1999)
17. Bierenbaum, I., Blümlein, J., Klein, S., Schneider, C.: Two–loop massive operator matrix elements for unpolarized heavy flavor production to $o(epsilon)$. Nucl.Phys. B **803**(1–2), 1–41 (2008). arXiv:hep-ph/0803.0273
18. Blümlein, J.: Algebraic relations between harmonic sums and associated quantities. Comput. Phys. Commun. **159**(1), 19–54 (2004). arXiv:hep-ph/0311046
19. Blümlein, J.: Structural relations of harmonic sums and Mellin transforms at weight w=6. In: Carey, A., Ellwood, D., Paycha, S., Rosenberg, S. (eds.) Motives, Quantum Field Theory, and Pseudodifferential Operators. Clay Mathematics Proceedings, vol. 12, pp. 167–186 (2010). DESY 08–206, SFB-CPP/09–002, arXiv:0901.0837 (math-ph)
20. Blümlein, J., Kurth, S.: Harmonic sums and Mellin transforms up to two-loop order. Phys. Rev. **D60**, 014018 (31p) (1999)
21. Blümlein, J., Kauers, M., Klein, S., Schneider, C.: Determining the closed forms of the $O(a_s^3)$ anomalous dimensions and Wilson coefficients from Mellin moments by means of computer algebra. Comput. Phys. Commun. **180**, 2143–2165 (2009)
22. Blümlein, J., Broadhurst, D.J., Vermaseren, J.A.M.: The multiple zeta value data mine. Comput. Phys. Commun. **181**, 582–625 (2010). arXiv:0907.2557 (math-ph)
23. Blümlein, J., Hasselhuhn, A., Schneider, C.: Evaluation of multi-sums for large scale problems. In: Proceedings of RADCOR 2011 PoS(RADCOR2011), Durham, vol. 32, pp. 1–9 (2012)

24. Blümlein, J., Klein, S., Schneider, C., Stan, F.: A symbolic summation approach to Feynman integral calculus. J. Symb. Comput. **47**, 1267–1289 (2012)
25. Blümlein, J., Hasselhuhn, A., Klein, S., Schneider, C.: The $O(\alpha_s^3 n_f T_F^2 C_{A,F})$ contributions to the gluonic massive operator matrix elements. Nucl. Phys. B **866**, 196–211 (2013)
26. Bronstein, M.: On solutions of linear ordinary difference equations in their coefficient field. J. Symb. Comput. **29**(6), 841–877 (2000)
27. Brown, F.: The massless higher-loop two-point function. Math. Phys. **287**, 925–958 (2009)
28. Chen, S., Kauers, M.: Order-degree curves for hypergeometric creative telescoping. In: van der Hoeven, J., van Hoeij, M. (eds.) Proceedings of ISSAC 2012, Grenoble, pp. 122–129 (2012)
29. Chu, W., Donno, L.D.: Hypergeometric series and harmonic number identities. Adv. Appl. Math. **34**(1), 123–137 (2005)
30. Chyzak, F.: An extension of Zeilberger's fast algorithm to general holonomic functions. Discret. Math. **217**, 115–134 (2000)
31. Cohn, R.: Difference Algebra. Interscience/Wiley, New York (1965)
32. Driver, K., Prodinger, H., Schneider, C., Weideman, J.: Padé approximations to the logarithm III: alternative methods and additional results. Ramanujan J. **12**(3), 299–314 (2006)
33. Eröcal, B.: Algebraic extensions for summation in finite terms. Ph.D. thesis, RISC, Johannes Kepler University, Linz (2011)
34. Gosper, R.: Decision procedures for indefinite hypergeometric summation. Proc. Natl. Acad. Sci. U.S.A. **75**, 40–42 (1978)
35. Hardouin, C., Singer, M.: Differential Galois theory of linear difference equations. Math. Ann. **342**(2), 333–377 (2008)
36. Hendriks, P., Singer, M.: Solving difference equations in finite terms. J. Symb. Comput. **27**(3), 239–259 (1999)
37. Hoeij, M.: Finite singularities and hypergeometric solutions of linear recurrence equations. J. Pure Appl. Algebra **139**(1–3), 109–131 (1999)
38. Horn, P., Koepf, W., Sprenger, T.: m-fold hypergeometric solutions of linear recurrence equations revisited. Math. Comput. Sci. **6**(1), 61–77 (2012)
39. Karr, M.: Summation in finite terms. J. ACM **28**, 305–350 (1981)
40. Karr, M.: Theory of summation in finite terms. J. Symb. Comput. **1**, 303–315 (1985)
41. Kauers, M.: The holonomic toolkit. In: Blümlein, J., Schneider, C. (eds.) Computer Algebra in Quantum Field Theory: Integration, Summation and Special Functions. Springer, Vienna (2013)
42. Kauers, M., Zimmermann, B.: Computing the algebraic relations of C-finite sequences and multisequences. J. Symb. Comput. **43**(11), 787–803 (2008)
43. Koornwinder, T.: On Zeilberger's algorithm and its q-analogue. J. Comput. Appl. Math. **48**, 91–111 (1993)
44. Koutschan, C.: A fast approach to creative telescoping. Math. Comput. Sci. **4**(2–3), 259–266 (2010)
45. Krattenthaler, C., Rivoal, T.: Hypergéométrie et fonction zêta de Riemann. Mem. Am. Math. Soc. **186**(875) (2007)
46. Levin, A.: Difference Algebra. Algebra and Applications, vol. 8. Springer, New York (2008)
47. Moch, S., Uwer, P., Weinzierl, S.: Nested sums, expansion of transcendental functions, and multiscale multiloop integrals. J. Math. Phys. **6**, 3363–3386 (2002)
48. Moch, S., Vermaseren, J., Vogt, A.: The three-loop splitting functions in QCD: the non-singlet case. Nucl. Phys. B **688**, 101 (2004). ArXiv:hep-ph/0403192v1
49. Nemes, I., Paule, P.: A canonical form guide to symbolic summation. In: Miola, A., Temperini, M. (eds.) Advances in the Design of Symbolic Computation Systems. Texts and Monographs in Symbolic Computation, pp. 84–110. Springer, Wien/New York (1997)
50. Nørlund, N.E.: Vorlesungen über Differenzenrechnung. Springer, Berlin (1924)
51. Osburn, R., Schneider, C.: Gaussian hypergeometric series and extensions of supercongruences. Math. Comput. **78**(265), 275–292 (2009)
52. Paule, P.: Greatest factorial factorization and symbolic summation. J. Symb. Comput. **20**(3), 235–268 (1995)

53. Paule, P., Riese, A.: A Mathematica q-analogue of Zeilberger's algorithm based on an algebraically motivated aproach to q-hypergeometric telescoping. In: Ismail, M., Rahman, M. (eds.) Special Functions, q-Series and Related Topics. Fields Institute Communications, vol. 14, pp. 179–210 AMS, Providence (1997)
54. Paule, P., Schneider, C.: Computer proofs of a new family of harmonic number identities. Adv. Appl. Math. **31**(2), 359–378 (2003)
55. Paule, P., Schorn, M.: A mathematica version of Zeilberger's algorithm for proving binomial coefficient identities. J. Symb. Comput. **20**(5–6), 673–698 (1995)
56. Pemantle, R., Schneider, C.: When is 0.999... equal to 1? Am. Math. Mon. **114**(4), 344–350 (2007)
57. Petkovšek, M.: Hypergeometric solutions of linear recurrences with polynomial coefficients. J. Symb. Comput. **14**(2–3), 243–264 (1992)
58. Petkovšek, M., Zakrajšek, H.: Solving linear recurrence equations with polynomial coefficients. In: Blümlein, J., Schneider, C. (eds.) Computer Algebra in Quantum Field Theory: Integration, Summation and Special Functions. Springer (2013, to appear)
59. Petkovšek, M., Wilf, H.S., Zeilberger, D.: $A = B$. A. K. Peters, Wellesley (1996)
60. Riese, A.: qMultisum – a package for proving q-hypergeometric multiple summation identities. J. Symb. Comput. **35**, 349–377 (2003)
61. Risch, R.: The problem of integration in finite terms. Trans. Am. Math. Soc. **139**, 167–189 (1969)
62. Schneider, C.: Symbolic summation in difference fields, Technical Report 01–17. Ph.D. thesis, RISC-Linz, J. Kepler University (2001)
63. Schneider, C.: A collection of denominator bounds to solve parameterized linear difference equations in $\Pi\Sigma$-extensions. An. Univ. Timişoara Ser. Mat.-Inform. **42**(2), 163–179 (2004)
64. Schneider, C.: The summation package Sigma: underlying principles and a rhombus tiling application. Discret. Math. Theor. Comput. Sci. **6**, 365–386 (2004)
65. Schneider, C.: Symbolic Summation with Single-Nested Sum Extensions. In: Gutierrez, J. (ed.) Proceedings of the ISSAC'04, Santander, pp. 282–289. ACM (2004)
66. Schneider, C.: Degree bounds to find polynomial solutions of parameterized linear difference equations in $\Pi\Sigma$-fields. Appl. Algebra Eng. Commun. Comput. **16**(1), 1–32 (2005)
67. Schneider, C.: Finding telescopers with minimal depth for indefinite nested sum and product expressions. In: Kauers, M. (ed.) Proceedings of the ISSAC'05, Beijing, pp. 285–292. ACM (2005)
68. Schneider, C.: A new Sigma approach to multi-summation. Adv. Appl. Math. **34**(4), 740–767 (2005)
69. Schneider, C.: Product representations in $\Pi\Sigma$-fields. Ann. Comb. **9**(1), 75–99 (2005)
70. Schneider, C.: Solving parameterized linear difference equations in terms of indefinite nested sums and products. J. Differ. Equ. Appl. **11**(9), 799–821 (2005)
71. Schneider, C.: Apéry's double sum is plain sailing indeed. Electron. J. Comb. **14** (2007)
72. Schneider, C.: Simplifying sums in $\Pi\Sigma$-extensions. J. Algebra Appl. **6**(3), 415–441 (2007)
73. Schneider, C.: Symbolic summation assists combinatorics. Sém. Lothar. Combin. **56**, 1–36 (2007). Article B56b
74. Schneider, C.: A refined difference field theory for symbolic summation. J. Symb. Comput. **43**(9), 611–644 (2008). arXiv:0808.2543v1
75. Schneider, C.: A symbolic summation approach to find optimal nested sum representations. In: Carey, A., Ellwood, D., Paycha, S., Rosenberg, S. (eds.) Motives, Quantum Field Theory, and Pseudodifferential Operators. Clay Mathematics Proceedings, vol. 12, pp. 285–308. American Mathematical Society (2010). ArXiv:0808.2543
76. Schneider, C.: Parameterized telescoping proves algebraic independence of sums. Ann. Comb. **14**(4), 533–552 (2010). arXiv:0808.2596
77. Schneider, C.: Structural theorems for symbolic summation. Appl. Algebra Eng. Commun. Comput. **21**(1), 1–32 (2010)
78. Vermaseren, J.: Harmonic sums, Mellin transforms and integrals. Int. J. Mod. Phys. **A14**, 2037–2976 (1999)

79. Wegschaider, K.: Computer generated proofs of binomial multi-sum identities. Master's thesis, RISC, J. Kepler University (1997)
80. Wilf, H., Zeilberger, D.: An algorithmic proof theory for hypergeometric (ordinary and "q") multisum/integral identities. Invent. Math. **108**, 575–633 (1992)
81. Zeilberger, D.: A holonomic systems approach to special functions identities. J. Comput. Appl. Math. **32**, 321–368 (1990)
82. Zeilberger, D.: The method of creative telescoping. J. Symb. Comput. **11**, 195–204 (1991)

Potential of FORM 4.0

Jos A.M. Vermaseren

Abstract I describe the main new features of FORM version 4.0. They include factorization, polynomial arithmetic, new special functions, systems independent. sav files, a complete ParForm, open source code and a forum for user communication. I also mention a completely new feature for code simplification.

1 Pre-introduction: Why FORM?

One may wonder why FORM is so important when there are programs like Mathematica and Maple. We were faced recently with a request for help in programming a gravity problem occurring in Type IIb superstrings [1]. The calculation to be done was in a 10-dimensional theory. It needed the evaluation of the Weyl tensor (W) and a tensor (T) with six indexes. Then 20 quartic invariants of the types WWWW, WWWT, WWTT, WTTT and TTTT had to be evaluated. People had been trying this with the RGTENSOR package in Mathematica. Especially the last category of invariants was very hard on Mathematica. There are 12 indexes in 10 dimensions that needed to be contracted (summed over) as in Fig. 1.

Most of the TTTT invariants took 2–3 days each but the last one was never completed. The program had to be halted after a week. The whole effort with Mathematica took the better part of a year. We took 3 days programming this from scratch in FORM after which the whole project ran in 1 h on my laptop. With a few extra optimizations and a few error corrections(!) it now runs in 40 min on the same laptop.

The above shows a combination of the strengths and weaknesses of FORM. Its strength is that it can be very flexible and very fast. Its weakness is that there

J.A.M. Vermaseren (✉)
Nikhef, Science Park 105, 1098 XG, Amsterdam, The Netherlands
e-mail: t68@nikhef.nl

C. Schneider and J. Blümlein (eds.), *Computer Algebra in Quantum Field Theory*, Texts & Monographs in Symbolic Computation, DOI 10.1007/978-3-7091-1616-6_15,

Fig. 1 Tensor structure of objects to be contracted. Each vertex represents a tensor and each *line* represents an index that should be summed over

are few packages and people like to use packages so that they do not have to think much. With FORM you do have to think. You have to build up expertise, which takes a while, but after that you can do calculations that are otherwise impossible.

2 Introduction

Over the past 12 years FORM was available as version 3.0–3.3 [2, 3]. This version was significantly more powerful than version 2, but also version 3 needed extensions. During the past few years there were a few opportunities to hire extra people for special contributions and development picked up speed. A number of much needed but very labour intensive projects were undertaken and completed. This marked a good point to clean up the whole program, make it open source and bring it out as a new version.

On March 29 version 4.0 was released. It took much more time than anticipated to prepare this release, because there are quite a few new features and the debugging was a slow procedure. In addition making the source code available and easy installable took much work.

Most work for version 4.0 has been put in by Jan Kuipers, Takahiro Ueda, Jens Vollinga and me. Other people (and also Jens) who worked on FORM in the past have left the field and the current team consists of Takahiro Ueda, Jan Kuipers and me. Jan will be leaving in the autumn of 2012. This means that the speed of advances will be slowed down until there are new opportunities to hire good people.

Jens Vollinga has added a number of very nice features to FORM, including systems independent .sav files, checkpoints (points from which a program can be restarted if it crashes), much documentation and the forum. He also designed much of the open source infrastructure.

Jan Kuipers has made the factorization, polynomial libraries and more recently a completely new method for output simplification [4,5] which will be part of a future release.

Takahiro Ueda has been hired by Karlsruhe on DFG money with the task to improve the parallelization of FORM. The project is to combine TFORM and ParPARFORM to make use of clusters of multicore computers. This project is still in its infancy, because his first task was to make ParPARFORM complete. This has been finished now. He has also taken over the task to manage the open source infrastructure.

This talk will take us through a number of new features and in the end give a few hints about what the future might bring.

The first thing the user will notice is the new header.

```
FORM 4.0 (Mar 29 2012) 64-bits
                                  Run: Wed Apr  4 14:23:50 2012
```

The mentioning of the 64/32-bits version is for version 4, because we are in a period that many people still use 32-bits computers or operating systems. Other headers are

```
TFORM 4.0 (Mar 29 2012) 64-bits 8 workers
                                  Run: Wed Apr  4 14:24:22 2012
```

or

```
ParFORM 4.0 (Mar 29 2012) 64-bits 8 workers
                                  Run: Wed Apr  4 14:32:45 2012
```

In all the examples in the following sections we will omit this header.

3 Factorization

The first feature we are going to look at is one that many people have asked for in the past.

It should be realized that factorization is a subject that many mathematicians have paid attention to. In addition big commercial programs have spent much effort on making good packages for this. Hence one should not expect to outperform other packages. The best model would be to look whether there are packages under the GNU license that have been created and are maintained by good mathematicians. Unfortunately we could not find any that deal with more than a single variable. The better packages are all closed source and part of a commercial system. This means we had to make our own. But then we could optimize it for what WE anticipate that the use will be (which is not artificial benchmarking).

```
Symbols x,y,z;
CFunction f;
Off Statistics;
Format nospaces;
Local F = f((x+y)*(x*y+4*z^2+3*y*z^4-7*x^2*y)*(x+z));
Print;
.sort

F=
   f(4*y*z^3+3*y^2*z^5+4*x*z^3+4*x*y*z^2+3*x*y*z^5+x*y^2*z+
   3*x*y^2*z^4+4*x^2*z^2+x^2*y*z+3*x^2*y*z^4+x^2*y^2-7*x^2*
   y^2*z+x^3*y-7*x^3*y*z-7*x^3*y^2-7*x^4*y);
FactArg,f;
Print;
.end
```

```
F=
   f(y+x,z+x,4*z^2+3*y*z^4+x*y-7*x^2*y);
```

The first example shows the factorization of function arguments. This is probably the most important use in complicated calculations. The `FactArg` statement is natural for this.

Factorization of expressions is a bit more complicated. How to (re)present the results?

```
#define MAX "5"
#define TERMS "6"
#define POW "3"
Symbols a1,...,a'MAX',j;
Off Statistics;
Format NoSpaces;
#do i = 1,'MAX'
Local F'i' = sum_(j,1,'TERMS',random_('TERMS')*
        <a1^random_('POW')/a1>*...*
        <a'MAX'^random_('POW')/a'MAX'>);
#enddo
Print;
.sort

F1=
   5*a2*a4+3*a2*a3^2*a4^2*a5+6*a1*a2*a5^2+5*a1*a2*a3^2*a4^2+
   6*a1*a2^2*a3*a4^2+a1^2*a4^2*a5;

F2=
   2*a3^2*a4^2*a5+2*a1*a5+2*a1^2*a3^2*a4^2+2*a1^2*a2^2*a5^2+
   4*a1^2*a2^2*a3*a4^2*a5^2+a1^2*a2^2*a3^2*a4*a5^2;

F3=
   6*a3^2*a4*a5^2+5*a2*a3*a4^2+a1*a2*a3^2*a5+a1^2*a3^2*a4^2*
   a5+4*a1^2*a2*a3*a4^2+a1^2*a2^2*a3^2*a4^2;

F4=
   6*a2^2*a4^2*a5+4*a2^2*a3^2*a4*a5^2+4*a1*a2*a3^2*a4^2+a1^2
   *a2*a3+6*a1^2*a2*a3^2*a5^2+a1^2*a2^2*a5^2;

F5=
   5*a2*a3^2*a4^2*a5+3*a1*a3^2*a4^2+2*a1*a2*a3*a4^2*a5+a1^2*
   a3+5*a1^2*a2^2*a3^2*a4*a5+a1^2*a2^2*a3^2*a4*a5^2;
On Statistics;
Drop;
Local F = F1*...*F'MAX';
.sort

Time =       0.02 sec    Generated terms =      7776
               F         Terms in output =      5540
                         Bytes used      =    158532
Factorize;
Print;
.end
```

```
Time =        0.02 sec    Generated terms =       5540
                F         Terms in output =       5540
                          Bytes used      =     158532

Time =        1.65 sec    Generated terms =         33
                F         Terms in output =         33
                factorize Bytes used      =       1684

   F=
       (a3)
      *(a3)
      *(6*a3*a4*a5^2+5*a2*a4^2+a1*a2*a3*a5+a1^2*a3*a4^2*a5+4*
         a1^2*a2*a4^2+a1^2*a2^2*a3*a4^2)
      *(2*a3^2*a4^2*a5+2*a1*a5+2*a1^2*a3^2*a4^2+2*a1^2*a2^2*
         a5^2+4*a1^2*a2^2*a3*a4^2*a5^2+a1^2*a2^2*a3^2*a4*a5^2)
      *(a2)
      *(5*a2*a4+3*a2*a3^2*a4^2*a5+6*a1*a2*a5^2+5*a1*a2*a3^2*
         a4^2+6*a1*a2^2*a3*a4^2+a1^2*a4^2*a5)
      *(6*a2*a4^2*a5+4*a2*a3^2*a4*a5^2+4*a1*a3^2*a4^2+a1^2*a3+
         6*a1^2*a3^2*a5^2+a1^2*a2*a5^2)
      *(5*a2*a3*a4^2*a5+3*a1*a3*a4^2+2*a1*a2*a4^2*a5+a1^2+5*
         a1^2*a2^2*a3*a4*a5+a1^2*a2^2*a3*a4*a5^2);
```

Factorization is considered a 'state' in which the expression exists. It is either factorized or unfactorized. Conversion takes place at the end of the module after the expression has been processed and sorted. Hence we have two output statistics. The second one refers to the factorization procedure. To store the factorized expression we use the FORM bracket system with the built in symbol `factor_`. This allows also a way to refer to the brackets.

The execution time depends critically on how complicated the expression is. If we raise the powers of the variables we can see the effect:

```
#define MAX "5"
#define TERMS "6"
#define POW "4"
Symbols a1,...,a'MAX',j;
Off Statistics;
Format NoSpaces;
#do i = 1,'MAX'
Local F'i' = sum_(j,1,'TERMS',random_('TERMS')*
        <a1^random_('POW')/a1>*...*
        <a'MAX'^random_('POW')/a'MAX'>);
#enddo
Print;
.sort

   F1=
      6*a1^2*a2*a3^2*a5+6*a1^2*a2*a3^3*a4^2*a5+3*a1^2*a2^2*a3^3
      *a5^2+a1^2*a2^3*a3*a4^3*a5^3+5*a1^3*a2^3*a3*a4^3*a5+5*
      a1^3*a2^3*a3^2*a4*a5^2;

   F2=
      a4^2*a5+2*a1*a2^3*a3^3*a5+4*a1^2*a2*a3*a5^2+2*a1^2*a2*
```

```
      a3^3*a4^3*a5^3+2*a1^2*a2^2*a4*a5+2*a1^3*a2^2*a3*a4^3*a5;

   F3=
      5*a2+6*a2^3*a4*a5^3+4*a1*a3*a4^3*a5+a1*a2^2*a4*a5+a1*a2^3
      *a3^3*a4^3*a5+a1^3*a3^3*a4^3;

   F4=
      6*a2*a3^2*a4^2*a5^2+a2^3*a3^3*a4^3*a5+4*a1*a2*a3^3*a4^2*
      a5^2+4*a1*a2^2*a3^3*a4^3*a5+a1^2*a2^2*a3*a5^2+6*a1^3*a2^2
      *a3*a5;

   F5=
      3*a1*a2^3*a3^3*a4^2+a1^2*a3*a4^2*a5^3+2*a1^2*a2*a4+a1^2*
      a2*a3*a4*a5^2+5*a1^2*a2^2*a4^2+5*a1^3*a3^2*a4^3;
    On Statistics;
    Drop;
    Local F = F1*...*F`MAX';
    .sort

Time =       0.01 sec    Generated terms =       7776
               F         Terms in output =       7125
                         Bytes used      =     213980
    Factorize;
    Print;
    .end

Time =       0.01 sec    Generated terms =       7125
               F         Terms in output =       7125
                         Bytes used      =     213980

Time =      77.67 sec    Generated terms =         41
               F         Terms in output =         41
               factorize Bytes used      =       2116

   F=
       (a5)
      *(a5)
      *(a5)
      *(a4)
      *(a4^2+2*a1*a2^3*a3^3+4*a1^2*a2*a3*a5+2*a1^2*a2*a3^3*a4^3
         *a5^2+2*a1^2*a2^2*a4+2*a1^3*a2^2*a3*a4^3)
      *(a3)
      *(a3)
      *(6*a3+6*a3^2*a4^2+3*a2*a3^2*a5+a2^2*a4^3*a5^2+5*a1*a2^2*
         a4^3+5*a1*a2^2*a3*a4*a5)
      *(6*a3*a4^2*a5+a2^2*a3^2*a4^3+4*a1*a3^2*a4^2*a5+4*a1*a2*
         a3^2*a4^3+a1^2*a2*a5+6*a1^3*a2)
      *(a2)
      *(a2)
      *(5*a2+6*a2^3*a4*a5^3+4*a1*a3*a4^3*a5+a1*a2^2*a4*a5+a1*
         a2^3*a3^3*a4^3*a5+a1^3*a3^3*a4^3)
      *(3*a2^3*a3^3*a4+a1*a3*a4*a5^3+2*a1*a2+a1*a2*a3*a5^2+5*a1
         *a2^2*a4+5*a1^2*a3^2*a4^2)
      *(a1)
```

```
*(a1)
*(a1);
```

It is also possible to put expressions in the input in factorized form:

```
    Symbols x,y,z;
    LocalFactor F = (x+1)*(x+y)*(z+2)^2*((x+2)*(y+2));
    Print;
    .sort

Time =       0.00 sec    Generated terms =         12
               F         Terms in output =         12
                         Bytes used      =        448

   F =
         ( 1 + x )
       * ( y + x )
       * ( 2 + z )
       * ( 2 + z )
       * ( 4 + 2*y + 2*x + x*y );

    id  x = -y;
    Print;
    .sort

Time =       0.00 sec    Generated terms =         12
               F         Terms in output =          8
                         Bytes used      =        300

   F =
         ( 1 - y )
       * ( 0 )
       * ( 2 + z )
       * ( 2 + z )
       * ( 4 - y^2 );

    UnFactorize F;
    Print;
    .end

Time =       0.00 sec    Generated terms =          8
               F         Terms in output =          8
                         Bytes used      =        300

Time =       0.00 sec    Generated terms =          2
               F         Terms in output =          2
             unfactorize Bytes used      =         84

Time =       0.00 sec    Generated terms =          0
               F         Terms in output =          0
             unfactorize Bytes used      =          4

   F = 0;
```

This example shows also that if during further processing a factor becomes zero, we still keep the expression and the other factors. If, on the other hand, we unfactorize the expression, we end up with zero of course.

Factorization of \$-expressions is yet another case. Here we do not have the bracket system. Neither do we have the possibility to store the factors as arguments. On the other hand, we are not limited by the maximum size of terms.

```
    #define MAX "5"
    #define TERMS "6"
    #define POW "3"
    Symbols a1,...,a'MAX',j;
    Off Statistics;
    Format NoSpaces;
    #do i = 1,'MAX'
    #$v'i' = sum_(j,1,'TERMS',random_('TERMS')*\
            <a1^random_('POW')/a1>*...*\
            <a'MAX'^random_('POW')/a'MAX'>);
    #enddo
    #$V = <$v1>*...*<$v'MAX'>;
    .sort
    #factdollar $V
    #write <> "Factors in $V: '$V[0]'";
Factors in $V: 8
    #do i = 1,'$V[0]'
    #write <> "  Factor 'i': %$",$V['i'];
  Factor 1: a3
    #enddo
  Factor 2: a3
  Factor 3: 6*a3*a4*a5^2+5*a2*a4^2+a1*a2*a3*a5+a1^2*a3*a4^2*a5+
4*a1^2*a2*a4^2+a1^2*a2^2*a3*a4^2
  Factor 4: 2*a3^2*a4^2*a5+2*a1*a5+2*a1^2*a3^2*a4^2+2*a1^2*a2^
2*a5^2+4*a1^2*a2^2*a3*a4^2*a5^2+a1^2*a2^2*a3^2*a4*a5^2
  Factor 5: a2
  Factor 6: 5*a2*a4+3*a2*a3^2*a4^2*a5+6*a1*a2*a5^2+5*a1*a2*a3^
2*a4^2+6*a1*a2^2*a3*a4^2+a1^2*a4^2*a5
  Factor 7: 6*a2*a4^2*a5+4*a2*a3^2*a4*a5^2+4*a1*a3^2*a4^2+a1^2*
a3+6*a1^2*a3^2*a5^2+a1^2*a2*a5^2
  Factor 8: 5*a2*a3*a4^2*a5+3*a1*a3*a4^2+2*a1*a2*a4^2*a5+a1^2+
5*a1^2*a2^2*a3*a4*a5+a1^2*a2^2*a3*a4*a5^2
    .end
  3.25 sec out of 3.25 sec
```

We refer to the factors as if they are array elements. The zero element tells the number of factors.

Of course \$-variables can be used in two ways: during compilation as shown above, and during execution:

```
    Symbols x,y,z;
    CFunction f;
    Off Statistics;
    Format nospaces;
    Local F = f((x+y)*(x*y+4*z^2+3*y*z^4-7*x^2*y)*(x+z));
    Print;
```

```
  .sort

F=
   f(4*y*z^3+3*y^2*z^5+4*x*z^3+4*x*y*z^2+3*x*y*z^5+x*y^2*z
   +3*x*y^2*z^4+4*x^2*z^2+x^2*y*z+3*x^2*y*z^4+x^2*y^2-7*x^2*
   y^2*z+x^3*y-7*x^3*y*z-7*x^3*y^2-7*x^4*y);
  id  f(x?$v) = 1;
  FactDollar,$v;
  do $i = 1,$v[0];
     Print "        Factor %$ in $v = %$",$i,$v[$i];
     $t = nterms_($v[$i]);
     Print "          There are %$ terms in factor %$",$t,$i;
  enddo;
  .end
      Factor 1 in $v = z+x
        There are 2 terms in factor 1
      Factor 2 in $v = 4*z^2+3*y*z^4+x*y-7*x^2*y
        There are 4 terms in factor 2
      Factor 3 in $v = y+x
        There are 2 terms in factor 3
```

Here we need an extra supporting facility: the do loop during execution. Its variable is a $-variable.

Internally the factorization algorithms work only with symbols and numbers. Yet we may use other objects as well. FORM will replace them temporarily by an internal set of symbols, called the "extra symbols". Then, after factorization these are replaced back. Hence the following example works properly.

```
  Symbols x,y,z;
  CFunction f,g;
  Off Statistics;
  Format nospaces;
  Local F = f((x+y)*(x+g(z))*(x*y+4*z^2+3*g(y)*z^4-7*x^2*y));
  Print;
  .sort

F=
   f(4*x*y*z^2+4*x^2*z^2+x^2*y^2+x^3*y-7*x^3*y^2-7*x^4*y+
   3*g(y)*x*y*z^4+3*g(y)*x^2*z^4+3*g(y)*g(z)*y*z^4+3*g(y)*
   g(z)*x*z^4+4*g(z)*y*z^2+4*g(z)*x*z^2+g(z)*x*y^2+g(z)*x^2
   *y-7*g(z)*x^2*y^2-7*g(z)*x^3*y);

  id  f(x?$v) = 1;
  FactDollar,$v;
  do $i = 1,$v[0];
     Print "        Factor %$ in $v = %$",$i,$v[$i];
     $t = nterms_($v[$i]);
     Print "          There are %$ terms in factor %$",$t,$i;
  enddo;
  .end
      Factor 1 in $v = 4*z^2+x*y-7*x^2*y+3*g(y)*z^4
        There are 4 terms in factor 1
      Factor 2 in $v = y+x
```

```
    There are 2 terms in factor 2
  Factor 3 in $v = x+g(z)
    There are 2 terms in factor 3
```

There are more things that can be said about the factorization, but the talks is supposed to be finite in time.

4 Rational Polynomials

Another important thing that was missing, was the capability to deal with rational polynomials. This has even led to the introduction of the external channels to use other programs like FERMAT [6] for this purpose. It would have been nice to have FERMAT in the form of a library, like zlib (compression) or the GMP [7] (for multiprecision calculations), but that was not to be. Now we have our own capabilities.

```
    Symbols x,y,z,a,b;
    CFunction rat;
    Format Nospaces;
    Local F = a*rat(x+1,y+1)+a*rat(x+z,y+z)+a*rat(y+z,y-z);
    Print;
    .sort

Time =       0.00 sec    Generated terms =          3
               F         Terms in output =          3
                         Bytes used      =        476

   F=
      rat(1+x,1+y)*a+rat(z+y,-z+y)*a+rat(z+x,z+y)*a;
    PolyRatFun rat;
    Print;
    .end

Time =       0.00 sec    Generated terms =          3
               F         Terms in output =          1
                         Bytes used      =        584

   F=
      a*rat(2*x*y^2-x*y*z+x*y-x*z^2-x*z+y^3+3*y^2*z+2*y^2+3*y*z
      -z^2,y^3+y^2-y*z^2-z^2);
```

Like many things in FORM this is of course limited to the maximum size of the terms. If this turns out to be a limitation, there are usually other ways to attack the problem. In that case the numerators and denominators are very big expressions and it is better to store them in separate expressions or \$-variables. These then can be used in the new functions `gcd_`, `div_`, `rem_` to obtain results.

The first application of the rational polynomials was to make a new version of the Mincer [8] library. This version works exact. This means that it does not use expansions in ϵ. All ϵ dependence is put inside the rational polynomial. The example shown is as in Fig. 2.

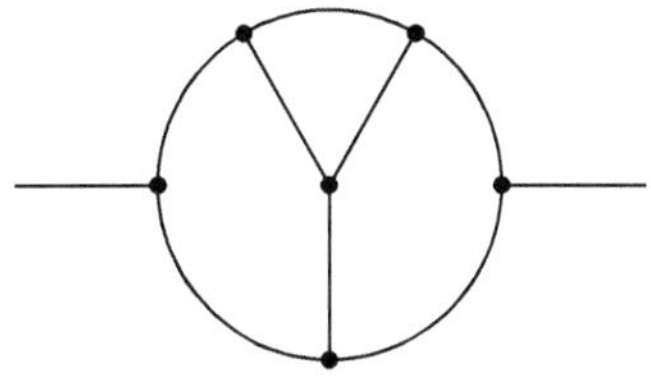

Fig. 2 A typical diagram of type BE

Fig. 3 The diagrams corresponding to the various master integrals

```
#include- minceex.h
Off Statistics;
Format nospaces;
.global
L    F = Q.Q^2/p1.p1/p2.p2/p3.p3/p4.p4/p5.p5/p6.p6/p7.p7
           /p8.p8;
#call integral(be,0)
Print +f +s;
.sort

F=
     +GschemeConstants(0,0)*BasicT1Integral*
         rat(6*ep^3-3*ep^2,2*ep+1)
     +GschemeConstants(0,0)^2*GschemeConstants(1,0)*
         rat(18*ep^2-15*ep+3,2*ep^2+ep)
     +GschemeConstants(0,0)^2*GschemeConstants(2,0)*
         rat(-128*ep^2+96*ep-16,6*ep^2+3*ep)
     +GschemeConstants(0,0)*GschemeConstants(1,0)*
         GschemeConstants(2,0)*
         rat(84*ep^2-49*ep+7,6*ep^2+3*ep)
     ;

  #call subvalues
~~~Answer in the Gscheme
  #call expansion(1)
~~~Answer in the Gscheme
  Print +f;
  .end

F=
     -2*ep^-1*z3-3*z4+12*z3+46*ep*z5+18*ep*z4-32*ep*z3;
```

As is shown above, there are some constants, which are basic one loop integrals with zero, one or two insertions and there is a two loop integral of type T1 with one insertion as shown in Fig. 3. There is one more constant which is the basic non-planar integral in three loops.

The first three integrals are known in terms of Γ-functions and can be expanded as far as wanted/needed.

The T1 integral can be expanded to any precision but that takes more and more time and runs eventually into the limitation that there are relations between the Multiple Zeta Values and these are known only to a certain weight [9]. Enough precision is built in for any practical calculations.

The fifth(NO) integral is more of a problem, but is known to sufficient precision for even four loop calculations.

The above program shows that this exact treatment is quite good because we do not have to worry about cancellations of powers of ϵ.

```
   L    F1 = GschemeConstants(0,0)^2*GschemeConstants(1,0)*
          rat(18*ep^2-15*ep+3,2*ep^2+ep);
   L    F2 = GschemeConstants(0,0)^2*GschemeConstants(2,0)*
          rat(-128*ep^2+96*ep-16,6*ep^2+3*ep);
   L    F3 = GschemeConstants(0,0)*GschemeConstants(1,0)*
          GschemeConstants(2,0)*
          rat(84*ep^2-49*ep+7,6*ep^2+3*ep);
   L    F4 = GschemeConstants(0,0)*BasicT1Integral*
          rat(6*ep^3-3*ep^2,2*ep+1);

   #call subvalues
~~~Answer in the Gscheme
   #call expansion(1)
~~~Answer in the Gscheme
   Print +f;
   .end

  F=
     -2*ep^-1*z3-3*z4+12*z3+46*ep*z5+18*ep*z4-32*ep*z3;

  F1=
     192+3*ep^-4-18*ep^-3+48*ep^-2-96*ep^-1-18*ep^-1*z3-27*z4+
     108*z3-384*ep-126*ep*z5+162*ep*z4-288*ep*z3;

  F2=
     -1024/3-16/3*ep^-4+32*ep^-3-256/3*ep^-2+512/3*ep^-1+256/3
     *ep^-1*z3+128*z4-512*z3+2048/3*ep+1024*ep*z5-768*ep*z4+
     4096/3*ep*z3;

  F3=
     448/3+7/3*ep^-4-14*ep^-3+112/3*ep^-2-224/3*ep^-1-154/3*
     ep^-1*z3-77*z4+308*z3-896/3*ep-546*ep*z5+462*ep*z4-2464/3
     *ep*z3;

  F4=
     -18*ep^-1*z3-27*z4+108*z3-306*ep*z5+162*ep*z4-288*ep*z3;
```

As one can see, there are quite a few terms cancelling between the terms with only one loop constants.

```
    #include- minceex.h
    Off Statistics;
    Format nospaces;
    .global
    L   F = Q.Q^3*Q.p2^2/p1.p1^2/p2.p2^2/p3.p3^2/p4.p4/p5.p5/
            p6.p6/p7.p7/p8.p8;
    #call integral(be,0)
    Print +f +s;
    .sort
   F=
      +GschemeConstants(0,0)*BasicT1Integral*rat(162*ep^8+729*
      ep^7+1008*ep^6+405*ep^5-60*ep^4-72*ep^3-12*ep^2,8*ep^4+44
      *ep^3+88*ep^2+76*ep+24)
      +GschemeConstants(0,0)^2*GschemeConstants(1,0)*rat(144*
      ep^9+960*ep^8+2418*ep^7+3051*ep^6+1620*ep^5-450*ep^4-690*
      ep^3-99*ep^2+54*ep+12,4*ep^7+34*ep^6+118*ep^5+214*ep^4+
      214*ep^3+112*ep^2+24*ep)
      +GschemeConstants(0,0)^2*GschemeConstants(2,0)*rat(-288*
      ep^9-2640*ep^8-7486*ep^7-9899*ep^6-5723*ep^5+821*ep^4+
      2179*ep^3+500*ep^2-112*ep-32,6*ep^7+51*ep^6+177*ep^5+321*
      ep^4+321*ep^3+168*ep^2+36*ep)
      +GschemeConstants(0,0)*GschemeConstants(1,0)*
      GschemeConstants(2,0)*rat(-1296*ep^10+16308*ep^9+47592*
      ep^8+43275*ep^7+2601*ep^6-20189*ep^5-9321*ep^4+2300*ep^3+
      1490*ep^2-84*ep-56,96*ep^8+720*ep^7+2088*ep^6+2844*ep^5+
      1584*ep^4-180*ep^3-528*ep^2-144*ep)
      ;

    #call subvalues
~~~Answer in the Gscheme
    #call expansion(1)
~~~Answer in the Gscheme
    Print +f;
    .end

   F=
    -2903/1296-1/18*ep^-2+125/216*ep^-1-1/3*ep^-1*z3-1/2*z4
    -5/18*z3+28541/7776*ep+23/3*ep*z5-5/12*ep*z4+467/108*ep*z3;
  0.41 sec out of 0.44 sec
```

The "Mincer Exact" package has been added to the FORM distribution.

5 New Functions

FORM has obtained many new functions. We name them here. Some names are selfevident:

`Random_`	`RanPerm_`	`Div_`	`Rem_`
`Gcd_`	`Inverse_`	`FirstTerm_`	`Prime_`
`ExtEuclidean_`	`MakeRational_`	`NumFactors_`	`Content_`
`ExtraSymbol_`			

A number of these functions are designed for use in future packages, like a package for Gröbner bases. Such bases can often be calculated faster when calculus is over a prime number and in the end the results over several prime number calculations are combined into a result modulus the product of these numbers.

Making a decent Gröbner basis package is a major undertaking. Again the better packages are not available for inclusion as a library and are part of a commercial product. There is much heuristics involved to take shortcuts and all of that is kept secret. This means that one has to develop ones own heuristics. For this reason we have only been experimenting a little bit with Gröbner bases.

```
#-
#include- groebner.h
Off Statistics;
ON HighFirst;
.global
Local Poly1 = x1^2 + x2*x3 - 2;
Local Poly2 = x1^2*x3 + x2^3 - 3;
Local Poly3 = x1*x2 + x3^2 - 5;
#$n = 3;
#write <> "The input polynomials are:"
Print +f;
.sort
#call groebner(Poly,n)
#write <> "The Groebner basis is:"
Print +f;
.end
```

The above shows what this should look like from the users perspective. The output of this program is:

```
    #-
The input polynomials are:

   Poly1 =
      x1^2 + x2*x3 - 2;

   Poly2 =
      x1^2*x3 + x2^3 - 3;

   Poly3 =
      x1*x2 + x3^2 - 5;

The Groebner basis is:

   Poly1 =
    4093136817253*x1 - 999056107380*x3^9 + 3162784725684*x3^8 +
    15617604960159*x3^7 - 52374677099676*x3^6 - 78004955176188*
    x3^5 + 303405612909504*x3^4 + 117232980911431*x3^3 -
    685255923260685*x3^2 - 3397254300818*x3 + 498469518662424;

   Poly2 =
    30*x3^10 - 9*x3^9 - 570*x3^8 + 222*x3^7 + 4173*x3^6
     - 1782*x3^5 - 14569*x3^4 + 5523*x3^3 + 24357*x3^2
```

```
      - 5721*x3 - 15553;

    Poly3 =
     372103347023*x2 + 159558175470*x3^9 - 307637902881*x3^8
     - 2539999354128*x3^7 + 5261771049639*x3^6 + 13722480161265*
     x3^5 - 31064912209032*x3^4 - 27045529790992*x3^3 +
     69969321110925*x3^2 + 14416898137155*x3 - 48858412479378;

  0.10 sec out of 0.11 sec
```

One can see here that this can run out of hand rather quickly. Of course the secret is in what is in the library groebner.h. It uses a large number of the new functions.

This is for example a routine that defines an S-polynomial:

```
#procedure Spoly(A,B,S);
*
*   Procedure defines the S-polynomial of the two polynomials
*   A and B. We work with $'s for the intermediate variables
*   because that is faster and that way we have to compute the
*   gcd only once and do the divisions only once.
*
    #$firstA = firstterm_('A');
    #$firstB = firstterm_('B');
    #$gcdfirst = gcd_($firstA, $firstB);
    #if ( isnumerical($gcdfirst) )
        Local 'S' = 0;
    #else
        #$nfirstA = $firstA/$gcdfirst;
        #$nfirstB = $firstB/$gcdfirst;
        Skip;
        Local 'S' = 'A'*$nfirstB - 'B'*$nfirstA;
    #endif
    .sort
#endprocedure
```

Note the use of the functions `firstterm_` and `gcd_`. Also the new option `isnumerical` in the preprocessor 'if statement' is being used. Another routine:

```
#procedure MakeInteger(IN,OUT)
*
*   Defines the polynomial 'OUT' as a multiple of 'IN' so that
*   all its coefficients are integer.
*
    #$MkI = content_('IN');
    #inside $MkI
        $cMkI = coeff_;
    #endinside
    Skip;
    Drop 'IN';
    Local 'OUT' = 'IN'/('$cMkI');
    .sort
*
#endprocedure
```

Here we use the function `content_` to eventually obtain the GCD of the numerators and the LCM of the denominators.

Now we hope for volunteers to make a good package.

6 Miscellaneous

The parallel versions TFORM and ParPARFORM are both fully functional now. Till about a year ago ParPARFORM was still missing much of the functionality and was also not very portable. This has all been rectified. It is part of the open source distribution. It does however need a proper MPI installation.

One of the complaints in the past was that .sav files of different executables of FORM were incompatible. This meant that a .sav file generated on one computer might not be usable on another. Also new versions would often need extra variables in the .sav file and hence the old files would be useless. Starting with version 4.0 we have made an attempt to solve these problems. The files should now be uniform, even between 32- and 64-bits versions. In addition we have left much spare space in the headers to allow for future extensions that would otherwise invalidate old files.

Of course, some files cannot be carried from a 64-bits computer to a 32-bits computer. If we use x^{123456}, the power is more than the maximum power allowed for symbols on 32-bits systems (which is 10000). Similarly one can exceed the total number of different objects used in all expressions together. But those are rather natural limitations and they would be encountered very rarely.

Checkpoints are selected points in a FORM program from which the program can be restarted. One can define many such checkpoints but the program will remember only the last one it has encountered. This facility allows the user to restart the program when external causes have halted execution, like a power outage. The user has to define these checkpoints. It is not done automatically.

There is now a forum for FORM users to communicate with each other. To post on the forum one needs to register. This involves answering an easy question and then waiting till one of the moderators approves of the registration. This procedure is needed because of SPAM attacks and organizations having automatic programs for trying to register on forums like this.

The forum is the proper way to report bugs, to ask questions about installation and versions, or to ask help with certain features or techniques.

For all features in FORM , TFORM and ParPARFORM holds that we have tried our best to make them running flawlessly on systems that are available to us. Yet it is not excluded that on other systems strange things happen. This may be due to errors in the other systems, unanticipated behaviour, or just insufficiently careful programming on our side. Whenever such a thing occurs we ask the user to report the problems by means of the forum.

7 What Brings the Future?

The one line answer to this question: "Hopefully something spectacular."

At the moment I have an ERC grantrequest[1] for a project in which I want to use game theoretical concepts to solve the systems of recursion relations that are encountered in multi-loop calculations of all Mellin moments in either DIS or Drell Yan processes. The idea is to use something called Monte Carlo Tree Search (MCTS) to work ones way through the enormous number of possibilities to combine equations in order to find either an acceptable solution, or to find the best among a large number of solutions.

Indications are that the chance of success for this method is quite good.

The automatic derivation of the formulas would eliminate the most time consuming part of this kind of calculations, even if the derivation by means of MCTS costs a significant amount of CPU time.

If one can derive solutions automatically and if there is more than one solution and if the derivation is reasonably fast, one can envision to make solutions that are optimal for a given diagram. This would make the application of the solution much faster.

Currently we (mainly Jan) are trying to make a system for rewriting outputs for numerical programs in a way that takes as few operations as possible. This is called simplification. At the highest level of optimization this system uses MCTS to determine a good ordering of the variables in a multivariate Horner scheme. At the moment we are obtaining exciting results [4, 5] that are considerably better than with existing systems [10–12]. Applied to the GRACE system for automatic one loop calculations [13, 14] it gives expressions that are much shorter than what was available in previous versions. The code will become available in version 4.1 of FORM, but here is already a small example:

```
Symbols x,y,z;
Off Statistics;
Local F = 6*y*z^2+3*y^3-3*x*z^2+6*x*y*z-3*x^2*z+6*x^2*y;
Format O1,stats=on;
Print;
.end
  Z1_=y*z;
  Z2_= - z + 2*y;
  Z2_=x*Z2_;
  Z3_=z^2;
  Z1_=Z2_ - Z3_ + 2*Z1_;
  Z1_=x*Z1_;
  Z2_=y^2;
  Z2_=2*Z3_ + Z2_;
  Z2_=y*Z2_;
  Z1_=Z2_ + Z1_;
```

[1] By now (publication of these proceedings) this request has been approved and the project will start in July 2013.

```
      F=3*Z1_;
*** STATS: original  1P 16M 5A : 23
*** STATS: optimized 0P 10M 5A : 15
```

The statistics show that we started with 23 operations (one power which counted double because it was a third power, 16 multiplications and 5 additions) and that we are left with 15 operations. Note that squares are counted like a single multiplication. If we run this program with the option O2 we obtain

```
      Z1_=z^2;
      Z2_=2*y;
      Z3_=z*Z2_;
      Z2_= - z + Z2_;
      Z2_=x*Z2_;
      Z2_=Z2_ - Z1_ + Z3_;
      Z2_=x*Z2_;
      Z3_=y^2;
      Z1_=2*Z1_ + Z3_;
      Z1_=y*Z1_;
      Z1_=Z1_ + Z2_;
      F=3*Z1_;
*** STATS: original  1P 16M 5A : 23
*** STATS: optimized 0P 9M 5A : 14
```

and with O3 (MCTS) we have

```
      Z1_=x + z;
      Z2_=2*y;
      Z3_=Z2_ - x;
      Z1_=z*Z3_*Z1_;
      Z3_=y^3;
      Z2_=x^2*Z2_;
      Z1_=Z1_ + Z3_ + Z2_;
      F=3*Z1_;
*** STATS: original  1P 16M 5A : 23
*** STATS: optimized 1P 6M 4A : 12
```

It is possible to obtain an even better decomposition, but this requires simplifications of the type $x^2 + xz + z^2 \to (x + z)^2 - xz$ which is not within the scope of the simplifications we apply. Similarly $x^2 + 2xz + z^2$ would not be seen as a square. Such simplifications require entirely different algorithms which should be executed before the FORM algorithms are applied.

References

1. Banerjee, N., Dutta, S., Lodato, I.: The fate of flat directions in higher derivative gravity. 36p (2013). e-Print: arXiv:1301.6773 [hep-th]
2. Vermaseren, J.A.M.: New features of FORM. arXiv math-ph/0010025
3. Tentyukov, M., Vermaseren, J.A.M.: The multithreaded version of FORM. Comput. Phys. Commun. **181**, 1419 (2010). arXiv hep-ph/0702279
4. Kuipers, J., Vermaseren, J.A.M., Plaat, A., van den Herik, H.J.: Improving multi-variate Horner schemes with Monte Carlo tree search. arXiv 1207.7079

5. Kuipers, J., Vermaseren, J.A.M.: Code simplification in FORM (In preparation)
6. Lewis, R.H.: Fermat. http://www.bway.net/~lewis/
7. GNU Multiple Precision Arithmetic Library. http://gmplib.org/
8. Mincer is a program for the evaluation of massless three loop propagator diagrams. The original paper is: Gorishnii, S.G., Larin, S.A., Surguladze, L.R., Tkachev, F.V.: Mincer: program for multiloop calculations in quantum field theory for the schoonschip system. Comput. Phys. Commun. **55**, 381–408 (1989). A much improved FORM version is included in the FORM distribution
9. Blümlein, J., Broadhurst, D.J., Vermaseren, J.A.M.: The multiple zeta value datamine. Comput. Phys. Commun. **181**, 582–625 (2010)
10. Kojima, M.: Efficient evaluation of polynomials and their partial derivatives in homotopy continuation methods. J. Oper. Res. Soc. Jpn. **51**, 29–54 (2008)
11. Leiserson, C.E., Li, L., Maza, M.M., Xie, Y.: Efficient evaluation of large polynomials. LNCS **6327**, 342–353 (2010)
12. Reiter, T.: Optimising code generation with haggies. Comput. Phys. Commun. **181**, 1301–1331 (2010). arXiv 0907.3714
13. Yuasa, F., et al.: Automatic computation of cross sections in HEP. Prog. Theor. Phys. Suppl. **138**, 18 (2000). arXiv hep-ph/0007053
14. Fujimoto, J., et al.: GRACE with FORM. Nucl. Phys. Proc. Suppl. **160**, 150–154 (2006)

Feynman Graphs

Stefan Weinzierl

Abstract In these lectures I discuss Feynman graphs and the associated Feynman integrals. Of particular interest are the classes functions, which appear in the evaluation of Feynman integrals. The most prominent class of functions is given by multiple polylogarithms. The algebraic properties of multiple polylogarithms are reviewed in the second part of these lectures. The final part of these lectures is devoted to Feynman integrals, which cannot be expressed in terms of multiple polylogarithms. Methods from algebraic geometry provide tools to tackle these integrals.

1 Feynman Graph Polynomials

The first part of these lectures is centred around two graph polynomials. We will give four different definitions of these two polynomials, each definition will shed a different light on the nature of these polynomials. The presentation in this section follows [8].

1.1 Graphs

Let us start with a few basic definitions: A graph consists of edges and vertices. We will mainly consider connected graphs. The valency of a vertex is the number of edges attached to it. Vertices of valency 0, 1 and 2 are special. A vertex of valency 0 is necessarily disconnected from the rest of graph and therefore not relevant for connected graphs. A vertex of valency 1 has exactly one edge attached to it. This

S. Weinzierl (✉)
PRISMA Cluster of Excellence, Institut für Physik, Johannes Gutenberg-Universität Mainz, D-55099 Mainz, Germany
e-mail: stefanw@thep.physik.uni-mainz.de

C. Schneider and J. Blümlein (eds.), *Computer Algebra in Quantum Field Theory*, Texts & Monographs in Symbolic Computation, DOI 10.1007/978-3-7091-1616-6_16,

edge is called an external edge. All other edges are called internal edges. In the physics community it is common practice not to draw a vertex of valency 1, but just the external edge. A vertex of valency 2 is called a mass insertion and is usually not considered. Therefore in physics it is usually implied that a genuine vertex has a valency of three or greater.

An edge in a Feynman graph represents a propagating particle. The edges are drawn in a way as to represent the different types of particles. For example, one uses lines with an arrow for fermions, wavy lines for photons or curly lines for gluons. A simple line without decorations is used for scalar particles. To each (orientated) edge we associate a D-dimensional vector q and a number m, describing the momentum and the mass of the particle. D is the dimensions of space-time.

Vertices of valency $n \geq 3$ represent interactions of n particles. At each vertex we have momentum conservation: The sum of all momenta flowing into the vertex equals the sum of all momenta flowing out of the vertex.

To each Feynman graph we can associate a new graph, obtained by replacing each propagator of the original graph by a scalar propagator. This new graph is called the underlying topology. This new graph does no longer carry the information on the type of the particles propagating along the edges. We will later associate to each Feynman graph an integral, called the Feynman integral of this graph. It turns out, that the Feynman integral corresponding to an arbitrary Feynman graph can always be expressed as a linear combination of Feynman integrals corresponding to Feynman graphs with scalar propagators. Therefore it is sufficient to restrict ourselves to the underlying topology and to restrict our study to Feynman graphs with scalar propagators.

Let us now consider a graph G with n edges and r vertices. Assume that the graph has k connected components. The loop number l is defined by

$$l = n - r + k. \tag{1}$$

If the graph is connected we have $l = n - r + 1$. The loop number l is also called the first Betti number of the graph or the cyclomatic number. In the physics context it has the following interpretation: If we fix all momenta of the external lines and if we impose momentum conservation at each vertex, then the loop number is equal to the number of independent momentum vectors not constrained by momentum conservation.

A connected graph of loop number 0 is called a tree. A graph of loop number 0, connected or not, is called a forest. If the forest hast k connected components, it is called a k-forest. A tree is a 1-forest.

1.2 Spanning Forests

Given an arbitrary connected graph G, a spanning tree of G is a subgraph, which contains all the vertices of G and which is a tree. In a similar way, given an arbitrary

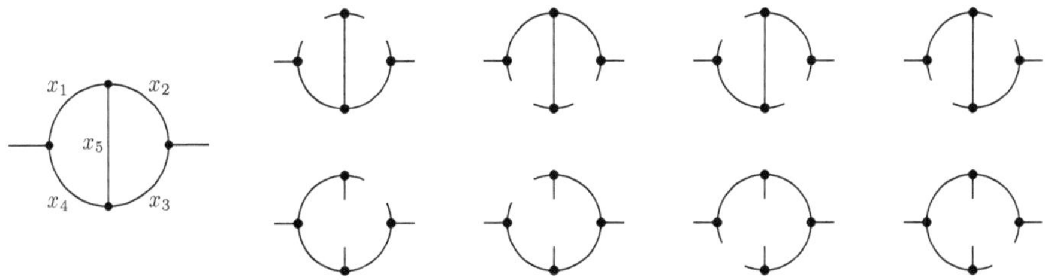

Fig. 1 An example of a Feynman graph and the associated set of spanning trees

connected graph G, a spanning k-forest of G is a subgraph, which contains all the vertices of G and which is a k-forest.

We have already associated to each edge a momentum vector and a mass. In addition we associate now to each internal edge e_j a real (or complex) variable x_j. The variables x_j are called Feynman parameters. For each graph we can define two polynomials $\mathcal{U}$ and $\mathcal{F}$ in the variables x_j as follows: Let G be a connected graph and $\mathcal{T}_1$ the set of its spanning trees. The first graph polynomial is defined by

$$\mathcal{U} = \sum_{T \in \mathcal{T}_1} \prod_{e_j \notin T} x_j. \tag{2}$$

This is best illustrated with an example. Figure 1 shows a Feynman graph decorated with the Feynman parameters x_1 to x_5, as well as the associated set $\mathcal{T}_1$ of spanning trees. For each spanning tree we take the Feynman parameters associated to the edges not belonging to this spanning tree. Summing over all spanning trees we obtain for this example

$$\mathcal{U} = x_1 x_2 + x_3 x_4 + x_1 x_3 + x_2 x_4 + x_2 x_5 + x_1 x_5 + x_4 x_5 + x_3 x_5. \tag{3}$$

$\mathcal{U}$ is also called the first Symanzik polynomial of the graph G. In mathematics, the Kirchhoff polynomial of a graph is better known. It is defined by

$$\mathcal{K} = \sum_{T \in \mathcal{T}_1} \prod_{e_j \in T} x_j. \tag{4}$$

The difference between the two definitions is given by the fact that in the case of $\mathcal{K}$ we consider all edges belonging to the spanning tree T, while in the case of $\mathcal{U}$ we consider all edges not belonging to T. There is a simple relation between the Kirchhoff polynomial $\mathcal{K}$ and the first Symanzik polynomial $\mathcal{U}$:

$$\mathcal{U}(x_1, \ldots, x_n) = x_1 \ldots x_n \mathcal{K}\left(\frac{1}{x_1}, \ldots, \frac{1}{x_n}\right). \tag{5}$$

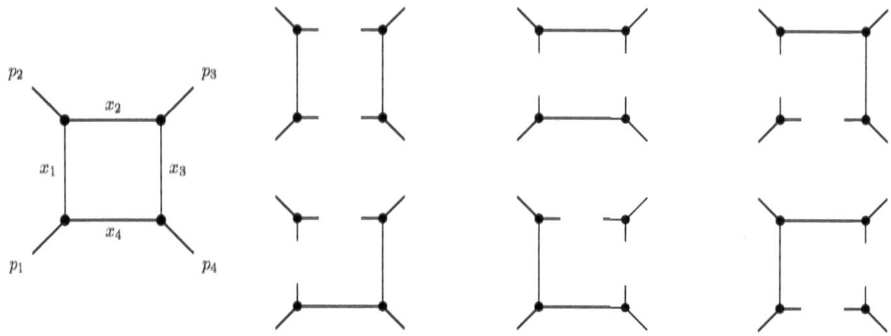

Fig. 2 An example of a Feynman graph and the associated set of spanning 2-forests

We now turn to the definition of $\mathcal{F}$. Let G be a connected graph and $\mathcal{T}_2$ the set of its spanning 2-forests. An element of $\mathcal{T}_2$ is denoted as (T_1, T_2). Let us further denote by P_{T_i} the set of external momenta of G attached to T_i. We first define a polynomial $\mathcal{F}_0$ by

$$\mathcal{F}_0 = \sum_{(T_1,T_2)\in\mathcal{T}_2} \left(\prod_{e_i \notin (T_1,T_2)} x_i \right) \left(\sum_{p_j \in P_{T_1}} \sum_{p_k \in P_{T_2}} \frac{p_j \cdot p_k}{\mu^2} \right). \tag{6}$$

Here, $p_j \cdot p_k$ is the Minkowski scalar product of two momenta vectors. μ is an arbitrary scale introduced to make the expression dimensionless. $\mathcal{F}$ is defined by

$$\mathcal{F} = \mathcal{F}_0 + \mathcal{U} \sum_{i=1}^{n} x_i \frac{m_i^2}{\mu^2}. \tag{7}$$

m_i denotes the mass of the i-th internal line. If all internal masses are zero, we have $\mathcal{F} = \mathcal{F}_0$. $\mathcal{F}$ is called the second Symanzik polynomial. Again, let us illustrate the definition of $\mathcal{F}$ with an example. Figure 2 shows a Feynman graph and the associated set $\mathcal{T}_2$ of 2-forests. For simplicity we assume that all internal masses are zero, therefore we have $\mathcal{F} = \mathcal{F}_0$. In Fig. 2 we have labelled the internal edges with the Feynman parameters x_1 to x_4. The external edges have been labelled with the external momenta p_1 to p_4. We orientate these edges such that p_1 to p_4 are all flowing outwards. With this choice momentum conservation reads

$$p_1 + p_2 + p_3 + p_4 = 0. \tag{8}$$

The Mandelstam variables s and t are defined by

$$s = (p_1 + p_2)^2, \; t = (p_2 + p_3)^2. \tag{9}$$

From the definition in Eq. (6) we find for the polynomial $\mathcal{F}$ for this example

$$\mathcal{F} = x_2x_4\frac{(-s)}{\mu^2} + x_1x_3\frac{(-t)}{\mu^2} + x_1x_4\frac{(-p_1^2)}{\mu^2} + x_1x_2\frac{(-p_2^2)}{\mu^2} + x_2x_3\frac{(-p_3^2)}{\mu^2}$$
$$+x_3x_4\frac{(-p_4^2)}{\mu^2}.$$

A few remarks on the two Symanzik polynomials are in order: Both polynomials are homogeneous in the Feynman parameters, $\mathcal{U}$ is of degree l, $\mathcal{F}$ is of degree $l+1$. The polynomial $\mathcal{U}$ is linear in each Feynman parameter. If all internal masses are zero, then also $\mathcal{F}$ is linear in each Feynman parameter. In expanded form each monomial of $\mathcal{U}$ has coefficient $+1$.

1.3 Feynman Integrals

Feynman graphs have been invented as a pictorial notation for mathematical expressions arising in the context of perturbative quantum field theory. Each part in a Feynman graph corresponds to a specific expression and the full Feynman graph corresponds to the product of these expressions. For scalar theories the correspondence is as follows: An internal edge corresponds to a propagator

$$\frac{i}{q^2 - m^2}, \tag{10}$$

an external edge to the factor 1. A vertex corresponds in scalar theories also to the factor 1. In addition, there is for each internal momentum not constrained by momentum conservation an integration

$$\int \frac{d^Dk}{(2\pi)^D}. \tag{11}$$

Let us now consider a Feynman graph G with m external edges, n internal edges and l loops. To each internal edge we associate apart from its momentum and its mass a positive integer number ν, giving the power to which the propagator occurs. (We can think of ν as the relict of neglecting vertices of valency 2. A number $\nu > 1$ corresponds to $\nu - 1$ mass insertions on this edge). The momenta flowing through the internal lines can be expressed through the independent loop momenta $k_1, \ldots, k_l$ and the external momenta $p_1, \ldots, p_m$ as

$$q_i = \sum_{j=1}^{l} \rho_{ij}k_j + \sum_{j=1}^{m} \sigma_{ij}p_j, \quad \rho_{ij}, \sigma_{ij} \in \{-1, 0, 1\}. \tag{12}$$

We define the Feynman integral by

$$I_G = \frac{\prod\limits_{j=1}^{n} \Gamma(\nu_j)}{\Gamma(\nu - lD/2)} \left(\mu^2\right)^{\nu - lD/2} \int \prod_{r=1}^{l} \frac{d^D k_r}{i\pi^{\frac{D}{2}}} \prod_{j=1}^{n} \frac{1}{(-q_j^2 + m_j^2)^{\nu_j}}, \tag{13}$$

with $\nu = \nu_1 + \ldots + \nu_n$. The prefactor in front of the integral is the convention used in this article. This choice is motivated by the fact that after Feynman parametrisation we obtain a simple formula. Feynman parametrisation makes use of the identity

$$\prod_{j=1}^{n} \frac{1}{P_j^{\nu_j}} = \frac{\Gamma(\nu)}{\prod\limits_{i=1}^{n} \Gamma(\nu_j)} \int_{\Delta} \omega \left(\prod_{i=1}^{n} x_i^{\nu_i - 1} \right) \left(\sum_{j=1}^{n} x_j P_j \right)^{-\nu}, \tag{14}$$

where ω is a differential $(n-1)$-form given by

$$\omega = \sum_{j=1}^{n} (-1)^{j-1} \, x_j \, dx_1 \wedge \ldots \wedge \widehat{dx_j} \wedge \ldots \wedge dx_n. \tag{15}$$

The hat indicates that the corresponding term is omitted. The integration is over

$$\Delta = \left\{ [x_1 : x_2 : \ldots : x_n] \in \mathbb{P}^{n-1} | x_i \geq 0, 1 \leq i \leq n \right\}. \tag{16}$$

We use Eq. (14) with $P_j = -q_j^2 + m_j^2$. We can write

$$\sum_{j=1}^{n} x_j (-q_j^2 + m_j^2) = -\sum_{r=1}^{l} \sum_{s=1}^{l} k_r M_{rs} k_s + \sum_{r=1}^{l} 2 k_r \cdot Q_r - J, \tag{17}$$

where M is a $l \times l$ matrix with scalar entries and Q is a l-vector with D-vectors as entries. After Feynman parametrisation the integrals over the loop momenta $k_1, \ldots, k_l$ can be done and we obtain

$$I_G = \int_{\Delta} \omega \left(\prod_{j=1}^{n} x_j^{\nu_j - 1} \right) \frac{\mathcal{U}^{\nu - (l+1)D/2}}{\mathcal{F}^{\nu - lD/2}}. \tag{18}$$

The functions $\mathcal{U}$ and $\mathcal{F}$ are given by

$$\mathcal{U} = \det(M), \quad \mathcal{F} = \det(M) \left(-J + Q M^{-1} Q \right) / \mu^2. \tag{19}$$

It can be shown that Eq. (19) agrees with the definition of $\mathcal{U}$ and $\mathcal{F}$ given in Sect. 1.2 in terms of spanning trees and spanning forests. Thus, Eq. (19) provides a second definition of the two graph polynomials. Equation (18) defines the Feynman integral of a graph G in terms of the two graph polynomials $\mathcal{U}$ and $\mathcal{F}$. A few remarks are in order: The integral over the Feynman parameters is a $(n-1)$-dimensional integral in projective space $\mathbb{P}^{n-1}$, where n is the number of internal edges of the graph. Singularities may arise if the zero sets of $\mathcal{U}$ and $\mathcal{F}$ intersect the region of integration. The dimension D of space-time enters only in the exponents of the integrand and the exponents act as a regularisation.

1.4 The Laplacian of a Graph

For a graph G with n edges and r vertices define the Laplacian L [38, 42] as a $r \times r$-matrix with

$$L_{ij} = \begin{cases} \sum x_k & \text{if } i = j \text{ and edge } e_k \text{ is attached to } v_i \text{ and is not a self-loop,} \\ -\sum x_k & \text{if } i \neq j \text{ and edge } e_k \text{ connects } v_i \text{ and } v_j. \end{cases}$$

We speak of a self-loop (or tadpole) if an edge starts and ends at the same vertex. In the sequel we will need minors of the matrix L and it is convenient to introduce the following notation: For a $r \times r$ matrix A we denote by $A[i_1, \ldots, i_k; j_1, \ldots, j_k]$ the $(r-k) \times (r-k)$ matrix, which is obtained from A by deleting the rows $i_1, \ldots, i_k$ and the columns $j_1, \ldots, j_k$. For $A[i_1, \ldots, i_k; i_1, \ldots, i_k]$ we will simply write $A[i_1, \ldots, i_k]$. The matrix-tree theorem relates the Laplacian of a graph to its Kirchhoff polynomial:

$$\mathcal{K} = \det L[i]. \tag{20}$$

A generalisation by the all-minor matrix tree theorem [13, 14, 31] leads to the following expressions for the graph polynomials $\mathcal{U}$ and $\mathcal{F}_0$: Starting from a graph G with n internal edges, r internal vertices $(v_1, \ldots, v_r)$ and m external legs, we first attach m additional vertices $(v_{r+1}, \ldots, v_{r+m})$ to the ends of the external legs and then associate the parameters $z_1, \ldots, z_m$ with the external edges. This defines a new graph $\tilde{G}$. We now consider the Laplacian $\tilde{L}$ of $\tilde{G}$ and the polynomial

$$\mathcal{W}(x_1, \ldots, x_n, z_1, \ldots, z_m) = \det\ \tilde{L}[r+1, \ldots, r+m]. \tag{21}$$

We then expand $\mathcal{W}$ in polynomials homogeneous in the variables z_j:

$$\begin{aligned} \mathcal{W} &= \mathcal{W}^{(0)} + \mathcal{W}^{(1)} + \mathcal{W}^{(2)} + \ldots + \mathcal{W}^{(m)}, \\ \mathcal{W}^{(k)} &= \sum_{1 \leq j_1 < \ldots < j_k \leq m} \mathcal{W}^{(k)}_{(j_1, \ldots, j_k)}(x_1, \ldots, x_n)\, z_{j_1} \ldots z_{j_k}. \end{aligned} \tag{22}$$

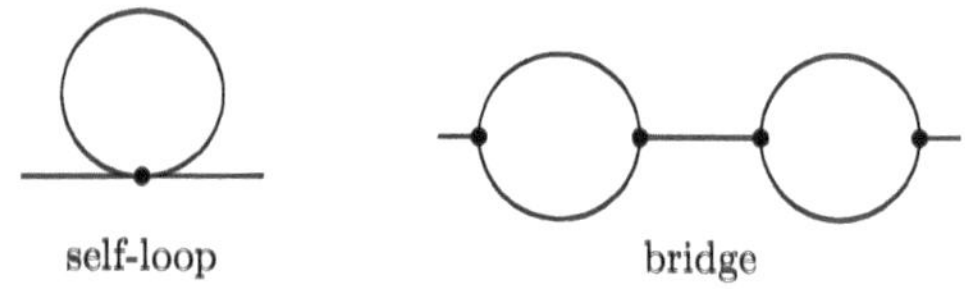

Fig. 3 Examples of graphs containing a self-loop (*left*) or a bridge (*right*)

We then have

$$\begin{aligned} 0 &= \mathcal{W}^{(0)}, \\ \mathcal{U} &= x_1 \dots x_n \, \mathcal{W}^{(1)}_{(j)} \left(\frac{1}{x_1}, \dots, \frac{1}{x_n} \right) \qquad \text{for any } j, \\ \mathcal{F}_0 &= x_1 \dots x_n \sum_{(j,k)} \left(\frac{p_j \cdot p_k}{\mu^2} \right) \cdot \mathcal{W}^{(2)}_{(j,k)} \left(\frac{1}{x_1}, \dots, \frac{1}{x_n} \right). \end{aligned} \tag{23}$$

This provides a third definition of the Feynman graph polynomials $\mathcal{U}$ and $\mathcal{F}$. This formulation is particularly well suited for computer algebra.

1.5 Deletion and Contraction Properties

Let us now consider a recursive definition of the two graph polynomials based on deletion and contraction properties. We first define a regular edge to be an edge, which is neither a self-loop nor a bridge. In graph theory an edge is called a bridge, if the deletion of the edge increases the number of connected components. Examples for graphs containing either a self-loop or a bridge are shown in Fig. 3. For a graph G and a regular edge e we define

$$\begin{aligned} G/e & \quad \text{to be the graph obtained from } G \text{ by contracting the regular edge } e, \\ G - e & \quad \text{to be the graph obtained from } G \text{ by deleting the regular edge } e. \end{aligned} \tag{24}$$

The operations of deletion and contraction are illustrated in Fig. 4. For any regular edge e_k we have

$$\begin{aligned} \mathcal{U}(G) &= \mathcal{U}(G/e_k) + x_k \mathcal{U}(G - e_k), \\ \mathcal{F}_0(G) &= \mathcal{F}_0(G/e_k) + x_k \mathcal{F}_0(G - e_k). \end{aligned} \tag{25}$$

The recursion terminates when all edges are either bridges or self-loops. These graphs are called terminal forms. If a terminal form has r vertices and l (self-) loops, then there are $(r-1)$ "tree-like" propagators, where the momenta flowing through these propagators are linear combinations of the external momenta p_i alone and independent of the independent loop momenta k_j. The momenta of the remaining l propagators are on the other hand independent of the external momenta and can be

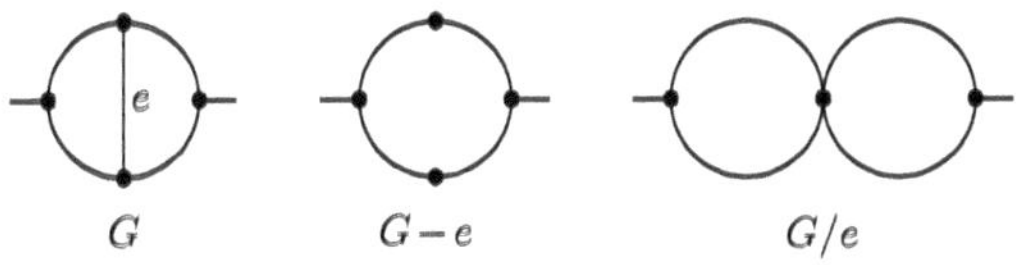

Fig. 4 A graph G, together with the graph $G-e$, where the edge e has been deleted and the graph G/e, where the edge e has been contracted

taken as the independent loop momenta k_j, $j = 1, \dots, l$. Let us agree that we label the $(r-1)$ "tree-like" edges from 1 to $r-1$, and the remaining l edges by $r, \dots, n$ with $n = r + l - 1$. We further denote the momentum squared flowing through edge j by q_j^2. For a terminal form we have

$$\mathcal{U} = x_r \dots x_n, \quad \mathcal{F}_0 = x_r \dots x_n \sum_{j=1}^{r-1} x_j \left(\frac{-q_j^2}{\mu^2}\right). \tag{26}$$

Equation (25) together with Eq. (26) provides a fourth definition of the graph polynomials $\mathcal{U}$ and $\mathcal{F}$.

Let e_a and e_b be two regular edges, which share a common vertex. We have the following factorisation theorems:

$$\begin{aligned}
&\mathcal{U}\left(G/e_a - e_b\right)\mathcal{U}\left(G/e_b - e_a\right) - \mathcal{U}\left(G - e_a - e_b\right)\mathcal{U}\left(G/e_a/e_b\right) = \left(\frac{\Delta_1}{x_a x_b}\right)^2, \\
&\mathcal{U}\left(G/e_a - e_b\right)\mathcal{F}_0\left(G/e_b - e_a\right) - \mathcal{U}\left(G - e_a - e_b\right)\mathcal{F}_0\left(G/e_a/e_b\right) \\
&+\mathcal{F}_0\left(G/e_a - e_b\right)\mathcal{U}\left(G/e_b - e_a\right) - \mathcal{F}_0\left(G - e_a - e_b\right)\mathcal{U}\left(G/e_a/e_b\right) = \\
&\quad 2\left(\frac{\Delta_1}{x_a x_b}\right)\left(\frac{\Delta_2}{x_a x_b}\right).
\end{aligned} \tag{27}$$

Δ_1 and Δ_2 are polynomials in the Feynman parameters and can be expressed as sums over 2-forests and sums over 3-forests, respectively [8]. If for all external momenta one has

$$\left(p_{i_1} \cdot p_{i_2}\right) \cdot \left(p_{i_3} \cdot p_{i_4}\right) = \left(p_{i_1} \cdot p_{i_3}\right) \cdot \left(p_{i_2} \cdot p_{i_4}\right), \tag{28}$$

then

$$\mathcal{F}_0\left(G/e_a - e_b\right)\mathcal{F}_0\left(G/e_b - e_a\right) - \mathcal{F}_0\left(G - e_a - e_b\right)\mathcal{F}_0\left(G/e_a/e_b\right) = \left(\frac{\Delta_2}{x_a x_b}\right)^2.$$

The factorisation theorems follow from Dodgson's identity [15, 48], which states that for any $n \times n$ matrix A one has

$$\det(A)\det(A[i,j]) = \det(A[i])\det(A[j]) - \det(A[i;j])\det(A[j;i]). \tag{29}$$

We recall that

$$
\begin{aligned}
A[i] &\ \text{ is obtained from } A \text{ by deleting the } i\text{-th row and column,}\\
A[i;j] &\ \text{ is obtained from } A \text{ by deleting the } i\text{-th row and the } j\text{-th column,}\\
A[i,j] &\ \text{ is obtained from } A \text{ by deleting the rows and columns } i \text{ and } j.
\end{aligned}
$$

The first formula of Eq. (27) is at the heart of the reduction algorithm of [11, 12].

2 Multiple Polylogarithms

Let us come back to the Feynman integrals defined in Eq. (18). A Feynman integral has an expansion as a Laurent series in the parameter $\varepsilon = (4 - D)/2$ of dimensional regularisation:

$$
I_G = \sum_{j=-2l}^{\infty} c_j \varepsilon^j. \tag{30}
$$

The Laurent series of an l-loop integral can have poles in ε up to the order $(2l)$. The poles in ε correspond to ultraviolet or infrared divergences. The coefficients c_j are functions of the scalar products $p_j \cdot p_k$, the masses m_i and (in a trivial way) of the arbitrary scale μ. An interesting question is, which functions do occur in the coefficients c_j.

2.1 One-Loop Integrals

The question, which functions occur in the coefficients c_j has a satisfactory answer for one-loop integrals. If we restrict our attention to the coefficients c_j with $j \leq 0$ (i.e. to c_{-2}, c_{-1} and c_0), then these coefficients can be expressed as a sum of algebraic functions of the scalar products and the masses times two transcendental functions, whose arguments are again algebraic functions of the scalar products and the masses.

The two transcendental functions are the logarithm and the dilogarithm:

$$
\begin{aligned}
\text{Li}_1(x) &= \sum_{n=1}^{\infty} \frac{x^n}{n} = -\ln(1-x),\\
\text{Li}_2(x) &= \sum_{n=1}^{\infty} \frac{x^n}{n^2}.
\end{aligned} \tag{31}
$$

2.2 The Sum Representation of Multiple Polylogarithms

Beyond one-loop an answer to the above question is not yet known. We know however that the following generalisations occur: From Eq. (31) it is not too hard to imagine that the generalisation includes the classical polylogarithms defined by

$$\text{Li}_m(x) = \sum_{n=1}^{\infty} \frac{x^n}{n^m}. \tag{32}$$

However, explicit calculations at two-loops and beyond show that a wider generalisation towards functions of several variables is needed and one arrives at the multiple polylogarithms defined by Goncharov [22, 23] and Borwein et al. [9]

$$\text{Li}_{m_1,\dots,m_k}(x_1,\dots,x_k) = \sum_{n_1>n_2>\dots>n_k>0}^{\infty} \frac{x_1^{n_1}}{{n_1}^{m_1}} \cdots \frac{x_k^{n_k}}{{n_k}^{m_k}}. \tag{33}$$

Methods for the numerical evaluation of multiple polylogarithms can be found in [44]. The values of the multiple polylogarithms at $x_1 = \dots x_k = 1$ are called multiple ζ-values [6, 9]:

$$\zeta_{m_1,\dots,m_k} = \text{Li}_{m_1,m_2,\dots,m_k}(1,1,\dots,1) = \sum_{n_1>n_2>\dots>n_k>0}^{\infty} \frac{1}{n_1^{m_1}} \cdot \ldots \cdot \frac{1}{n_k^{m_k}}. \tag{34}$$

Important specialisations of multiple polylogarithms are the harmonic polylogarithms [21, 36]

$$H_{m_1,\dots,m_k}(x) = \text{Li}_{m_1,\dots,m_k}(x, \underbrace{1,\dots,1}_{k-1}), \tag{35}$$

Further specialisations leads to Nielsen's generalised polylogarithms [34]

$$S_{n,p}(x) = \text{Li}_{n+1,1,\dots,1}(x, \underbrace{1,\dots,1}_{p-1}). \tag{36}$$

2.3 The Integral Representation of Multiple Polylogarithms

In Eq. (33) we have defined multiple polylogarithms through the sum representation. In addition, multiple polylogarithms have an integral representation. To discuss the integral representation it is convenient to introduce for $z_k \neq 0$ the following functions

$$G(z_1, \ldots, z_k; y) = \int_0^y \frac{dt_1}{t_1 - z_1} \int_0^{t_1} \frac{dt_2}{t_2 - z_2} \cdots \int_0^{t_{k-1}} \frac{dt_k}{t_k - z_k}. \tag{37}$$

In this definition one variable is redundant due to the following scaling relation:

$$G(z_1, \ldots, z_k; y) = G(xz_1, \ldots, xz_k; xy) \tag{38}$$

If one further defines $g(z; y) = 1/(y - z)$, then one has

$$\frac{d}{dy} G(z_1, \ldots, z_k; y) = g(z_1; y) G(z_2, \ldots, z_k; y) \tag{39}$$

and

$$G(z_1, z_2, \ldots, z_k; y) = \int_0^y dt \; g(z_1; t) G(z_2, \ldots, z_k; t). \tag{40}$$

One can slightly enlarge the set and define $G(0, \ldots, 0; y)$ with k zeros for z_1 to z_k to be

$$G(0, \ldots, 0; y) = \frac{1}{k!} (\ln y)^k . \tag{41}$$

This permits us to allow trailing zeros in the sequence $(z_1, \ldots, z_k)$ by defining the function G with trailing zeros via Eqs. (40) and (41). To relate the multiple polylogarithms to the functions G it is convenient to introduce the following shorthand notation:

$$G_{m_1,\ldots,m_k}(z_1, \ldots, z_k; y) = G(\underbrace{0, \ldots, 0}_{m_1 - 1}, z_1, \ldots, z_{k-1}, \underbrace{0 \ldots, 0}_{m_k - 1}, z_k; y) \tag{42}$$

Here, all z_j for $j = 1, \ldots, k$ are assumed to be non-zero. One then finds

$$\text{Li}_{m_1,\ldots,m_k}(x_1, \ldots, x_k) = (-1)^k G_{m_1,\ldots,m_k}\left(\frac{1}{x_1}, \frac{1}{x_1 x_2}, \ldots, \frac{1}{x_1 \ldots x_k}; 1\right). \tag{43}$$

The inverse formula reads

$$G_{m_1,\ldots,m_k}(z_1, \ldots, z_k; y) = (-1)^k \; \text{Li}_{m_1,\ldots,m_k}\left(\frac{y}{z_1}, \frac{z_1}{z_2}, \ldots, \frac{z_{k-1}}{z_k}\right). \tag{44}$$

Equation (43) together with Eqs. (42) and (37) defines an integral representation for the multiple polylogarithms. As an example, we obtain from Eqs. (43) and (38) the integral representation of harmonic polylogarithms:

$$H_{m_1,\ldots,m_k}(x) = (-1)^k\, G_{m_1,\ldots,m_k}(1,\ldots,1;x)\,. \tag{45}$$

The function $G_{m_1,\ldots,m_k}(1,\ldots,1;x)$ is an iterated integral in which only the two one-forms

$$\omega_0 = \frac{dt}{t},\; \omega_1 = \frac{dt}{t-1} \tag{46}$$

corresponding to $z = 0$ and $z = 1$ appear. If one restricts the possible values of z to zero and the n-th roots of unity one arrives at the class of cyclomatic harmonic polylogarithms [1].

2.4 Shuffle and Quasi-shuffle Algebras

Multiple polylogarithms have a rich algebraic structure. The representations as iterated integrals and nested sums induce a shuffle algebra and a quasi-shuffle algebra, respectively. Shuffle and quasi-shuffle algebras are Hopf algebras. Note that the shuffle algebra of multiple polylogarithms is distinct from the quasi-shuffle algebra of multiple polylogarithms.

Consider a set of letters A. The set A is called the alphabet. A word is an ordered sequence of letters:

$$w = l_1 l_2 \ldots l_k. \tag{47}$$

The word of length zero is denoted by e. Let K be a field and consider the vector space of words over K. A shuffle algebra $\mathscr{A}$ on the vector space of words is defined by

$$(l_1 l_2 \ldots l_k) \cdot (l_{k+1} \ldots l_r) = \sum_{\text{shuffles } \sigma} l_{\sigma(1)} l_{\sigma(2)} \ldots l_{\sigma(r)}, \tag{48}$$

where the sum runs over all permutations σ, which preserve the relative order of $1, 2, \ldots, k$ and of $k+1, \ldots, r$. The name "shuffle algebra" is related to the analogy of shuffling cards: If a deck of cards is split into two parts and then shuffled, the relative order within the two individual parts is conserved. A shuffle algebra is also known under the name "mould symmetral" [16]. The empty word e is the unit in this algebra:

$$e \cdot w = w \cdot e = w. \tag{49}$$

A recursive definition of the shuffle product is given by

$$\begin{aligned}(l_1 l_2 \ldots l_k) \cdot (l_{k+1} \ldots l_r) = {} & l_1 \left[(l_2 \ldots l_k) \cdot (l_{k+1} \ldots l_r)\right] \\ & + l_{k+1} \left[(l_1 l_2 \ldots l_k) \cdot (l_{k+2} \ldots l_r)\right].\end{aligned} \tag{50}$$

It is a well known fact that the shuffle algebra is actually a (non-cocommutative) Hopf algebra [37]. In a Hopf algebra we have in addition to the multiplication and the unit a counit, a comultiplication and a antipode. The unit in an algebra can be viewed as a map from K to A and multiplication in an algebra can be viewed as a map from the tensor product $A \otimes A$ to A (e.g. one takes two elements from A, multiplies them and gets one element out). The counit is a map from A to K, whereas comultiplication is a map from A to $A \otimes A$. We will always assume that the comultiplication is coassociative. The general form of the coproduct is

$$\Delta(a) = \sum_i a_i^{(1)} \otimes a_i^{(2)}, \tag{51}$$

where $a_i^{(1)}$ denotes an element of A appearing in the first slot of $A \otimes A$ and $a_i^{(2)}$ correspondingly denotes an element of A appearing in the second slot. Sweedler's notation [39] consists in dropping the dummy index i and the summation symbol:

$$\Delta(a) = a^{(1)} \otimes a^{(2)} \tag{52}$$

The sum is implicitly understood. This is similar to Einstein's summation convention, except that the dummy summation index i is also dropped. The superscripts $^{(1)}$ and $^{(2)}$ indicate that a sum is involved. Using Sweedler's notation, the compatibility between the multiplication and comultiplication is expressed as

$$\Delta\left(a \cdot b\right) = \left(a^{(1)} \cdot b^{(1)}\right) \otimes \left(a^{(2)} \cdot b^{(2)}\right). \tag{53}$$

The antipode S is map from A to A, which fulfils

$$a^{(1)} \cdot S\left(a^{(2)}\right) = S\left(a^{(1)}\right) \cdot a^{(2)} = e \cdot \bar{e}(a). \tag{54}$$

With this background at hand we can now state the coproduct, the counit and the antipode for the shuffle algebra: The counit $\bar{e}$ is given by:

$$\bar{e}\left(e\right) = 1, \quad \bar{e}\left(l_1 l_2 \dots l_n\right) = 0. \tag{55}$$

The coproduct Δ is given by:

$$\Delta\left(l_1 l_2 \dots l_k\right) = \sum_{j=0}^{k} \left(l_{j+1} \dots l_k\right) \otimes \left(l_1 \dots l_j\right). \tag{56}$$

The antipode S is given by:

$$S\left(l_1 l_2 \dots l_k\right) = (-1)^k \, l_k l_{k-1} \dots l_2 l_1. \tag{57}$$

The shuffle algebra is generated by the Lyndon words. If one introduces a lexicographic ordering on the letters of the alphabet A, a Lyndon word is defined by the property $w < v$ for any sub-words u and v such that $w = uv$.

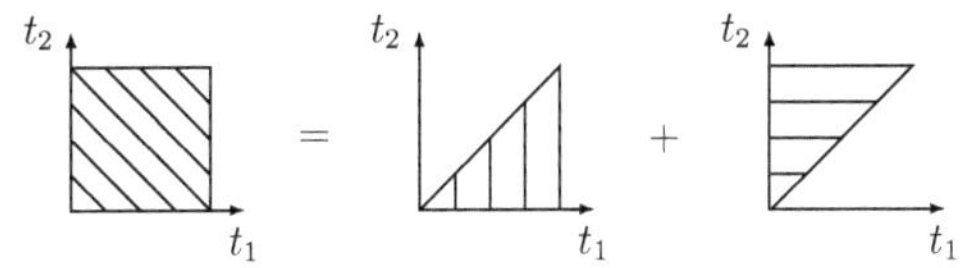

Fig. 5 A shuffle algebra follows from replacing the integral over the *square* by an integral over the *lower triangle* and an integral over the *upper triangle*

An important example for a shuffle algebra are iterated integrals. Let $[a, b]$ be a segment of the real line and f_1, f_2,... functions on this interval. Let us define the following iterated integrals:

$$I(f_1, f_2, \dots, f_k; a, b) = \int\limits_a^b dt_1 f_1(t_1) \int\limits_a^{t_1} dt_2 f_2(t_2) \dots \int\limits_a^{t_{k-1}} dt_k f_k(t_k) \qquad (58)$$

For fixed a and b we have a shuffle algebra:

$$I(f_1, f_2, \dots, f_k; a, b) \cdot I(f_{k+1}, \dots, f_r; a, b) = \sum_{\text{shuffles } \sigma} I(f_{\sigma(1)}, f_{\sigma(2)}, \dots, f_{\sigma(r)}; a, b), \qquad (59)$$

where the sum runs over all permutations σ, which preserve the relative order of $1, 2, \dots, k$ and of $k + 1, \dots, r$. The proof is sketched in Fig. 5. The two outermost integrations are recursively replaced by integrations over the upper and lower triangle. The definition of multiple polylogarithms in Eq. (37) is of the form of iterated integrals as in Eq. (58). Therefore it follows that multiple polylogarithms obey a shuffle algebra. An example for the multiplication is given by

$$G(z_1; y) G(z_2; y) = G(z_1, z_2; y) + G(z_2, z_1; y). \qquad (60)$$

Let us now turn to quasi-shuffle algebras. Assume that for the set of letters we have an additional operation

$$\begin{aligned} (.,.) : A \otimes A &\to A, \\ l_1 \otimes l_2 &\to (l_1, l_2), \end{aligned} \qquad (61)$$

which is commutative and associative. Then we can define a new product of words recursively through

$$\begin{aligned} &(l_1 l_2 \dots l_k) * (l_{k+1} \dots l_r) \\ &\quad = l_1 \left[(l_2 \dots l_k) * (l_{k+1} \dots l_r) \right] + l_{k+1} \left[(l_1 l_2 \dots l_k) * (l_{k+2} \dots l_r) \right] \\ &\qquad + (l_1, l_{k+1}) \left[(l_2 \dots l_k) * (l_{k+2} \dots l_r) \right]. \end{aligned} \qquad (62)$$

This product is a generalisation of the shuffle product and differs from the recursive definition of the shuffle product in Eq. (50) through the extra term in the last line. This modified product is known under the names quasi-shuffle product [25], mixable shuffle product [24], stuffle product [9] or mould symmetrel [16]. Quasi-shuffle algebras are Hopf algebras. Comultiplication and counit are defined as for the shuffle algebras. The counit $\bar{e}$ is given by:

$$\bar{e}\,(e) = 1, \quad \bar{e}\,(l_1 l_2 \dots l_n) = 0. \tag{63}$$

The coproduct Δ is given by:

$$\Delta\,(l_1 l_2 \dots l_k) = \sum_{j=0}^{k} \left(l_{j+1} \dots l_k\right) \otimes \left(l_1 \dots l_j\right). \tag{64}$$

The antipode S is recursively defined through

$$S\,(l_1 l_2 \dots l_k) = -l_1 l_2 \dots l_k - \sum_{j=1}^{k-1} S\left(l_{j+1} \dots l_k\right) * \left(l_1 \dots l_j\right), \quad S(e) = e. \tag{65}$$

An example for a quasi-shuffle algebra are nested sums. Let n_a and n_b be integers with $n_a < n_b$ and let $f_1, f_2, \dots$ be functions defined on the integers. We consider the following nested sums:

$$S(f_1, f_2, \dots, f_k; n_a, n_b) = \sum_{i_1=n_a}^{n_b} f_1(i_1) \sum_{i_2=n_a}^{i_1-1} f_2(i_2) \dots \sum_{i_k=n_a}^{i_{k-1}-1} f_k(i_k). \tag{66}$$

(The letter S denotes here a function, and not the antipode.) For fixed n_a and n_b we have a quasi-shuffle algebra:

$$\begin{aligned}
S(f_1, f_2, \dots, f_k; n_a, n_b) * S(f_{k+1}, \dots, f_r; n_a, n_b) = \\
\sum_{i_1=n_a}^{n_b} f_1(i_1)\; S(f_2, \dots, f_k; n_a, i_1 - 1) * S(f_{k+1}, \dots, f_r; n_a, i_1 - 1) \\
+ \sum_{j_1=n_a}^{n_b} f_k(j_1)\; S(f_1, f_2, \dots, f_k; n_a, j_1 - 1) * S(f_{k+2}, \dots, f_r; n_a, j_1 - 1) \\
+ \sum_{i=n_a}^{n_b} f_1(i) f_k(i)\; S(f_2, \dots, f_k; n_a, i - 1) * S(f_{k+2}, \dots, f_r; n_a, i - 1)
\end{aligned} \tag{67}$$

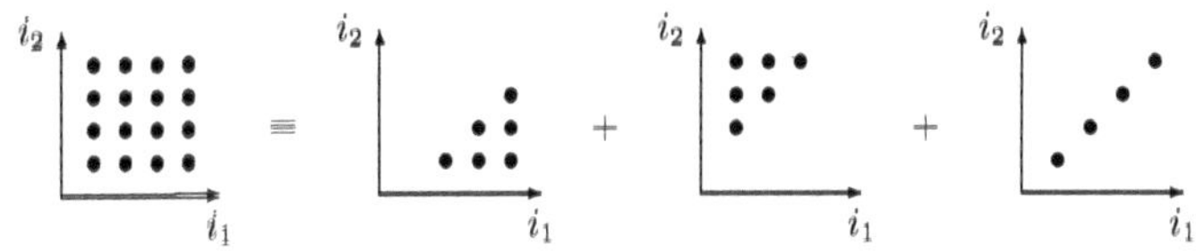

Fig. 6 A quasi-shuffle algebra follows from replacing the sum over the *square* by a sum over the *lower triangle*, a sum over the *upper triangle* and a sum over the diagonal

Note that the product of two letters corresponds to the point-wise product of the two functions:

$$(f_i \cdot f_j)(n) = f_i(n) f_j(n). \tag{68}$$

The proof that nested sums obey the quasi-shuffle algebra is sketched in Fig. 6. The outermost sums of the nested sums on the l.h.s of (67) are split into the three regions indicated in Fig. 6. The definition of multiple polylogarithms in Eq. (33) is of the form of nested sums as in Eq. (66). Therefore it follows that multiple polylogarithms obey also a quasi-shuffle algebra. An example for the quasi-shuffle multiplication is given by

$$\text{Li}_{m_1}(x_1)\text{Li}_{m_2}(x_2) = \text{Li}_{m_1,m_2}(x_1, x_2) + \text{Li}_{m_2,m_1}(x_2, x_1) + \text{Li}_{m_1+m_2}(x_1 x_2). \tag{69}$$

2.5 *Mellin-Barnes Transformation*

In Sect. 1.3 we saw that the Feynman parameter integrals depend on two graph polynomials $\mathcal{U}$ and $\mathcal{F}$, which are homogeneous functions of the Feynman parameters. In this section we will continue the discussion how these integrals can be performed and exchanged against a (multiple) sum over residues. The case, where the two polynomials are absent is particular simple:

$$\int_\Delta \omega \left(\prod_{j=1}^{n} x_j^{\nu_j - 1} \right) = \frac{\prod_{j=1}^{n} \Gamma(\nu_j)}{\Gamma(\nu_1 + \ldots + \nu_n)}. \tag{70}$$

With the help of the Mellin-Barnes transformation we now reduce the general case to Eq. (70). The Mellin-Barnes transformation reads

$$(A_1 + A_2 + \ldots + A_n)^{-c} = \frac{1}{\Gamma(c)} \frac{1}{(2\pi i)^{n-1}} \int_{-i\infty}^{i\infty} d\sigma_1 \ldots \int_{-i\infty}^{i\infty} d\sigma_{n-1}$$
$$\times \Gamma(-\sigma_1) \ldots \Gamma(-\sigma_{n-1}) \Gamma(\sigma_1 + \ldots + \sigma_{n-1} + c)\, A_1^{\sigma_1} \ldots A_{n-1}^{\sigma_{n-1}} A_n^{-\sigma_1 - \ldots - \sigma_{n-1} - c}. \tag{71}$$

Each contour is such that the poles of $\Gamma(-\sigma)$ are to the right and the poles of $\Gamma(\sigma + c)$ are to the left. This transformation can be used to convert the sum of monomials of the polynomials $\mathcal{U}$ and $\mathcal{F}$ into a product, such that all Feynman parameter integrals are of the form of Eq. (70). As this transformation converts sums into products it is the "inverse" of Feynman parametrisation. With the help of Eq. (70) we may perform the integration over the Feynman parameters. A single contour integral is then of the form

$$I = \frac{1}{2\pi i} \int\limits_{-i\infty}^{+i\infty} d\sigma \, \frac{\Gamma(\sigma + a_1) \dots \Gamma(\sigma + a_m)}{\Gamma(\sigma + c_2) \dots \Gamma(\sigma + c_p)} \frac{\Gamma(-\sigma + b_1) \dots \Gamma(-\sigma + b_n)}{\Gamma(-\sigma + d_1) \dots \Gamma(-\sigma + d_q)} \, x^{-\sigma}. \tag{72}$$

The contour is such that the poles of $\Gamma(\sigma + a_1), \dots, \Gamma(\sigma + a_m)$ are to the right of the contour, whereas the poles of $\Gamma(-\sigma + b_1), \dots, \Gamma(-\sigma + b_n)$ are to the left of the contour. We define

$$\alpha = m + n - p - q, \qquad \beta = m - n - p + q,$$

$$\lambda = \mathrm{Re}\left(\sum_{j=1}^{m} a_j + \sum_{j=1}^{n} b_j - \sum_{j=1}^{p} c_j - \sum_{j=1}^{q} d_j \right) - \frac{1}{2}(m + n - p - q). \tag{73}$$

Then the integral Eq. (72) converges absolutely for $\alpha > 0$ [17] and defines an analytic function in

$$|\arg x| < \min\left(\pi, \alpha \frac{\pi}{2}\right). \tag{74}$$

The integral Eq. (72) is most conveniently evaluated with the help of the residuum theorem by closing the contour to the left or to the right. Therefore we need to know under which conditions the semi-circle at infinity used to close the contour gives a vanishing contribution. This is obviously the case for $|x| < 1$ if we close the contour to the left, and for $|x| > 1$, if we close the contour to the right. The case $|x| = 1$ deserves some special attention. One can show that in the case $\beta = 0$ the semi-circle gives a vanishing contribution, provided $\lambda < -1$. To sum up all residues which lie inside the contour it is useful to know the residues of the Gamma function:

$$\mathrm{res}\ (\Gamma(\sigma + a), \sigma = -a - n) = \frac{(-1)^n}{n!}, \quad \mathrm{res}\ (\Gamma(-\sigma + a), \sigma = a + n) = -\frac{(-1)^n}{n!}.$$

In the general case, the multiple integrals in Eq. (71) lead to multiple sums over residues.

2.6 Z-Sums

The multiple sums over the residues can be expanded as a Laurent series in the dimensional regularisation parameter ε. For particular integrals the coefficients of the Laurent series can be expressed in terms of multiple polylogarithms. To see this, we first introduce a special form of nested sums, called Z-sums [29, 30, 45, 46]:

$$Z(n; m_1, \ldots, m_k; x_1, \ldots, x_k) = \sum_{i_1 > i_2 > \ldots > i_k > 0}^{n} \frac{x_1^{i_1}}{i_1^{\,m_1}} \cdots \frac{x_k^{i_k}}{i_k^{\,m_k}}. \tag{75}$$

k is called the depth of the Z-sum and $w = m_1 + \ldots + m_k$ is called the weight. If the sums go to infinity ($n = \infty$) the Z-sums are multiple polylogarithms:

$$Z(\infty; m_1, \ldots, m_k; x_1, \ldots, x_k) = \mathrm{Li}_{m_1,\ldots,m_k}(x_1, \ldots, x_k). \tag{76}$$

For $x_1 = \ldots = x_k = 1$ the definition reduces to the Euler-Zagier sums [5, 7, 18, 43, 47]:

$$Z(n; m_1, \ldots, m_k; 1, \ldots, 1) = Z_{m_1,\ldots,m_k}(n). \tag{77}$$

For $n = \infty$ and $x_1 = \ldots = x_k = 1$ the sum is a multiple ζ-value:

$$Z(\infty; m_1, \ldots, m_k; 1, \ldots, 1) = \zeta_{m_1,\ldots,m_k}. \tag{78}$$

The Z-sums are of the form as in Eq. (66) and form therefore a quasi-shuffle algebra. The usefulness of the Z-sums lies in the fact, that they interpolate between multiple polylogarithms and Euler-Zagier sums. Euler-Zagier sums appear in the expansion of the Gamma-function:

$$\begin{aligned} \Gamma(n+\varepsilon) &= \Gamma(1+\varepsilon)\Gamma(n) \\ &\times \left[1 + \varepsilon Z_1(n-1) + \varepsilon^2 Z_{11}(n-1) + \varepsilon^3 Z_{111}(n-1) + \ldots + \varepsilon^{n-1} Z_{11\ldots1}(n-1)\right]. \end{aligned} \tag{79}$$

The quasi-shuffle product can be used to reduce any product

$$Z(n; m_1, \ldots; x_1, \ldots) \cdot Z(n; m_1', \ldots; x_1', \ldots) \tag{80}$$

of Z-sums with the same upper summation index n to a linear combination of single Z-sums. The Hopf algebra of Z-sums has additional structures if we allow expressions of the form [30]

$$\frac{x_0^n}{n^{m_0}} Z(n; m_1, \ldots, m_k; x_1, \ldots, x_k), \tag{81}$$

e.g. Z-sums multiplied by a letter. Then the following convolution product

$$\sum_{i=1}^{n-1} \frac{x^i}{i^m} Z(i-1;\ldots) \frac{y^{n-i}}{(n-i)^{m'}} Z(n-i-1;\ldots) \tag{82}$$

can again be expressed in terms of expressions of the form (81). In addition there is a conjugation, e.g. sums of the form

$$-\sum_{i=1}^{n} \binom{n}{i} (-1)^i \frac{x^i}{i^m} Z(i;\ldots) \tag{83}$$

can also be reduced to terms of the form (81). The name conjugation stems from the following fact: To any function $f(n)$ of an integer variable n one can define a conjugated function $C \circ f(n)$ as the following sum

$$C \circ f(n) = \sum_{i=1}^{n} \binom{n}{i} (-1)^i f(i). \tag{84}$$

Then conjugation satisfies the following two properties:

$$\begin{aligned} C \circ 1 &= 1, \\ C \circ C \circ f(n) &= f(n). \end{aligned} \tag{85}$$

Finally there is the combination of conjugation and convolution, e.g. sums of the form

$$-\sum_{i=1}^{n-1} \binom{n}{i} (-1)^i \frac{x^i}{i^m} Z(i;\ldots) \frac{y^{n-i}}{(n-i)^{m'}} Z(n-i;\ldots) \tag{86}$$

can also be reduced to terms of the form (81).

With the help of these algorithms it is possible to prove that the Laurent series in ε of specific Feynman integrals contains only multiple polylogarithms [3].

3 Beyond Multiple Polylogarithms

Although multiple polylogarithms form an important class of functions, which appear in the evaluation of Feynman integrals, it is known from explicit calculations that starting from two-loop integrals with massive particles one encounters functions beyond the class of multiple polylogarithms. The simplest example is given by the two-loop sunset diagram with non-zero masses, where elliptic integrals make their appearance [28]. Differential equations provide a tool to get a handle on these integrals [2, 19–21, 26, 27, 33, 35].

3.1 The Two-Loop Sunset Integral with Non-zero Masses

In this subsection we review how ideas of algebraic geometry can be used to obtain a differential equation for the Feynman integral [32]. We first focus on the example of the two-loop sunrise integral. The two-loop sunrise integral is given in D-dimensional Minkowski space by

$$S\left(D, p^2\right) = \frac{\left(\mu^2\right)^{3-D}}{\Gamma\left(3-D\right)} \int \frac{d^D k_1}{i\pi^{\frac{D}{2}}} \frac{d^D k_2}{i\pi^{\frac{D}{2}}} \frac{1}{\left(-k_1^2+m_1^2\right)\left(-k_2^2+m_2^2\right)\left(-k_3^2+m_3^2\right)}, \tag{87}$$

with $k_3 = p - k_1 - k_2$. Here we suppressed on the l.h.s. the dependence on the internal masses m_1, m_2 and m_3 and on the arbitrary scale μ. It is convenient to denote the momentum squared by $t = p^2$. In terms of Feynman parameters the integral reads

$$S\left(D, t\right) = \int\limits_{\Delta} \omega \frac{\mathcal{U}^{3-\frac{3}{2}D}}{\mathcal{F}^{3-D}}, \tag{88}$$

where the two Feynman graph polynomials are given by

$$\mathcal{F} = \left[-x_1 x_2 x_3 t + \left(x_1 m_1^2 + x_2 m_2^2 + x_3 m_3^2\right)\mathcal{U}\right]\mu^{-2}, \quad \mathcal{U} = x_1 x_2 + x_2 x_3 + x_3 x_1.$$

It is simpler to consider this integral first in $D = 2$ dimensions and to obtain the result in $D = 4 - 2\varepsilon$ dimensions with the help of dimensional recurrence relations [40, 41]. In two dimensions this integral is finite, depends only on the second Symanzik polynomial $\mathcal{F}$ and is given by

$$S\left(2, t\right) = \int\limits_{\Delta} \frac{\omega}{\mathcal{F}}. \tag{89}$$

From the point of view of algebraic geometry there are two objects of interest in Eq. (89): On the one hand the domain of integration Δ and on the other hand the algebraic variety X defined by the zero set of $\mathcal{F} = 0$. The two objects X and Δ intersect at the three points $[1:0:0]$, $[0:1:0]$ and $[0:0:1]$ of the projective space $\mathbb{P}^2$. This is shown in Fig. 7 on the left. We blow-up $\mathbb{P}^2$ in these three points and we denote the blow-up by P. We further denote the strict transform of X by Y and the total transform of the set $\{x_1 x_2 x_3 = 0\}$ by B. With these notations we can now consider the mixed Hodge structure (or the motive) given by the relative cohomology group [4]

$$H^2\left(P \backslash Y, B \backslash B \cap Y\right). \tag{90}$$

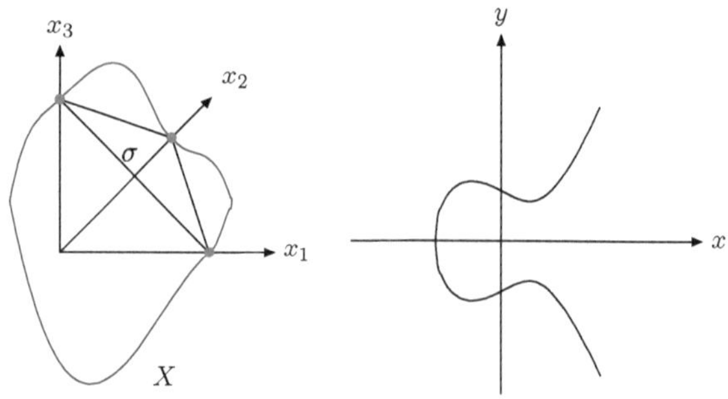

Fig. 7 The intersection of the domain of integration Δ with the zero set X of the second Symanzik polynomial (*left*) and the *elliptic curve* $y^2 = x^3 - x + 1$ (*right*)

In the case of the two-loop sunrise integral considered here essential information on $H^2(P\backslash Y, B\backslash B\cap Y)$ is already given by $H^1(X)$. We recall that the algebraic variety X is defined by the second Symanzik polynomial:

$$-x_1x_2x_3t + \left(x_1m_1^2 + x_2m_2^2 + x_3m_3^2\right)(x_1x_2 + x_2x_3 + x_3x_1) = 0. \quad (91)$$

This defines for generic values of the parameters t, m_1, m_2 and m_3 an elliptic curve. The elliptic curve varies smoothly with the parameters t, m_1, m_2 and m_3. By a birational change of coordinates this equation can brought into the Weierstrass normal form

$$y^2z - x^3 - a_2(t)xz^2 - a_3(t)z^3 = 0. \quad (92)$$

The dependence of a_2 and a_3 on the masses is not written explicitly. In the chart $z = 1$ this reduces to

$$y^2 - x^3 - a_2(t)x - a_3(t) = 0. \quad (93)$$

The curve varies with the parameter t. An example of an elliptic curve is shown in Fig. 7 on the right. It is well-known that in the coordinates of Eq. (93) the cohomology group $H^1(X)$ is generated by

$$\eta = \frac{dx}{y} \text{ and } \dot{\eta} = \frac{d}{dt}\eta. \quad (94)$$

Since $H^1(X)$ is two-dimensional it follows that $\ddot{\eta} = \frac{d^2}{dt^2}\eta$ must be a linear combination of η and $\dot{\eta}$. In other words we must have a relation of the form

$$p_0(t)\ddot{\eta} + p_1(t)\dot{\eta} + p_2(t)\eta = 0. \quad (95)$$

The coefficients $p_0(t)$, $p_1(t)$ and $p_2(t)$ define the Picard-Fuchs operator

$$L^{(2)} = p_0(t)\frac{d^2}{dt^2} + p_1(t)\frac{d}{dt} + p_2(t). \quad (96)$$

Applying the Picard-Fuchs operator to our integrand gives an exact form:

$$L^{(2)}\left(\frac{\omega}{\mathscr{F}}\right) = d\beta. \tag{97}$$

The integration over Δ yields

$$L^{(2)} S(2,t) = \int_{\Delta} d\beta = \int_{\partial\Delta} \beta \tag{98}$$

The integration of β over $\partial\Delta$ is elementary and we arrive at

$$\left[p_0(t)\frac{d^2}{dt^2} + p_1(t)\frac{d}{dt} + p_2(t)\right] S\,(2,t) = p_3(t). \tag{99}$$

This is the sought-after second-order differential equation. The coefficients are given in the equal mass case by Laporta and Remiddi [28] and Broadhurst et al. [10]

$$\begin{aligned} p_0(t) &= t\left(t-m^2\right)\left(t-9m^2\right), & p_2(t) &= t-3m^2, \\ p_1(t) &= 3t^2-20tm^2+9m^4, & p_3(t) &= -6\mu^2. \end{aligned} \tag{100}$$

The coefficients for the unequal mass case can be found in [32].

3.2 Differential Equations

The ideas of the previous subsection can be generalised to arbitrary Feynman integrals. For a given Feynman integral let us pick one variable t from the set of the Lorentz invariant quantities $(p_j + p_k)^2$ and the internal masses squared m_i^2. Let us write

$$\omega_t = \omega\left(\prod_{j=1}^{n} x_j^{\nu_j-1}\right)\frac{\mathscr{U}^{\nu-(l+1)D/2}}{\mathscr{F}^{\nu-lD/2}}. \tag{101}$$

The subscript t indicates that ω_t depends on t through $\mathscr{F}$. The Feynman integral is then simply

$$I_G = \int_{\Delta} \omega_t \tag{102}$$

We seek an ordinary linear differential equation with respect to the variable t for the Feynman integral I_G. We start to look for a differential equation of the form

$$L^{(r)}\omega_t = d\beta, \tag{103}$$

where

$$L^{(r)} = \sum_{j=0}^{r} p_j \left(\mu^2 \frac{d}{dt}\right)^j \tag{104}$$

is a Picard-Fuchs operator of order r. Suppose an equation of the form as in Eq. (103) exists. Following the same steps as in Sect. 3.1 we arrive at

$$L^{(r)} I_G = \int_{\partial\Delta} \beta. \tag{105}$$

The right-hand side corresponds to simpler Feynman integrals, where one propagator has been contracted. The coefficients of the Picard-Fuchs operator and the coefficients of the form β can be found by solving a linear system of equations [33].

References

1. Ablinger, J., Blümlein, J., Schneider, C.: Harmonic sums and polylogarithms generated by cyclotomic polynomials. J. Math. Phys. **52**, 102301 (2011)
2. Argeri, M., Mastrolia, P.: Feynman diagrams and differential equations. Int. J. Mod. Phys. **A22**, 4375–4436 (2007)
3. Bierenbaum, I., Weinzierl, S.: The massless two-loop two-point function. Eur. Phys. J. **C32**, 67–78 (2003)
4. Bloch, S., Esnault, H., Kreimer, D.: On motives associated to graph polynomials. Commun. Math. Phys. **267**, 181 (2006)
5. Blümlein, J.: Algebraic relations between harmonic sums and associated quantities. Comput. Phys. Commun. **159**, 19–54 (2004)
6. Blümlein, J., Broadhurst, D.J., Vermaseren, J.A.M.: The multiple zeta value data mine. Comput. Phys. Commun. **181**, 582 (2010)
7. Blümlein, J., Kurth, S.: Harmonic sums and mellin transforms up to two-loop order. Phys. Rev. **D60**, 014018 (1999)
8. Bogner, C., Weinzierl, S.: Feynman graph polynomials. Int. J. Mod. Phys. **A25**, 2585–2618 (2010)
9. Borwein, J.M., Bradley, D.M., Broadhurst, D.J., Lisonek, P.: Special values of multiple polylogarithms. Trans. Am. Math. Soc. **353**(3), 907 (2001)
10. Broadhurst, D.J., Fleischer, J., Tarasov, O.: Two loop two point functions with masses: asymptotic expansions and Taylor series, in any dimension. Z. Phys. **C60**, 287–302 (1993)
11. Brown, F.: The massless higher-loop two-point function. Commun. Math. Phys. **287**, 925–958 (2008)
12. Brown, F.: On the periods of some Feynman integrals. arXiv:0910.0114 [math.AG] (2009)

13. Chaiken, S.: A combinatorial proof of the all minors matrix tree theorem. SIAM J. Algebr. Discret. Methods **3**, 319–329 (1982)
14. Chen, W.K.: Applied Graph Theory, Graphs and Electrical Networks. North Holland, Amsterdam (1982)
15. Dodgson, C.L.: Condensation of determinants. Proc. R. Soc. Lond. **15**, 150–155 (1866)
16. Ecalle, J.: Ari/gari, la dimorphie et l'arithmétique des multizêtas: un premier bilan. Journal de Théorie des Nombres de Bordeaux **15**, 411 (2003)
17. Erdélyi, A., Magnus, W., Oberhettinger, F., Tricomi, F.: Higher Transcendental Functions, vol. I. McGraw Hill, New York (1953)
18. Euler, L.: Meditationes circa singulare serierum genus. Novi Comment. Acad. Sci. Petropol. **20**, 140 (1775)
19. Gehrmann, T., Remiddi, E.: Differential equations for two-loop four-point functions. Nucl. Phys. **B580**, 485–518 (2000)
20. Gehrmann, T., Remiddi, E.: Two-loop master integrals for gamma* → 3 jets: the non- planar topologies. Nucl. Phys. **B601**, 287–317 (2001)
21. Gehrmann, T., Remiddi, E.: Two-loop master integrals for gamma* → 3 jets: the planar topologies. Nucl. Phys. **B601**, 248–286 (2001)
22. Goncharov, A.B.: Multiple polylogarithms, cyclotomy and modular complexes. Math. Res. Lett. **5**, 497 (1998)
23. Goncharov, A.B.: Multiple polylogarithms and mixed Tate motives. math.AG/0103059 (2001)
24. Guo, L., Keigher, W.: Baxter algebras and shuffle products. Adv. Math. **150**, 117 (2000)
25. Hoffman, M.E.: Quasi-shuffle products. J. Algebr. Comb. **11**, 49 (2000)
26. Kotikov, A.V.: Differential equation method: the calculation of n point Feynman diagrams. Phys. Lett. **B267**, 123–127 (1991)
27. Kotikov, A.V.: Differential equations method: new technique for massive Feynman diagrams calculation. Phys. Lett. **B254**, 158–164 (1991)
28. Laporta, S., Remiddi, E.: Analytic treatment of the two loop equal mass sunrise graph. Nucl. Phys. **B704**, 349–386 (2005)
29. Moch, S., Uwer, P.: Xsummer: transcendental functions and symbolic summation in form. Comput. Phys. Commun. **174**, 759–770 (2006)
30. Moch, S., Uwer, P., Weinzierl, S.: Nested sums, expansion of transcendental functions and multi-scale multi-loop integrals. J. Math. Phys. **43**, 3363–3386 (2002)
31. Moon, J.: Some determinant expansions and the matrix-tree theorem. Discret. Math. **124**, 163–171 (1994)
32. Müller-Stach, S., Weinzierl, S., Zayadeh, R.: A second-order differential equation for the two-loop sunrise graph with arbitrary masses. Commun. Number Theory Phys. **6**, 203–222 (2012)
33. Müller-Stach, S., Weinzierl, S., Zayadeh, R.: Picard-Fuchs equations for Feynman integrals. arXiv:1212.4389 [hep-ph] (2012)
34. Nielsen, N.: Der Eulersche Dilogarithmus und seine Verallgemeinerungen. Nova Acta Leopoldina (Halle) **90**, 123 (1909)
35. Remiddi, E.: Differential equations for Feynman graph amplitudes. Nuovo Cim. **A110**, 1435–1452 (1997)
36. Remiddi, E., Vermaseren, J.A.M.: Harmonic polylogarithms. Int. J. Mod. Phys. **A15**, 725 (2000)
37. Reutenauer, C.: Free Lie Algebras. Clarendon Press, Oxford (1993)
38. Stanley, R.P.: Spanning trees and a conjecture of Kontsevich. Ann. Comb. **2**, 351–363 (1998)
39. Sweedler, M.: Hopf Algebras. Benjamin, New York (1969)
40. Tarasov, O.V.: Connection between Feynman integrals having different values of the space-time dimension. Phys. Rev. **D54**, 6479–6490 (1996)
41. Tarasov, O.V.: Generalized recurrence relations for two-loop propagator integrals with arbitrary masses. Nucl. Phys. **B502**, 455–482 (1997)
42. Tutte, W.T.: Graph Theory. Encyclopedia of Mathematics and Its Applications, vol. 21. Addison-Wesley, Menlo Park (1984)

43. Vermaseren, J.A.M.: Harmonic sums, mellin transforms and integrals. Int. J. Mod. Phys. **A14**, 2037 (1999)
44. Vollinga, J., Weinzierl, S.: Numerical evaluation of multiple polylogarithms. Comput. Phys. Commun. **167**, 177 (2005)
45. Weinzierl, S.: Symbolic expansion of transcendental functions. Comput. Phys. Commun. **145**, 357–370 (2002)
46. Weinzierl, S.: Expansion around half-integer values, binomial sums and inverse binomial sums. J. Math. Phys. **45**, 2656–2673 (2004)
47. Zagier, D.: Values of zeta functions and their applications. In: Joseph A. et al. (eds.) First European Congress of Mathematics, Paris, 1992 vol. II, pp. 497–512. Birkhauser, Boston (1994)
48. Zeilberger, D.: Dodgson's determinant-evaluation rule proved by two-timing men and women. Electron. J. Comb. **4**(R22), 2 (1997)

Index

C. Schneider and J. Blümlein (eds.), *Computer Algebra in Quantum Field Theory*, Texts & Monographs in Symbolic Computation, DOI 10.1007/978-3-7091-1616-6,

MIX
Papier aus verantwortungsvollen Quellen
Paper from responsible sources
FSC® C105338

If you have any concerns about our products,
you can contact us on
ProductSafety@springernature.com

In case Publisher is established outside the EU,
the EU authorized representative is:
Springer Nature Customer Service Center GmbH
Europaplatz 3, 69115 Heidelberg, Germany

Printed by Libri Plureos GmbH
in Hamburg, Germany